Time- and Strata-Bound Ore Deposits

Edited by
D.D. Klemm and
H.-J. Schneider

With 160 Figures and 29 Tables

Springer-Verlag
Berlin Heidelberg New York 1977

Professor Dr. Dietrich D. Klemm
Institut für Allgemeine und Angewandte Geologie
der Universität, Luisenstraße 37, 8000 München 2, FRG

Professor Dr. Hans-Jochen Schneider
Institut für Angewandte Geologie der Freien Universität
Wichernstraße 6, 1000 Berlin 33

ISBN 3-540-08502-5 Springer-Verlag Berlin Heidelberg New York
ISBN 0-387-08502-5 Springer-Verlag New York Heidelberg Berlin

Library of Congress Cataloging in Publication Data. Main entry under title:
Time- and strata-bound ore deposits. Bibliography: p. Includes index. 1. Ore-
deposits. 2. Geological time. 3. Geology, Stratagraphic. I. Klemm, Dietrich D., ·
1933- . II. Schneider, Hans Jochen, 1923- . QE390.T55. 553'.1. 77-21293.

Printing and bookbinding: Beltz Offsetdruck, Hemsbach/Bergstr.
2132/3130-543210

This book is dedicated to
ALBERT MAUCHER
the always helpful and stimulating teacher
and friend, the unflagging advocate of the
idea on time- and strata-bound ore deposits

Preface

The problem of time- and strata-bound formation of ore deposits has during the past decade become one of the most debated topics in current international discussion. Due to the amazing results of modern mineral exploration and world-wide geophysical research, the mutual relationship between the complex geological history of a crustal segment and the development of distinct metallogenic provinces (ore belts) has received much interest. Reviewing the earth's history in this light one can now recognize metallogenic epochs even of global range which document the existence of world-wide time-bound ore enrichments.

The knowledge of these metallogenetic processes has been growing step by step for several decades. It began with simple observations and sceptic interpretations, which at first threw heretical spot lights on to the edifices of the prevailing theories on granitic differentiation as the favoured source of ore deposits. It was obvious that the new ideas at first referred to ore enrichments in sedimentary sequences, nowadays summarized under the term strata-bound, and mainly interpreted as stratiform or sedimentary ore deposits. Moreover, the modern term "strata-bound" also includes ore mineralizations which are bound to distinct units of layered (intrusive or extrusive) igneous complexes as a general descriptive term without genetical restriction!

Albert Maucher is one of the representatives of the initial era who discussed these genetical questions critically in the decade before the 2nd World War. Even in his first papers on the genesis of metamorphic polymetallic deposits of Northern Bavaria (1932–1939) he supported the theory of a strata-bound character and sedimentary origin of this mondial type of metamorphic kies ore. However, at that time such ideas were still "too modern" and did not attract general attention.

After a longer interruption caused by the war, *Albert Maucher* again devoted his interest to the genetical problems of strata-bound and stratiform (sedimentary) ore deposits, beginning in the fifties with the lead-zinc ores in carbonate rocks of the Alps and in Turkey. With the increasing number of his students and research proceedings at the rebuilt institute of the University of München, a distinct critical common sense in scientific discussion developed amongst the team, which has been called by outsiders the "Munich School".

At the beginning of the sixties he initiated a research scheme on stratiform Sb-Hg-W deposits in European Paleozoic and abroad. These activities culminated in the strategically prepared prospection and discovery of the great Felbertal (Mittersill) scheelite deposit in the Austrian Alps in 1967.

The scientific base for this success was the genetical concept on time- *and* strata-bound formation of Sb-Hg-W ore deposits, which has been advocated by *Albert Maucher* in his papers since 1965. By his characteristic willingness for frank and critical discussions, at any time he has stimulated a great number of his students, friends, and colleagues during the past decades, influencing research activities also outside Germany and Europe.

In the sixties, in addition to his research engagements, he began, at first together with *G. Rehwald*, the editing of the well-known *Card Index of Ore Photomicrographs.*

The title of the present book comprises the prevalent scope of *Albert Maucher*'s scientific work, showing at the same time the simultaneous international development of these prolific hypotheses at different centres of the world. As the circle of authors was limited to the personal sphere of activity of *Albert Maucher*, the list of competent colleagues is understandably incomplete. Nevertheless, the various contributions to the volume reflect the huge range of a modern geoscientific topic which has arisen only within the last two decades. They indicate additionally the multifarious mutual relations between physico-chemical, geochemical, petrological, and microscopical investigations, as well as geological and geophysical field observations, partly in close connection with plate tectonic aspects, revealing the recent progress in these fields in all continents.

Naturally the authors were interested in presenting personal results of their actual research or even fundamental reviews; the book cannot therefore cover all details summarized recently under the general theme, even though there are stimulating citations for nearly all topical questions.

In view of the comprehensive title of the book, the editors have aimed at a subdivision into the following six sections according to the different topics:

1. General Topics.	Papers stating the problem, including general information or regional reviews.
2. Precambrian Deposits	As examples of time- and strata-bound ore formation in different geological eras and regions.
3. Paleozoic Deposits	
4. Mesozoic Deposits	
5. Strata-Bound Intrusive Deposits	

6. Geochemical and Minerogenetical Problems	Papers related to the general topics with special results or representative details of the adjacent fields.

In order to offer the volume for a greater international readership, the manuscripts are presented in the English language, certainly for some of the authors an additional labour. The editors are indepted to all contributors for their kind co-operation which has obviated many technical difficulties.

We are especially happy to notice contributions from *Paul Ramdohr*, one of the academic teachers of *Albert Maucher*, and *Hermann Borchert*, one of his oldest friends since the early scientific activities.

Our cordial thanks are due to Dr. *Konrad F. Springer* and his team at Springer-Verlag, Heidelberg, for their patient assistance in raising the technical quality of the book.

Amongst the list of authors a great number of colleagues, friends, and students of *Albert Maucher* are missing, who were not in a position to contribute a paper because of temporary hindrance or engagement in other fields of geological research. Many of them expressed their regret at not being able to participate in this volume.

On behalf of them all we edit this book as a dedication to

Albert Maucher,

on occasion of his 70th birthday.

Berlin and München, October 1977   Dietrich D. Klemm
Hans-J. Schneider

Contents

Paleozoic Deposits

Mesozoic Deposits

Strata-Bound Intrusive Deposits

Geochemical and Minerogenetical Problems

Contributors

Alonso, F. Fdez., Avenida San Luis 40, Madrid–33, Spain

Amstutz, G.C., Mineralogisch-Petrographisches Institut der Universität Heidelberg, Im Neuenheimer Feld 236, 6900 Heidelberg, FRG

Borchert, H., Altenauer Str. 16, 3392 Clausthal-Zellerfeld, FRG

Brigo, L., Istituto di Mineralogia, Università degli studi Milano, Via Botticelli 23, 20133 Milano, Italy

Brodtkorb, A., Paso 258 – 9A, 1640 Martinez, Prov. Buenos Aires, Argentina

Brodtkorb, M.K. de, Paso 258 – 9A, 1640 Martinez, Prov. Buenos Aires, Argentina

Burchard, U., Urangesellschaft Canada Ltd., Suite 3100, 2 Bloor Street East, Toronto, Ontario M4W 1A8, Canada

Chukrov, F.V., Institute of Ore Geology, Staromonetnypereulok 35, 109017 Moskau, USSR

Degens, E.T., Geologisch-Paläontologisches Institut der Universität Hamburg, Bundesstraße 55, 2000 Hamburg 13, FRG

Derkmann, K., Lehrstuhl für Lagerstättenforschung und Rohstoffkunde der TU Clausthal, Adolf-Römer-Str. 2A, 3392 Clausthal-Zellerfeld, FRG

Dora, O.Ö., Ege Üniversitesi, FEN Fakültesi, Jeoloji Kürsüsü, Izmir /Bornova, Turkey

Eichmann, R., Max-Planck-Institut für Chemie, Saarstr. 23, 6500 Mainz, FRG

Ermilova, L.P., Institute of Ore Geology, Staromonetnypereulok 35, 109017 Moskau, USSR

Grafenauer, S., Lehrstuhl für Mineralogie und Petrographie der Universität, Aškerceva 20, 61000 Ljubljana, Yugoslavia

Gruenewaldt, G. von, Institute for Geological Research on the Bushveld Complex, University of Pretoria, Hillcrest, Pretoria 0002, Rep. S. Africa

Höll, R., Institut für Allgemeine und Angewandte Geologie der Universität, Luisenstr. 37, 8000 München 2, FRG

Kantor, J., Dionyz Stur Institute of Geology, Mlanskà dolina 1, 80940 Bratislava, CSSR

Klemm, D.D., Institut für Allgemeine und Angewandte Geologie der Universität München, Luisenstr. 37, 8000 München 2, FRG

Kostelka, L., Bleiberger Bergwerks-Union, Radetzkystr. 2, 9010 Klagenfurt, Austria

Kralik, M., Institut für Geologie und Lagerstättenlehre, Montanistische Hochschule, 8700 Leoben/Stmrk., Austria

Kräutner, H.G., Institutul de Geologie si Geofizica, Str. Caransebes, Nr. 1, Sect. 8, Bucuresti 32, Romania

Lawrence, L.J., The University of New South Wales, P.O. Box 1, Kensington, New South Wales, Australia 2033

Lehmann, B., Institut für Angewandte Geologie der Freien Universität Berlin, Wichernstr. 16, 1000 Berlin 33, FRG

Madel, J., Saarberg-Interplan, Stengelstr. 1, 6600 Saarbrücken, FRG

Nosik, L.P., Institute of Ore Geology, Staromonetnypereulok 35, 109017 Moskau, USSR

Omenetto, P., Istituto di Mineralogia e Petrografia, Corso Garibaldi 9, 35100 Padova, Italy

Pereira, J., 1 Kensington Palace Green, London W 8, England

Petrascheck, W.E., Geologisches Institut, Erzh.-Johann-Str. 10, 8700 Leoben, Austria

Ramdohr, P., Kaiserstr. 39, 6940 Weinheim-Hohensachsen, FRG

Ranzenbacher, A., Institut für Geologie und Lagerstättenlehre, Montanistische Hochschule, 8700 Leoben/Stmrk., Austria

Saupe, F., Centre National de la Recherche Scientific, 15, Rue N.-D. des Pauvres, 54500 Vandoeuvre, France

Schidlowski, M., Max-Planck-Institut für Chemie, Saarstr. 23, 6500 Mainz, FRG

Schneider, H.-J., Institut für Angewandte Geologie der Freien Universität Berlin, Wichernstr. 16, 1000 Berlin 33, FRG

Schroll, E., Bundesversuchs- und Forschungsanstalt Arsenal, Geotechnisches Institut, Dirmoserstr. 8, Obj. 210, 1030 Wien 3, Austria

Schulz, O., Institut für Mineralogie und Petrographie der Universität Innsbruck, Universitätsstr. 4, 6020 Innsbruck, Austria

Smirnov, V.J., Department of Geology, State University, Moskau B-234, 117234, USSR

Snethlage, R., Institut für Allgemeine und Angewandte Geologie der Universität, Luisenstr. 37, 8000 München 2, FRG

So, Chil-Sup, Department of Geology, Korea University, 1, Anam-Dong, Sungbuk-Ku, Seoul, Korea

Söhnge, P.G., Department of Geology, University of Stellenbosch, 7600 Stellenbosch, Rep. S. Africa

Sözen, A., M.T.A. Ege Bölge Müdürü, P.K. 1, Iszmir-Bornova, Turkey

Stoffers, P., Institut für Sedimentforschung der Universität Heidelberg, Im Neuenheimer Feld 236, 6900 Heidelberg, FRG

Strauss, G.K., Companía de Azufre y Cobre de Tharsis, Tharsis (Huelva), Spain

Štrucl, I., Rudniki svinca in topilnica Mežica, 62392 Mežica, Jugoslavia

Vavtar, F., Institut für Mineralogie und Petrographie der Universität,
Universitätsstr. 4, 6020 Innsbruck, Austria

Weber-Diefenbach, K., Institut für Allgemeine und Angewandte Geologie der Universität, Luisenstr. 37, 8000 München 2, FRG

Zuffardi, P., Istituto di Mineralogia, Università degli studi di Milano,
Via Botticelli 23, 20133 Milano, Italy

General Topics

Factor of Time in Formation of Strata-Bound Ore Deposits

V. J. SMIRNOW, Moscow

With 3 Figures

Summary

A long period of formation during which the feature of the ore mineralization changes is character-istic for all stratiform ore deposits.

Kies ore deposits are especially clear examples of a repeated ore formation over long intervals of time. The age and genesis of many deposits cannot be attributed to only one cycle of formation but at least two stages must be considered. The total time of formation for kies ore deposits lies between 50 and 250 my. The examples of Karatau and Sardinia prove a theory according th which the forma-tion of Pb/Zn deposits in carbonate rocks cannot only be attributed to a primary syngenetic genesis on the one hand or an epigenetic hydrothermal genesis on the other, but that rather an evolution of the formation has to be considered, which means that a time of formation of up to 300 my must be taken into account.

Geochemical aureoles of stratiform ore deposits have so far hardly been studied from the view-point of a long formation time and evolution of the deposit; taking these facts into consideration, however, gives a better understanding of the deposit. Whereas in one deposit a younger ore forma-tion can usually be distinguished from an older one, it is impossible to attribute a cordant or discor-dant formation, or the amount of deposited ore mineralization, for example, clearly to either the younger or older stage. Both stages can take a long period of time and develop in different substages.

The genesis of stratiform ore deposits can only be understood by taking into account that the deposits are polychronous and polygenous, i.e., the long period of formation and the evolution of the mineralization must be considered.

Many kies deposits in early stages of geosynclinal formation are typical representatives of stratiform ore deposits, for example lead-zinc deposits in carbonate rocks, and copper deposits in sandstones and schists in later stages of geosynclinal development during folding in the upper part of the crust. All kinds of stratiform ore deposits have in com-mon a long period of formation and a radical change in mineralization between the first and the final stage. This becomes especially evident not only when considering the various ore provinces, but also within ore deposits of stratiform character.

Kies Ore Deposits

Kies ore deposits differ considerably from other ore formations in their contents, geo-logic positions, conditions of formation, and genesis. Remarkable among them are pyrite and chalcopyrite deposits, linked to basaltic-andesitic rocks and polymetallic (Pb, Zn) sulphidic ores which are connected with andesitic-liparitic rocks.

The formation of many provinces of kies ores includes long periods of time, during which the deposits were repeatedly reformed. In Causasia (USSR), for example, kies ore deposits are known which were formed during Caledonian, Hercynian, Kimerian and Alpidic orogeneses; in the Ural (USSR) deposits of Hercynian, Caledonian and Proterozoic times are found; in Canada, ore fileds of Proterozoic, Hercynian and Kimerian times, and in Japan deposits have been dated as Hercynian, Kimerian and Alpidic. Within a single cycle the kies ore deposits could have been formed several times. In the Ural for example, deposits are known which are Hercynian, upper Silurian, middle Devonian and lower Carboniferous..

For single ore fields, ore deposits – and even ore bodies of kies mineralization – a constant ore feeding is characteristic, beginning with a volcanic stage and ending with a post-volcanic stage which finally characterizes the deposits. The very long period of formation of kies ore deposits is illustrated in Erzaltei, in Asasu in Kasachstan and Osernego in Buratia, USSR.

Erzaltei. The discussions on the genesis of polymetallic kies ore deposits are well known. For a long time the traditional opinion was held, which postulates the classical hydrothermal genesis, connected with derivatives of the Smeinogorsk granite complex of upper paleozoic age (*P. Burow, I. Grigoriew, W Nechoroschew, W. Popow* and others). In the years 1940–1950 there was an opposing view which postulated a close genetic, spatial, and short-lived connection of those deposits with Devonian volcanic rocks which are found in most parts of the ore deposits (*Weiz*, 1959; *Scherba*, 1957, 1968; *Derbikow*, 1966; *Jakowlew*, 1975, and others).

The arguments for these two geological views have been published more than once and need not be repeated here. It may suffice to say; a Paleozoic age and the plutonic-hydrothermal hypothesis is supported by emplacement of the Devonian effusives through several orebodies and the existence of veins of polymetallic composition which cut through Smeinogorsk granites (ore deposits of Pariginskol); a Devonian age and a volcanic hydrothermal hypothesis is supported by the layered texture of the ores and the form of most ore layers which lie conformably in Devonian volcanic rocks. Recently polymetallic ores were found and described in the area of Ridder-Sokolnogo of volcanic-sedimentary origin and of Devonian age (*Pokrowskaia* and *Kowrigo*, 1970). These ores form the socalled second Ridder ore deposit which lies in volcanic sedimentary depositions of the Eifel-Devonian.

Parallel orebodies of this deposit have a layered form and lie conformably in Alewro pelites of the centre part of the Krjukowskoi deposits.

They consist of successive layers of schists with dense fine-grained ores such as sphalerite, galena, pyrite, chalcopyrite, Bleklih ores, dolomite and other nonore minerals.

A synsedimentary ore deposition is proved by the following:

1. sedimentary rhythm of the ore deposit;
2. specific form of the ore layers with linear lying foot margin and nonlinear hanging foot margins;
3. characteristic deformation of the ore and host rocks, simultaneous sedimentation;
4. simultaneous folding of the ore layers and schists after dislocation.

In the kies-polymetallic ore deposits of the Rudnogo Altai, orebodies were found connected with effusives of middle Devonian on the one hand, and with intrusives of middle Carboniferous on the other. This can only be explained by the long period of formation of the ore deposits in this province, which originated in the middle Paleozoic with Devonian volcanism, when the metasomatic and volcano-sedimentary layers of the kies-polymetallic ores were formed, and terminated in the upper Paleozoic with granitic magmatism, when the formation of veins and possibly other forms of sulphidic ore-bodies took place. The whole period of ore formation took nearly 50 my.

Atasu in Kasachstan. The ore deposits of this region, the most famous of which are: Dscheirem, Uschkatan, Dschumart, Karadschal and Bestiube lie in the area of the Dscheil-minsk syncline which consists of upper Devonian volcano-sedimentary rocks.

Layered Fe-Mn and chalcopyrite-sphalerite ores which are contemporaneous with interlayered schistose rocks are here combined with overlying Pb/Zn baryte ores which were formed later. *A. Betedrlin, L. Pustowalov* and others regarded these ore deposits as of sedimentary origin which were later metamorphosed. *I. Jagowkin, M. Rusakow, K. Satpaew* and others regarded them as of hydrothermal metasomatic origin.

For the first theory, the following reasons are valid: fixed conformable layering of the orebodies, their rhythmic layered texture and simple mineralogic content. For the second theory, the characteristics of hydrothermal origin were taken into consideration, i.e., break-through character of some orebodies or parts of them, characteristics of meta-somatic ore formation with formation of albite, dolomite, quarzite and barite in the surrounding rocks.

However, already in 1938 *Streiss* pointed out the close resemblance of the character-istics of sedimentary and hydrothermal ore formation, and advanced a theory of two-stage formation of the ore, an earlier volcano-sedimentary and a later hydrothermal stage. Later *Scherba* (1964) confirmed this theory in all details and specified the Atas-niskiy type of polymetallic ore deposits in central Kasachstan. According to him this type of deposit was formed as follows: In the basin of the upper Devonian sea volcano-sedimentary rocks accumulated, consisting of tuffs, tuffites, argillites, calcareous rocks and dolomites cut by dikes of diabase-andesites and trachidacite-liparites. In the upper Gamennian, siderite, hematite, magnetite, pyrolisite, psilpnomelan, and braunite were contemporareously deposited between Carboniferous cherty carbonate rocks. These ore minerals now form the sedimentary Fe-Mn ore deposits of this region. When this miner-als were changed through addition of the thin globular pyrite and sphalerite, syngenetic layered Fe-Mn-Zn ores were formed. The metal content for the formation of these ores came to the sea floor from volcanic hydrothermal sources. Later, the layered ores were covered, but the hydrothermal processes continued. Accordingly, layered and lens-like bodies containing barite-sphalerite-galena exist in upper Gamennian rocks.

Scherba is of the opinion that the hydrothermal-metasomatic ores of Atasu were formed after the sedimentary ones, with a gradational change of sedimentary origin to metasomatic origin. Other geologists regard the time of their formation as the post-folding period of the Hercynian cycle. In that case 100 my passed from the beginning of the sedimentary ore accumulation to the end of the metasomatic ore formation.

Osernoe, Buratia. Sedimentary, pyroclastic and effusive formations, together with their co-magmatic rocks, are folded in a synclinal fold which is broken through faults. In the fold are ten layers of kies-polymetallic and sideritic ores with mainly layered and brecciform sedimentary structures imprinted with metasomatic veins which were formed during diagenesis and the following hydrothermal processes: upper ore bodies reach just shortly below the land surface, others up to 1300 m under the surface. Layers lying on top of each other are separated by sedimentary and volcanic rocks. The thickness of the sedimentary and volcanic rocks reaches from 1–10 m (Fig. 1).

According to *Distanow* et al. (1972) the primary amount of the kies-polymetallic ores was deposited contemporaneously with other rocks in a hydrothermal-sedimentary way. The local zones of overlying hydrothermal metasomatic mineralization, which include mainly siderite and siderite-barite with beds of sulphides are later formations. In the light of these facts, there is no doubt as to the formation of the primary ores according to the theory of volcano-sedimentary origin in early Cambrian times.

The actual time of hydrothermal-metasomatic mineralization is not clear. It is possible that its formation followed the early Precambrian volcano-sedimentary ore formation, or that it is linked to locally known early Paleozoic granites or that it is even linked to Permo-Triassic granosyenite-porphyries. In the latter case the average time of formation of the ore deposits in Buratia took 250 my.

The main period of iron formation in Oservoe is dated as early Cambrian, but it lasted over a long period, since the orebodies were formed one after the other in rocks with a thickness of 1.5 km.

Pb/Zn Deposits in Carbonate Rocks

Stratiform ore deposits of phalerite and galena often associated with pyrite, barite, fluorite and other minerals are characteristic for this mineralization. They are connected with certain stratigraphic horizons in thick carbonate rocks, especially dolomites. No magmatic rocks occur in the area of these ore deposits, so that they cannot be post-magmatic derivatives.

The best-known representatives of this type of ore deposit in the USSR are the deposits of the mountain range of Karatau, Kasachstan; others are the deposits of the Mississippi valley in the US, Paint Point in Canada, Upper Silesia (Poland), in the Northern, Eastern and Western parts of Africa, and elsewhere. These ore deposits are the subject of intensive discussion among researching geologists.

Some geologists consider them as of primary sedimentary syngenetic origin, deposited by deeper hot mineral-carrying waters after the formation of their host rocks. These mineral-carrying waters can be magmatic, metamorphic, sedimentary and meteoric (*A. Betechtin, E. Sacharow, I. Knjasew, W. Kreiter, Tsch. Tschere, T. Galkewitsch*, and others). The author also agrees with this hypothesis, which seems to be at least partly true.

The second hypothesis is supported by the fact that apart from layered ore bodies, also crossing ore bodies are found, furthermore there is characteristic metasomatic ore formation and metamorphism of the rocks and a rather high temperature of mineralization at 70–200° C.

In recent times, the long time of formation of the stratiform ore deposits of Pb and Zn in carbonate rocks has been stressed. *Cannon* (1973) proved that the large galena crystals of the Mississippi Valley ore deposit grew over a period of 100–300 my. *Juschko* (1969) found six generations of galena in the ore deposits of Karatau. The first

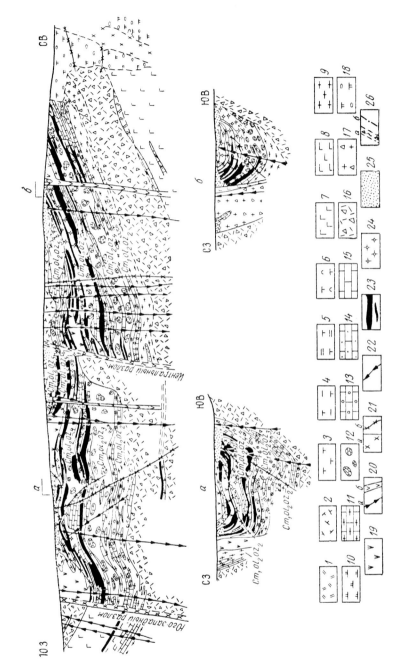

Fig. 1. Vertical and horizontal section through the polymetallic ore deposit of Osernoe/Buriatia, USSR (compiled after *Distanow* et al., 1972). *1–15*: Volcanogenic-sedimentary complex of Oldindinskoi Formation. *1*: Rhyolitic-dacitic lavas (cm_1), *2*: Acid crystalloclastic tuffs, *3*: Acid heteroclastic tuffs, *4*: Acid medium-clastic tuffs (1–5 mm), *5*: Acid coarse-clastic tuffs (5–30 mm), *6*: ignimbrites, *7*: Andesitic-dacitic lavas, *8*: Intermediate tuffs, *9*: Calcareous tuffs, *10*: Carbonaceous tuffs, *11*: Fine lamination of tuff and limestone, *12*: Calcareous breccias with tuffitic-calcareous cement, *13*: Calcareous-tuffitic graywackes, *14*: Calcareous-tuffitic sandstones, *15*: Limestone, *16–20*: Cambrian subvolcanic complex, *16*: Rhyolitic-dacitic magmatic breccias, *17*: Dacitic magmatic breccias, *18*: Agglomerates of breccia pipe facies, *19*: Diabase porphyries altered into greenstone, *20*: Quartz-diabase porphyries, *21*: Trachydacitic and granosyenitic porphyries, *22*: Dolerites and trachy-dolerites, *23*: Polymetallic orebodies, *24*: Siderite orebodies, *25*: Oxidized zone, *26*: Faults

generation is contemporaneous with upper Devonian dolomites, whereas the last generation was formed on crystals of cerussite in the oxidation zone during the Mesozoic or
Cainozoic. Thus the theory of an evolution of ore formation was formed, proposing
original syngenetic sedimentary processes which are later replaced by epigenetic hydrothermal processes. Such views are also held by *Laffite* (1966), *Iuschko* (1969), and
others. The following gives examples of ore deposits of Karatau (according mainly to
Juschko) and of Sardinia (according to material by *P. Zuffardi*).

Karatau, Kasachstan. The geologic basement of the mountain range of Karatau consists of thick layers of carbonate rocks of upper devonian and lower Carboniferous,
which are folded and broken. Among the Pb/Zn deposits of this region two main types
are discerned which are named after deposits in Mirgalisainskij and Atschisaiskij. Deposits of the Mirgalisainskij-type are formed with layered bodies between dolomites, in the
so-called second band horizon in the Gamennian. They consist of thin layers of dolomite and ore minerals, among which the most important are galena, less sphalerite,
pyrite, sulphidic copper, barite, quartz, ankerite and calcite. Bituminized dolomites have
oolithic structures. Collomorphous, metacolloidal structures and kidney-like oolithic
textures, which do not imply a syngenetic formation with ore-carrying dolomites, are
characteristic for ore minerals in the earliest stages of ore formation in Karatau.

Ore deposits of the Atschisaiskij-type which have the form of conformable lenses and
large breaking-through bodies are common in carbonate rocks of Devonian and Carboniferous times. They consist of massive pyrite with galena, sphalerite and other minerals
which were deposited after the first generation. These ore bodies are surrounded by
aureoles of hydrothermal dolomitization of the rocks. The formation of the Pb/Zn
deposits of Karatau is as follows:

During an earlier stage in the late Devonian the foundations of the layered sedimentary ores of the Mirgalisaimskij-type were formed. Later, during the earlier carboniferous time and maybe even later, ore formation is continuing under the influence of circulation of hot underground mineral waters. At that time a change of the primary sedimentary ore concentration and formation of new orebodies could take place, with all
the characteristics of hydrothermal origin. The process from lead formation in Karatau
to the formation of an ore deposit lasted some 160 my.

The Island of Sardinia, Italy. The Pb/Zn-baryte ore deposits are in the South-West
of the island which is built up of thick carbonate rocks, dolomites, sandstones and
schists of Cambrian, Ordovician and Silurian times. These rocks have been folded by
Caledonian, Hercynian and Alpidic orogeneses and are broken by many faults. Hercynian orogenesis is accompanied by granites, and Aplpidic orogenesis by mafic-felsic volcanites. On Sardinia three groups of ore deposits are found (*Zuffardi*, 1966):

1. Synsedimentary layers in so-called ore-bearing horizons in carbonate rocks and
dolomites of the Cambrian period and less in deposits of the Ordovician and Silurian
periods. In the ore layers are the following minerals: chalcopyrite with usually framboid structure, marcasite, chalcopyrite, lightly colored sphalerite, galena, baryte, carbonates. These have a thin-banded, rhythmic-layered texture, collomorphous structure,
and have been deposited, as sulphur isotopes show, by biogene factors. These ores are a
product of lower Paleozoic sea-sediment formation.

2. Breaking-through, epigenetic layers of the ores of analogous contents, in fissures
of Caledonian orogenesis are regarded as rearrangements of sedimentary formations
through chemically active amagmatic hot waters circulating in deeper parts.

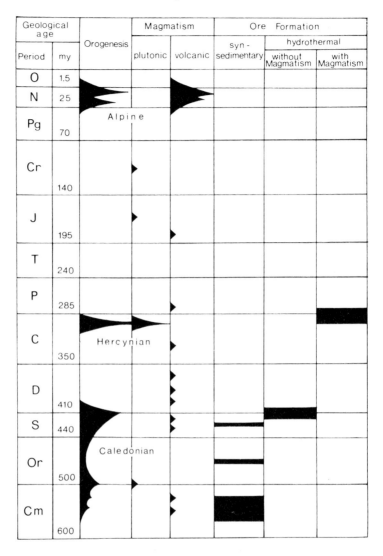

Fig. 2. Synoptic scheme of formation of strata-bound Pb/Zn deposits in Sardinia. (Compiled after *Tamburrini* and *Zuffardi*, 1967)

3. Breaking-through, epigenetic layers of the ores surrounding Hercynian granite massives, with zonar changes from skarn to medium-temperature hydrothermal formations. In the composition of these ore layers epidote, chlorite, quartz, barite, pyrite, phyrrhotite, magnetite, hematite, dark sphalerite, and galena are found.

The history of formation of ore deposits in Sardinia is described by *Zuffardi* as follows: During the early Paleozoic the accumulation of the syngenetic disseminated sulphides of Fe, Zn, Pb and $BaSO_4$ with the carbonate rocks took place. During the Caledonian folding a partial mobilization of earlier deposited ores and their tansformation to a magmatic hydrothermal ore deposit through the effects of hot waters circulating in deeper parts occurred. During Mesozoic and Caenozoic times a hypergene transformation of the ores took place (Fig. 2). From the beginning of the deposition of the ores in the sediments of the Cambrian to the formation of a Hercynian hydrothermal ore deposit ca. 300 my passed.

Copper-Ore Deposits in Sandstones and Schists

Bedded and other forms of ore deposits of pyrite, chalcopyrite and bornite in layered bodies of terrigene rocks have long been known as an important source of copper ores. Their most famous representatives in the USSR are Dschiskasgan, Kasachstan and Udokan, Siberia. Other ore deposits are Zambia, Mansfeld (GDR), Lower Silesia (Poland) and many others.

In many of these deposits not only the characteristics of sedimentary syngenetic, but also of epigenetic hydrothermal genesis are found, to that discussions on the origin of these deposits have been continuing for decades.

Those who support a syngenetic sedimentary origin of these deposits (*W. Popow, D. Saposchnikow, Ju. Bogdanow, W. Garlik, P. Putie, H. Schneiderhöhn* and others) offer as arguments a bedded form of the orebodies, lying conformably in sedimentary rocks, their stratigraphic and lithologic context and having no connection with magmatic rocks; defenders of an epigenetic hydrothermal genesis of these ores present the "break-through character", single ore layers, stage-wise ore formation and the characteristics of metasomatic change of the rocks. The characteristics of a long period of formation are less accentuated in these deposits than in the two mentioned earlier. *Darnley* (1960), however, presents a theory on a primary concentration of ore-forming material in the sediments with a secondary complication by high-temperature hydrothermal solutions.

Some particularities of these deposits regarding the formation are described in the examples of the ore fields of Udokan, Siberia, Zambia and Mansfeld.

Udokan. The ore deposits of copper sandstone in West Sabaikalie, USSR, are in the sedimentary rocks of the early Proterozoic, which have been metamorphosed to greenschist facies. Copper mineralization was found in sediments of six layers. Especially important ore mineralization is in the three layers with stratigraphic intervals between 2500 and 1500 m.

All higher concentrations of copper are connected with fine-grained diagonally layered sandstones, deposited under water, in surface deltas, and in lagoons (*Bogdanov* et al., 1966). The most important deposit, Udokan, lies in a shallow syncline with a length of 25 km. It is broken, sometimes has dikes of gabbro-diabase and granite-porphyries of the upper part of the lower Proterozoic and syenite-porphyries of Mesozoic age.

The orebody forms a horizon in the sandstones with a thickness of 330 m and lies conformably between the hanging and footwall rocks. The ore consists of chalcopyrite, chacosine, bornite and pyrite, with other rarer minerals, with the following zoning: pyrite–chalcopyrite–bornite–chalcosine. Typical textures of the ores are stripes and diagnonal bedding which are characteristic of sedimentary formation. In Udokan break-through veins of quartz or quartz/carbonate can also be found with the same sulphides as in layered bodies, but richer in copper.

The formation took place in two stages. In the first stage syngenetic layers of copper mineralization are bedded between the thick series of delta sands, which were later changed during diagenesis and katagenesis. In the second stage, during the intrusion of magmatic dikes under the influence of the hot mineral waters, the epigenetic ore bodies were formed.

Detailed information on the interval during the two stages are not available to us, but it must be at least several tens of my.

Zambia. According to *Mendelsohn* (1961) and others, the known deposits of Africa form ore belts; that in Sambia/Katanga has a length of 700 km and a width of up to 40 km. This copper belt is in folded, weakly metamorphosed argillites and marly dolomites with orebodies of a thickness of 20–70 m. It belongs to the lower part of the Katanga system of upper Proterozoic age.

Most of the orebodies in this belt have a layered form, lying conformably in the schists and dolomites. They are characterized by the thin-layered stictolothic structure of the ores which have important sedimentary features (diagnonal bedding, slump breccias, etc.). Besides the stratigraphic, the lithologic context is remarkable: ore layers cut out at the transition from argillites and marly dolomites to conglomerates, arkose sandstones and sandy dolomites.

The main ore minerals are pyrite, chalcopyrite, bornite, sometimes linnaeite and other Co-minerals, and uranium minerals. In the copper belt epigenetic orebodies are also known, for example Kipushi and Shinkolobwe. Among these are quartz-carbonate-feldspar rocks with sulphide veins in crystalline rocks of the pre-Katanga system. It has been noted that these veins disappear upwards where the rock changes to conglomerates, sandstones, schists and dolomites of the Katanga series, where sulphidic impregnation predominates.

Although many geologists consider the hydrothermal veins in the cristalline basement, which is supposed to be the main source for the strata-bound ore mineralization, as belonging to the Katanga transgression, no proof is available that there are younger veins as well. Such a possibility was given by radiometric dating of the ore mineralization at Brannerit of Kibushi and Uranirite of Shinkolobwe, which showed an age of 500– 620 my, i.e., lower Paleozoic.

All research scientists of the Zambia copper belt agree that the ore minerals in the strata-bound layers are remobilized by metamorphism after precipitation. The above facts do not exclude hydrothermal influences as described by *Darnley* (1960) and *Garlik* (1964).

Mansfeld. The synclines of Mansfeld, and associated regions in the GDR have an area of several 100 km^2, and consist of middle and upper Permian sediments. The basement rocks are diagonally layered sandstones of the "Rotliegendes", on which the lower Zechstein conglomerates lie from 1–2 to several tens of m thick. On the latter lie the the copper-ore bearing layers, 20–40 cm thick, covered by limestones of the same age. The profile ends with anhydrites of the middle Zechstein.

The ore layer consists of bituminous thin-layered marl. As mounds within it are a few layered ore minerals, mainly bornite, sphalerite and chalcosine. Secondary are chalcopyrite, galena and elementary silver. This layer has all the characteristics of syngenetic sedimentary formation in Permian times.

These characteristics are known and need not be repeated; it is, however, also known that barite veins are in faults which cut through this layer, as well as through the over- and underlying rocks. Where they cut through the copper ore layer, the barite veins are enriched with cobalt arsenides, only little nicke, and also chalcopyrite, bornite, bismuth and pitchblende, which occur at the place of crossing about 15–20 m above the copper layer. They are regarded as later hydrothermal formations. The overall composition of the Manfeld ore deposits of a combination of the syngenetic sediment layer of the copper schists and the cutting-through epigenetic hydrothermal veins. The time interval between the formation of the copper schists and veins is not definite. However, considering the well-dated age of the ore layer, i.e., middle Permian, and according to *Schneiderhöhn* (1958: "All veins are younger than the variszic orogenesis"), this time interval could be several millions of years.

Ore Clasts in the Roof Rocks

If it is true that the older part of the ores is syngenetic and synchronous in the stratiform layers, not only clasts of the surrounding rocks but also disintegrated ore clasts must be found in the roof rocks which were formed during a period of destruction of the ore-bearing layers. Formation of Pb/Zn ore deposits in carbonate rocks, copper schists and sandstones and schists took place under quiet conditions with a gradual covering of the ore horizons with roof rocks. The author does not know of any discoveries of ore clasts in the roof rocks of these deposits. Another matter are the kies ore deposits, which were formed under dynamic conditions of volcanic eruptions which bore ore clasts and deposited them in the pyroclastica of the roof.

Such ore clasts, which prove that the ore sedimentation took place before the formation of the roof rocks, are no rarity in a series of kies ore deposits. Their form is angular, rounded, or irregular. The size of the clasts corresponds to the size of the clasts of the surrounding rocks and reaches from several mm to several tens of cm.

Mineralogical and chemical composition of the ore clasts corresponds completely to the underlying orebody, if the latter did not undergo post-depositional changes (Fig. 3).

Such clasts differ distinctly from metasomatic sulphidic aggregations between volcanic breccia and so-called compactions of colloidal matter, which some geologists have noticed in the hanging rocks. Their dispositional characteristics have been described in many publications (*Smirnov*, 1968).

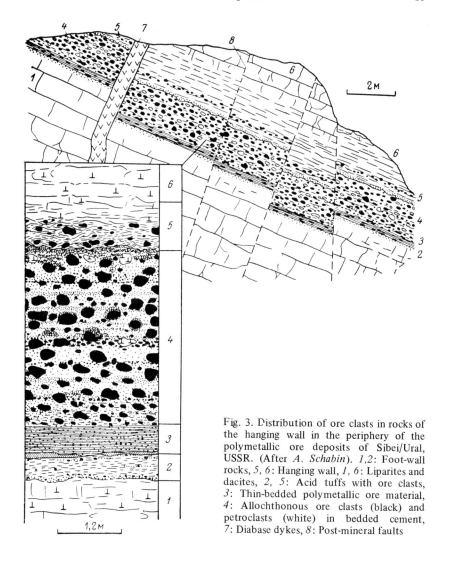

Fig. 3. Distribution of ore clasts in rocks of the hanging wall in the periphery of the polymetallic ore deposits of Sibei/Ural, USSR. (After *A. Schabin*). *1,2*: Foot-wall rocks, *5, 6*: Hanging wall, *1, 6*: Liparites and dacites, *2, 5*: Acid tuffs with ore clasts, *3*: Thin-bedded polymetallic ore material, *4*: Allochthonous ore clasts (black) and petroclasts (white) in bedded cement, *7*: Diabase dykes, *8*: Post-mineral faults

Dikes and Orebodies

The stratiform ore deposits of Pb and Zn in carbonate rocks and copper ores in sandstones and schists are accompanied by dykes, among them dykes of diabase, lesser gabbro-porphyries, albitophytes and plagiogranites.

In the many publications which consider the dykes as post-ore formations, the following observations are given as proofs:

1. dykes breaking through orebodies cut the hanging wall where there is no ore formation;
2. dykes break through faults which cut off ore bodies.

3. dykes cement ore clasts;
4. thin apophyses of the dykes penetrate the ore along the cleavage;
5. the margins of the ore minerals are cut off by the selvages of the dykes;
6. dykes cut through the ribbon structure of the ore.
7. As the contact of dykes with kies ores, pyrrhotine, hematite, magnetite and anto-
   phyllite are formed; bornite is replaced by chalcopyrite and the latter by leaf-struc-
   tured carbonate. Also partial solution and recrystallzation of pyrite, quartz and other
   minerals takes place.

There are, however, also publications on the syngenetic origin of the diabase and porphyry dykes in some kies ore deposits of the Ural. They maintain that such dykes
intruded after the deposition of most of the pyrite, but before the precipitation of the
main content of sulphides of nonferrous metals (*Borodaewskaia*, 1964).

It could be so that a part of the magmatic dykes could have intruded shortly after
deposition of the ore-forming material.

Geochemical Aureoles

Up to now the aureoles within the altered rocks and the distributions of the chemical
elements around the orebody have been studied without considering the long period of
ore formation and evolution of the ore-forming processes. Very interesting results
regarding the geochemic aureoles of kies ore deposits were gained, for example, by
Owtschinikow and *Baranowa* (1970).

The chemical content of such aureoles was determined; considerable enrichment of
metals in them which are not counted to in the reserves of the orebodies and a regular
zonar distribution of the metals in the aureoles was found. However, no attempt was
made to analyze the formation of the aureoles of altered rocks and geochemical aureoles,
taking into consideration the time of formation and evolution of the stratiform ore
deposits. Without this the results of such a research are considerably devalued. By considering the time factor, the following problems could be solved.

In an early stage during sedimentation of ore-forming material of continental origin,
layers in nonaltered rocks must have formed. During that time geochemically detectable
aureoles could have formed in the footwall as introductory formation, and in the hanging wall as final formation. Such aureoles are characteristic for copper schists and
copper sandstones; they probably can also be found in other stratiform deposits, which
have passed through an earlier stage of sedimentary formation, whereas in hydrothermal
ascendant ore transportation into sedimentary layers of stratiform ore deposits, this
type of aureole can only be found in the footwall, as has been proven, for example, for
the kies-ore deposits of Chudes, Caucasia. The same is true for the deposits of Osernoe,
Buratia and others.

Aureoles in the hanging wall of this type of deposit can only be formed by resedimentation of the orebody. An example for this is the deposit at Filistshei, Caucasia.
Furthermore, aureoles and geochemical distribution all around the orebody can occur in
the altered rocks by hydrothermal processes.

Many stratiform ore deposits can only be understood by considering the long time of formation, just as the accompanying aureoles can only be correctly interpreted if one considers the evolution of ore formation in time.

Folding, Metamorphism and Ore Formation

The investigation of the interactions of folding, metamorphism and ore formation is one of the main problems of genesis of stratiform ore deposits, especially of those in Proterozoic and Paleozoic rocks which are all more or less dislocated and altered. Those geologists who mainly consider the characteristics of the ore body before folding and metamorphism belong to the group favoring a syngenetic sedimentary formation. Other geologists who regard mainly the characteristics which occur during folding and metamorphism belong to that favoring an epigenetic hydrothermal formation.

However, if this problem is studied in the light of the evolution of stratiform ore formation in time, many contradictions can be solved.

The first stage of ore formation obviously takes place before folding and metamorphism, as is well proven for the products of the primary ore-formation period of Atasu and others. The ore formation of later stages can take place by tectonic deformation and in some cases also after metamorphism of different intensities, for example, the Dschusinskoe deposit in the Ural, which was formed in two stages.

In the early Devonina, before folding and metamorphism, pyrite and chalcopyrite ores were formed, associated with formations of quartz and sericite in the rocks. In the late Carboniferous, after folding and metamorphism, lead and zinc minerals were formed accompanied by alteration of the wall rocks to quartz, sericite, carbonate and chlorite. Thus the polymetallic kies deposits were formed with complicated aureoles of the altered rocks.

Connections between Younger and Older Ore Formation

The earlier stage formation of many stratiform ore deposits, as well as the later stage, can take a long time and can develop in different substages. The long duration of ore formation in the earlier stage is proven by the stockwerk column of layers in Osernoe, Buratia, which was repeatedly formed within a km-thick layer of lower Cambrian tuffs and schists. Examples for a stage-wise ore accumulation during later periods are even more abundant.

It should be noted that a concordant or discordant position within the layered rocks gives no clues as to an earlier or later stage. In the earlier stage during deposition of ore-forming material out of hydrothermal solutions in concordant layers, discordant bodies near the footwall can also be formed, as can often be seen in kies ore deposits. On the other hand, pseudoconcordant metasomatic layers can be formed simultaneously with discordant bodies during later stages, in Atasu, for example.

The products of earlier and later stages or ore formation occur in different ways.

In a first case, when chemically active solutions which do not add new ore-forming elements affect earlier deposited ores, only a rearrangement on a large or small scale can take place. Examples of this are the enriched parts and the veins breaking-through in deposits of copper sandstones and slates. During such processes no additional ore minerals are formed.

In a second case, with chemically active solutions which contain not sulphur, but nonferrous metals, an enrichment of these ores by copper, lead, zinc, and other elements can take place through displacement of iron and of parts of its sulphides which were formerly abundant; some polymetallic kies ore deposits were formed in this way (*Lawring*, 1961).

In a third case, primary syngenetic layers can be remobilized by hydrotherms after being covered by hanging-wall sediments and then form epigenetic bodies, as observed in Atasu and possibly also in Osernoe.

In a fourth case, primary ores, influenced by later magmatism, e.g., granite intrusives, can be re-arranged and concentrated by post-magmatic processes, as in Sardinia.

The amount of ores formed during the earlier or later stage varies. in some deposits in copper sandstones and slates, the amount of the syngenetic ores is very large, whereas in some kies deposits, such as Atasu or Leninogorsk, epigenetic ores of the later hydrothermal stage predominate.

Supply with Ore-Forming Material

In the eralier stage ore-forming material for syngenetic strata-bound deposits could enter the sediment by continental erosion or by ascendant volcanic hydrothermal solutions. As examples of the first type many scientists quote the copper sandstones and slates, for the second, certain layers in the kies ore mineralization. In both cases a large variation of sulphur isotopes can be noted, which indicates a participation of biogenetic processes due to the remarkable fractionation of this element.

The supply of metals during a later stage was caused by deep circulation of hot mineralized hydrothermal solutions, in which the metals originate from an autochthonous source either of earlier deposited material or from far away. If such a source was connected with a deeper magmatic focus, the sulphur isotopic ratios would be close to the meteoritic standard. A remarkable geochemical stability during the whole formation of stratiform ore deposits is noted. During the whole process, however long it may have taken, a repetition of the same metals — copper, lead and zinc in the form of sulphides, and also iron as oxide, carbonate or sulphide — can be observed. Sometimes additional elements like gold, silver, uranium and others can be found. The ratio of the main elements and the variation of this ratio within time for different deposits can vary, but the general consistency is stable. As a rule, stratiform ore formation is not disturbed by the occurrence of other metals.

Some General Reflections

Geochemical stability and formation over long periods of time is not only characteristic of stratiform ore deposits where it is stressed by radical evolution of ore deposition.

With intensive research it can also be determined for other types of ore deposit. It has been found in areas of uranium deposits, which are mostly known as sedimentary or sedimentogene-infiltrated, but also as hydrothermal formations (*Smirnow*, 1963).

Hydrothermally altered uranium deposits can, for example, be destroyed and then form sedimentary uranium deposits.

Returning to the main object of this discussion – the stratiform ore deposits –, the fundamental error in the research of their genesis is, that those geologists who regard syngenetic features, consider these deposits as sedimentary formations, while those who regard epigenetic features, consider them as hydrothermal.

They are, however, polychronous and polygenous formations which can only be understood when considering the long period of formation and the evolution of the ore-forming processes from the syngenetic deposition of the material during the formation period of ore-contaning layers up to the subsequent epigenetic enrichment of the ores taking place after formation and alteration of the layers.

The geologic age of these deposits cannot be determined only by the time of precipitation of the ore-forming products, but the geologic position of layers in which the ore formation originated and of layers in which it terminated has also to be considered.

References

Bogdanow, J.B., Kotschin, G.G., et al.: Copper deposits of Olemko-Witimskoi district. Moscow: Nedra (in Russian) 1966

Borodaewskaia, M.B.: Relation of polymetallic mineralization with magmatism and questions of origin of polymetallic ore deposits of Southern Ural. 22nd Intern. Geol. Congr., Reports of Soviet Geol. Moscow: Nedra (in Russian) 1964

Cannon, R.S.: Lead isotope variation with crystal zoning in a galena crystal. Science **142** (3592) (1963)

Darnley, A.G.: Petrology of some Rhodesian copperbelt ore bodies and associated rocks. Trans. Inst. Mineral. Metall. **69** (1960)

Derbikow, I.W.: To the problem of origin of sulfide ores of Smeinogorsk district, Rudnogo Altaja. Geol. Ore Deposits **6** (in Russian) (1966)

Distanow, E.G., Kowalew, K.P., Tarasowa, P.S.: Geology and genesis of the polymetallic Pb-Zn ore deposits (W-Sabaikalia). Geol. Ore Deposits **2** (in Russian) (1972)

Eremin, N.I.: The significance of schistose and fractured zones for the localization of polymetallic ore deposits as demonstrated by the Dschuschinskoe ore deposits, S-Ural. Bestnik MGU **6** (in Russian) (1963)

Garlick, W.G.: Criteria for recognition of syngenetic sedimentary mineral deposits and veins formed by their remobilisation. Zambia, 1964

Iuschko, S.A.: Mineralogy of stratiform Pb-Zn ore deposits of Southern Kasachstan. Moscow: Nedra (in Russian) 1969

Jakowlew, G.F., Chisamutdinow, M.G., Demin, J.I.: Polygene and polychrone polymetallic ore deposits of Rudnogo Altaia. Geol. Ore Deposits **3** (in Russian) (1975)

Laffitte, P.: La métallogénie de la France. Bull. Soc. Geol. France **8** (1966)

Lawring, T.S.: Sulfide ores from solutions undersaturated in sulfur. Physico-Chemical Problems in Petrogenesis. Vol. I. Moscow: Akad. Nauk USSR (in Russian) 1961

Mendelsohn, F. (ed.): The Geology of the Northern Rhodesian Copperbelt. London: McDonald & Co. Ltd. London 1961, p. 396–399

Owtschinnikow, L.N., Baranow, E.N.: The endogenous geochemical halos of polymetallic ore deposits. Geol. Ore Deposits **2** (in Russian) (1970)

Pokrowskaia, I.W., Kowrigo, O.A.: On the volcanogene-sedimentary origin of the stratified ore deposit of Ridder-Sokolnogo. Geol. Ore Deposits **3** (in Russian) (1970)

Scherba, G.N.: Geology of Leninogorskogo ore district. In: The Polymetallic Ore Deposits of Rudnogo Altaia. Moscow: Gosgeoltechisdat (in Russian) 1957

Scherba, G.N.; Some special features of ore deposits of the Atasuiskij type. Report Akad. Nauk SSR Kasachstan. Geol. Ser. 5 (in Russian) (1962)

Scherba, G.N.: Problems of origin of the polymetallic ore deposits of Rudnogo Altaia. Sow. Geol. 6 (in Russian) (1968)

Schneiderhöhn, H.: Ore Deposits. Moscow: Inostr. Lit. Publ. (in Russian) 1958

Skriptschenko, N.S., Roschnow, A.A., Litkin, W.A.: The zonality of the ore deposits of Scheiremskoe group, Central Kasachstan. Geol. Ore Deposits 5 (in Russian) (1971)

Smirnow, W.I.: Metallogeny of uranium. In: Problems of Applied Radiogeology. Moscow: Gosatomisdat (in Russian) 1963

Smirnow, W.I.: Polymetallic ore deposits. In: Genesis of Endogenous Ore Deposits. Moscow: Nedra (in Russian) 1968

Smirnow, W.I.: On the origin of endogenous ore deposits. Isw. Akad. Nauk, USSR, Geol. Ser. 3 (in Russian) (1969)

Tamburrini, D., Zuffardi, P.: Ulteriori sviluppi delle conoscenze e delle ipotesi sulla metallogenesi sarda. Giornata Studi Geominerari, Agordo, 7 Oct. 1967, 21 p.. Trento: 1967

Weitz, B.I.: Mineralogy of polymetallic ore deposits of Rudnogo Altaia. Alma Ata (in Russian) (1959)

Plate Tectonics and Mineralization in China

J. PEREIRA, London

With 1 Figure

Contents

Summary

The mineral resources of China have not yet been put into full action. A knowledge of plate tectonics and this link with mineralization is essential before the major works of exploration can begin. This outline suggests how the deficiencies in China's actual mineral resources could be extended and place China on an independent basis. The link between plate movement and mineralization began as a largely empirical series of observations, but is rapidly becoming fundamental to exploration.

1. Introduction

For the purpose of this work it is necessary to use the Chinese general geological map of China 1:4 million and the Geological Society of America (A.G.S.) Tectonic map of China and Mongolia 1:5 million (1974) in conjunction. Using the Tectonic map, areas referred to as "stable blocks, platforms" etc are more or less synonyms with "plates", while fold belts are "interplate zones" or "plate edges". This is an interpretation that is too simple, but is needed to avoid complex geological detail. The Chinese map is essential in order to pick out the main magmatic or migmatite masses. There are also variations in the distribution of Cenozoic volcanics and the Ultramifics, some of which are probably ophiolites and important to drawing plate boundaries. The resulting map is based on information from both these sources.

It is assumed that Proto China and China proper represent a collision state. A further assumption is that the "spreading area" lies along the coast of China – the Island Areas of which Japan is a part, the edges of the Phillipines plate, and close to China, the island of Taiwan.

It should also be mentioned that there are problems in placing the Sinian in its correct relationship to the Phanerozoic time scale and the Proterozoic.

2. The Main Plates and Interplate Zones

Sparce paleomagmatic data (*McElhinning*, 1965) Sesomic evidence (*Savarasky* and *Gohnbourg*, 1969) and Cenozoic volcanics suggest that the greater part of China is a wedge-shaped mass, driven into Tibet and Mongolia in a north-easterly direction. The outline of the wedge appears to be still fairly active, with the main mass itself composed of a number of plates that appear to have been in existence in Sinian times.

3. Outline of Main or Tethyian Wedge

Starting in the south the "interplate zone" runs north along the Salween/Mekong fold belt, between the Indian plate and the South China plate, then swings north-east and finally east along the border of Tibet which itself appears to be a plate or stable area of long standing.

It then makes a big loop to the north-east of Lanchow. This is well-marked by a series of granitic intrusions that can only be seen on the Chinese map. They are "filon"-shaped as in Brittany, and their form suggests the turning point of the loop.

The interplate zone then runs west along the boundary between the Ala Shan stable block and the Ordos Shanshi region (itself a plate). From here it swings north east through a series of Cenozoic volcanics and into Russia, along the line of the Verhoyansk and Sikhote-Alin mountains.

Areas shown on the A.G.S. map as ultramafic rocks, including serpentinities etc are of importance, as one would suspect that they are ophiolites. In the south below the L of the Mekong Foldbelt and by the A of Hsing-an volcanic belt, it is not of importance to examine the ophiolites in the field; it is more important to see if they are placed where one would expect them to occur.

Fig. 1. Plate tectonic pattern and distribution of mineral deposits in China (*Pereira*, 1976)
1: Musan (Fe) 42° 128°, *2*: Liao-ning (Fe) 41° 123°, *3*: Kaiping Haicheng (Fe) 40° 123°, *4*: Tchahak (Fe) 40° 115°, *5*: Houpeh (Fe) 30° 114°, *6*: Kewichow (Fe) 26° 107°, *7*: Sud Kiangsi (Wo) 1 (26° 115°), *8*: Sining Ho (Au, Cu) 36° 103°, *9*: Chang Tong (Ba) 37° 117°, *10*: Chang Tong (Fe) 36° 117°, *11*: Fusin (Zn) 42° 122°, *12*: Hei Lung Kiang (An) 47° 129°, *13*: Kochiu shan (Pb) 23° 103°, *14*: Kungshan (Fe) 26° 103°, *15*: Hsikuang Hsihua (Sn) 26° 112°, *16*: Shui Kou Shan (Pb, Zn) 27° 114°, *17*: Sining HO (Au, Cu) 37° 101°, *18*: Taiyuan (Fe) 41° 122°, *19*: Anshan (Fe) 26° 105°, *20*: Penhsi (Fe) 41° 124°, *21*: Wuhan (Fe) 40° 117°, *22*: Paotou (Fe) 41° 110°, *23*: Chenghao (Fe) 36° 115°, *24*: Lungyen (Fe) 25° 117°, *25*: Funshun (Al) 23° 111°, *26*: Changling (Al) 44° 124°, *27*: Hungtoushan (Cu) 42° 124°, *28*: Chingyuan (Cu) 42° 125°, *29*: Tung Hua (Cu) 42° 126°, *30*: Yashan (Cu) 37° 115°, *31*: Shuikoushan (Pb/Zn) 42° 124°, *32*: Chinghengtzu (Pb/Zn) 42° 124°, *33*: Hsiuyen (Pb/Zn) 24° 110°, *34*: Chunanan (U) 25° 114°, *35*: Nanshan Mts (Fe) 38° 95°, *36*: Kueiping (Mn) 23° 110°, *37*: Tao Chung (Fe) 29° 100°, *38*: Taoling (Pb, Zn) 29° 113°

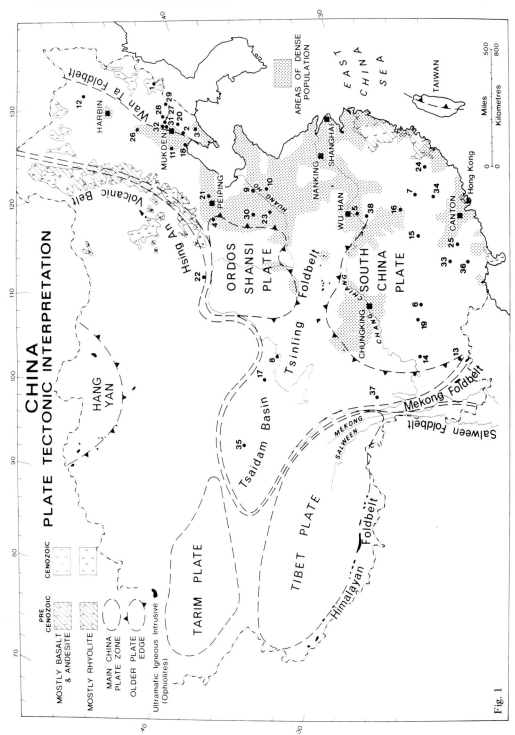

Fig. 1

4. The Interior Plates

The South China plate seems distinctly separated from the Ordos/Shanshi plate by the Tsinling fold belt running roughly east to west towards Shanghai. This is also marked by ultramafics (Ophiolites) just by the F and B of Tsinling Fold Belt (see Fig. 1). The ultramafics are also shown on the Taiwan map, but appear more specific and larger. As in the outline of the wedge area bounding the main China wedge, this zone is also marked by Cenozoic volcanics.

The Coastal Regions. In the north-east, the Wan Ta fold belt suggests a plate-collision situation with the Sikahote area and Korea moving as mentioned in the preceding paragraph.

In the south-east various authors place Taiwan as a spreading area (Cenozoic), marked on the mainland of China by a volcanic belt of Palaeozoic age, supposing continued activity along plate edges of old origin.

5. General Factors

Palaeogeography and Tectonics. Grabau's (1924) Palaeogeography of China outline all the marine incursions and periods of orogeny from the pre-Sinian to the present day and from his description of events, it appears that there is little evidence of major marine deposition after the Permian, in fact from the middle of the Palaeozoic upwards, marine deposition appears to have been more and more confined to the south of China. He also points out that important volcanic areas in the Cretaceous and Jurassic were succeeded in the Tertiary by basaltic flows, in particular the Hsing-an Volcanic area in the north and in the south through Yunnan and into Laos. One must add that in all probability the marine incursions were not marine incursions. Proto-China was coming into existence.

A much more helpful general guide to the Geology of China is T.K. Huang (1945), who gives an excellent account of the main tectonic events and units. It is clear that the picture presented by the author is, as it is intended to be, a simplification. It seems to confirm that the Himalayan orogeny produced a good deal of plate-against-plate movement, but no major change to the form of the mainland continent.

Seismic Evidence. The best seismic data I can find appears to support the plates outlined in section 3. Naturally seismic data tends to emphasize only the present stresses.

Paleomagnetic Data. Very little is available except for the north-east, where a number of measurements suggest a collision state along the line of the Sikhote mountains in the Cretaceous.

Mineralization. China's known mines appear to be insufficient for the needs of this rapidly growing country. Inron ore is worked at grades which are far below those in current world production (Layfitte and Rouverol, 1964).

There are no reported porphery copper mines, though I would suspect that Sining (Youngbushand, 1925) is such a site. The lead zinc mines which are exploited are below normal grades, as are also the Sinian Iron mines which are reported to be oolitic. Aluminium also appears to be in short supply, although this is difficult to believe that bauxite is not available in vast quantities, especially in a country which has given the world the name Kaolin. There is an interesting reference to diamonds in the Tunhuang region, which supports the tectonic make-up of that area, but according to *Needham* (in press) the report dates back to AD 277. It is anomalous that one of the only big exports is antimony, apparently only matched in scale by the Rhodesian deposits. In brief, the apparent mineral resources of China cannot be equated with the potential that a country of this size and importance should have.

It is suggested that the deficiencies are not so much a real lack of resources, but a reflection of the historical distribution of the Chinese population. For detailed information *Tregear* (1965) *A Geography of China* is recommended. The distribution of the plates suggests that some of the most promising areas have not yet been explored or exploited.

A.B. Ikonnikov's (unpubl.) highly detailed study of the mineral resources of China was not available to the public at the time of writing, but his maps may be tabulated as follows:

Map No.	Elements	No. of sites located
7	Mn and Cu	148
8	Pb/Zn Sn Ni Hg Sb As Co Cd Bi	112
9	Al, Mg ti and other light metals	79
10	Au, Ag, Pt	87
11	Wo, Mo, U and others	116

This is too many to plot on the map in this paper, so the only ones shown are well-known sites listed in Appendix A; nonetheless, the highly detailed information tabulated above still gives the same distribution pattern with the addition of sites in the Altai Shan area, the most north-westerly plate marked.

Certain metals are specific plate edge indicators, notably Pt, Cr, and Ni. In this connection one must mention Pt in the Hsing-an belt, Cr in the Tsidan basin and Ni in the Mekong fold belt.

6. Conclusions

Most of mainland China appears to have become more or less of one piece in Sinian times and *Grabau* (1924) strongly stresses the contrast between the Sinian and the immensely altered pre-Sinian systems. It is difficult to suggest major Continental movement after the Permian, so the time when one seeks major changes appears to be confined to shifts which in Tethyian terms are relatively minor: a kind of hinging action beginning in the north-east where palaeomagnetic evidence suggests a Cretaceous collision, closing to the south with movement right through Yunnan and to the south across the border. Probably the southern part moved in a sinestralsence against the India Plate.

It is also important to note that over 90% of the population of China lives along the coastal belt in only some 11% of the total area.

If one assumes the kind of picture shown by *Dixon* and *Pereira* (1974) or *Garson* and *Mitchell* (1976) where the mineralization of the Tethyian is well known, it appears from mines that are known that (as in Brazil), they tend to be concentrated in areas of high or moderate population density and that China has a vast potential along the inner plate zones outlined, probably unexplored on a thorough basis. Undoubtedly inaccessability adds to China's exploration problems.

7. Discussion

Evans (1975), discussing the data of mineralization characteristic of Tethyan subduction zones in Switzerland, suggests very convincingly that lateral movements between plates produce geosynclinal development that is poorly mineralzed and not typical of the region as a whole.

On this basis one must add to the previous paragraphs on plates in China that much of the plate-against-plate movement may have been lateral. If this is the case, then the northern areas of what has been termed the Tethyan wedge should be more favourable than the southern areas such as Yunan, since strong lateral movement is suspected along this region and to the south in Cambodin and Laos.

References

Bebb, J.T., Harland, B.H.: Personal communication
Burk, C.A., Drake, C.L. (eds.): The Geology of Continental Margins. Berlin–Heidelberg–New York: Springer 1974
Chinese Academy of Geological Sciences: 1:4,000,000 geological map 1976
Dixon, C.J., Pereira, J.: Plate tectonics and mineralization in the Tethyan region. Mineralium Deposita 9, 185–198 (1974)
Evans, A.M.: Mineralisation in geosynclines – the Alpine enigma. Mineralium Deposita 10, 254–260 (1975)
Garson, M.S., Mitchell, A.H.: Mineralization at Destructive Plate Boundaries. Volcanic Processes in Ore Genesis. Imm. Symposium 1976, pp. 81–97
Geological Society of America: 1:5.000.000 Tectonic Map of China and Mongolia (1974)
Geological Survey of Taiwan: 1:4.000.000 General Geology of China. Unpubl. Note, this map is a copy of the official Chinese map 1:2 million which is not available outside China
Grabau, A.W.: Paleogeography of China. Vols. I, II (1924)
Huang, T.K.: On major tectonic forms of China. Geol. Surv. China Memoir *1* (1945)
Laffite, P., Rouveyrol, P.: Carte minière du globe à 1:20.000.000, sur fond tectonique. Orléans: BRGM (1964)
McElhinning, M.W.: The South Pacific. USA, 1965
Needham, J.: History of Science in China, Vol. III. Cambridge: University Press, in press
Savarasky, R., Gohnbourg, J.: Seismicity of continental Asia and the region of the Sea of Okhotsk 1953–65. Am. Geophys. Union Monogr. 13, 902–910 (1969)
Tregear, T.R.: A geography of China. Cambridge: University Press 1965
Younghusband, F.: Peking to Lhasa. Journeys Made by Brigadier General G. Pereira. London, 1925

Phase Boundaries as an Instrument for Metal Concentration in Geological Systems

E.T. DEGENS, Hamburg, and P. STOFFERS, Heidelberg

With 11 Figures

Contents

Summary

Hutton's principle of the uniformity of process, which holds that the present is the key to the past, is seen in a new light. The well-mixed oxygenated ocean of today seems not to be the model environment for the past 600 my. Instead, we are dealing with a sea which alternates between two opposing states: stratified and nonstratified. Depending on local or regional hydrographic settings, which may effect deep-water circulation and the hydrodynamics of water movement vertically and horizontally, one may encounter all sorts of transitions.

At present, some areas in the open ocean are already anaerobic, and indicate the delicate balance between the oxic and anoxic states. All these findings lead us to conclude that a number of former interpretations regarding the origin of sedimentary cycles, the history of the oceans, the workings of evolution, and the origin of strata-bound ore deposits have to be re-examined.

1. Introduction

The system of air, water, life and soil is in an almost balanced state. At their respective boundaries, the phases interact and many significant geological and biological reactions occur. For instance: on air–soil contact the process of weathering and erosion takes place; at the sediment–water interface we witness almost in slow motion the gradual formation of sediments with all the intriguing facets of texture and structure; at the air—water interface the exchange kinetics of gases is controlled, and the interplay between a planktonic cell and the environment is regulated by the outer membrane of the organism. It thus appears that the physicochemical phenomena established at the boundary between two different states characterize these two states; or expressed differently, the physicochemical properties of the environment are "written" on the boundary layer, which can be a soil, a sediment, a membrane, or any interface present in the open environment.

In the following study, we will examine in which way the flux of metal ions in aqueous systems is controlled at phase boundaries.

2. Background

Upon weathering in tropical regions, metals such as iron and aluminum may concentrate in the form of bauxites and laterites. These processes take place at the air—soil interface and are well understood (*Valeton*, 1972).

In contrast, metal enrichment at the water—sediment interface are questions of immense importance which prove to have tremendously large uncertainties (e.g., *Lynn* and *Bonatti,* 1965; *Presley* et al., 1967; *Price* and *Calvert*, 1970; *Edington* and *Callender,* 1970; *Bischoff* and *Ku,* 1971; *Glass* and *Podolski*, 1975). This is so, because exchange across the interface is controlled by a variety of factors such as: (1) mineral—water equilibria, (2) sorption processes, (3) ion exchange, (4) redox potential, (5) biological activities, and (6) water depth. With respect to the last factor it could be shown (*L.W. Wood*, 1975) that in shallow water, bioturbation is more critical for material transport across a mud—water interface than ordinary diffusion, whereas at greater water depth it is the molecular diffusion (*Berner*, 1976).

At the entrance of a river into a lake or the ocean, changes in chemistry, water velocity, and temperature may cause a rapid deposition of minerals; coastlines too can act as an interface, and placer deposits may form. The physical and chemical principles at work in this regime are already part of textbook geology. Less profound is our knowledge on density structures, that may arise in a lake or the open sea in the form of a quasi-horizontal stratification in mid-water, and their effect on metal distribution. Such phase boundaries, termed pycnoclines, are a consequence of temperature or salinity fluctuations and are referred to as thermoclines or haloclines, respectively. Stratification may last for hours, days, months and even thousands of years. In addition, the interface can move up and down a water column in response to variations in salt content and water temperature (*Degens* and *Stoffers*, 1976).

The boundary between a biological cell and the environment is a membrane. The flow of metal ions in and out of a cell is regulated by this barrier. Organisms are known to concentrate certain metal ions in substantial amounts. Biomineralization is a consequence of such activities (*Degens*, 1976). Other metal ions can be introduced into the organism which are physiologically harmful, and in the field of pollution chemistry one is presently trying very hard to understand the uptake characteristics of metals by organisms (e.g., *J.M. Wood*, 1975; *Goldberg*, 1976).

In this article we will focus on mechanisms at work in the concentration of metals at: (1) the water–sediment interface, (2) pycnoclines, and (3) organic membranes. We have selected a few critical areas to illustrate the regulation principles at or close to phase boundaries.

3. Sediment–Water Interface

3.1 Fractionation of Iron and Manganese

Most crystalline rocks have Fe/Mn ratios close to 25. Weathering processes will mobilize the two elements at different rates, as will also retrograde weathering occurring during diagenesis. As a result, sediments show a wide range in Fe/Mn ratio. Low-temperature mineral equilibria (*Krauskopf*, 1957; *Panel on Orientations for Geochemistry*, 1973; *Garrels* et al., 1975) are principally responsible for the fractionation of iron and manganese. For instance, the two elements may become separated because iron oxides and hydroxides precipitate at lower oxidation potentials than the comparable manganese compounds at any given pH. Furthermore, manganese sulphides are more soluble than iron sulphides and the "right" Eh/pH relationships for the precipitation of a manganese sulphide are practically nowhere met within the exogenic cycle. Thus, under anoxic conditions, much of the Fe^{2+} is precipitated as FeS, which rapidly transforms into pyrite or marcasite, whereas manganese stays in solution.

If a water contains a certain high level of carbonate ions, precipitation of manganosiderite may occur (*Calvert* and *Price*, 1972), and the geochemical implications will be discussed later in our discussion (see sect. 4.3).

Abundance of dissolved SiO_2 may promote the formation of nontronite as found in Lake Chad and Lake Malawi (*Lemoalle* and *Dupont*, 1973; *Müller* and *Förstner*, 1973). In Lake Malawi precipitation of nontronite and limonite is assumed to occur at the contact sediment–water. The dissolved Fe^{2+} required for the formation of nontronite is leached from the sediment under reducing conditions at a pH below 7; SiO_2 is thought to be derived from geothermal discharge (*Müller* and *Förstner*, 1973). Alternatively, dissolution of diatom frustules which are abundant in the lake may supply the required SiO_2.

The mode of nontronite formation, as suggested by *Müller* and *Förstner* (1973), is shown in Figure 1. In addition to nontronite and limonite, vivianite is also present in Lake Malawi sediments. The formation of vivianite is explained by dissolution (pH > 7) of calcium phosphates within the sediment and redeposition as iron phosphate in the uppermost sediment layer under reducing but slightly alkaline conditions. It is of note that the wide-spread marine phosphorite deposits are believed to be a product of metasomatosis of a calcium carbonate-precursor mineral present at the sediment–water interface. The phosphorus is thought to be derived from the decomposition of planktonic debris (e.g., *Ames*, 1959).

The occurrence of recent glauconite and chamosite in Loch Etive, Scotland (*Rohrlich* et al., 1969) and in front of the Niger Delta (*Porrenga*, 1966) is obviously also triggered by processes taking place in the topmost strata. The mode of formation of these iron-silicates, however, is still in doubt.

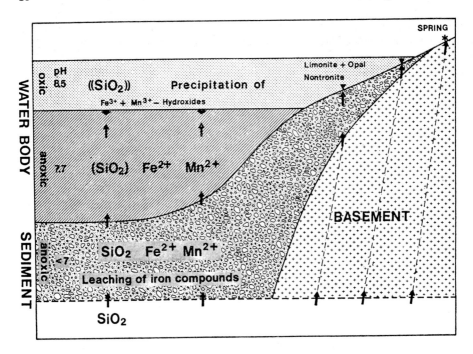

Fig. 1. Reaction scheme of iron, manganese and silica precipitation in a stratified Lake Malawi (*Müller* and *Förstner*, 1973)

3.2 Manganese Nodules

Some of the most interesting metal enrichments found at the sediment—water interface are the manganese nodules which are scattered in great numbers over large parts of the sea floor. For the Pacific Ocean alone, about 200 million t of nodules are estimated.

Ferromanganese nodules and crusts were first discovered at the sea bed during the Challenger Expedition of the 1870s (*Murray* and *Renard*, 1891). Since then many papers have appeared on the origin and geochemistry of these nodules (e.g., *Goldberg*, 1954; *Mero*, 1965; *Manheim*, 1965; *Seibold*, 1973). In recent years interest in manganese nodules has increased greatly with the recognition of the economic value of the major accessory metals in the nodules such as cobalt, nickel, and copper.

It has been suggested that sediment pore water is perhaps the most important source for manganese, iron, and the accessory metals found in the crusts and nodules. However, the separation during continental weathering, as well as the release by ocean ridge volcanism, may also play an inportant part. Under reducing conditions a considerable amount of the manganese and iron in the sediment is dissolved and concentrated in the pore water. As much as tens of mg/l manganese is found in pore waters from the Baltic Sea (*Hartmann*, 1964), and an up to 150-fold enrichment for manganese is reported from Saanich-Inlet-Fjord, B.C. (*Presley* et al., 1972). An upward migration of the diagenetically derived metals through the sediment is suggested. The upward diffusion of the reduced metals is caused by concentration gradients or sediment compaction.

Above the sediment-water interface there are profound changes in Eh and pH which cause iron and manganese ions to be oxidized and precipitate as hydrated ferric hydroxides and hydrated manganese oxides.

Ferromanganese nodules and crusts are not restricted to the marine environment. They have been widely reported from numerous freshwater lakes (*Neumann*, 1930; *Ljungren*, 1953; *Gorham* and *Swain*, 1965; *Cronan* and *Thomas*, 1970; *Dean*, 1970). However, in contrast to deep-sea nodules, a low trace metal content is characteristic of the ferromanganese deposits of freshwater lakes. The high concentration of trace metals found in the marine ferromanganese nodules is due to the fact that hydrated oxides and hydroxides of iron and manganese can pick up certain anions and cations (*Goldberg*, 1954). The efficiency of the metal ion "scavenging" process depends on the flux of heavy metals in sediment or water, and the time available for fixation to the Mn/Fe mineral substrate placed at the sediment—water interface.

4. Stratified Water

4.1 Feedback Mechanism

The world ocean is fully oxygenated except for some local areas in regions of upwelling or stagnation. Climate-controlled physical factors turn over the ocean's water in a matter of a few hundred to a few thousand years. At this rate, molecular oxygen is recharged must faster in the deep sea than it is consumed by the oxidation of organic matter at greater water depth.

During prolonged warmer climatic stages or when land-locked seas develop, thermo- or haloclines may become so stabilized that they only move up and down in response to seasonal changes, tectonic activities, of some other major perturbations in the environment: but they rarely break up entirely. In such conditions molecular oxygen will remain abundant in the euphotic zone, but will gradually drop to zero below the density boundary. This will cause the development of a euxinic environment in which no higher forms of life can exist and molecular oxygen is replaced by hydrogen sulfide.

The speed at which anoxic conditions develop, once a permanent stratification is achieved, is rather short if one takes data from the Black Sea as a standard of reference. Each g-atom of carbon in plankton requires 1.3 mol of oxygen for its oxidation (*Redfield* et al., 1963). Since the amount of organic matter that falls through the present O_2-H_2S interface in the Black Sea is fairly well known (*Deuser*, 1971, 1974), the "decay" constant of the oxygen reservoir can be calculated. From this number, the upward progression of the O_2-H_2S interface can be derived. Calculations show that the thermohaline boundary requires ~ 3000 years to rise from the 2200 m-deep abyssal plain to the present pycnocline located at a water depth of about 200 m. Using these figures, one can infer that geologically speaking a stratified global sea should become anoxic within a short time. On the other hand, a Precambrian stratified ocean may contain plenty of oxygen in its upper layer for extended periods of time, even if primary productivity was lower than today. This has considerable consequence for the origin and evolution of life and atmosphere in the Precambrian (*Degens*, 1977).

A significant aspect of this problem, which has received little attention in the past, concerns a possible feedback mechanism between oxidizing and reducing environments, which may result in the formation of specific mineral deposits. We will focus attention on this feedback question by examining a few representative modern stratified basins.

4.2 Black Sea

The Black Sea is the largest anoxic water body in the world ocean, measuring almost half a million km^3. The O_2-H_2S interface will restrict molecular exchange (e.g., *Craig*, 1969a; *Spencer* and *Brewer*, 1971). Vertical advection and diffution will move deep water through a density boundary to the surface layer. For the main pycnocline in the Black Sea this results in a vertical eddy-diffusion coefficient of $\sim 0.014\ cm^2\ s^{-1}$ (*Brewer* and *Spencer*, 1974). From this value one can readily calculate the net upward flux of dissolved species such as iron, manganese, or hydrogen sulphide. Against the upward advective gradient there is a downward diffusion gradient of oxygen. These two opposing fluxes will generate a wide range in redox potentials within a narrow segment of the water column. Some elements and minerals will respond to the Eh gradient by becoming reduced or oxidized, and precipitated or dissolved, respectively. In turn, minerals can form at one point, but by sinking through an Eh gradient they may redissolve. Depending on the speed of this reaction and the general hydrographic setting, the process of precipitation and dissolution can continue over and over again and lead to substantial concentrations of certain elements and minerals close to the pycnocline.

Models using a constant eddy-diffusion coefficient for the vertical transfer of passive properties in a continuously layered medium are only applicable for periods of hours and days (e.g., *Riley* et al., 1949; *Radach* and *Maier-Reimer*, 1974), because density boundaries tend to rise and fall, and many other physical pertubations exist (*Kraus* and *Turner*, 1967; *Hasselmann* et al., 1973; *Kitaigorodskii* and *Mirolpol'skii*, 1970; *Denman*, 1973). If the interface is lowered rapidly, an almost spasmodic mineralization in the upper layer will take place, resulting in massive precipitation of a series of minerals. In contrast, a rise of the interface will dissolve the same minerals. Figure 2 shows calcite which has precipitated in the upper layer. In Figure 2a, the calcite remained in the surface layer, whereas in Figure 2b, it sank into the anoxic water. It is of note that all carbonates, including dolomite and siderite, will eventually dissolve if exposed to highly reducing environments.

Beneath the density boundary, an up to 2000 m thick anoxic water layer is established in which dissolved suphide species are present. The lower water layer is fed by detritus and tripton settling through the pycnocline. On the other hand, disssolved organic and inorganic material released from the compacting strata at the sediment—water interface will remineralize the deep water. Some of these constitutents, for instance copper, lead or zinc, will immediately precipitate in the form of sulphides, whereas others, such as manganese, remain in solution as the high redox values (-100 to -200 mV) hinder precipitation in the form of manganese sulphide. At the O_2-H_2S interface, a manganese oxide phase is formed by in situ precipitation (*Brewer* and *Spencer*, 1974). Once formed, the particles sink through the interface, redissolve, and upward progression will again move the ions through the phase boundary into the oxygenated environment, and the

Fig. 2 a and b. Scanning electron micrographs of chemically precipitated calcites, sedimented above O_2 –H_2S interface (a), and below interface (b). Size of individual crystals $5-10$ μm

cycle turns on again. The behavior of some metal ion species in Black sea water profiles is depicted in Figure 3.

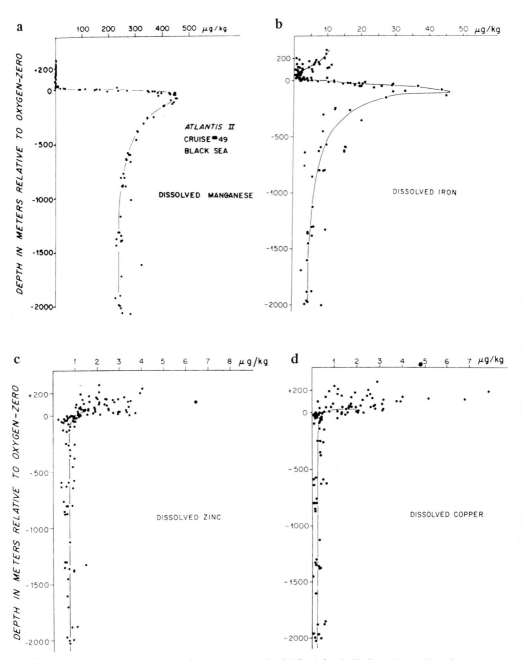

Fig. 3 a–d. Water stations occupied during cruise 49 of R/V *Atlantis II* of the Woods Hole Oceanographic Institution in the Black Sea in spring 1969. Vertical distribution of dissolved: manganese (a); iron (b); zinc (c); and copper (d). (After *Brewer* and *Spencer*, 1974)

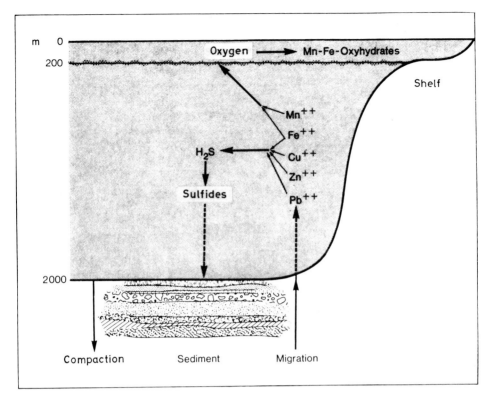

Fig. 4. Reaction scheme of metal ions precipitated above and below the pycnocline in the Black Sea (*Degens*, 1970)

The relation between the various migration and concentration processes for heavy metals at work in the Black Sea environment is schematically shown in Figure 4. Mn-Fe oxyhydrates precipitated close to the pycnocline can be physically carried to the shelf environment and may accumulate there. In contrast, the sulphides will settle to the sea floor. If the supply of detritus is kept at a minimum, ore deposits may form. We suggest that the process of metal enrichments of economic value may occur only if the supply of detritus is greatly reduced.

This mechanism of compaction, migration, and concentration of certain metals at phase boundaries is considered to be responsible for the formation of the large Eocene sedimentary manganese deposits of Tschiaturi Kutais and Nikopol, Ukraine, found in the vicinity of the Black Sea. These processes may also have been involved in the generation of the Jurassic Minette ore deposits and the Permian Kupferschiefer.

4.3 Lake Kivu

The deep lakes of East Africa — Tanganyika, Malawi and Kivu — are thermally stratified. Lake Kivu has a stable thermohaline density structure with an interface at ~ 70 m

(*Degens* et al., 1973). The position and nature of the density boundary influence the occurrence of distinct mineral and chemical facies observed in sediment cores. Most notably affected by the position of the interface is the distribution of Fe-oxides, Fe-silicates, Fe-phosphates, Fe-sulphides, Mn-siderites and a series of Ca- and Ca--Mg-carbonates (*Hecky* and *Degens*, 1973; *Müller* and *Förstner*, 1973; *Stoffers*, 1975).

The impact of an oscillating stratification is illustrated by the distribution of manganese, and related facies in a core form the deep part of Lake Kivu, taken at a water depth of ~ 500 m (Fig. 5). The observed Mn^{2+} oscillations are interpreted as resulting from the following processes (*Degens* and *Hecky*, 1974):

Thermal stratification allows Mn^{2+} and Fe^{2+} to accumulate in the anoxic water of relatively low pH below the thermocline. Vertical advection and diffusion transfer both ions through the $O_2 - H_2 S$ interface where they instantaneously precipitate as manganosiderite. The newly formed mineral phase sinks back into deep water, dissolves, and the cycle starts over again. Its incorporation into the sediment will only be possible near the thermocline. The manganosiderite in Lake Kivu occurs, however, as discrete layers, rather than as a mixture with the regular detritus. This suggests a more powerful transfer mechanism than advective eddy-diffusion: the rapid lowering of the thermocline and a resultant massive precipitation of the manganosiderite. Fluctuations in Mn^{2+} content shown in Figure 5 reflect oscillations of the thermocline which shifted the area of manganosiderite deposition away from or back toward the location of the core site. High water stands are indicated by the emergence of Fe-Ni-sulphides and specific diatom assemblages in an organic-rich facies free of carbonates.

4.4 Regions of Hydrothermal Activity

Phase boundaries also play an important role in metal enrichments resulting from submarine hydrothermal exhalations in connection with local volcanism, rifting, or postvolcanic activities (*Degens* and *Ross*, 1976). These processes, accompanied by halmyrolysis, can favor metal accumulations as known from Vulcano, Italy (*Honorez* et al., 1973), Santorini, Greece (*Puchelt*, 1973), Matupi, New Guinea (*Ferguson* and *Lambert*, 1972) and from areas close to the East Pacific Rise (*Boström*, 1970, 1973). In general, these deposits are related to gaseous exhalations and thermal springs composed of CO_2, H_2, $H_2 S$ which are rich in leached metals. In contact with the sea water, precipitation of metals as sulphides or hydroxides/oxides occur. In the case of the East Pacific Rise the metal enrichment is obvioulsy controlled by tholeitic basalts which result from sea-floor spreading.

There are, however, hardly any commercial aspects to these deposits. The lack of convenient trap structures for the metals leads to a dilution by normal detrital sedimentation or to a simple mixing with the bottom waters. An exception are the brine deeps – especially the Atlantis II, Chain, and Discovery deeps – in the central Red Sea rift zone, which are characterized by heavy metal deposits rich enough to encourage commercial exploitation. These metal-rich hot brines were discovered in the 1960s and since then have been studied in great detail (e.g., *Miller* et al., 1966; *Degens* and *Ross*, 1969; *Brewer* and *Spencer*, 1969; *Bäcker* and *Richter*, 1973; *Ross* et al., 1973; *Schoell*, 1974). Metal-bearing brine discharge is active and still increasing in intensity. The origin of

Fig. 5. Distribution of manganese in Kivu sediments deposited during the time interval between 13,700 and 11,000 yr BP (before present). Mineral facies and reconstructed thermocline oscillations are shown. High water stands are indicated by the emergence of Fe-Ni sulphides and specific diatom assemblages. (After *Hecky* and *Degens*, 1973)

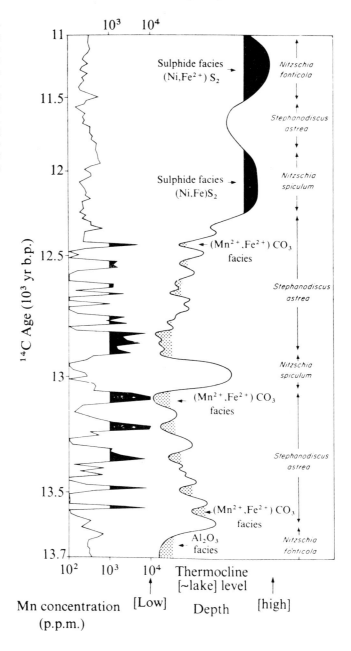

these brines is a matter of controversy; opinions range from long-distance migration from the southern end of the Red Sea (*Craig*, 1969b), to discharge from the flanks (*Manheim*, 1974). Whether the heavy metal-rich brines can be directly attributed to normal sea-floor spreading, or are derived by a leaching process from local evaporitic and pre-evaporitic strata is also still in discussion (*Manheim*, 1974). The heating of the brine is definitely due to the volcanic activity in the area. Obviously there is a relation between the brine pools and the crop-out of the S-reflector representing Miocene evaporites (*Ross* et al., 1973).

In general, the deeps are filled with brines differing in temperature and salt concentrations. This leads to a stratification and the development of transition zones. In 1972 the deepest brine in the Atlantis II deep had a temperature of 60° C with a chlorinity of 156.5 ‰. It was topped by a cooler and less concentrated brine (49.8° C, 82 ‰ Cl). While the transition zone between these two brines is small, a broad transition exists between the upper bine and the normal Red Sea water (*Bäcker* and *Richter*, 1973). The physicochemical conditions within the different brine layers form individual sediment facies such as limonite, hematite, manganite, sulphide and nontronite facies (*Bischoff*, 1969; *Bäcker* and *Richter*, 1973). According to *Hartmann* (1973), the redox potential in the 60° C brine is so low that Fe and Mn are kept in solution. Manganese and iron hydroxides sinking from the upper brine through the interface into the lower brine also become dissolved. Under reducing conditions and in the presence of dissolved sulphides, Fe, Zn, and Cu-sulphides are precipitated. The 50° C brine is characterized by iron hydroxides with little Zn and Cu. Manganese is still kept in solution. In the transition zone to the normal Red Sea water, Mn and Fe-hydroxides are formed basically. A general review on the subject matter of strata-bound metalliferous deposits found in or near active spreading centres has been prepared by *Degens* and *Ross* (1976).

4.5 Present as a Key to the Past

In the modern fully oxygenated ocean, sands and clays are deposited in the shallow sea (Fig. 6a) and carbonates are found at greater water depth. On stratification of the water, anoxic conditions develop below the interface and euxinic sediments form in the H_2S zone, whereas carbonates may develop in the shallow oxic zone (Fig. 6b). It is essential to know that these two contrasting environments not only differ in oxygen and hydrogen sulphide but in a number of chemical ingredients. As a common rule, the anoxic waters are enriched in (1) mineral nutrients such as nitrate, phosphate, ammonia and silica, (2) some common elements and trace elements, and (3) dissolved gases.

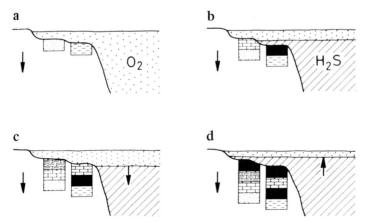

Fig. 6 a–d. Formation and evolution of stratified waters (see text). (After *Degens* and *Stoffers* 1976)

Lowering the interface (Fig. 6c) will result in the transfer of dissolved chemical species into the upper layer. Primary productivity will be stimulated and series of minerals will precipitate. An upward progression of the interface (Fig. 6d) will extend euxinic conditions to shallower parts of the basin, with established benthic communities becoming extinct, and carbonates starting to dissolve because the level of carbonate compensation will in time be identical with the level of the $O_2 - H_2S$ interface.

Sediment facies will mirror these events through distinct cycles and mineral assemblages. Heavy metals, notably those that are affected by changes in Eh/pH conditions, may become concentrated or removed from the water column or sediment–water interface, depending on the kind of water chemistry established in the environment. Eventually this may lead to ore accumulations of economic value, providing that low sedimentation rates prevail. In addition, upwelling or overturning of water masses may cause an almost spasmodic release of some of the dissolved constitutents, and phosphorites, carbonates, and Fe-Mn oxides and hydroxides are a few of the major phases that may come into existence.

This type of background might, for example, explain the origin of the extended Permian and Paleocene phosphorite deposits or the Precambrian banded iron formation. The pathway of the two critical elements Fe^{2+} and Si^{4+} through geological time is depicted in Figure 7. The concentration level for the two elements in the deep sea is shown at the far right. The main features are as follows:

Fe^{2+} becomes oxidized in the euphotic zone. Lowering of the $O_2 - H_2S$ interface will release Fe^{2+} to the upper layer, causing the formation of siderite, a series of Fe-oxides and hydroxides, and Fe-silicates (e.g., *Floran* and *Papike*, 1975). Local upwellings or perturbations of the water column due to surface currents may yield a similar effect.

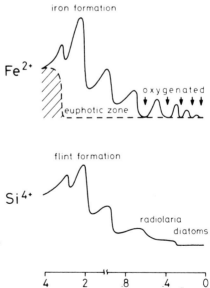

Fig. 7. Chemical evolution curves of Fe^{2+} and Si^{4+} levels in the deep sea (schematic). The Fe^{2+} content in the upper layer drops to zero early in the Precambrian as soon as the O_2-generation reaches a critical level. (After *Degens* and *Stoffers*, 1976)

In the anoxic sea, Fe^{2+} is extracted as sulphide or silicate. Towards the end of the Precambrian the ocean water turned over, fully oxygeneting the deep sea for the first time, but went back to the anoxic state during the Silurian. This sequence has been repeated several times throughout geological time. We consider some of the Silurian and Jurassic iron ores (thuringite, chamosite etc.) as a consequence of such a development.

In the Precambrian, the water was most likely saturated with silica below the interface. When released to the upper layer, it precipitated on evaporation in shallow-water environments as amorphous silica. Biogenic silica extraction has been the dominant release mechanism in more recent times (*Degens*, 1976).

In conclusion, we suggest that many strata-bound ore deposits of present and past have originated in a stratified sea. Heavy metals were derived from: (1) compacting sediments, (2) hydrothermal discharge, (3) reworked detritus, or (4) river run-off. The sediment—water interface and density boundaries in the water column provided the mechanism for selection and concentration of commercially interesting elements.

5. Organic Membranes

5.1 Cellular

Organisms utilize a series of heavy metals for a wide variety of physiological functions. Typical examples are iron in haemoglobin, or zinc metalloenzymes such as alkaline phosphatase and carbonic anhydrase. Apart from the beneficial aspect of heavy metals in cellular systems there are a number of heavy metals which are detrimental to life (*Wood*, 1975; *Goldberg*, 1976). Organisms can often protect themselves against metal "poisoning" by forming chelating compounds which they excrete into the environment or use internally for the efficient removal or neutralization of heavy metals. The Golgi apparatus, for instance, produces distinct proteins and mucopolysaccharides for this purpose. Some of the operation principles are described in *Degens* (1976a).

Figure 8 depicts the ultrastructure of a coccolith in which the organic matrix of the calcite structure has been naturally stained by heavy metals, revealing the growth pattern of the coccolith plates. The metal staining may have occurred as a result of normal physiological activities of the planktonic cell. Metal ions became removed from within the cell by Golgi-elaborated structures such as vesicles, which eventually transferred interfering metals to the plasmalemma, i.e., the outer membranes. Uranium in recent Black Sea coccolith oozes which were deposited over the past 1000 years is thought to have been concentrated in this fashion; the U_3O_8 values are close to 100 ppm. Some organisms may thus be scavengers for certain heavy metals. In the case of the Black Sea uranium, the concentration factor between water and sediment is about 10,000; the top 1 m layer contains about 5 million t of U_3O_8 (*Degens* et al., 1977).

5.2 Extracellular

Dissolved organic matter in aqueous systems can be derived in the following ways: (1) excretion by organisms, (2) remineralization of cellular material in the water column,

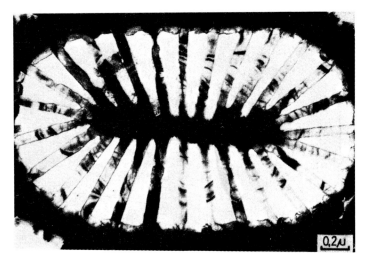

Fig. 8. Electron micrograph of coccolith *Umbilicosphaera* sp. Note staining pattern which is the result of uptake of heavy metals by shell proteins and mucopolysaccharides from the environment or the sediment

Fig. 9 a–c. Electron micrographs of hollow spheres termed resin globules suspended in Lake Kivu. The about 1 μ sized spheres are shown in various enlargements (a–c). Note the "bubbly" surface of the globules and the tiny sphalerite crystals (5–50 Å) that appear as black dots. (After *Degens* et al., 1972)   a

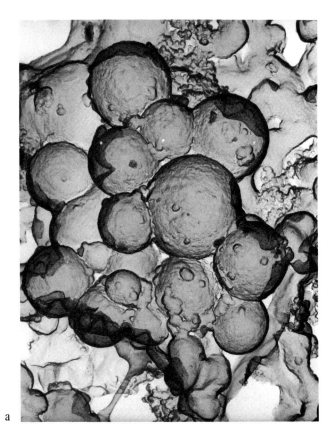

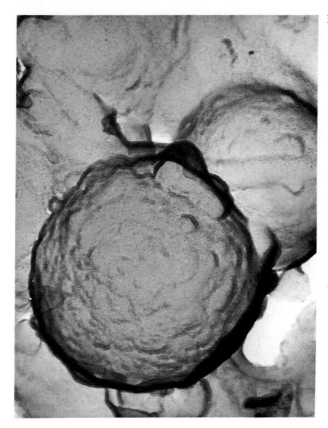

Fig. 9b (legend see p. 39)

(3) release from compacting sediment, and (4) pollution. We will draw attention to a few examples in which metal—organic interactions are displayed.

In 1971, we discovered zinc anomalies in the waters of Lake Kivu (*Degens* et al., 1973). The zinc is associated with micron-sized spheres. It can be shown that the hollow spheres with a wall thickness of 500 Å consist of a complex resinous material which has little functionality, except for hydroxyl groups (*Degens* et al., 1972). The spheres arise in the process of degassing of water samples at depth. Tiny gas bubbles, about 1 μ in size, act as scavengers of dissolved resinous material. The newly created resinous membrane covering the surface of the gas bubble promotes the selective coordination of zinc dissolved in the water column. In the prevailing H_2S regime, formation of sphalerite crystals is induced. The size range of the crystals, 5 to 50 Å, corresponds to 1 to 10 unit cells of ZnS and suggests that the resinous membrane also acts as a template in sphalerite growth processes. The source of the zinc and dissolved gases (CO_2, CH_4, H_2S, H_2) is hydrothermal springs emanating from the lake bottom into the basin.

Average Kivu water contains 2 ppm zinc. Thus, 1 million t of zinc are presently contained in Lake Kivu waters in the form of sphalerite. The zinc-containing globules are

Fig. 9c (legend see p. 39)

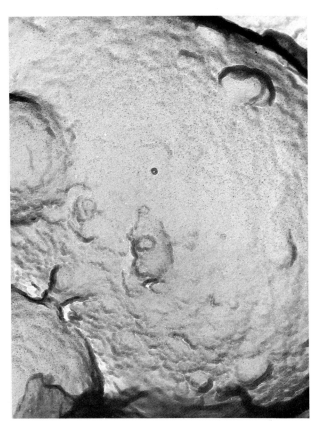

shown in Figure 9; the individual sphalerite crystals can be recognized as black dots. By means of energy-dispersive X-ray analysis, the presence of zinc, sulphur and traces of iron could be established (*Degens* et al., 1972).

During the Chain cruise to the Red Sea in spring 1971, several cores from the Atlantis II deep were taken and analyzed for their heavy metal content. Zinc- and copper-containing spheres were discovered in the sediment material retrieved from the deep which are similar in morphology to the Lake Kivu spheres described above (Fig. 10). We suggest a similar origin of formation, except that in the case of the Red Sea samples, in addition to zinc, copper is also scavenged. The high yield in sulphur suggests that the two elements occur in the form of sulphides.

Much work has been done on the origin of the so-called framboids which are pyrite aggregations. Such aggregates may from thin layers (Fig. 11), but are also found as suspended material. It is conceivable that they have an origin similar to the μ-sized zinc and copper globules previously described.

Acknowledgments: We gratefully acknowledge the financial support of the Deutsche Forschungsgemeinschaft, Bad Godesberg.

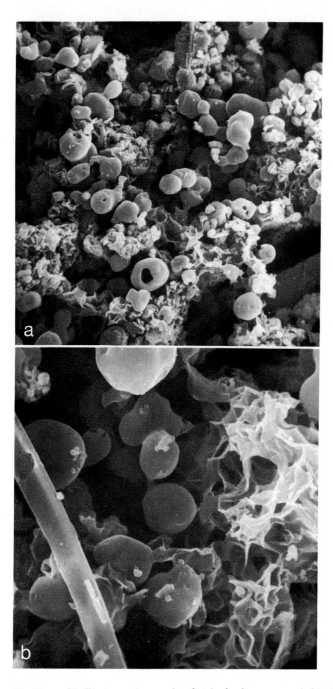

Fig. 10 a and b. Electron micrographs of μ sized spheres suspended in the Red Sea

Fig. 11. Electron micrographs of "framboid" layer in Black Sea sediments. Indentical layers are found in Lake Kivu sediments. The individual pyrite crystals have a size of about half a μ

References

Ames, L.L., Jr.: The genesis of carbonate apatites. Econ. Geol. **54**, 829–841 (1959)

Bäcker, H., Richter, H.: Die rezente hydrothermal-sedimentäre Lagerstätte Atlantis-II-Tief im Roten Meer. Geol. Rdsch. **62**, 697–737 (1973)

Berner, R.A.: The benthic boundary from the viewpoint of a geochemist. In: The Benthic Boundary Layer. McCave, I.N. (ed.). New York: Plenum Press 1976, pp. 33–55

Bischoff, J.L.: Red Sea geothermal brine deposits: Their mineralogy, chemistry, and genesis. In: Hot Brines and Recent Heavy Metal Deposits in the Red Sea. Degens, E.T., Ross, D.A. (eds). New York–Heidelberg–Berlin: Springer 1969, pp. 368–401

Bischoff, J.L., Ku, T.L.: Pore fluids of recent marine sediments: II, Anoxic sediments of 35° to 45° N Gibraltar to Mid-Atlantic Ridge. J. Sedim. Petrol. **41**, 1008–1017 (1971)

Boström, K.: Submarine volcanism as a source for iron. Earth Planet. Sci. Lett. **9**, 348–354 (1970)

Boström, K.: Origin and fate or ferromanganoan active ridge sediments. Stockholm Contr. Geol. **27**, 149–242 (1973)

Brewer, P.G., Spencer, D.W.: A note on chemical composition of the Red Sea brines. In: Hot Brines and Recent Heavy Metal Deposits in the Red Sea. Degens, E.T., Ross, D.A. (eds.). New York–Heidelberg–Berlin: Springer 1969, pp. 174–179

Brewer, P.G., Spencer, D.W.: Distribution of some trace elements in the Black Sea and their flux between dissolved and particulate phases. In: The Black Sea – Geology, Chemistry, and Biology. Tulsa, OK: Am. Assoc. Petrol. Geol. Mem. **20**, 137–143 (1974)

Calvert, S.E., Price, N.B.: Diffusion and reaction profiles of dissolved manganese in the pore waters of marine sediments. Earth Planet. Sci. Lett. **16**, 245–249 (1972)

Craig, H.: Abyssal carbon and radiocarbon in the Pacific. J. Geophys. Res. **74**, 5491–5506 (1969a)

Craig, H.: Geochemistry and origin of the Red Sea brines: In: Hot Brines and Recent Heavy Metal Deposits in the Red Sea. Degens, E.T., Ross, D.A. (eds.). New York–Heidelberg–Berlin: Springer 1969b, pp. 208–242

Cronan, D.S., Thomas, R.L.: Ferromanganese concentrations in Lake Ontario. Can. J. Earth Sci. **7**, 1346–1349 (1970)

Dean, W.E.: Fe-Mn oxidate crusts in Oneida Lake, New York. Proc. 13th Conf. Great Lakes Res. 217–226 (1970)

Degens, E.T.: Sea floor spreading: Lagerstättenkundliche Untersuchungen im Roten und im Schwarzen Meer. Umschau in Wissenschaft und Technik **70**, 268–274 (1970)

Degens, E.T.: Carbonate, phosphate, and silica deposition in the living cell. Topics Curr. Chem. **64**, Inorganic Biochemistry. Berlin–Heidelberg–New York: Springer 1976a, pp. 1–112

Degens, E.T.: Physikalische und chemische Steuerungsmechanismen der Evolution. Gesellschaft für Strahlen- und Umweltforschung mbH München Nr. 340, 10–46 (1977)

Degens, E.T., Hecky, R.E.: Paleoclimatic reconstruction of Late Pleistocene and Holocene based on biogenic sediments from the Black Sea and a tropical African lake. Coll. Intern. C.N.R.S. **219**, 13–24 (1974)

Degens, E.T., Herzen, R.P. von, Wong, H.K., Deuser, W.G., Jannasch, H.W.: Lake Kivu: Structure, chemistry and biology of an East African rift lake. Geol. Rdsch. **62**, 245–277 (1973)

Degens, E.T., Okada, H., Honjo, S., Hathaway, J.C.: Microcrystalline sphalerite in resin globules suspended in Lake Kivu, East Africa. Mineral. Deposita **7**, 1–12 (1972)

Degens, E.T., Ross, D.A. (eds.): Hot Brines and Recent Heavy Metal Deposits in the Red Sea. New York–Heidelberg–Berlin: Springer 1969, p. 600

Degens, E.T., Ross, D.A.: Strata-bound metalliferous deposits found in or near active spreading centers. In: Handbook of Strata-bound and Stratiform Ore Deposits. Wolf, K.H. (ed.). Amsterdam: Elsevier 1976

Degens, E.T., Stoffers, P.: Stratified waters as a key to the past. Nature (London) **263**, 22–27 (1976)

Degens, E.T., Khoo, F., Michaelis, W.: Uranium anomaly in Black Sea sediments. Nature (London) **269** (1977)

Denman, K.L.: A time-dependent model of the upper ocean. J. Phys. Oceanogr. **3**, 173–184 (1973)

Deuser, W.G.: Organic-carbon budget of the Black Sea. Deep-Sea Res. **18**, 995–1004 (1971)

Deuser, W.G.: Evolution of anoxic conditions in Black Sea during Holocene. In: The Black Sea – Geology, Chemistry, and Biology. Tulsa, OK: Am. Assoc. Petrol. Geol. Mem. **20**, 1974, pp. 133–136

Edington, D.N., Callender, E.: Minor element geochemistry of Lake Michigan ferromanganese nodules. Earth Planet. Sci. Lett. **8**, 97–100 (1970)

Ferguson, J., Lambert, I.B.: Volcanic exhalations and metal enrichments at Matupi Harbor, New Britain, T.P.N.G. Econ. Geol. **67**, 25–37 (1972)

Floran, R.J., Papike, J.J.: Petrology of the low-grade rocks of the Gunflint Iron Formation, Ontario–Minnesota. Geol. Soc. Am. Bull. **86**, 1169–1190 (1975)

Garrels, R.M., Mackenzie, F.T., Hunt, C.: Chemical Cycles and the Global Environment. Los Altos, CA: William Kaufmann, Inc. 1975, p. 206

Glass, G.E., Podolski, J.E.: Interstitial water components and exchange across the water sediment interface of Western Lake Superior. Verh. Intern. Verein. Limnol. **19**, 405–420 (1975)

Goldberg, E.D.: Marine geochemistry. I. Chemical scavengers of the sea. J. Geol. **62**, 249–265 (1954)

Goldberg, E.D.: The Health of the Oceans. Paris: The UNESCO Press 1976, p. 172

Gorham, E., Swain, D.J.: The influence of oxidizing and reducing conditions upon the distribution of some elements in lake sediments. Limnol. Oceanogr. **10**, 268–279 (1965)

Hartmann, M.: Zur Geochemie von Mangan und Eisen in der Ostsee. Meyniana **14**, 3–20 (1964)

Hasselmann, K., Barnett, T.P., Bouws, E., Carlson, H., Cartwright, D.E., Henke, K., Ewing, J.A., Gienapp, H., Hasselmann, D.E., Kruseman, P., Meerburg, A., Müller, P., Obers, D.J., Richter, K., Sell, W., Walden, H.: Measurement of wind-wave growth and swell decay during the Joint North Sea Wave Project (JONSWAP). Ergänzungsh. Deut. Hydrol. Z., R. A (8°) 12 (1973)

Hecky, R.E., Degens, E.T.: Late Pleistocene-Holocene chemical stratigraphy and paleolimnology of the rift valley lakes of Central Africa. Woods Hole Oceanogr. Inst. Techn. Rep. 73–28, 1–114 (1973)

Honorez, J., Honorez-Guerstein, B., Valette, J., Wauschkuhn, A.: Present day formation of an exhalative sulfide deposit at Vulcano (Thyrrhenian Sea), Part. II: Active crystallization of fumarolic sulfides in the volcanic sediments of the Baia di Levante. In: Ores in Sediments. Amstutz, G.C., Bernard, A.J. (eds.). Berlin–Heidelberg–New York: Springer 1973, pp. 139–166

Kitaigorodskii, S.A., Miropol'skii, Yu.Z.: On the theory of the open-ocean active layer. Izv. Akad. Sci. USSR. Atm. Ocean Phys. 6, 97–102 (1970)

Kraus, E.B., Turner, J.S.: A one-dimensional model of the seasonal thermocline, II. The general theory and its consequences. Tellus 19, 98–106 (1967)

Krauskopf, K.B.: Separation of manganese from iron in sedimentary processes. Geochim. Cosmochim. Acta 12, 61–84 (1957)

Lemoalle, J., Dupont, B.: Iron-bearing oolites and the present conditions of iron sedimentation in Lake Chad. In: Ores in Sediment. Amstutz, G.C., Bernard, A.J. (eds.). Berlin–Heidelberg–New York: Springer 1973, pp. 167–178

Ljungren, P.: Some data concerning the formation of manganiferous and ferriferous bog ores. Geol. Fören. Förhandl. 75, 277–297 (1953)

Lynn, D.C., Bonatti, E.: Mobility of manganese in diagenesis of deep sea sediments. Mar. Geol. 3, 457–474 (1965)

Manheim, F.T.: Manganese iron accumulations in the shallow marine environment. Symp. Mar. Geochim., Occas. Publ. 3, Narragansset Marine Lab. Univ. Rhode Island, Kingston, R.I., 217–276 (1965)

Manheim, F.T.: Red Sea geochemistry. In: Initial Rep. DSDP 23, 1974, pp. 975–999

Matheja, J., Degens, E.T.: Structural Molecular Biology of Phosphates. Stuttgart: Fischer 1971, p. 180

Mero, J.L.: The Mineral Resources of the Sea. Amsterdam: Elsevier, 1965, p. 312

Miller, A.R., Densmore, C.D., Degens, E.T. et al.: Hot brines and recent iron deposits in deeps of the Red Sea. Geochim. Cosmochim. Acta 30, 341–359 (1966)

Müller, G., Förstner, U.: Recent iron ore formation in Lake Malawi, Africa. Mineral. Deposita 8, 278–290 (1973)

Murray, J., Renard, A.F.: Report on Deep Sea Deposits Based on Specimens Collected During the Voyage of HMS Challenger, London 1891, p. 525

Neumann, E.: Einführung in die Bodenkunde der Seen. Stuttgart: Schweizerbart 1930, p. 126

Panel on Orientations for Geochemistry. Washington, D.C.: Natl. Acad. Sci. 1973, p. 122

Porrenga, D.H.: Glauconite and chamosite as depth indicators in the marine environment. Mar. Geol. 5, 495–501 (1966)

Presley, B.J., Brooks, R.R., Kaplan, I.R.: Manganese and related elements in the interstitial water of marine sediments. Science 158, 906–909 (1967)

Presley, B.J., Kolodny, Y., Nissenbaum, A., Kaplan, I.R.: Early diagenesis in a reducing fjord, Saanich Inlet, British Columbia. II. Trace element distribution in interstitial water and sediment. Geochim. Cosmochim. Acta 36, 1073–1090 (1972)

Price, N.B., Calvert, S.E.: Compositional variation in Pacific Ocean ferromanganese nodules and its relationship to sediment accumulation rates. Mar. Geol. 9, 145–171 (1970)

Puchelt, H.: Recent iron sediment formation at the Kameni Islands, Santorini (Greece). In: Ores in Sediments. Amstutz, G.C., Bernard, A.J. (eds.). Berlin–Heidelberg–New York: Springer 1973, pp. 227–245

Radach, G., Maier-Reimer, E.: The vertical structure of phytoplankton growth dynamics – a mathematical model. Mem. Soc. Sci. Liège, 6th Liège Coll. (1974)

Redfield, A.C., Ketchum, B.H., Richards, F.A.: The influence of organisms on the composition of sea-water. In: The Sea. Hill, M.N. (ed.). New York: Interscience, Vol. II, 1963, pp. 26–77

Riley, G.A., Stommel, W., Bumpus, D.F.: Quantitative ecology of the plankton of the western North Atlantic. Bull. Bingham Oceanogr. Coll. 12, 1–169 (1949)

Rohrlich, V., Price, B., Calvert, S.E.: Chamosite in recent sediments of Loch Etive, Scotland. J. Sediment Petrol. 39, 624–631 (1969)

Ross, D.A. et al.: Red Sea drillings. Science 179, 377–380 (1973)

Schoell, M.: Valdivia: VA 01/03 Rotes Meer – Golf von Aden. Hydrographie II/III. Hannover: Bundesanst. Bodenforsch. 1974, p. 1063

Seibold, E.: Rezente submarine Metallogenese. Geol. Rdsch. 62, 641–684 (1973)

Spencer, D.W., Brewer, P.G.: Vertical advection diffusion and redox potentials as controls on the distribution of manganese and other trace metals dissolved in waters of the Black Sea. J. Geophys. Res. 76, 5877–5892 (1971)

Stoffers, P.: Sedimentpetrographische, geochemische und paläoklimatische Untersuchungen an ostafrikanischen Seen. Thesis, Ruprecht-Karl-Univ. Heidelberg (1975), p. 117

Valeton, I.: Bauxites. Amsterdam: Elsevier 1972, p. 226

Wood, J.M: Biological cycles for elements in the environment. Naturwissenschaften 8, 357–365 (1975)

Wood, L.W.: Role of oligochaetes in the circulation of water and solutes across the mud-water interface. Verh. Intern. Verein. Limnol. 19, 1530–1533 (1975)

The Syngenetic-epigenetic Transition
An Australian Example

L.J. LAWRENCE, Kensington

With 4 Figures

Contents

Summary

The Mount Morgan pyritic copper-gold deposit of central eastern Queensland, Australia, exhibits both syngenetic stratiform processes and epigenetic hydrothermal processes. These facies of mineralization lie vertically one above the other, thus facilitating a chronological study of ore-forming sequences. Remnants of a former stratiform sulphide horizon are followed by a massive sulphide breccia pipe and this is intruded by a latite stock which has yielded sulphide-bearing quartz veins injected into the ore of the breccia pipe.

1. Introduction

Ore genesis concepts, over the last 15 to 20 years, have shown a number of marked changes, among which has been a reconciliation between the ideas of syngenesis and those of epigenesis. This has been aided, inter alia, by the expanding research work on younger deposits, including those of active island arc areas such as the South-West Pacific region and the Kuroko deposits of Japan, where both stratiform and transgressive vein-type ores occur as parts of the same deposit. Further evidence has come from the realization that discordant igneous intrusions such as stocks are an integral part of volcanism and terms such as "sub-volcanic plutonism" are now widely accepted.

The Mount Morgan pyritic gold-copper orebody of central eastern Queensland, Australia, located at longitude 150°22′E and latitude 23°38′S, bears evidence of a sequence of events ranging from submarine sulphide exhalations through to vein formation, presumably generated by a shallow intrusion which has been penetrated at the base of the open cut. This intrusion is injected in part into the orebody itself.

2. Geology of the Mount Morgan Area

The oldest rocks of the Mount Morgan area are referred to as the Mine Corridor Complex: they are of Devonian age and are surrounded by younger Permian rocks with a cover of terrestrial sandstones of Cretaceous age.

Mine geologists of the operating company (Peko-Wallsend Ltd.) have divided the Devonian rocks of the mine area from top to bottom as follows: Calliungal Porphyries, Banded Formation and Mount Morgan Porphyries.

The Calliungal Porphyries consist of massive and sometimes fragmented quartz-feldspar porphyry and quartz porphyry. The fragmental porphyry contains, additionally, copious fragments of chert, jasper and occasional recrystallized limestone – one boulder measuring 5 m across. These porphyries are considered to be extrusive volcanic agglomerates with some admixed sedimentary material (*Frets* and *Balde*, 1974). The thickness is not in excess of 200 m.

The Banded Formation comprises thin beds of quartz-feldspar porphyry and of quartz porphyry alternating with thinly laminated and massive cherts and jaspers and some haematite-magnetite layers. The jaspers show structures suggesting colloidal depositions and contain minute phenocrystal grains derived, presumably, from the hardening and incipient crystallization of a predominantly siliceous gel in the manner proposed by *Elliston* (1968). Flow folding is common in this formation and the various beds pinch and swell or form discontinuous lenses. Total thickness is from 150 to 200 m.

The Mount Morgan Porphyries are lowermost in the succession and consist of massive quartz-feldspar porphyry, which is the host rock for a large part of the Mount Morgan orebody. The thickness is in excess of 730 m.

The above rocks are intruded, in the mine area, by dykes and irregular shaped masses – presumably the upper portion of a stock-like body – of quartz latite. In the bottom of the open cut this rock can be seen also intruding the orebody. The quartz latite is a greenish-grey melanocratic rock of variable texture and composition and somewhat altered along its outer surface. Mineralogically the rock consists of fine oligoclase laths sometimes with a trachytic texture, a pale green amphibole and interstitial quartz.

All the above rocks: the Calliungal Porphyries, the Banded Formation, the Mount Morgan Porphyries and the ensuing quartz latite intrusion belong to the so-called *Mine Corridor Complex*, which has been subsequently intruded by the *Mount Morgan Igneous Complex* of Upper Devonian age. This complex includes substantial areas of tonalite and quartz diorite near the open cut with gabbro and diorite also occurring to the southwest.

The area was finally intruded (including the orebody itself) by a host of basic to intermediate dykes of late Permian age.

The Mine Corridor rocks show strikes and dips that indicate a local doming prior to the later intrusions. The area is also faulted and the orebody is cut by some of these faults.

On the western side the margin of the orebody tends to parallel rocks of the Banded Formation producing a degree of conformability. On the north and east, however, the margin of the orebody is distinctly disconformable with respect to both the Calliungal Porphyries and the Banded Formation.

3. Ore Mineralogy

The Mount Morgan orebody is essentially of massive pyrite with variable accessory amounts of chalcopyrite, magnetite, pyrrhotite and sometimes sphalerite. Minor amounts of gold, petzite, hessite and calaverite are present mainly as small inclusions in chalcopyrite but some gold is recovered also from the pyritic ore. Other minerals, some only in trace amounts, include bismuthinite, bornite, arsenopyrite, coloradoite, cubanite, galena, cassiterite, molybdenite, tellurobismuthite, tetradymite and tetrahedrite. The ore mineralogy has been described in some detail by *Lawrence* (1967).

Some of the massive pyrite exhibits a vague banding in the form of alternating coarse and fine grain size. Banding is also seen in masses of highly silicified material consisting of microcrystalline silica alternating with silica plus diffuse magnetite with sparsely disseminated pyrite.

Most of the ore, including the silica, has been subject to annealing recrystallization (*Lawrence*, 1972). Causes of the stress and temperature elevation needed to recrystallize a pyritic ore are not known with certainty. It was originally thought that dilation due to dyke emplacement, together with the attendant temperature elevation, induced the annealing but there is now reason to believe that this process occurred earlier and is related in some way to the forces and accompanying temperature rise during a period of vein formation toward the close of the overall mineralization process (see following section; *Frets*, 1974).

Chalcopyrite has two main modes — as narrow films interstitial to closely fitting polygonal pyrite grains and in narrow later-formed quartz veins and stockworks transsecting the massive ore. The distribution of chalcopyrite in the former case may stem from its relative surface tension whereby it has migrated to the pyrite grain boundaries during annealing of that mineral.

The orebody is a pipe-like mass with a sharply discordant contact on its eastern and northern margins. It is shaped like a boot and is surrounded by a zone of silification. The ore within the pipe appears to be heavily brecciated and this has led to the belief that it is a breccia pipe, i.e., a "brecciated massive sulphide deposit" (*Frets*, 1974). However, the fragments of sulphide breccia often do not show any relative displacement and at least some of the brecciation results from the introduction of silica into stockwork fractures causing an apparent breaking up of the pyritic ore.

4. Mineralization Processes

Several papers have dealt with the origin of the Mount Morgan orebody viz. *Cornelius* (1967, 1968, 1969), *Hawkins* (1967), *Lawrence* (1967, 1972, 1974), *Paltridge* (1967), *Frets* and *Balde* (1974), *Frets* (1974). In summary, these various publications range from a simple epigenetic origin with the mineralization arising from the intrusive Mount Morgan Igneous Complex — particularly the Mount Morgan tonalite — through a straightforward syngenetic (stratiform) origin to an epigenetic origin from a source postulated to lie below the present level of exposure. Each worker presented evidence in support of his thesis. The most comprehensive studies are those of *Cornelius* (1968) and *Frets*

(1974). In 1967 *Lawrence* suggested a volcanogenic origin for the orebody and extended this in 1974 to include both epigenic and syngenetic processes.

Field studies by the staff of Peko-Wallsend Ltd. have confirmed the view that the Mount Morgan Tonalite (Mount Morgan Igneous Complex) is younger than the orebody which it intrudes in places. There is also confirmation of an intrusion older than the tonalite and sequentially closely related to the orebody, i.e., the latite intrusion.

In 1965, on number 4A Bench South (toward the top of the open cut), remnants of two distinctive ore horizons were encountered. The first was a mass of stratified pyritic ore (Fig. 1) measuring 2 m x 1.5 m with a 20° dip away from the pit; the second, lying below the stratified pyrite, was in the form of fragmented masses of a black "shale" with interstratified pyrite layers exhibiting, in places, soft sediment slumping (Fig. 2).

Fig. 1. Bedded pyrite outcrop on No. 4 Bench South. Mount Morgan, Queensland

Within the open cut the ore is traversed by numerous sulphide-bearing quartz veins which pass downwards and increase in frequency toward the base of the pit. These veins sometimes form stockworks (Fig. 3) and they are occasionally cut off by the latite intrusion, the upper surface of which is seen in the bottom of the pit. The veins and their sulphide ore content are thought to belong to the hydrothermal stage of the upwelling latite which now intrudes them, as well as the massive sulphide breccia of the main ore pipe.

Fig. 2. Pyritic black shale with minor soft-sediment slumping. Mount Morgan, Queensland (x 1)

Fig. 3. Massive pyritic ore traversed by quartz veins which are truncated by intrusive latite (*dark area bottom left*), Mount Morgan, Queensland

The exact nature of the massive ore pipe is not clear. It originates either from the downward collapse of massive stratiform ore, originally higher in the stratigraphic sequence, following an explosive episode, or it stems from the upwelling of ore fluid implicitly of an early hydrothermal discharge from the advancing latite.

There are therefore both syngenetic stratiform and epigenetic hydrothermal manifestations in the overall mineralization process at Mount Morgan.

The following reconstruction is proposed (Fig. 4):

Stage 1. Upwelling of acidic magma at depth with doming of the sea floor and eventual extrusion of porphyritic rhyolites, i.e., the Lower Mine Porphyries.

Stage 2. Continued extrusion of acidic magma with periods of quiescence where colloidal silica was generated alternating with further thin lava flows. Limestone reefs developed around the flanks of the domed area and were periodically destroyed yielding calcareous sediment. Intervening expulsion of silica gel (eventually cherts and jaspers). Some slumping of unstable sediments occurred, i.e., the Banded Mine Sequence.

Stage 3. Explosive eruption and deposition of agglomerates comprising porphyritic acid volcanics, cherts and jaspers, limestone; larger fragments settled out first including some enormous limestone blocks, i.e., the Upper Mine Porphyries (actually explosive agglomerates).

Stage 4. In respect to the banded sulphide ore it is postulated that the explosive activity was followed by a period of quiescence whence the following processes ensued.

a) Within a basin-shaped depression, possibly caused by the explosive activity and in-filled by the Upper Mine Porphyries, colloidal volcanic sediment was deposited perhaps by re-working of fine colloidal ash.

b) Through the agglomeratic pile rose gases and vapours perhaps of heavy metal halides together with sulphurous vapours leading to the precipitation of a colloidal sulphide mud interstratified with ultrafine sediment.

c) Deposition of a black pyritic shale with some slumping and progressive lithification and grain growth.

Stage 5. Here the massive pyritic pipe was developed. It is not clear whether the main ore pipe resulted from the upwelling of hydrothermal fluids within the pipe-like structure or whether the pipe was filled by the downward collapse of the overlying banded stratiform ore as suggested in Figure 4.

Stage 6. Progressively during this time the plug-like mass of latite, now visible at the base of the pit, was slowly rising and fracturing the hood of rock and ore above it. Into these fractures was injected quartz, chalcopyrite and accessory molybdenite and gold minerals stemming from the hydrothermal discharge of the latite plug which itself carries minor disseminated pyrite and chalcopyrite.

In some instances the quartz veins, contained as they are in breccia pipe ore, are truncated by the latite from which they issued. It is not improbable that the disseminated sulphides, sparsely developed in the outer margins of the latite intrusion, originated from the assimilated vein sulphides, i.e., a type of "redigestion" process!

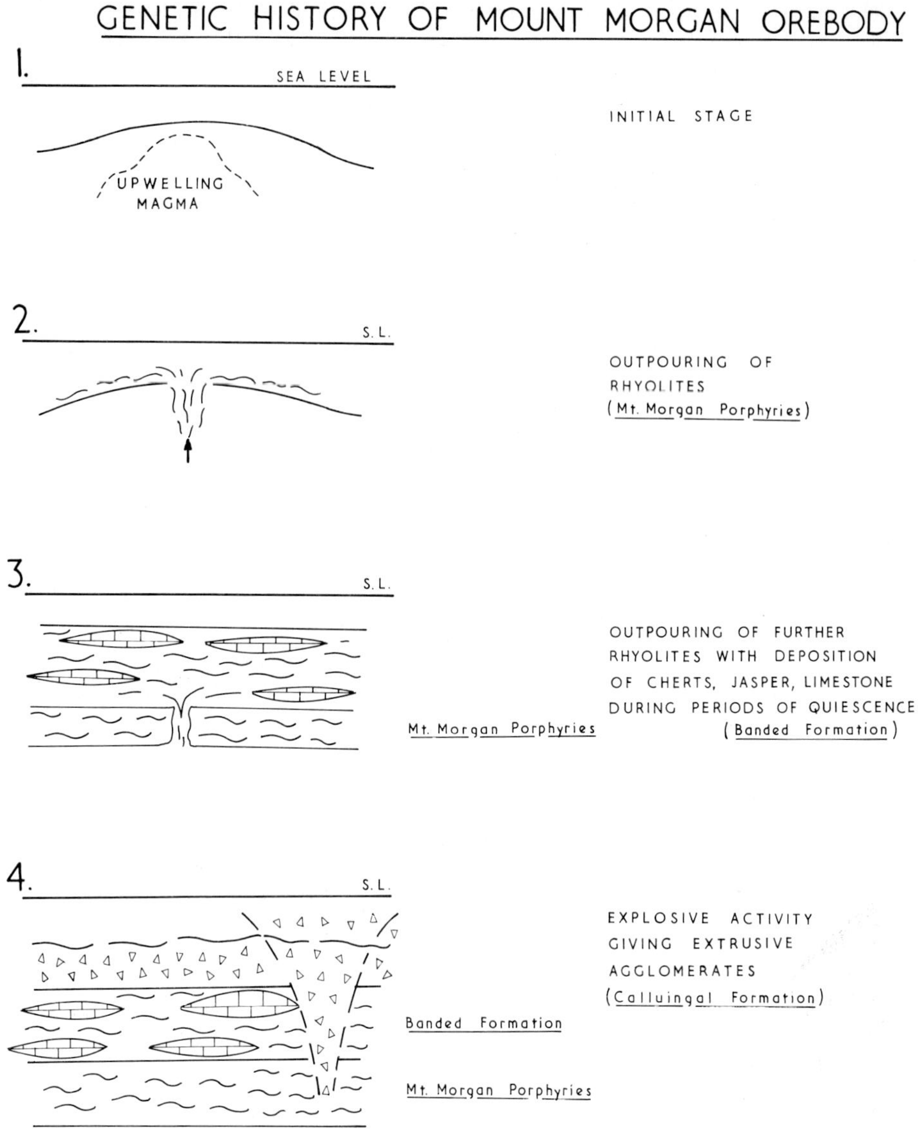

GENETIC HISTORY OF MOUNT MORGAN OREBODY

Fig. 4. Postulated sequence of ore forming processes at Mount Morgan, Queensland

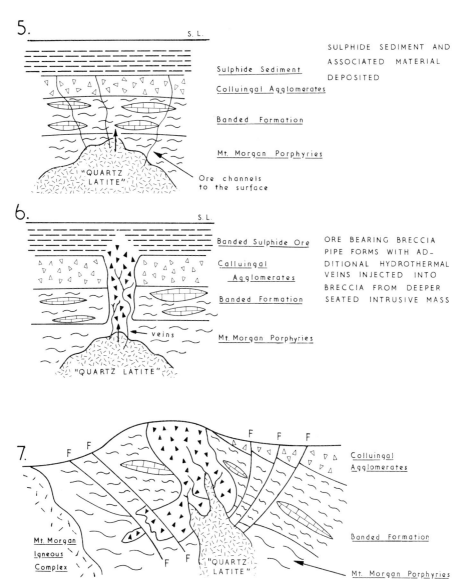

5.

S.L.

Sulphide Sediment

Colluingal Agglomerates

Banded Formation

Mt. Morgan Porphyries

"QUARTZ LATITE"

Ore channels to the surface

SULPHIDE SEDIMENT AND ASSOCIATED MATERIAL DEPOSITED

6.

S.L.

Banded Sulphide Ore

Colluingal Agglomerates

Banded Formation

veins

Mt Morgan Porphyries

"QUARTZ LATITE"

ORE BEARING BRECCIA PIPE FORMS WITH AD-DITIONAL HYDROTHERMAL VEINS INJECTED INTO BRECCIA FROM DEEPER SEATED INTRUSIVE MASS

7.

F   F

F   F   F

Colluingal Agglomerates

Mt. Morgan Igneous Complex

F   F

"QUARTZ LATITE"

Banded Formation

Mt. Morgan Porphyries

CONTINUED INTRUSION OF QUARTZ LATITE DYKES INTO BRECCIA PIPE. FOLDING AND FAULTING. EROSION TO PRESENT SURFACE.

Fig. 4

5. Conclusions

The Mount Morgan ore deposit provides a unique opportunity to study an unusually complete sequence of mineralization ranging from syngenetic stratiform processes through a breccia pipe filling to epigenetic vein-forming processes. It warrants emphasis that although the veins are epigenetic, the time gap with respect to the early stratiform ore, only small remnants of which still remain, may not be great.

It would appear that all ore "facies" are related to the latite plug or stock, the upper portion of which was penetrated in the deeper levels of the open cut. At some depth below the surface (somewhere in excess of 1000 m) the latite, or a differentiate thereof, issued ore-bearing fluids which reached the surface to form the (now virtually eroded) stratified sulphide ore. As the latite rose still further – according to one line of thought – the massive ore was emplaced hypogenetically within a breccia pipe (alternatively the stratified ore above collapsed into the pipe). At still shallower levels of intrusion, the latite discharged a second pulse of ore fluid which filled fractures and stockworks in the ore within the pipe. Eventually the latite intrusion rose to levels sufficiently shallow to truncate and possibly digest sulphide-bearing quartz veins that it had earlier discharged. Specimens have been observed where quartz veins up to 2 cm wide end abruptly at the outer margin of the latite which is intrusive into massive brecciated sulphide ore.

Many stratiform ore deposits are associated with intrusive rocks to which they are genetically related. Such mineralization is apparently a long protracted process with an early syngenetic cum volcanogenic mineralization leading to an epigenetic cum hydrothermal vein-type mineralization. Mount Morgan is a unique example in so far as the requisite amount of erosion (leaving only a vestige of the overlying stratiform ore), together with open cut mining, has revealed these various features which here lie vertically above one another.

References

Cornelius, K.D.: Breccia pipe associated with epigenetic mineralization, Mount Morgan, Queensland, Econ. Geol. **62**, 853–860 (1967)

Cornelius, K.D.: The ore deposit and general geology of the Mount Morgan area. Ph.D. thesis (unpubl.), Univ. Queensland (1968)

Cornelius, K.D.: The Mount Morgan Mine, Queensland. A massive gold-copper pyritic replacement deposit. Econ. Geol. **64**, 885–902 (1969)

Elliston, J.N.: Retextured sediments, Rep. Intern. Geol. Congress, 23rd Sess. Czech. Proc. Sect. 8, Genesis and classification of sedimentary rocks, 84–104 (1968)

Frets, D.C.: Rock relationships and mineralization at Mount Morgan, Centr. Queensland, Proc. Australas. Inst. Mineral. Met. Ann. Conf. 425–439 (1974)

Frets, D.G., Balde, R.: The copper-gold ore deposits of Mount Morgan. In: Economic Geology of Australia and Papua New Guinea – Metals Volume. Knight, C.L. (ed.). Melbourne: Australas. Inst. Mineral. Met. 1974

Hawkins, B.W.: Breccia pipe associated with epigenetic mineralization, Mount Morgan, Queensland. Econ. Geol. **62**, 1000–1001 (1967)

Lawrence, L.J.: A mineragraphic study of Mount Morgan copper-gold ore. Proc. Australas. Inst. Mineral. Met. **223**, 29–47 (1967)

Lawrence, L.J.: The thermal metamorphism of a pyritic sulfide ore. Econ. Geol. **67**, 487–496 (1972)

Lawrence, L.J.: The nature and origin of the ore minerals of Mount Morgan. Centr. Queensland: Proc. Australas. Inst. Mineral. Met. Ann. Conf. 417–424 (1974)

Paltridge, I.M.: Breccia pipe mineralization at Mount Morgan – discussion. Econ. Geol. **62**, 861 (1967)

Exploration Practice for Strata-Bound Volcanogenic Sulphide Deposits in the Spanish-Portuguese Pyrite Belt: Geology, Geophysics, and Geochemistry*

G.K. STRAUSS, Tharsis, J. MADEL, Saarbrücken, and F. FDEZ. ALONSO, Madrid

With 14 Figures

Contents

Summary

The time- and strata-bound volcanogenic sulphide deposits of the Paleozoic Spanish-Portuguese Pyrite Belt have been the object of intensive exploration activities during the past 20 years. As a result of these efforts, some 500 million t of new ore reserves were delineated, of which some 180 million t were pure geological and geophysical discoveries.

This paper describes shortly the exploration history in the Pyrite Belt and the different exploration targets, massive pyritic sulphide ores, disseminated copper and copper-zinc mineralizations, all of them within a volcanic-sedimentary environment. It is emphasized that exploration geology blocks out specific areas with characteristic features known to be present in mineralized areas, such as specific facies of acid and basic volcanic rocks, hydrothermal alteration of host rocks and structural control, among others. Airborne geophysical surveys in Spain are enumerated briefly and their failure is discussed. Ground geophysics, namely different varieties of earth resistivity combined with gravimetry, have proved to be the most efficient methods in prospecting for hidden sulphide ores, as is shown by several case histories. Classical copper-lead-zinc soil geochemistry did not prove to be of much practical use, because of lack of soils and contamination produced by 3000 years of mining and smelting activities in the Pyrite Belt. "Mercury in soil gas" and "mercury in soil" test surveys over different sulphide orebodies are presented and the efficiency of the latter method discussed within an integrated exploration programme for new hidden sulphide mineralizations.

* This paper is based to a great part on the results of geological, geophysical and geochemical exploration work by the authors and on behalf of THE THARSIS SULPHUR & COPPER COMPANY Ltd. The authors would like therefore to express their sincere gratefullness to this company.

1. Introduction

The Spanish-Portuguese Pyrite Belt represents one of Europe's most prominent metallo-genic provinces, as well as one of its oldest mining districts, with a record of over 3000 years of mining activities. It extends approximately E-W over 230 km with an average width of 35 km and covers some 7800 km^2 (Fig. 1).

The total mineral reserves of the Pyrite Belt are estimated to have been in excess of 1000 million t of massive sulphide ores before modern mining started. To date some 250 million t have been mined since the middle of the last century. Other ore types, geologically closely associated in time and space, include low-grade copper and copper-zinc disseminations of only recently recognized economic importance, and numerous small-sized siliceous manganese mineralizations, which were worked in the past.

Average massive sulphide ore grades range from 44–48% S, 39–44% Fe, 2–6% Cu + Pb + Zn, 0.3–0.5% As, 0.2–1.5 g/t Au and 5–30 g/t Ag. Economic copper dissemina-tions have overall average values of 0.7–1.2% Cu. One marginally exploitable dissemin-ated copper-zinc deposit is fairly low-grade with only 0.6% Cu, 0.4% Zn and 10g/t Ag. It follows from these figures that the metal concentration due to sulphide ores is, at least, of the order of some 47 kg Fe and 5 kg base metals per m^2 of the Pyrite Belt.

Mining and smelting in ancient times was recorded by the Tartessians, Phoenicians and Romans. Gold and silver were produced from the extensive massive gossan caps, silver from the soft yellow, jarositic layer underlying the gossans, and some minor cop-per from the secondarily enriched cementation zone. Silver mining and smelting by the Romans during some 350 years of continuous industrial activity have produced most of the 30 million t of slag at different points of the Pyrite Belt.

Following this period, mining was limited to some discontinuous activities, centred mainly on the Rio Tinto deposit, until the sudden resurgence of mining in the middle of the 19th century, when the production of copper became the principle objective. This was achieved by direct smelting of high-grade ores or open-air roasting of cupreous pyrite ores with subsequent leaching and recovery of the copper in launders by the use of scrap iron. Later, following the depletion of the Sicilian sulphur deposits, the huge Spanish pyrite ore reserves became the only alternative source of raw materials for the ever-growing sulphuric acid production of the European chemical industry. The base metal contents of the pyrite cinders, mainly copper and zinc, were then recovered by wet chemical processes in different Northern and Central European plants with the remaining iron ("purple") ore being used as blast furnace feed stock.

Today, the annual output of some 3 million t of massive pyritic ores (sulphur ores) stems from seven Spanish and two Portuguese mines. Furthermore, some 20,000 t of copper, 25 t of silver and 3 t of gold are produced by the Cerro Colorado operation at Riotinto. Also at Riotinto, the new Alfredo deposit is scheduled to come on-stream with an additional 8500 t of copper production per year. Shortly the new Aznalcollar mine will produce 12,800 t of copper, 21,000 t of lead and 45,200 t of zinc in concen-trates.

To a great extent, the exploration effort of the past 20 years must be credited for the profound change of some of the long-established mine areas in the Pyrite Belt, to name only three: (1) Aljustrel (Portugal), with virtually no ore reserves in the early 1950s, has now well over 170 million t of massive pyritic ore, some with important copper and zinc

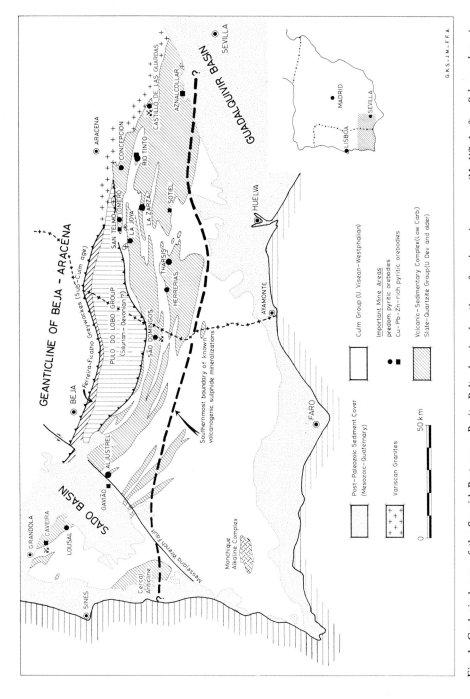

Fig. 1. Geological map of the Spanish-Portuguese Pyrite Belt, showing location of major mine areas. (Modified after *Schermerhorn* in *Carvalho* et al., 1975)

grades. (2) The Riotinto area (Huelva Province), that in over 100 years had produced more than 130 million t of cupreous pyrites, now looks to the future with reserves of some 60 million, or more t of medium-to-low-grade disseminated copper ore overlain by some 20 million t of gold-and silver-bearing massive gossan. (3) Finally, the nearly depleted Aznalcollar mine (Sevilla Province) has been revived after a successful drilling programme that disclosed some 80 million t of massive Cu, Pb, Zn sulphides, together with associated low-grade Cu-Zn disseminations.

In two earlier papers (*Strauss* et al., 1974a, b) the authors have already stressed the coincidence between the latest geological theories about ore formation and the exploration successes achieved thereby in the Pyrite Belt. This paper will discuss and summarize the most relevant data on exploration history and techniques, as the authors see it now, based on an extensive working experience in both the Portuguese and Spanish parts of the belt.

2. Exploration History and Exploration Targets

Since modern mining started, the exploration activities can be subdivided approximately into four phases:

The *first phase* (mid 19th century) is marked by the systematic reexamination of those sites where Roman workings were found, most of them having remained virtually untouched since the fourth century AD. In this way, mines such as Rio Tinto, Tharsis ans Sao Domingos were rediscovered and put in operation.

Practically all mines operating today were worked previously by the Romans, with shafts, galleries and stopes following the soft, silver-rich jarosite layer underneath the gossan caps. Modern mine workings, as opencasts and exploration shafts went deeper, then discovered new huge reserves, first of secondarily enriched copper ores and then of the primary, massive pyritic protore.

The *second phase* started approximately after 1945–1950. In Spain, an extensive drilling programme explored surface showings around operating and abandoned mines. This task was performed by the state organization Piritas Españolas over many years and, despite the effort, the results were negligible, partly because little was known about the geology of the ores and/or the erroneous concept of an epigenetic, hydrothermal origin of massive sulphide ores was postulated. In Portugal this exploration phase saw the beginning of the systematic use of geolphysical methods, namely electromagnetics, in all exploration programmes.

The *third phase* was characterized by airborne geophysical surveys in Spain and Portugal, starting approximately in 1957. These surveys were originated primarily by state organizations and, later, by private mining companies, all employing North-American exploration practice. The results were again negative and are reviewed in Section 6.1.

The *fourth exploration phase* started sometime around 1960. By that time a completely new geological approach concerning massive sulphide ore genesis had taken place. It was the switch away from the epigenetic-hydrothermal model to one favouring a syngenetic, synsedimentary volcanogenic origin, combined with a strata-bound control of the Pyrite Belt sulphide ores and their associated manganese mineralizations. Geological mapping suddenly became a much more important component in any exploration programme, and this was accompanied by extensive use of long-established ground geophysical methods, particularly earth-resistivity methods and gravimetry. The remarkable results of this exploration phase are evident from Table 1.

The fourth phase is still continuing and its principal targets are the most promising and rewarding areas around operating or abandoned mines. Altogether these areas add up to some 300–400 km², or only some 4% of the whole Pyrite Belt surface.

Since about 1970 a fifth phase is in progress, directed mainly towards virgin areas. Special attention is being paid to approximately 1800 km² of ground, associated with the ore-bearing Volcanic-Sedimentary Complex, and to an estimated further 1000–1500 km² where this complex is covered by 50–200 m of Paleozoic (Culm) sediments or Tertiary Basin fill (Fig. 1).

It becomes evident that this latter task needs detailed geological, stratigraphical and structural control and the first spectacular result, the discovery of the Gaviao ore bodies (Fig. 1), under some 60–80 m Tertiary sediments in 1970 (*Freire d'Andrade* and *Schermerhorn*, 1971) 4 km W of Aljustrel (Portugal), is to be credited entirely to geological methods that guided the exploration boreholes into the Paleozoic ore-bearing structures underlying the Sado Basin sediments. Other discoveries are those at Aznalcollar – "Anomaly 3" (1973), Nueva Almagrera and Cantareras, both at Tharsis (1965 and 1973).

In view of the huge reserves of pyritic ores (sulphur ores) the exploration targets are nowadays both massive, base-metal rich sulphide deposits and/or disseminated copper or copper-zinc sulphides. Geologically both types are of volcanogenic origin and closely related with each other in space and time.

As evidenced by its name, the primary Pyrite Belt ores normally have a very high pyrite content and rather low grades of Cu, Pb, Zn and Ag, when compared by world standards. From a statistical point of view, maximum average grades in massive ores may be expected to reach 1.5% Cu, 4% Zn and 2% Pb, however only in low tonnage deposits of the 500,000 t – 5 million t range. Marginally economical base metal grades of large deposits that may warrant selective flotation are best characterized by the recently developed 45 million t Aznalcollar massive orebody with overall grades of 0.44% Cu, 1.74% Pb, 3.33% Zn and 67g/t Ag (*San Pedro Querejeta*, 1975). The main obstacle is that nearly all such massive ores are very fine-grained and this requires grinding to < 400 mesh for economical base metal recoveries through flotation.

There are two different types of sulphide disseminations: (1) stockwork ore (= stringer ore, *Sangster*, 1972), always bound to footwall rocks of massive pyritic ore masses, and normally accompanied by intense hydrothermal chloritization (*Strauss* and *Madel*, 1974; *Aye* and *Strauss*, 1975; *Williams* et al., 1975); (2) synsedimentary disseminations, normally in the extensions along strike and/or in the hanging wall, like at Aznalcollar.

Most of these ores are predominantly pyritic with copper grades of only 0.1–0.2% Cu. Exceptions are deposits at Riotinto, where two large stockwork ore masses in chloritized acid footwall volcanics have overall grades of 0.7% Cu (Cerro Colorado, *Pryor* et al., 1972) and appromimately 1.2% Cu (Alfredo Deposit). The only workable synsedimentary dissemination deposit is currently being developed jointly with underlying massive ore at Aznalcollar and has low grades of 0.6% Cu, 0.4% Zn and 10 g/t Ag. Here, the sulphide minerals occur in tuffaceous shales.

Table 1. Exploration Results 1955–1974: 470–500 million t

Year	Mine Area	Orebody	Tonnage and Grades	Geology	Discovery Method	Additional Methods	Remarks
1955	Tharsis (Huelva)	San Guillermo	> 50 M. tons mass. pyr. ore	30–60°N dipping ore lens	exploration galleries and cross–cuts		Orebody open with depth
1956	Aljustrel (Portugal)	Carrasco and Moinho	> 70 M. tons mass. pyr. ore	3 steeply dipping ore–lenses. 50 m below surface	exploration gallery	Gravimetry Turam	
1962	Riotinto (Huelva)	San Antonio	12 M. tons Cu + Pb + Zn – pyr. ore	20° E dipping ore sheet. 100–300 m below surface	exploration drilling	Gravimetry and EM	
1963	Aljustrel (Portugal)	Feitais	> 70 M. tons mass. pyr. ore	50° N dipping orebody. 100 m below surface	Gravimetry (1.2 milligals)	diam. drilling	Orebody open with depth
1965–	Tharsis (Huelva)	Nueva Almagrera	3 M. tons mass. pyr. ore	vertical dip. No surface indication	Resistivity	Turam Gravimetry	Orebody open with depth
1966	Lousal (Portugal)	Massa Antonio	2 M. tons mass. Cu–pyr. ore 1,5 % Cu	vertically dipping ore lens	underground mine geology	exploration cross–cuts underground diam drilling	
1966–	Riotinto (Huelva)	Cerro Colorado	>40 M. tons copper stockwork ore: 0,8 % Cu	cpy veinlets in chloritized acid pyroclastics. Overlain by mass.gossan	grid drilling	pitting, trenching,exploration galleries	Additional 20 M. tons of mass. gossan 2.5 g Au/ton 47 g Ag/ton
1968	Aljustrel (Portugal)	Estaçao	>30 M. tons mass. pyr. ore	60° N dipping ore lens. 300 m below surface	Gravimetry (0.4 milligals)	diam. drilling	in exploration stage

Year	Location	Name	Tonnage / ore	Ore geometry	Geology	Geophysics / drilling	Metals / stage
1970	4 Km West of Aljustrel (Portugal)	Gavião	25 M. tons mass. Cu + Zn-pyr. ore	2 vertically dipping orebodies below 60–100 m of Tertiary sedim. cover	Geology based on grid drilling into Paleozoic of Sado Basin	diam. drilling	small residual anomaly (0.14 milligals)
1971	Sotiel (Huelva)	Sotiel	41 M. tons mass. Cu + Pb + Zn pyrite ore	3 ore lenses 45° N dipping	grid drilling under old mine workings	exploration shaft, galleries and cross-cuts	Total combined base metals: 255.000 t Cu, 550.000 t Pb, 1.750.000 t Zn
1971	Aznalcollar (Sevilla)	Higuereta	45 M. tons mass. Cu + Pb + Zn pyr. ore. 35 Mio tons diss. Cu-Zn sulph.	45° N dipping ore lenses	grid drilling under old mine-workings		Total combined base metals: 400.000 t Cu, 775.000 t Pb, 1.600.000 t Zn
1973	Tharsis (Huelva)	Cantareras	> 5 M. tons mass. pyr. ore	steeply dipping ore lens. No surface indication	Resistivity	Gravimetry Mise à la Masse	
1973–1975	Riotinto (Huelva)	Alfredo	> 20 M. tons 1,2 % Cu stockwork ore	cpy-pyr. veinlets in footwall chlorit. acid volcanics adjacent to San Dionisio mass. pyr. ore deposit	underground mine geology	grid drilling	in development stage
1973	Aznalcollar (Sevilla)	"Anomaly 3"	>10 M. tons. mass. Cu + Pb + Zn-pyr. ore	steeply N dipping ore lens. Under 25 m Tertiary cover sediments	Mise à la Masse at 700 m distant abandoned shaft	Resistivity gravimetry	in exploration stage
1974	Paymogo area (Huelva)	Romanera	> 5 M. tons mass. Cu + Pb + Zn-pyr. ore	2 steeply dipping ore lenses	Resistivity showed extension of abandoned orebody	Mise à la Masse grid drilling	in exploration stage
1974	Lousal (Portugal)	Massa José Massa Fernando	≈ 4 M. tons mass. pyr. ore	vertically dipping lenses below 460 m level	underground resistivity	underground gravimetry	in development stage

3. General Geological Setting

The Pyrite Belt forms the central part of a Silurian (?) — Carboniferous Geosyncline (Fig. 1) termed the "South-Portuguese Zone" by *Lotze* (1945) and which represents the southernmost tectonic segment of the Hercynian Hesperic Massif (= "Iberian Meseta").

To the north of the Pyrite Belt, the Beja-Aracena Geanticline is made up of older, metamorphic rocks of Upper Precambrian to Upper Paleozoic ages. This area is considered to be the source of clastic materials of the Pyrite Belt sediments, as well as of, towards the north, the Ferreira-Ficalho greywackes and the "Pulo do Lobo" group composed of phyllites, quartzites, greywackes and minor volcanics (Fig. 1).

A general description of the Pyrite Belt geology has been given, among others, by *Strauss* (1965, 1970) and *Schermerhorn* (1971). According to the former author *the rocks of this belt may best be grouped into three units (from bottom to top):*

1. The basal *Slate-Quartzite Group* ("Phyllite-Quartzite Group", *Schermerhorn*, 1971) consists principally of slates and quartzites with minor quartzitic conglomerate occurrences at its top (Fig. 3). The thickness of this group is unknown, but it certainly exceeds several 1000 m. Also at its top and in a few places some limestone lenses have been dated by conodonts as of Upper Devonian (Famenne) age (*Höllinger*, 1958; *v.d. Boogaard*, 1967; *v.d. Boogard* and *Schermerhorn*, 1975).

2. The *Volcanic-Sedimentary Complex* ("Volcanic-Siliceous Complex", *Schermerhorn*, 1971) overlies conformably the "Slate-Quartzite Group", with a variable thickness of a few tens of m up to some 800 m. This is the only group with acid and basic submarine volcanic rocks and it contains exclusively all the strata-bound, volcanogenic sulphide and manganese mineralizations. Its age ranges from Tournai to Lower Visé (Lower Carboniferous), as determined by *Schermerhorn* (1971) and *v.d. Boogaard* and *Schermerhorn* (1975). However, *Carvalho* (1975) advocates an Upper Devonian age for the first acid volcanics at the Cercal Anticline (Fig. 1) at the southermost part of the Pyrite Belt and a volcanic activity lasting until Upper Visean times in the northern part of the Portuguese Belt.

3. At the top of the Paleozoic stratigraphic sequence, and again with a conformable junction to the Volcanic-Sedimentary Complex lies the *Culm*. Its composition is of alternating slates and greywackes, and it represents the flysch stage of the Pyrite Belt Geosyncline from Upper Visé to Lower Westphalian. In the western, Portuguese part of the belt, its thickness reaches several thousands of m (*Schermerhorn* and *Stanton*, 1970), whereas at its eastern part, in the Huelva Province, it certainly does not exceed some 500 m (according to results of electric-sounding surveys and some exploration boreholes).

During the Hercynian Orogeny (Middle Westphalian, *Schermerhorn*, 1971) this rock sequence was intensively folded. In Spain a general E-W direction prevails, changing into SE-NW directions in Portugal (Fig. 1). Mostly narrow, isoclinal and S-SW overturned fold structures were developed with accompanying thrust faults. Synchronously, a coaxial slaty clevage was produced. However, for the northern part of the belt *Schermerhorn* and *Stanton* (1970) and *Carvalho* et al. (1975) have shown good evidence of important overthrusts from north toward south, with transport of up to several km, prior to the development of the main slaty cleavage. Following the main folding phase, transverse faults have preferred approximate NE-SW strikes with important displacements, principally in Portugal.

The great sinistral Messejana Wrench Fault (Fig. 1), with an extension of some 500 km across the Iberian Peninsula, seems to be related with the early stages of the opening of the North Atlantic Ocean. In part it is marked by tholeitic dolerite dykes of Jurassic age (*Carvalho* et al., 1975).

A low-grade Hercynian regional metamorphism took place during the final stage, and still somewhat after the main folding phase. In the northern part of the belt a low-green-schist facies developed, grading into the prehnite-pumpellyite facies farther to the south (*Schermerhorn*, 1975).

Post-Paleozoic cover rocks are predominantly Tertiary sediments in the Sado Basin and the Guadalquivir Basin.

Out of the approximate time span of at least 60 my of recorded Pyrite Belt strati-graphy (Upper Devonian to Lower Westphalian) it is the Volcanic-Sedimentary Complex representing some 15–20 my, that is the only host of the strata-bound, volcanogenic polymetallic sulphide ores and its closely associated siliceous manganese mineralizations (*Strauss*, 1965, 1970).

The *Volcanic-Sedimentary* Complex is a heterogenous rock sequence with a rapidly varying thickness from place to place and a quickly changing volcano-sedimentary facies in both lateral and vertical directions. A very much generalized lithostratigraphy is given in Figure 2.

The volcanics are felsites and mafites, with occasionally some intermediate rocks. The felsic volcanics, by volume, outnumber the basic ones.

Principally, the *acid volcanics* are pyroclastics of all grain sizes, and most frequently interbedded with shales, indicating a submarine deposition. Relatively few, mostly auto-brecchiated lavas are found around eruptive centres. A few places (e.g., La Zarza, see Fig. 5) have been mapped with acid lavas exhibiting columnar jointing, that might indi-cate either subaerial or extremely shallow submarine deposition.

Massive acid pyroclastics are grouped around eruptive centres lined up along fissural volcanic lineaments (*Strauss* and *Madel*, 1974) parallel to the general strike. These eruptive centres are outlined by volcanic breccias, agglomerates, isolated lava flows and, occasionally, volcanic chimneys have been recognized (*Madel* and *Lopera*, 1975; *Lopera* et al., 1975). Laterally these volcanics interfinger rapidly with sediments such as black shales, radiolarian chert, etc, and/or decrease in grain size to form well-bedded medium- to fine-graded tuffs or tuffaceous shales.

The felsic volcanism is predominantly of soda or potash-rich quartz-keratophyre or sodic rhyolite composition with albite being the only or principal felspar phase. Potash felspar megacryst bearing pyroclastics are only recognized at Aljustrel. Other, but minor, rock occurrences are quartz-latites, dacites, and even quartz-andesites at some places.

The *basic volcanism* is represented in its great majority by intrusive albite diabases, in lesser proportion by spilites and a few spilitic pyroclastics. According to *Schermer-horn* (1970), the silica contents for both the intrusive and the extrusive mafites may vary between 40–66% SiO_2, hence indicating intermediate compositions. Ultrabasic cumulates (approx. 50% serpentinized olivine and 50% clinopyroxene) within major diabase intrusions are known from the southern part of the Spanish Pyrite Belt at Thar-sis, and from the northern and southern flank of the Devonian Puebla de Guzmán Anti-cline (*Lopera* et al., 1975).

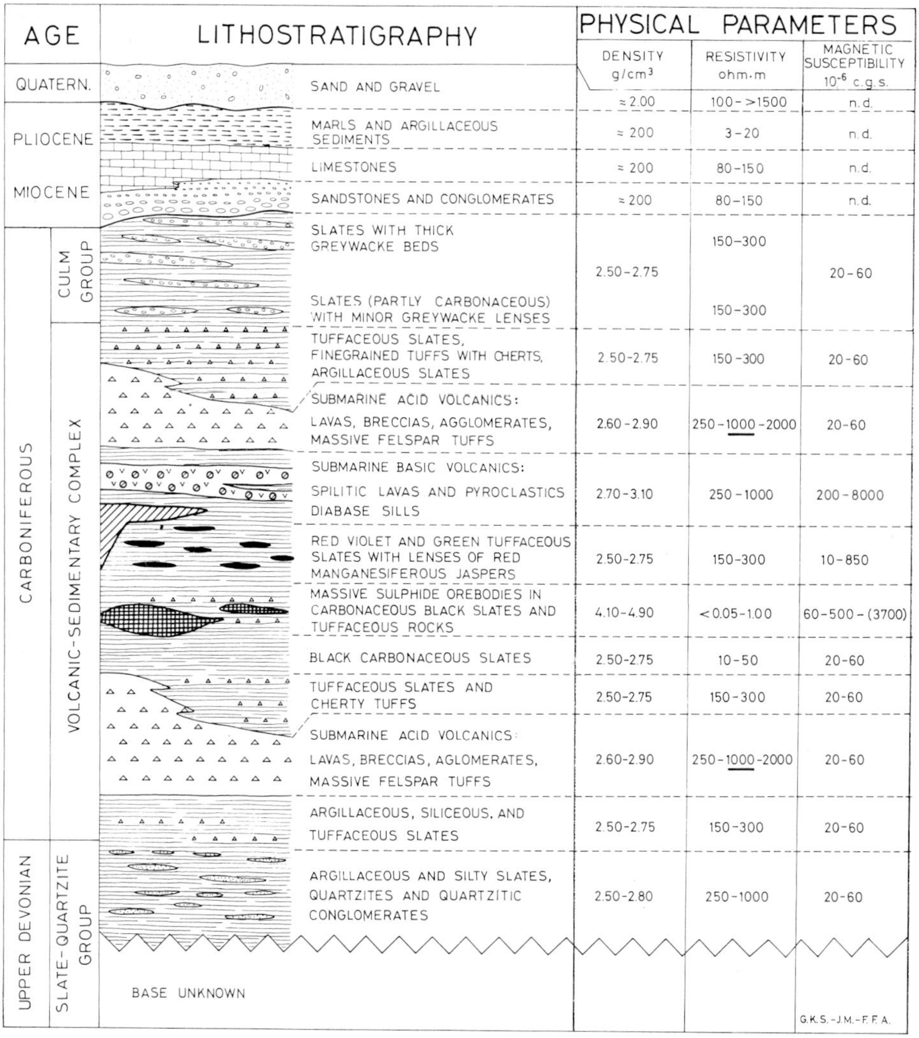

AGE		LITHOSTRATIGRAPHY	PHYSICAL PARAMETERS		
			DENSITY g/cm³	RESISTIVITY ohm·m	MAGNETIC SUSCEPTIBILITY 10⁻⁶ c.g.s.
QUATERN.		SAND AND GRAVEL	≈ 2.00	100->1500	n.d.
PLIOCENE		MARLS AND ARGILLACEOUS SEDIMENTS	≈ 200	3-20	n.d.
		LIMESTONES	≈ 200	80-150	n.d.
MIOCENE		SANDSTONES AND CONGLOMERATES	≈ 200	80-150	n.d.
CARBONIFEROUS	CULM GROUP	SLATES WITH THICK GREYWACKE BEDS	2.50-2.75	150-300	20-60
		SLATES (PARTLY CARBONACEOUS) WITH MINOR GREYWACKE LENSES		150-300	
	VOLCANIC-SEDIMENTARY COMPLEX	TUFFACEOUS SLATES, FINEGRAINED TUFFS WITH CHERTS, ARGILLACEOUS SLATES	2.50-2.75	150-300	20-60
		SUBMARINE ACID VOLCANICS: LAVAS, BRECCIAS, AGGLOMERATES, MASSIVE FELSPAR TUFFS	2.60-2.90	250-1000-2000	20-60
		SUBMARINE BASIC VOLCANICS: SPILITIC LAVAS AND PYROCLASTICS DIABASE SILLS	2.70-3.10	250-1000	200-8000
		RED VIOLET AND GREEN TUFFACEOUS SLATES WITH LENSES OF RED MANGANESIFEROUS JASPERS	2.50-2.75	150-300	10-850
		MASSIVE SULPHIDE OREBODIES IN CARBONACEOUS BLACK SLATES AND TUFFACEOUS ROCKS	4.10-4.90	<0.05-1.00	60-500-(3700)
		BLACK CARBONACEOUS SLATES	2.50-2.75	10-50	20-60
		TUFFACEOUS SLATES AND CHERTY TUFFS	2.50-2.75	150-300	20-60
		SUBMARINE ACID VOLCANICS: LAVAS, BRECCIAS, AGLOMERATES, MASSIVE FELSPAR TUFFS	2.60-2.90	250-1000-2000	20-60
		ARGILLACEOUS, SILICEOUS, AND TUFFACEOUS SLATES	2.50-2.75	150-300	20-60
UPPER DEVONIAN	SLATE-QUARTZITE GROUP	ARGILLACEOUS AND SILTY SLATES, QUARTZITES AND QUARTZITIC CONGLOMERATES	2.50-2.80	250-1000	20-60
		BASE UNKNOWN			

G.K.S.-J.M.-F.F.A.

Fig. 2. Geological column of the most important lithological formations and their physical parameters

Extrusive spilites are always known to occur in the immediate vicinity of acid volcanic lineaments. Their lithostratigraphic position is, almost everywhere, on top of a preceding acid volcanic eruption cycle (*Bernard* and *Soler*, 1974; *Strauss* and *Madel*, 1974). Intrusive diabases, either as stocks or, most frequently, as sills, are spaced at greater distances from acid pyroclastic accumulations, and have intruded all formations of the Volcanic-Sedimentary Complex.

The *sedimentary facies* of the Volcanic-Sedimentary Complex is largely dominated by shales, frequently black shales with a high content of carbonaceous matter, by siliceous and tuffaceous shales, radiolarian chert and by a few minor outcrops of calcarenitic limestones. At the central part of the Pyrite Belt, on both sides of the Spanish-Portuguese border, an important and thick formation of slates with impure quartzites and quartz-wackes (partly associated with nodular manganese mineralizations), occupies a litho-stratigraphical position between the first two acid volcanic eruptive cycles (*v.d. Boogard,* 1967; *Lopera* et al., 1975).

The *strata-bound massive sulphide deposits* are always connected, in time, to the different submarine, explosive acid volcanic rocks, and specifically to the final, waning stage of each eruptive cycle. In space they are confined to the vicinity of the volcanic centres, either deposited directly on massive acid pyroclastics (autochtonous orebodies) or sometimes deposited at some distance and interbedded with black shales and tufs (allochthonous ores).

The *manganese ores* are associated with red ferruginous jaspers and purple-violet tuffaceous shales that make up the "manganese formation". This formation is either contemporaneous with the sulphide deposits or, as in most cases, somewhat later. This ore facies (silicates, carbonates and oxides) was formed during greater time spans and occupies more extensive areas than the sulphide facies.

According to *Strauss* and *Madel* (1974) nearly all submarine acid eruptive phases are associated with a corresponding manganese formation. This excellent marker formation is, in most cases, the only key to a further subdivision of the heterogenous Volcanic-Sedimentary Complex into different acid extrusive formations which is, otherwise, extremely difficult in view of the tectonic complexity. Recently, *Bernard* and *Soler* (1974) maintained that the purple-violet shale formation (their "polvo-hematites") serves as a synchronous marker over the whole Pyrite Belt terrain. However, at some places like Tharsis, there exist two manganese formations including purple shales, separated by one spilitic and one acid volcanic eruptive cycle (Fig. 7).

4. Geology of the Sulphide Orebodies

As has been stated above there are three types of sulphide mineralizations:

1. The syngenetic-sedimentary massive polymetallic pyritic ore with 35–51% S, or 66–96% pyrite equivalent.

2. The syngenetic-sedimentary disseminated polymetallic pyritic ore with less than 35% S (66% pyrite equivalent). This ore is mostly associated along strike with the massive facies.

3. The epigenetic stockwork pyritic ore, predominantly in the highly chloritized footwall rocks of the massive ore. Its sulphide grade may vary from approximately 10–50% pyrite equivalent (5–25% S), only at some places, such as the Planes mine at Riotinto (*Rambaud Perez*, 1969), reaching as high as 95% pyrite equivalent.

The discussion of the ore formation of the massive sulphides has been reviewed exhaustively by *Strauss* (1965, 1970) and *Rambaud Perez* (1969). In contrast to the previous hypothesis of an epigenetic origin, there is now an unanimous acceptance of a syngenetic, synsedimentary exhalative (volcanogenic) origin of these ores (*Kinkel*, 1962; *Williams*, 1962; *Strauss*, 1965; *Schermerhorn*, 1971; *Bernard* and *Soler*, 1974).

More recently, the pyritic stockwork ores were interpreted, upon geological evidence, as of epigenetic origin with respect to their host rocks and as the hydrothermal feeder channels of the sedimentary sulphides (*Williams*, 1966; *Rambaud Perez*, 1969; *Bernard* and *Soler*, 1974; *Strauss* and *Madel*, 1974; *Garcia Palomero*, 1975; *Williams* et al., 1975). Mineralogical arguments backing the hydrothermal hypothesis have been given by *Aye* and *Strauss* (1975).

With respect to the *geological environment and formation of the sulphide ores*, particularly of the massive facies, the most important facts for exploration purposes have been summarized by *Strauss* and *Madel* (1974), as follows:

1. The massive sulphide ores and the manganese mineralizations occur exclusively in rocks of the Volcanic-Sedimentary Complex.

2. Both types of ore are sedimentary and are closely linked with the submarine explosive, acid alkaline volcanism.

3. In the case of multiple, acid eruptive cycles (up to three in the southern part of the belt) along one volcanic lineament, it is common experience that only the first cycle is "productive" with regard to economical sulphide mineralizations. The manganese-bearing formation may be associated with the first two eruptive phases.

4. The sedimentary sulphide concentrations, both massive and disseminated, are always located near the borders of the volcanic extrusive centres, and at a geologically well-defined distance, where black slates are interbedded with acid pyroclastic rocks. The manganese mineralizations, both laterally and vertically, are distributed more widely than the more restricted sulphide facies. The mangenese ores, normally, are somewhat later than the sulphides.

5. The concentration of sulphides into massive ore lenses is explained by the inflow of sulphide mud and/or detrital sulphide material into topographical depressions around submarine volcanoes. Those orebodies composed of different ore layers reflect the existence of separate inflows of ore, either under different facies conditions or from different exhalative sources, or both.

6. Sulphide ores with massive, fine-grained textures, lacking a layered internal composition, were generally deposited on coarse porphyritic pyroclastics, and hence in the immediate vicinity of the corresponding extrusive centre. Sulphide orebodies with a layered internal structure and detrital sedimentary textures (synsedimentary breccias, graded bedding, etc) were deposited in a more shaly environment.

7. Clastic sedimentary ore textures indicate resedimentation of sulphide material. This implies that previously formed, already solidified, pyritic deposits were submitted to submarine erosion and its material transported into newly formed depressions. Due to this process the ore was successively carried farther away from the corresponding volcanic centres into a more sedimentary, shaly environment, often becoming diluted in this way. These redepositions of sulphides into newly formed throughs may be explained by repeated volcanotectonic movements of the seafloor.

8. All big sulphide ore concentrations are confined to synclinal structures, sometimes the orebodies themselves being folded intensively. The only exceptions are the large pyritic sulphide masses of Riotinto on both flanks of the Cerro Colorado anticline, but linked together by the huge Cerro Colorado gossan (*Rambaud Perez*, 1969).

Regarding the *size and shape of the sulphide orebodies*, a recent statistical evaluation has been published by *Fernandez Alvarez* (1974) using all available data about some 60 different massive sulphide deposits of the Huelva Province. A combined total tonnage of approximately 750 million t was classified into four categories as follows:

1. Very big: > 20 million t; 850 m long, 80 m wide, 350 m deep:
   combined total = 475 million t
2. Big: 5–20 million t; 400 m long, 40 m wide, 260 m deep:
   combined total = 170 million t;
3. Medium: 1–5 million t; 350 m long, 25 m wide, 170 m deep;
   combined total = 95 million t
4. Small: < 1 million t; 200 m long, 12 m wide, 75 m deep:
   combined total = 10 million t

Although *Fernandez Alvarez* states that the three possible ore grades, viz. pyritic, cupreous-pyritic, and Cu + Pb + Zn-rich ores are not confined significantly to one or more of the four size categories, *Carvalho* et al. (1975) point to the fact that all deposits of above 30 million t and a true width exceeding 30–40 m are essentially pyritic with an overall primary copper grade of approx. 0,7% Cu. Significantly high overall grades of approx. 1.5% Cu and more are always restricted to tonnages below some 30 million t, as listed by some examples:

1. San Telmo (Huelva): approx. 10 million t; 1.5% Cu, 5% Zn
2. Gaviao (Portugal): approx. 25 million t; 1.5% Cu, 3.5% Zn
3. San Antonio (Riotinto): approx. 12 million t; 1.5% Cu, high Pb and Zn
4. Ntra. Sra. del Carmen (Paymogo, Huelva): approx. 0.3 million t, 3% Cu

5. Geological Exploration

The tasks of exploration geology in the Pyrite Belt are centred around two main objectives:

1. To explore working or abandoned mine areas with limited extensions of, say 1–3 km^2, mainly to guide exploration drilling. Mapping is done preferentially at scales of 1:5000 or greater.

2. Mapping of virgin ground with areas of 100 km^2 or more, in order to block out favourable areas in which to employ ground geophysical methods. Geological maps at scales of 1:25,000 or 1:10,000 proved appropriate for the display of sufficient details, e.g., volcanic facies and tectonics. Maps at scales 1:10,000, furthermore, may serve as a first basis for the geological interpretation of geophysical data.

As has been outlined in Section 2, at present it is the second task that is becoming of ever-growing importance. Therefore, geological mapping should stress those mappable features that may distinguish between mineralized areas and/or areas with possible mineral potential and barren ground. In short, *the following four geological features, along with structural control, should be sought:*

a) Facies of acid volcanics.
b) Facies of basic volcanics.
c) Facies of the manganese formations.
d) Hyrothermal alteration.

5.1 Acid Volcanics

The close link between sulphide mineralization and the acid, submarine and explosive volcanism requires geological maps showing volcano-stratigraphy and volcanic facies distribution. In the southern part of the Pyrite Belt (e.g., Tharsis) up to three different acid volcanic phases have been recognized along one volcanic lineament. In the authors' experience (see also *Strauss* et al., 1974) in any given exploration area of, say 1−2 km², economic mineralizations are always synchronous with the final stage of the very first acid explosive volcanism. The second acid volcanic phase may be accompanied by mineralizations, but only in the case of the previous phase being lacking at that place.

An example is the discontinuous acid volcanic lineament along the northern flank of the Devonian "Puebla de Guzman" anticline, which connects the mine areas of Tharsis to the east and Herrerias some 15 km to the west (*Lopera* et al., 1975):

At Tharsis the first phase "Tharsis 1" has been "productive" whilst the second phase "Tharsis 2" is known to show only minor sulphide disseminations at some places, but a highly "productive" manganese formation as deduced from many abandoned manganese mines. At Herrerias (Fig. 6) massive sulphide mineralizations are linked with acid volcanics that correspond to the "Tharsis 2" volcanism, because there the "Tharsis 1" phase is not present and thus "Tharsis 2" represents the first acid volcanic phase on that segment of the volcanic lineament, followed by the sterile "Tharsis 3" volcanics.

In space, all sulphide mineralizations are located on top of or nearby the acid extrusive centres (Fig. 3), lined up along volcanic lineaments, and marked by coarse massive pyroclastics, tuff-breccias, breccias, agglomerates and, occasionally lava flows. Similar rocks have been termed "mill rocks" by *Sangster* (1972). Consequently detailed volcanic facies maps are required (Figs. 4, 5 and 6). To the authors' knowledge all sulphide mineralization falls into a radius of approximately 1000 m around such an eruptive centre.

There are a few cases like that of Herrerias (Fig. 6) where the „productive" acid volcanism outcrops only as a few m of bedded acid pyroclastics (not mappable in Fig. 6) with its eruptive centre(s) being covered by Culm sediments to the ENE of by the outcropping eruptive centre of the second sterile acid volcanics towards the E-NE of E. However the Herrerias deposits are located on a volcanic lineament with, in this case, extensive basic, spilitic extrusives, and again within approximately 1000 m of one of the few mappable acid eruptive centres.

5.2 Basic Volcanics

Although the basic volcanism is not believed to be linked directly with sulphide ore formation, some mappable facies differences are recognized for mineralized and nonmineralized areas.

The albite diabases have intruded the upper horizons of the Devonian Slate-Quartzite group as well as practically all formations of the Volcanic-Sedimentary Complex. The youngest intrusions (e.g., at Herrerias) are contemperaneous with the last acid volcanic phase. Albite diabase intrusions preferentially accumulate in the "basin facies" of the Volcanic-Sedimentary Complex at some distance from the acid volcanic lineaments whereas the spilite lava flows (pillow lavas, amygdaloidal lavas) and (fewer) spilite tuffs are spaced closely to acid volcanic lineaments, occupying there relatively narrow throughs (Fig. 7).

The cases at Herrerias (Fig. 6) and the Tharsis area (Figs. 3, 7) illustrate well the distribution of the different basic volcanics with regard to sulphide mineralizations. These facts can be extended to all other mine areas as far as we know, and *some general conclusions for practical exploration purposes may be derived, as follows:*

1. Intrusive diabases accumulate mostly at greater distances from acid volcanic lineaments. Extrusive spilitic volcanics are lined up along "productive" acid volcanic lineaments.

2. In many cases the relatively restricted extrusive spilites are within a few hundred m of the orebodies, or occupy important parts of the specific mine areas, within the mine workings (opencast or underground) as for instance, in the Filon Norte opencast at Tharsis (Fig. 7).

3. The only basic volcanics within the dimensions of any massive sulphide orebody are extrusive spilitic rocks.

4. There is practically no chance of finding hidden sulphide mineralizations in a given exploration block, of say 1 km^2, where intrusive diabases make up more than approx. 20% of its surface.

5. From above it follows that extrusive spilites may serve as a possible guide to ore, whereas abundant diabase intrusions (stocks and sills) indicate sterile ground.

5.3 Manganese Formation

This formation, principally, is made up of purple-red tuffaceous shales ("polvo hematites" of *Rambaud Perez*, 1969; *Bernard* and *Soler*, 1974), and interbedded with predominantly red, ferruginous jasper lenses with primary manganese silicate and/or carbonate mineralizations (*Strauss*, 1965). The purple shales may grade laterally into greenish-grey tuffaceous shales. Less frequently, the manganese formation is represented by green and red versicoloured pyroclastics with interbedded jaspers (e.g., La Zarza, Figs. 5 and 8).

The manganese formation of the Pyrite Belt is comparable with similar formations in other metallogenic provinces with volcanogenic sulphide mineralizations. It certainly corresponds to the "cherty iron formation" (*Sangster*, 1972) or "ferruginous quartz zone" (*Roscoe*, 1971) associated with Precambrian and Paleozoic Canadian deposits, and to the "tetsusekiei bed" of the Tertiary Kuroko deposits of Japan (*Horikoshi*, 1969).

As has been argued above (Sect. 3 and 4), the red-purple shales and jaspers form excellent marker horizons that separate different acid volcanic eruptive phases. On the other hand the *lateral facies changes of the manganese formation seem to be related to some extent to sulphide mineralizations:*

1. The manganese formation, stratigraphically, is mostly later than the sulphide facies. Laterally, it may replace the sulphide facies or even precede it, as in the case of Almagrera (Fig. 9).

2. In the vicinity of sulphide deposits, the manganese formation thins out, that is purple shales turn into black shales or greenish tuffaceous shales with only very few small jasper lenses of very restricted thickness and extension (in the order of a few m). Such cases are documented at Lousal/Portugal, the Filon Norte opencast at Tharsis (Fig. 7) and at La Zarza, where important jasper concentrations are located peripherally around the very large orebody in all lateral directions (Fig. 5).

3. Our field experience is that ferruginous jaspers in the vicinity of sulphide mineralizations, up to a distance of approximately 200 m, may bear significant visible pyrite impregnations (1–5% pyrite). Jasper lenses in sterile areas have not been observed to have visible sulphide contents.

5.4 Hydrothermal Alteration

Hydrothermal alteration, mostly in the form of strong pretectonic chloritization, has been observed in many mine areas. Formerly this had been an argument favouring the hydrothermal, epigenetic formation of the massive sulphides (*Bateman*, 1927; *Williams*, 1934), although the chloritization accompanied by pyritization had only affected the footwall rocks of the massive sulphide ore lenses. Lately, some authors have attributed this type of alteration to the fumarolic activities related to the formation of the sedimentary ores (*Strauss* and *Madel*, 1974; *Aye* and *Strauss*, 1975; *Williams* et al., 1975).

Mapping of hydrothermal alterations as an exploration tool was reported for the first time by *Carvalho* (1975) from the Cercal-Odemira region (Portugal): several imperfect concentrical, elongated ellipsoidal alteration haloes of up to 0.5 km^2 extension could be mapped with increasing alteration grades, from the margins towards the centre, of: (1) argillaceous minerals and sericite; (2) argillaceous minerals, sericite, chlorite, pyrite and quartz; (3) abundant sericite and chlorite, pyrite and quartz, and (4) abundant sericite and chlorite, abundant pyrite, quartz and siderite.

Carvalho described that out of five mapped hydrothermal alteration haloes, four were located on acid volcanic agglomerates, that is, on eruptive centres. Furthermore, one of these anomalies was within a geochemical soil anomaly (Cu + Zn) and a Bouguer gravity anomaly. Follow-up diamond drilling intersected acid pyroclastics with disseminated sulphide mineralizations in the form of small veinlets of chalcopyrite, pyrite, sphalerite, galena and ankeritic carbonates.

5.5 Structural Control

The general structural pattern of the Pyrite Belt is southward-overturned fold structures with coaxial slaty cleavage planes dipping to the north. On a regional basis, in the northern part of the belt folding and accompanying thrust faulting have developed more extensively, and have mostly produced imbricated structures. Towards the south, open concentric folds are common features.

Within this general picture there are locally important breaks caused by the heterogeneous lithology of the Volcanic-Sedimentary Complex, and hence produced by local paleogeographic conditions: "volcanic rises" (acid volcanic piles of some 400–500 m or

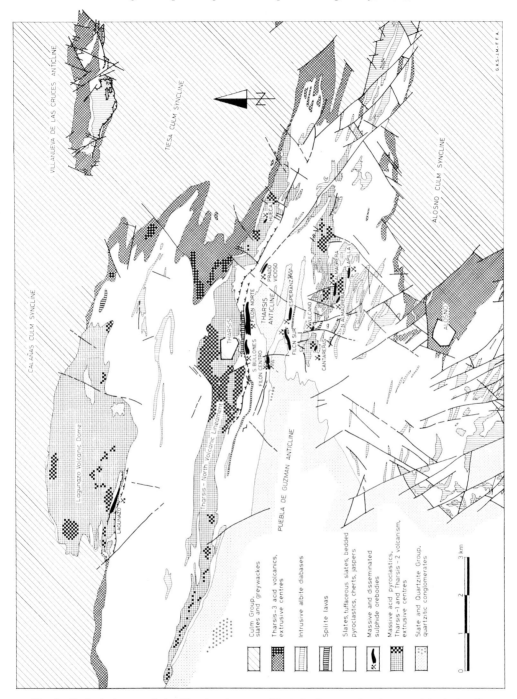

Fig. 3. Geological map of the Tharsis mine area (Huelva). Observe the close spatial relationship between sulphide mineralizations (*black*) and the eruptive centres (*block rectangles*) of three acid volcanic lineaments. Geology of Villanueva de las Cruces Anticline by *Aye* (1974). (Geology of the Alosno area compiled from unpublished maps courtesy of Messrs. *Lecolle* and *Roger*, Univ. Pierre et Marie Curie, Paris)

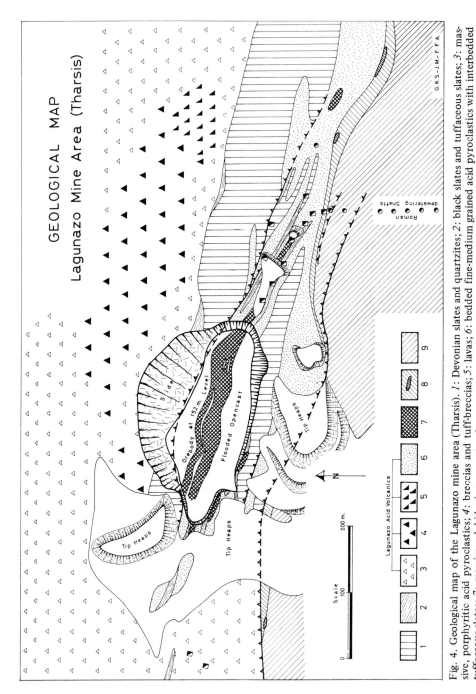

Fig. 4. Geological map of the Lagunazo mine area (Tharsis). *1*: Devonian slates and quartzites; *2*: black slates and tuffaceous slates; *3*: massive, porphyritic acid pyroclastics; *4*: breccias and tuff-breccias; *5*: lavas; *6*: bedded fine-medium grained acid pyroclastics with interbedded tuffaceous slates; *7*: massive sulphides in opencast and gossan outcrops of the Lagunazo orebody; *8*: purple slates with occasional ferruginous and manganesiferous jasper lenses; *9*: tuffaceous slates and minor bedded acid pyroclastics

Fig. 5. Geological map of the La Zarza mine area (Silos de Calañas, Huelva). *1*: massive acid porphyr- ▶ itic pyroclastics; *2*: breccias and tuff-breccias; *3*: lavas with columnar jointing; *4*: massive pyritic ore- body; *5*: black slates, partly tuffaceous; *6*: extrusive spilites and spilitic pyroclastics; *7*: massive and occasionally bedded, versicoloured, acid pyroclastics with breccias and tuff-breccias (*black circles*); *8*: ferruginous, manganesiferous jasper lenses; *9*: purple slates; *10*: intrusive albite diabases; *11*: fine grained, bedded acid pyroclastics and tuffaceous slates; *12*: Culm black slates with few greywackes beds

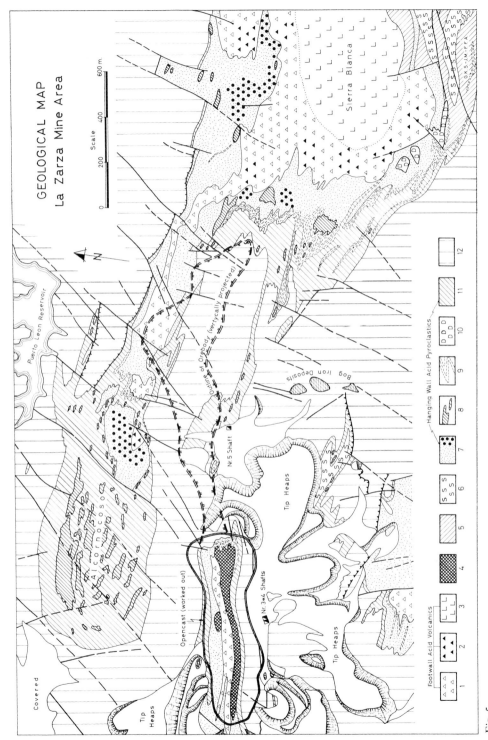

GEOLOGICAL MAP
La Zarza Mine Area

Fig. 5

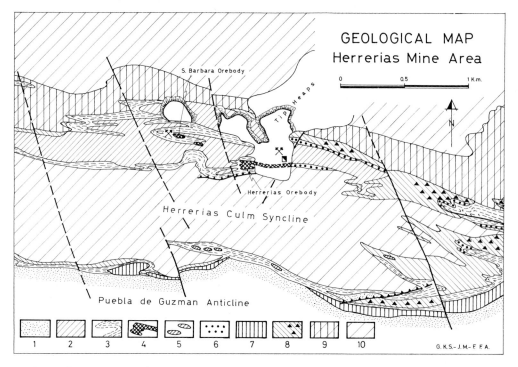

Fig. 6. Geological map of the Herrerias mine area (Huelva). *1*: Devonian slates and quartzites; *2*: extrusive spilites, predominately pillow lavas; *3*: black slates with few acid pyroclastic beds; *4*: massive cupreous pyritic ore and gossans; *5*: ferruginous, manganesiferous jasper lenses; *6*: tuffaceous purple slates; *7*: intrusive albite diabases; *8*: upper bedded acid pyroclastics, massive breccia occurrences (*black triangles*); *9*: Culm black slates without greywackes; *10*: Culm slates with predominately greywacke facies. *Note* that the Herrerias orebody is shown as projection to surface. The Santa Barbara gossans represent remnants of an eroded massive pyritic orebody. At present copper carbonates in decomposed spilites and slates are mined and leached at Santa Barbara. The lower acid volcanism contemporaneous with the formation of the sulphide ores is present as thin, not mappable beds of fine-medium grained pyroclastics on top of the spilite pillow lavas around the Herrerias orebody. (Geology compiled from *Lopera* et al., 1975)

more) are rigid blocks that later have been turned into broad anticlinal structures of monoclinic antiforms and "sedimentary basins", formed between the volcanic rises (with accumulations of shaly tuffaceous sediments and well-sorted and bedded pyroclastics) are characterized by dominating open concentric fold structures. The narrow border zone between both structural facies is the location of most of the massive and disseminated sulphide mineralizations. Here, coarse massive acid volcanic rocks grade laterally into medium-grained bedded pyroclastics with black carbonaceous shales and cherty, tuffaceous sediments.

The mineralized zones, due to the very heterogenous rock sequence, exhibit narrow, isoclinal, overturned folds that are thrust-faulted in most cases. Another mappable tectonic guide is the frequent and rapid change of axial plunge to both directions within a few hundred m along strike. From Herrerias (Fig. 6) *Febrel* (1972) has reported axial plunge of up to $60°$, whereas at la Zarza (Fig. 5) plunge of fold axes reaches $30-50°$

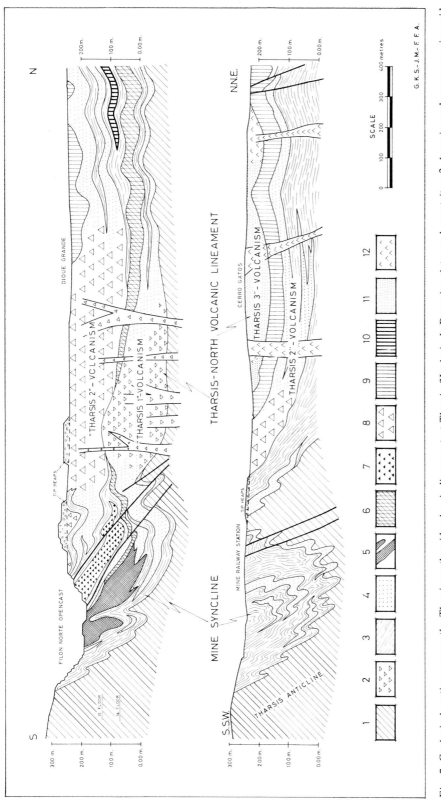

Fig. 7. Geological sections across the Tharsis-north acid volcanic lineament at Tharsis (Huelva). *1*: Devonian slates and quartzites; *2*: breccias and coarse massive acid pyroclastics of the Tharsis-1 volcanism; *3*: silty slates, black slates and tuffaceous slates; *4*: medium-fine grained acid pyroclastics, partly massive, partly well bedded; *5*: massive pyritic orebodies Filon Norte and San Guillermo; *6*: lower manganese formation with purple slates and very few ferruginous jaspers; *7*: extrusive spilite lavas; *8*: breccias, flow breccias and coarse, massive, acid pyroclastics of the Tharsis-2 volcanism; *9*: upper manganese formation with purple slates and abundant ferruginous jaspers; *10*: intrusive diabase sill; *11*: bedded acid pyroclastics, and *12*: breccias and coarse, acid, intrusive porphyries of the Tharsis-3 volcanism. *Note*: The *upper section is located approx. 1 km W from the *lower*

to the east and west around the mineralized zone, both at surface and in the underground mine workings.

Summarizing, it is safe to say that structural features such as those described will prove very helpful in blocking out promising ground in Volcanic-Sedimentary terrain. Furthermore we believe that some of these structures may trace hidden "volcanic rises" under Culm sediment cover of, say, some 100—200 m. In this way the area to cover with the far more expensive ground geophysical methods can be reduced significantly — and safely.

6. Geophysical Exploration

As early as 1919 *Jubes* and *Carbonell* discussed the possible application of the Eötvös torsion balance to explore subsurface massive pyritic ores in the Huelva Province. Then, some ten years later, a first company report is recorded that deals with a geoelectrical survey.

Sine 1944, systematic geophysical surveys have been carried out in the Portuguese part of the Pyrite Belt, mostly using electromagnetic methods (EM) in mine areas such as Sao Domingo, Aljustrel, Panoias, Lousal and Caveira (*Rocha Gomes* and *Da Silva*, 1955).

Starting from 1960, geophysics were integrated in nearly all mine exploration programmes in the Pyrite Belt. Nowadays, in both the Spanish and Portuguese parts of this Belt there are some five companies doing geophysical exploration continuously. The results are obvious from Table 1: geophysical discoveries, mainly by gravimetry and earth-resistivity methods have increased the ore reserves significantly, in the order of some 130 million t.

In the course of the last 15 years there has been considerable dispute about the practical use and efficiency of different geophysical methods. The main questions were: (1) the usefulness of airborne EM methods; (2) EM methods or Earth Resistivity methods; (3) Earth Resistivity methods or Induced Polarization?

At the same time there appear to exist no doubts about the extreme usefulness of gravimetry and the futility of employing magnetic surveys.

6.1 Airborne Geophysics

Table 2 lists airborne geophysical exploration programmes and test flights performed in the Spanish portion of the Pyrite Belt between 1957 and 1959 (*Pinedo Vara*, 1963) and later work carried out by two Canadian consulting companies.

It is a safe conclusion that all these airborne "blanket coverage" surveys have proved to be of no practical use to exploration. The numerous EM anomalies produced were, in most cases. outcrops of carbonaceous black shales and/or wet ground, as could be deduced by later follow-up investigations, including diamond drilling. Furthermore, many EM anomalies that were found by one company did not match with EM anomalies observed by later airborne surveys covering the same ground (*Pinedo Vara*, 1963).

Table 2. Airborne geophysics in the Spanish Pyrite Belt

Company	Methods	Area and data	Aircraft	Exploration results
Hunting Geophysics Ltd. (Great Britain)	Electromagnetic, magnetic (total field), radioactivity	300 km^2 Central Huelva Province 1957	Plane	Negative
Lundberg Exploration Ltd. (Sweden)	Electromagnetic (ground transmitter, aircraft receiver), magnetic (vert. comp.), radioactivity, gravimetry (tests)	400 km^2 Central Huelva Province 1957	Plane	Negative EM anomalies do not coincide with anomalies detected by Hunting
Hunting Geophysics Ltd. (Great Britain)	Electromagnetic (rotary field), test flights	Huelva Province 1959	2 planes	Negative
Seigel Associates Ltd. (Canada)	Electromagnetic (Scintrex HEM 701), magnetic	Huelva and Sevilla Provinces, total area of Pyrite Belt 1967–1968	Helicopter	Negative
Geoterrex Ltd. (Canada)	Electromagnetic, "Barringer INPUT" (Mark V "INPUT" System, 6 channels), test flights	Western Huelva Province Paymogo area 1973	Plane	Negative

Finally, the results of the 1973 "Barringer Input" test flight were disappointing, in as much as this new method hardly yielded anomalies in the second channel over various outcropping massive sulphide orebodies (Nuestra Sra. del Carmen, Trimpancho, Vuelta Falsa). On the other hand, it did produce even six-channel anomalies over terrain with abundant carbonaceous shales.

The airborne magnetic maps showed a perfect outline of the Volcanic-Sedimentary Complex with special enhancement of basic volcanics (*Armengot* and *Fdez. Alonso*, 1970). To the best of the authors' knowledge there are only the orebodies of Cueva de la Mora, Concepción and Castillo de las Guardas with minor pyrrhotite grades and/or magnetite lenses associated with the pyritic ore that could cause local, weak magnetic anomalies.

Summarizing this subject, it can be said that the airborne EM surveys with an excellent performance in other countries and mainly in Precambrian terrain, are not suitable in the exploration of the Pyrite Belt. In part this is due to the low depth penetration (certainly less than 30 m) under the local geological conditions and the fairly rough topography of the area.

In this region, nearly all orebodies that could be detected by this method, have some surface showings (gossans, etc.), and hence had been located already by Roman prospectors. Furthermore, the Tertiary sediment cover, such as the Sado and Guadalquivir Basins, made up principally of argillaceous sediments represents a thick electrical conductor and thus an impenetrable layer for any proper EM response of the underlying Paleozoic rocks.

6.2 Ground Geophysics

The relevant physical parameters of the most important rock units are listed in Figure 2. From this it is easily concluded that the best "direct" geophysical methods for Pyrite Belt ores are: (1) gravimetry; (2) direct current earth resistivity methods (ERM, different varieties): (3) electromagnetic methods (EM, e.g., Turam).

At present in Spain, because of the prevailing topographical conditions of the terrain, it is common practice first to explore virgin ground with ERM and check anomalies by gravimetry. *Ground EM Surveys* using Turam were the most important components within every exploration programme until approximately 1965, when they were progressively replaced by ERM due to the following reasons:

1. Limited depth penetrations compared with ERM (skin effect). The author's experience (F.F.A.) shows that a depth penetration of some 50 m is the utmost under existing geological conditions.

2. EM methods produce too many "non-ore" anomalies caused by very small resistivity contrast (e.g., faults and local cover). Consequently, very much follow-up work is involved, mainly checking by gravimetry and diamond drilling.

3. EM surveys, as compared to ERM (e.g., Electrical Trenching), do not provide a depth distribution of apparent resistivities. This in its turn provides very much desired geological information, as may be deduced for example from the La Zarza section (Fig. 8).

4. Electrical Trenching is an accurate mapping method of geological subsurface structures (*Strauss* et al., 1974).

Induced Polarization (IP), both time and frequency domain, has been employed only in a limited way until recently. From an exploration point of view, and considering the Pyrite Belt conditions, it can be argued that the IP method has several technical and economical disadvantages in comparison with ERM surveys:

1. IP is ideally suited to detect disseminated sulphide mineralizations. Although in the Pyrite Belt there are numerous disseminated pyritic mineralizations, those few of economic importance due to their base metal grades (mostly copper, in less proportion zinc and some lead) are always found in close contact to massive sulphide ores (e.g., Riotinto, Aznalcollar). Furthermore, these economic mineralizations form in most cases a network of sulphide veinlets (stockwork) which are electrical conductors detectable by ERM (see the Prado Vicioso example in Fig. 12).

2. The overall abundant carbonaceous slates (partly graphitic when faulted) originate the same amount of IP anomalies as ERM.

3. ERM depth penetration is far greater than that of IP methods, when similiar and practical electrode array dimensions are compared. Moreover, the more favourable "geological depth resolution" of the ERM (Fig. 8) is another argument in favour of the later method.

4. One of the most important facts, however, is that IP surveys are somewhat 50% slower than ERM (e.g., Electrical Trenching), and are thus far less economical.

The *Self-Potential method (SP)* has seen only limited applications during the past. Within the Pyrite Belt this method is of no practical use because only near-surface orebodies could be detected due to the shallow depth of the water table (varying between 10 and 50 m) above which oxidation processes take place. Recently, SP test surveys over the Almagrera and Cantareras orebodies at Tharsis have proved the SP method to be dependent on the local topography, and sensitive to climatic changes. No reproduceable anomalies could be measured during different seasons in the course of one year.

D.C. Earth Resistivity Methods (ERM). Of the different varieties of the ERM, the *Electrical Trenching Method (ETM)* a line-electrode mapping method, with the Schlumberger asymmetric three-electrode array is used most frequently (*Blokh,* 1962). Traverses are staked 100 or 200 m apart and perpendicular to the general strike with stations every 20 m. At each station, two different theoretical depths are measured simultaneously, mostly at 100 and 200 m (*Strauss* et al., 1974).

Another less common ETM variety is labelled the *Resistivity Block Method (RBM)*, as described by *Kunetz* (1966), among others. The difference is that with the RBM both current electrodes (A and B) remain fixed and only the potential probes M and N are moved along the traverse from station to station. As compared to the above-mentioned Schlumberger three-electrode-array the RBM yields a greater depth penetration, but at the expense of the "geological" depth resolution. On the other hand, this method is somewhat faster during field work and neeeds only a crew of three men.

Good comparison between the RBM and four different Schlumberger three-electrode-arrays can be observed at the traverse above the La Zarza orebody (Fig. 8): here, the RBM yields the best anomaly, but it must be taken into account that this is a somewhat "combined anomaly" of pyritic ore and Culm shales, whereas the different Schlumberger arrays yield a clear anomaly with theoretical depth penetrations of 320 and 420 m, although the inverse faults at the north side are overenhanced and mask part of the orebody.

Another variety of the ERM is the *Vertical Electric Sounding Method (VESM)*, a common method in water exploration. VESM can be extremely useful in deciphering large concentrical fold structures or delimiting extensions of nearly flat dipping ore-bodies as shown by the example of the Almagrera orebody as illustrated in Figure 10. Here, resistivity values could be converted directly into geological data using the parameters listed in Figure 2.

It becomes obvious that the future role of VESM will be predominantly to determine the thickness of the overlying sterile Culm sediments with the outline of "economical depths" in which to explore the underlying ore bearing Volcanic-Sedimentary Complex in the large Culm synclines.

Mise à la Masse. The Mise à la Masse method (MaM) is an appropriate tool for one of the main exploration tasks in the Pyrite Belt: investigation of abandoned mine areas or prospects of relatively small size ($1-2$ km^2).

The use of MaM needs an electrical contact with an orebody, either by a borehole that has intersected ore, an ore outcrop (e.g., opencast), or a flooded shaft where acid mine waters provide an electrical connection with the massive sulphide ore.

In this way MaM is best suited to investigate small prospects at relatively high speed and low cost, providing a good approximation of the lateral extension of an orebody, as well as the possibility of detecting any other electrical conductors in the vicinity of the "electrically charged" mineralization.

Figure 11 of the Almagrera-Lapilla mine area at Tharsis gives an example of the practical use of MaM. The Almagrera orebody was charged electrically through borehole DDH 170 and the equipotential contours outline perfectly the flatly north-dipping Almagrera orebody with approximately the same precision as the Bouguer gravity anomaly (see also Fig. 9 for the geophysical section). The Nueva Almagrera and Lapilla orebodies, which are not connected electrically with the Almagrera orebody, have caused deflections of the equipotential lines.

The Nueva Almagrera orebody had been drilled in 1969, based on ERM, Turam and gravimetric anomalies. The MaM survey was done in 1971. There is no doubt that MaM would have discovered this new orebody if this method had been included in the earlier exploration programme. The very first discovery of a new massive sulphide deposit by use of MaM through an abandoned flooded shaft was achieved by one of the authors (F.F.A.) at the Aznalcollar mine area in 1974.

The MaM survey grid is the same as for the ERM (Fig. 11). The electrode configuration has been described by *Poldini* (1947) and *Zaborovskii* (1963).

Gravimetry. This low speed—high cost method is essential in checking geoelectrical or electromagnetic anomalies. In this case it is common practice to measure stations every 10 m over the anomalies and at $40-100$ m intervals at greater distance from the anomalous points in order to access properly the regional field. When gravimetry is used as a "blanket coverage" method at the initial stage of an exploration programme, the grid of 50 x 50 m or 100 x 40 m has proved to be efficient enough to detect massive sulphide mineralizations of even subeconomic tonnages, provided the surface topography is relatively smooth.

A 100 x 100 m gravimetric grid is standard practice as part of any exploration programme of the Serviço Fomento Minero, a Portuguese state agency. In Aljustrel (Portugal), ore reserves have been increased in the order of some $100-150$ million t by systematic gravimetric grid measurements.

Gravimetry can also be employed successfully in underground mine exploration with stations staked along galleries and crosscuts. In this way recently, and in combination

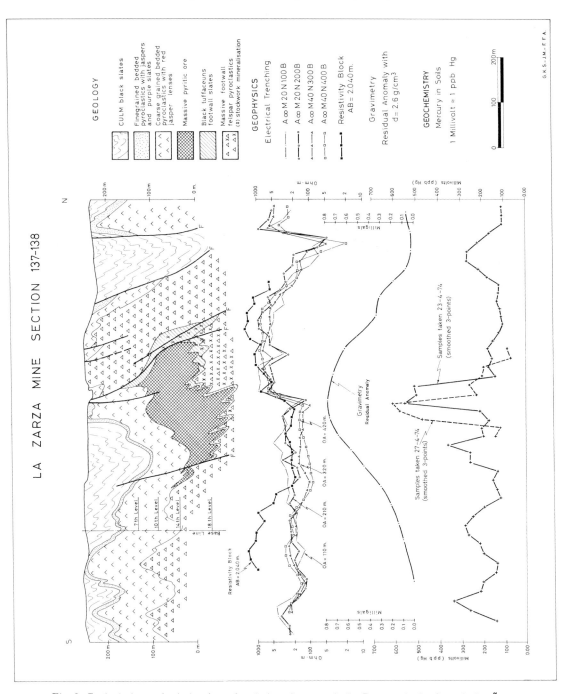

Fig. 8. Geological, geophysical and geochemical section over the La Zarza orebody (Silos de Calañas, Huelva). Mine section 137–138, geophysical traverse 300E. *Note* the depth distribution of electrical resistivities as marked by different Schlumberger electrode arrays (Electrical trenching). The orebody is marked by electrical trenching OA 320 m and OA 420 m, and particularly by the resistivity block method and gravimetry. However, the most striking, steep electrical anomaly is only due to faulted black slates, as confirmed by gravimetry

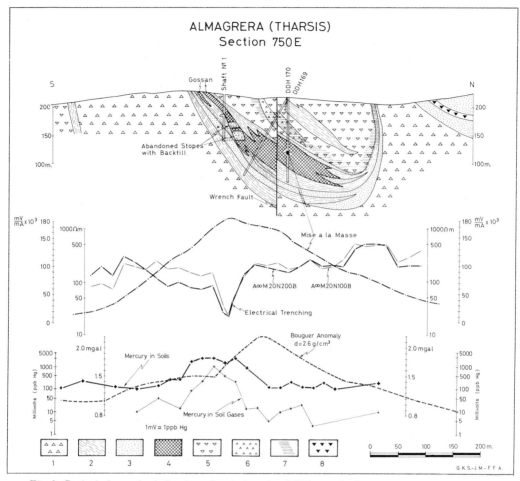

Fig. 9. Geological, geophysical and geochemical section 75OE over the Almagrera orebody (Tharsis).
1: Massive, coarse, porphyritic acid pyroclastics of the Almagrera-1 volcanism; *2*: black and/or tuffaceous slates; *3*: medium to fine grained bedded acid pyroclastics; *4*: massive pyritic ore; *5*: massive
coarse, porphyritic acid pyroclastics and *6*: medium grained, porphyritic pyroclastics of the Alma-
grera-2 volcanism; *7*: distribution cf very small ferruginous jasper lenses and cherts; *8*: intrusive
diabase sills. *Note* the clear electrical and gravimetric anomalies over the orebody. The anomalies are
nevertheless slightly displaced due to the flatly north dipping orebody

with vertical Electric Soundings, important new ore tonnages could be indicated and
drilled below the 460 m level at the Lousal mine (Portugal).

The low speed and high cost of gravimetric surveys stem from the need of accurate
topographical maps at scales of 1:5000 or 1:10,000 and the painstaking labour of the
terrain corrections which have to be done by hand by use of *Hammer*'s (1939) method.

Accurate gravimetric maps are the best means of outlining a massive sulphide mineralization and of providing preliminary estimates of tonnages as long as some basic figures about the density distribution of both the ore and host rocks are reasonably well
known. A good example is the Bouguer gravity map of the Tharsis mine area (Fig. 13),
where ore concentrations of some 100 million t (Filon Norte, San Guillermo and Sierra

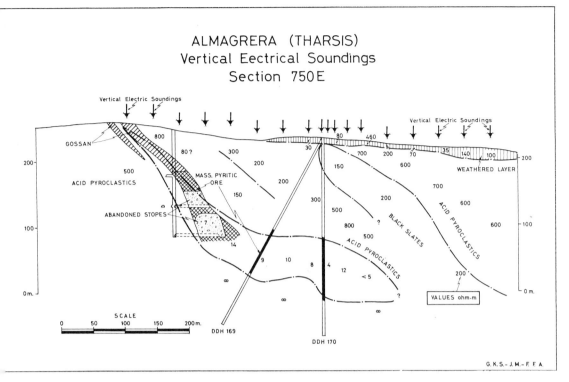

Fig. 10. Vertical electric soundings over the Almagrera orebody (Tharsis). VES stations are marked by *arrows*. The resulting electrical resistivities are shown in ohm. m. These have been converted into the corresponding rock formations by use of the data listed in Figure 2. The diamond drill holes DDH 169 and 170 have intersected massive pyritic ores as shown in black. Compare with geological section in Figure 9. *Note* the nearly perfect coincidence between VES-anomalous zone and the dimensions of the orebody, as confirmed by later diamond drilling.

Bullones) contrast with far smaller ones of the 3–6 million t range (e.g., Nueva Almagrera and Cantareras).

With this method, even relatively small disseminated sulphide mineralizations in chloritized host rocks with tonnages from 1–4 million t (Fig. 12) have been disclosed.

It is safe to conclude that gravimetry will, in future, play the most essential part as a "blanket coverage" method in any exploration programme directed towards those possible ore reserves covered by either Culm sediments or Tertiary basin fill.

7. Geochemical Exploration

With the advent of rapid analytical methods for trace elements, firstly organic solvent (dithizone) colourometric and, later, atomic absorption methods, geochemical exploration of soils and stream sediments for copper, lead and zinc has been integrated into many exploration programmes in the Pyrite Belt.

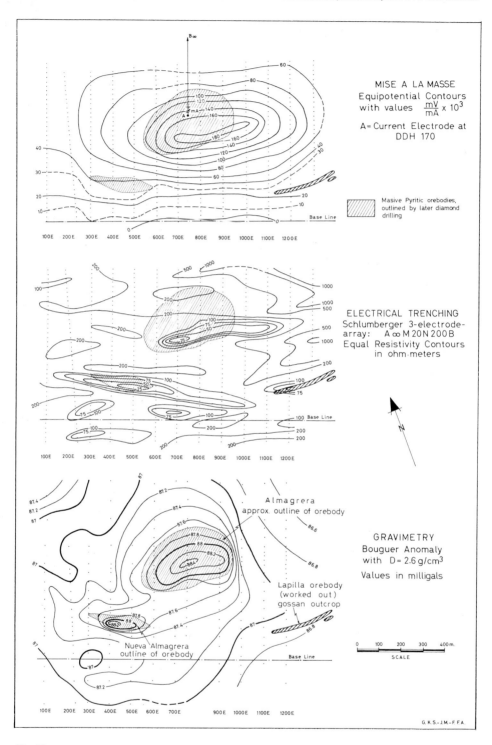

Fig. 11

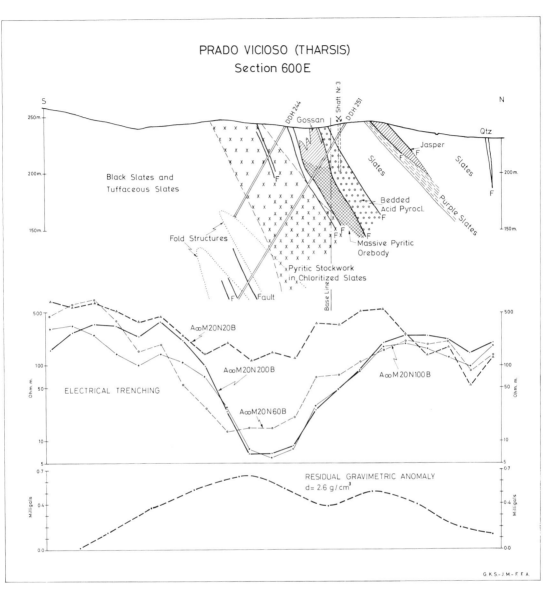

Fig. 12. Geological and geophysical section 600E over the Prado Vicioso orebody (Tharsis). Observe the space relationship between stockwork ore (footwall) and the massive, pyritic Prado Vicioso orebody. The resistivity anomaly is marking mainly the stockwork ore with its mineralization grade increasing with depth. This is evidenced by the response of different electrode arrays with theoretical depth penetrations of 30 70, 110 and 210 m, respectively. The residual gravimetric anomaly separates the two ore types

◄ Fig. 11. Geophysical surface map of the Almagrera–Nueva Almagrera and Lapilla mines area (Tharsis). Mise à la Masse, Electrical Trenching, Gravimetry. *Note* the clear anomalies in the geophysical surface maps which led to the discovery of important new ore tonnages. Some of the smaller electrical anomalies (ET) are caused by carbonaceous slates and were eliminated by the results of the gravimetric survey

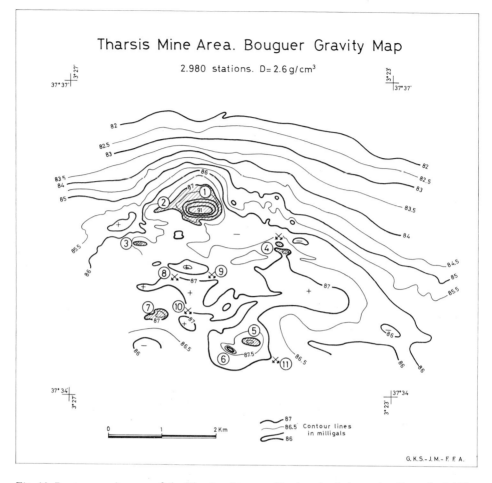

Fig. 13. Bouguer gravity map of the Tharsis mine area. *Numbers in circles* are locations of sulphide orebodies: *1*: Filon Norte and San Guillermo, representing approx. 90 million t; *2*: Sierra Bullones (partly worked out) with approx. 12 million t; *3*: Filon Centro, 3 million t; *4*: Prado Vicioso, massive and disseminated pyritic ore, total approx. 4 million t; *5*: Almagrera, approx. 10 million t; *6*: Nueva Almagrera, 3 million t; *7*: Cantareras, approx. 6 million t; *8*: Filon Sur, worked out; *9*: Esperanza, dissem. copper deposit, worked out; *10*: Vulcano, worked out; *11*: Lapilla, partly worked out, with estimated remaining reserves 0.5 million t

The Portuguese Serviço Mineiro includes soil geochemistry (Cu, Pb, Zn) in all exploration programmes using a standard sample grid of 100 x 100 m, simultaneously with gravimetric measurements. In Spain the Instituto Geológico y Minero de España (IGME) has recently performed several pilot programmes in different key areas of Cu, Pb, Zn determination in soils and in stream sediments (*IGME*, 1976). In the Huelva Province, at the present time only one major exploration group still systematically employs soil geochemistry of Cu, Pb, Zn as well as the recently introduced "mercury in soil" method. Here, soil samples are taken from traverses at 250 m distance from each other, and at stations every 20 m, where geophysical profiling is done simultaneously by Electrical Trenching and Induced Polarization.

The use of the *classical Cu, Pb, Zn geochemistry* is seriously limited due to the strong contamination produced over 3000 years of mining and smelting activities. It is common experience in the Pyrite Belt that the many local copper smelters and sites of open-air roasting of cupreous pyrites, that operated until the end of the last century, have caused geochemical contamination haloes with radii of some 10–15 km. Furthermore, there is a lack of appropriate soils in many areas, that have mostly developed only an imperfect B-horizon.

The recent geochemical programme carried out by the IGME (1976) in different parts of the Huelva and Sevilla provinces illustrates the truth of this statement. In the area of Paymogo-Puebla de Guzman (Herrerias) a total of 5260 stream sediment samples has been processed with the following results:

Copper: 30–40 ppm background, anomalies   70 ppm, max. 5000 ppm
Lead: 30–45 ppm background, anomalies   60 ppm, max. 9000 ppm
Zinc: 60 ppm background, anomalies 130 ppm, max. 6000 ppm

Their report admits that most anomalies can be explained as produced by pollution through abandoned mines (waste dumps, smelter stacks, mine rail tracks), and only one anomaly was centred around an acid volcanic eruptive centre. Furthermore, 612 soil samples taken on a grid 25 x 25 m in the near vicinity of the Herrerias mine showed only a few point anomalies without any practical information for exploration purposes.

The need for a more efficient geochemical method to suit the conditions of this old mining province has been satisfied only recently by the relatively long known *mercury method* in combination with a new rapid, reliable and low-cost analytical procedure for this element:

The use of mercury is based upon the assumption that a dynamic equilibrium of mercury vapour is developed around the channelways (e.g., faults) that connect the surface with buried sulphide orebodies. Mercury vapour is produced by oxidation processes of sulphides containing Hg trace amounts, e.g., sphalerite, pyrite, chalcopyrite and others, as well as through other processes taking place below the oxidation zone (*McNerney* and *Buseck*, 1973). Furthermore, mercury grades of the Paleozoic Pyrite Belt ores are relatively high, in the order of 10–200 ppm Hg, with high Hg grades corresponding to higher than 2% Zn grades. For comparison, Precambrian Canadian massive sulphides have only some 2–20 ppm Hg associated with higher than 10% Zn grades.

In the spring of 1974 the Huelva Chamber of Mines (Cámara Oficial Minera de Huelva) organized geochemical test work over different orebodies of this province by use of a portable SCINTREX HGG-3 mercury analyzer, as described by *Robbins* (1973). The mercury tests were carried out over seven outcropping and buried orebodies in different geological and mining environments, in order to study the feasibility and economics of "mercury in soil gas" and "mercury in soil" surveys, the response of base metal-poor vs. base metal-rich ores, and to assess the influence of contamination. Some geochemical sections are reproduced in Figures 8, 9 and 14 from the final report (unpubl., 1974).

The "soil gas" measurements were done directly in the field with a dynamic sampling procedure pumping soil gas from a 30–50 cm deep hole through a conical gas collector (to avoid dust) into the battery-powered mercury analyzer. At each station three holes were spiked and within 5–10 min three measurements could be realized in the field.

The "mercury in soil" analyses were done in a central laboratory with the minus-100 mesh fraction of air-dried samples of soils or weathered rock (approx. 200–500 g composite samples of three or four points around each sampling station). The time necessary for a duplicate analysis of each sample (approx. 10 mg) was only 1–2 min.

The *"soil gas"* measurements have not proved to be of much use under local conditions. Some sulphide deposits did not produce clear anomalies, as for example the Cantareras deposit (Fig. 14). In most cases, the "soil gas" method has yielded smaller (peak/background ratio) and narrower anomalies than the "mercury in soil" method, with the exception of Almagrera (Fig. 9). This certainly is due to the poorly developed, thin soil cover not suitable for storing enough mercury in the vapour phase under semiarid climatic conditions.

In contrast, the *"mercury in soil"* test measurements have always produced significant anomalies over a regional background in the order of 100–300 p.pb Hg, as determined in various parts of the province (Tharsis, Riotinto, Paymogo area). There has been good evidence of the interdependence of base metal grades and mercury haloes produced. Those ores with lower than normal base metal grades (e.g., Cantareras, Fig. 14) showed only fairly narrow and small anomalies, whereas ore deposits of similar size and depth, but with higher than normal base metal values (> 5%), had developed wide mercury haloes with peak/background ratios as high as 40 and 100.

The La Zarza example (Fig. 8) is particularly interesting, because this orebody with normal base metal grades (approx. 3% combined Cu, Pb, Zn) produced a significant anomaly of approximately 80 m width, that coincides well with thrust and transverse faults connecting the 150 m deep buried ore mass with the surface. The "noisy" background is likely to depend on the numerous directional and transverse faults that intersect the ore horizon with some disseminated sulphides (see surface map of La Zarza, Fig. 5). Proof of this is the almost perfect coincidence of most of the mercury peaks with resistivity lows that mark the faults, in particular the Culm "graben" to the north of the La Zarza orebody.

Also at La Zarza, mercury samples taken on two different days produced anomalies which did not match exactly and did not show the same peak/background ratios. This fact must be borne in mind. Mercury surveys performed on different days may yield somewhat differing absolute values, however the anomaly pattern should be essentially the same.

As to industrial contamination, the mercury test work allowed the conclusion that this danger exists only in the very near vicinity of a few tens of m around mine tailing and waste dumps.

To summarize, it is concluded that soil-mercury exploration is by far the best geochemical method for the Pyrite Belt conditions. Mercury surveys should give reliable results when employed during a first reconnaissance of geologically favourable areas, and prior to ground geophysical work. An optimum soil sample grid would be with stations every 10 m along traverses perpendicular to the general strike and 150 or 200 m distance between traverses.

An exploration block of approximately 1 km^2 is sampled and analyzed within three days, whereas a resistivity survey (including staking of traverses) needs 10–15 days, depending on the terrain conditions. Last but not least, a comparison of costs for one mercury sample station vs. one geophysical station results very much in favour of the

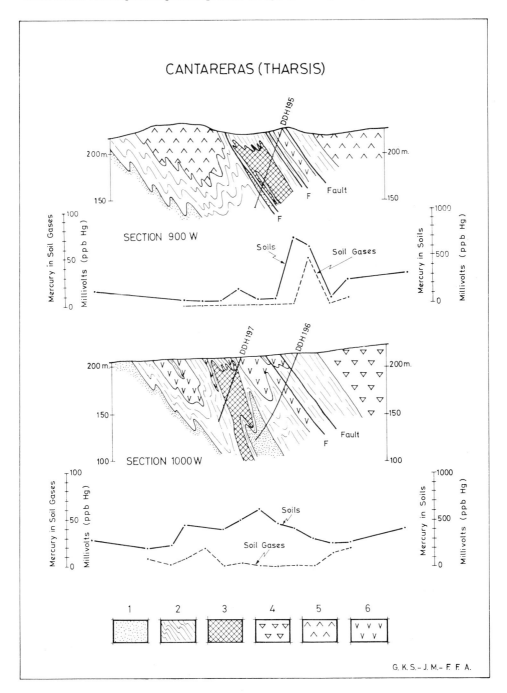

Fig. 14. Geological and geochemical (mercury) sections over the Cantareras orebody (Tharsis). *1*: Chloritized massive footwall pyroclastics; *2*: black slates and tuffaceous slates; *3*: massive pyritic ore; *4*: acid breccias; *5*: massive, coarse and *6*: medium grained porphyritic pyroclastic of the Almagrera-1 volcanism. *Note* that scale of "mercury in soil gases" is 0.10 of that of "mercury in soils". For "mercury in soils" 1 millivolt is equivalent to 1 p.p.b. Hg, approximately

mercury method: 20–25% of the cost for resistivity, and only some 5% of that of gravimetry. From the above, it becomes obvious that the integration of mercury geochemistry within any exploration programme would increase speed and might reduce costs significantly, with the more expensive ground geophysical profiling being centred mainly to restricted anomalous areas.

8. Conclusions

The exploration activities in the Pyrite Belt have been quite successful in the past 20 years. Out of the 470–500 million t of newly established ore reserves some 50 million t can be labelled geological, and approximately 130 million t geophysical discoveries. The balance has been discovered by exploration drifts and, principally, by systematic grid drilling around mine workings. However, as some authors state, it was "fresh geological views and optimism" that have contributed significantly to the initiation of these grid drilling programmes (as was the case with the Cerro Colorado copper-gold-silver deposit, *Prior* et al., 1972).

Most new discoveries have been made in promising areas around operating or abandoned mines, and these take up only 300–400 km^2 of the 7800 km^2 of the whole Pyrite Belt. There are still potential target areas of some 1800 km^2 of Volcanic-Sedimentary Complex and another approximately 2000 km^2 where the mineral-bearing Volcanic-Sedimentary Complex is covered by Culm sediments and Tertiary basin fill-up to an "economical" exploration depth of 100–200 m. Exploration progress and discovery rates in these 3800 km^2 will doubtless be slower than in the past 20 years.

From the practical experience of the last 10–20 years a successful exploration programme covering large areas should be performed in the following order:

1. Geological reconnaissance mapping (including photogeology) at 1:25,000 or greater.
2. Detailed geological mapping 1:5000 of favourable areas.
3. Eventually mercury soil geochemistry of the favourable areas delineated by detailed geological mapping.
4. Resistivity profiling of favourable areas, in part mercury anomalies.
5. Checking of resistivity and/or mercury anomalies by gravimetry.
6. Diamond drilling of the geophysical anomalies.

Prospective areas with considerable thickness of Culm or Tertiary sediments above the Volcanic-Sedimentary Complex, preferentially, should be explored first by "blanket coverage" gravimetry and the anomalies checked by deep resistivity profiling. If the terrain conditions were not suitable for gravimetry, then only deep resistivity surveys (e.g., the Resistivity Block method) could be employed, as well as, eventually, mercury soil geochemistry if the ground showed sufficient directional and transversal faults.

Smaller mine areas and small prospects of some 1–2 km^2 can be assessed in a preliminary way by use of the Mise à la Masse method and, above all, by gravimetry.

Finally, and although this sounds only common sense, we wish to stress the need for intimate collaboration between geologists and geophysicists: At every stage of any exploration programme both disciplines should establish a continuous flow of technical information and feed-back of relevant data. Geological interpretation of any exploration ground should determine the selection of the most appropriate geophysical method, or any combination of methods. Geophysical data and interpretations must be used to correct the earlier geological interpretations, and thus, finally, the necessary follow-up steps can be decided — be it to abandon the prospect, apply a different geophysical or geochemical method, or to start drilling.

Acknowledgments. Much of the data and conclusions presented here have arisen from discussions and field trips with many colleagues of state organizations, mining companies and universities from Spain and Portugal. To all of them, too numerous to name here, goes our gratitude. We are most thankful to our collaborators with the Tharsis Company, in particular to *Eusebio Lopera* and *Americo Santos* (geology), *Mariano González* and *Francisco Moreno* (geophysics) and *Gines Paez* for the drawings, among many others. Little would have been done properly without their enthusiastic, efficient work and the lively discussions. Mr. *Kenneth G. Gray* (Tharsis) cheerfully improved the English text, as so many times before.

Finally, we want to express our sincere gratitude to the directors of the Tharsis Sulphur and Copper Company Ltd. and Minera Balboa S.A. for permission to publish geological und geophysical information from company files. Messrs. *Michel Lécolle* and *Guy Roger* (Paris) were kind enough to allow us to copy part of their unpublished map of the Alosno area.

References

Armengot, J., Fdez. Alonso, F.: Aplicación de los métodos geológicos y geofísicos en la investigación de las piritas de Huelva. IV Jornadas Minero-Metalúrgicas de Cartagena (1970)

Aye, F.: Geologie et gites metallifères de la moyenne vallée de l'Odiel (Huelva-Espagne). Thesis 3e Cycle. Lab. Geol. Appliquée. Univ. Pierre et Marie Curie, Paris (1974)

Aye, F., Strauss, G.K.: Identification de cymerite (Ba, Al, SiO_3O_8OH) dans la province pyriteuse sud-iberique. Mineraux associés (La Zarza, Huelva, Espagne). C.R. Acad. Sci. (Paris) 281 (1975)

Bateman, A.M.: Ore deposits of the Rio Tinto (Huelva) district, Spain. Econ. Geol. **22**, 569–614 (1927)

Bernard, A.J., Soler, E.: Aperçu sur la Province Pyriteuse Sud-Ibérique. Cent. Soc. Géol. Belg. Gisements Stratiformes et Provinces Cuprifères, Liège (1974)

Blokh, I.M.: Elektroprofilirovanie Metodon Soprotivleniy. Moscow: Gosgeoltekhizdat 1962

Boogard, M. van den: Geology of the Pomarao region (Southern Portugal). Doct. thesis, Univ. Amsterdam (1967)

Boogard, M. van den, Schermerhorn, L.J.G.: Conodont faunas from Portugal and Southwestern Spain. Scripta Geologica, 28, Leiden (1975)

Carvalho, D.: Consideraçoes sobre o vulcanismo da regiao de Cercal-Odemira. Suas relaçoes com a Faixa Piritosa. III Reuniao de Geologia do Sudoeste Peninsular (Huelva-Beja) (in press)

Carvalho, D., Conde, L., Hernandez Enrile, J., Oliveira, V., Schermerhorn, L.J.G.: Faixa Piritosa Iberica. Guia da excursao da III Reuniao de Geologia do Sudoeste Peninsular. Huelva-Beja (1975)

Febrel, T.: Estructura de la masa de pirita de la mina de Herrerías, Puebla de Guzmán (Huelva). Bol. Geol. Minero. 83, 81–87 (1972)

Fernandez Alvarez, G.: Los yacimientos de sulfuros polimetálicos del Suroeste Ibérico y sus métodos de prospección. Stud. Geol. IX, 65–102, Salamanca (1974)

Freire d'Andrade, R., Schermerhorn, L.J.G.: Aljustrel e Gaviao. Principais jazigos minerais do Sul de Portugal. Livro-Guia Excursao No. 4. Congreso Hispano-Americano de Geología Economíca, Lisbon (1971)

Garcia Palomero, F.: Estudio geológico de la masa pirítica de San Antonio (Rio Tinto). Bilbao: Jornadas Minero-Metalúrgicas 1975

Hammer, S.: Terrain corrections for gravimeter stations. Geophysics 4, 184–194 (1939)

Höllinger, R.: Beitrag zur Kenntnis der Geologie im SW der Provinz Huelva. Doct. thesis, Univ. Münster, 168 p. (1958)

Horikoshi, E.: Volcanic activity related to the formation of the Kuroko-type deposits in the Kosaka district, Japan. Mineral. Deposita 4, 321–345 (1969)

IGME: Reserva Zona de Huelva: Investigación geoquímica en las hojas 936 (Paymogo), 939 (Castillo de las Guardas), 958 (Puebla de Guzmán) y 961 (Aznalcollar). Informe 11–936/1–939/1–958/ 1–961/1. Serie Colección-Informe, Instituto Geológico y Minero de España, Madrid (1976)

Jubes, E., Carbonell, A.: Estudio de los yacimientos de pirita ferrocobriza de la zona de la mina La Rica, sitos en los terrenos de Cabezas Rubias y El Cerro, Huelva. Oficial Minería y Metalurgia, 20, 21, Madrid (1919)

Kinkel, A.: Observations on the pyrite deposits of the Huelva district, Spain, and their relation to volcanism. Econ. Geol. 57 (1962)

Kunetz, G.: Principles of Direct Current Resistivity Prospecting. Berlin: Borntraeger 1966

Lopera, E., Madel, J., Santos, A.: Cartografía geológica 1:10.000 del flanco Norte del Anticlinal de Puebla de Guzmán, Provincia de Huelva. Unpubl. report Minera Balboa, S.A., Huelva (1975)

Lotze, F.: Zur Gliederung der Variziden der Iberischen Meseta. Geotekton. Forsch. 6 (1945)

Madel, J., Lopera, E.: Geología de la zona minera de Tharsis. Guia da excursao da III Reuniao de Geologia do Sudoeste Peninsular. Huelva-Beja (1975)

McNerney, J.J., Buseck, P.R.: Geochemical exploration using mercury vapor. Econ. Geol. 68, 8 (1973)

Pinedo Vara, I.: Piritas de Huelva. Madrid: Summa 1963

Poldini, E.: La prospection electrique du Sous-sol. Lausanne: F. Rouge et Co. 1947

Pryor, R.N., Rhoden, H.N., Villalon, M.: Sampling of Cerro Colorado, Rio Tinto, Spain. Trans. Inst. Mining Metallurgy, Sect. A, 788 (1972)

Rambaud Perez, F.: El sinclinal carbonífero de Rio Tinto (Huelva) y sus mineralizaciones asociadas. Mem Instituto Geológico y Minero de España, 71, Madrid (1969)

Robbins, J.L.: Zeeman spectrometer for measurement of atmospheric mercury vapour. Geochem. Explor. 1972, p. 315–323. Inst. Min. Metall. (1973)

Rocha Gomes, A.A., Da Silva, F.S.: Prospeccao de piritas no Baixo Alemtejo. Est Notas e Trabalhos Serviço de Fomento Mineiro, Oporto (1955)

Roscoe, W.E.: Geology of the Caribou Deposit, Bathurst, New Brunswick. Can. J. Earth Sci. 8, 1125 (1971)

Sangster, D.F.: Precambrian volcanogenic massive sulphide deposits in Canada: a review. Geol. Surv. Can. Paper 72, 22 (1972)

San Pedro Querejeta, F.: Inauguración del Complejo Minero de Aznalcollar (Sevilla). Industria Minera, 157, Madrid (1975)

Schermerhorn, L.J.G.: Mafic geosynclinal volcanism in the Lower Carboniferous of South Portugal. Geol. Mijnbouw 49 (1970)

Schermerhorn, L.J.G.: An outline stratigraphy of the Iberian Pyrite Belt. Bol. Geologico y Minero 82, Madrid (1971)

Schermerhorn, L.J.G.: Pumpellyite-facies metamorphism in the Spanish Pyrite Belt. Pétrologie, Nancy (1975)

Schermerhorn, L.J.G., Stanton, W.I.: Folded overthrusts at Aljustrel (South Portugal). Geol. Mag. 106,(1969)

Strauss, G.K.: Zur Geologie der SW-Iberischen Kiesprovinz und ihrer Lagerstätten, mit besonderer Berücksichtigung der Pyritgrube Lousal/Portugal. Doct. thesis, Univ. München (1965)

Strauss, G.K.: Sobre la Geología de la Provincia Piritífera del Suroeste de la Península Ibérica y de sus yacimientos, en especial sobre la mina de pirita de Lousal (Portugal). Mem. Instituto Geológico y Minero de España, 77, Madrid (1970)

Strauss, G.K., Madel, J.: Geology of Massive Sulphide Deposits in the Spanish-Portuguese Pyrite Belt. Geol. Rdsch. 63 (1), (1974)

Strauss, G.K., Madel, J., Fdez. Alonso, F.: La Faja Pirítica Hispano-Portuguesa y el papel de la Geología en su Investigación Minera. Industria Minera, 149, Madrid (1974a)

Strauss, G.K., Fdez. Alonso, F., Madel, J.: La Faja Pirítica Hispano-Portuguesa y el papel de la Geofísica en su Investigación Minera. Industria Minera, 150, Madrid (1974b)

Williams, D.: The geology of the Rio Tinto Mines, Spain. Bull. Inst. Min. Metall. **355** and **362** (1934)
Williams, D.: Further reflections on the origin of the porphyries and ores of Rio Tinto, Spain. Trans. Inst. Min. Metall. **71**, 265–266 (1962)
Williams, D.: Volcanism and Ore Deposits. Leipzig: Freiberger Forsch. C120, 1966
Williams, D., Stanton, R.L., Rambaud Perez, F.: The Planes-San Antonio pyritic deposit of Rio Tinto, Spain: Its nature, environment and genesis. Trans. Min. Metall. Sect. B., 84 (1975)
Zaborovoskii, A.J.: Elektrorazvedka. Moscow: Gostoptekhizdat 1963

On the Formation of Lower Cretaceous Potassium Salts and Tachhydrite in the Sergipe Basin (Brazil) with Some Remarks on Similar Occurrences in West Africa (Gabon, Angola etc.)

H. BORCHERT, Clausthal-Zellerfeld

With 3 Figures

Contents

Summary

Within an area of about 30 x 40 km^2 of the Sergipe state in the north-eastern coastal region of Brazil there are three main oil field anticlines (Carmópolis, Siririzinho and Riachuelo) with similar structures in the environment and off-shore continuations into the Atlantic Ocean. Between these more or less dome-like structures we find three salina sub-basins with potash salts, the most important of them being the basin of Taquari-Vassouras and the basin of Santa Rosa de Lima.

Figures 1 and 2 show the general geological conditions with the Lower Cretaceous (Aptian) sequence of the Muribeca Saline Formation especially in the Ibura Series, followed by the Albian Riachuelo Formation, the Cotinguiba Formation of Turonian and Santonian and the Piaçabuçu Lower Tertiary Formation. The sedimentation during the last 1000 my — and already in earlier Pre-Cambrian times — has often been interrupted because the principal tendency of the Brazilian Shield was directed upwards and resulted in uplifting processes of many km. During these dominant movements, tensional stress tectonics created several developments of horst/graben systems and also of some larger epicontinental sea basins with salinar facies, for instance even during the Pennsylvanian (Central Amazonas Basin) and also during the Permian in North–East Brazil.

Stronger dilatational movements with Continent drifting processes have proceeded since Jurassic and especially during Lower Cretaceous, forming mainly the salinar basins of Sergipe and Alagoas north of the San Francisco river. Within the Aptian Muribeca Formation, the principal salt sequences have been deposited in the Ibura Series, and these were so uniform that the sections of dozens of boreholes on the slopes of the three oil field domes could easily be parallelled (cf. scheme of Fig. 3). Therefore, there can be no doubt that the primary sequence of salinar crystallization products with 18 Markers of mainly foliated shales and bituminous clays and anhydritic beds have been homogenous in the whole basin of Sergipe.

At present, however, we find a systematic lateral zoning. Above the three oil field anticilines of Carmópolis, Siririzinho and Riachuelo, all the highly soluble chloridic salts have been leached away completely. Near the tops of the slopes we then find only rock salt, followed downwards by intercalations of sylvinitic layers, whereas the deeper regions of the sub-basins generally show predominantly carnallitic and tachhydritic salt masses. Similar zonal arrangements around oil fields, and dome-like structures are also found on the eastern side of the Atlantic Ocean in Central Africa from Gabon in the north to Angola in the south.

All observations and arguments suggest that the principal cause of the recent lateral zoning of the different salt formation types cannot be primary, but has been produced by secondary leaching and replacement processes, mainly by descending atmospheric waters, but also by slow percolation and impregnation of deeper ground waters and brines. These solutions were wandering mainly along and between impermable foliated shales and/or bituminous clays, but they also penetrated the salt formations locally across the bedding planes by fissure and fault zones. Thus, collapse structures are often found. "Oil field waters" of the underlying Carmopolis Series may also have ascended from below and may have also contributed their share to more or less local "salt horses" (Vertaubungen). Reaction processes between such oil field waters, anhydrite and carnallite decomposing brines which have also enriched carnallitites into sylvinites often with contents of approximately 25% K_2O are responsible also for the formation of great masses of tachhydrite.

In the two main sub-basins of Taquari – Vassouras and Santa Rosa de Lima the thorough investigations of the Compania de Pesquisa de Recursos Minerais (CPRM) and the Departemento Nacional da Produção Mineral (DNPM), together with special Potash Projects of PETROBRÁS and supported by dozens of systematically arranged boreholes, have proved the existence of two principal sylvinite areas along the slopes of the main oil field anticlines. The mineable quantities of sylvinite in the layers of more than 3 m thickness – in some areas more than 6–9 m of rich sylvinite – have been calculated to be 525×10^6 t, containing an average of 20–25% K_2O. These quantities and qualities would be sufficient for Brazil's steadily increasing home consumption for many decades, with a yearly production of 500,000 t or even 1 million t of potash fertilizers with 60% K_2O.

1. Introduction

During the past decade an essential revolution in thought has taken place in many fields regarding the genesis of ore deposits, the principal points of dispute being are they magmatic or sedimentary, syngenetic or epigenetic? These controversies were, however, not really new. Since *Abraham Gottlieb Werner* and his scholars of Freiberg, there have been "plutonists" and "neptunists", and similar discussions about ore lodes and layers continued throughout the 19th century. Towards the end of this epoch an overwhelming number of authorities inclined to ascendent solutions of magmatic origin as the principal factor of ore genesis and – for layered deposits of the Huelva and Rammelsberg type – very many authors preferred "hydrothermal replacement" as the most important genetic process, especially in the US where even the ordinary "true" pegmatites (cf. *Schneiderhöhn, 1961*), as mainly "closed systems" with a predominantly regular sequence of crystallization, were believed to be governed by metasomatic replacement and enrichment usually occurring later, of rare elements by solutions of unknown source and from unknown depths.

However, it should be borne in mind that many so-called modern ideas of ore genesis had been advanced much earlier by German mineralogists and economic geologists, for example, the synsedimentary origin of the cupriferous pyrite deposits of the Huelva district by *Klockmann* (1900) or the submarine exhalative genesis of iron ores of the Lahn-Dill type by *Harbort* (1903), followed by endless monographs and other publications on this type of geosynclinal ore deposits (cf. *Borchert*, 1960), before *Oftedahl* (1958), with his *Theory of exhalative-sedimentary ores*, spread the ideas to Anglo-American and other countries.

My friend *Albert Maucher* and the scholars of Munich, Innsbruck and Clausthal made great efforts from 1950 to 1970 (*Maucher*, 1954, 1965) to elucidate the first synsedimentary enrichment of Pb–Zn ores in the marine environment of the limestone and dolomitic chert bituminous facies in the northern and southern Alps, the Mississippi-Missouri district, etc (type Upper Silesia). Up to present, especially *Maucher's Antimony–Wolfram–Quicksilver Formation and its Relation to Magmatism and Geotectonics* (1965) has led to many controversies (cf. *Höll*, 1975).

Contrary to these and several other types of metallic ore deposits there has been general agreement for many decades that most of the thicker occurrences of rock salt, together with potassium and magnesium minerals, are principally synsedimentary crystallization products, mainly due to evaporation processes in more or less restricted epicontinental marine basins. Nevertheless, there are still controversial opinions on "primarists" and "secondarists", and whether the salt minerals and their textures are truly primary in the strictest sense, or what may have happened during the diagenesis or later geothermal metamorphism. Very problematic also is the role and importance of primary or "epigenetic" lateral and vertical facies variations of saline formations in connection with the original morphological basin conditions (synsedimentary subsiding etc) or later tectonics with leaching and secondary replacement processes. These are often combined with "oil field waters" ascending from the ordinarily deeper and earlier beds of the restricted basins with typical enrichments of organic matter (cf. the recent and earlier Black Sea, the Miocene Mediterranean and many other examples in other regions of other ages up to Cambrian time).

A very inteesting object related to such genetical and also economic questions is the deposit of large masses of potassium salts and tachyhydrite in north-eastern Brazil in the state of Sergipe. In spring 1973, in connection with a feasibility study of Krupp Industrie- und Stahlbau/Rheinhausen (Dept. for Techniques with Raw Materials), I had the opportunity to investigate these occurrences which had been discovered in 1964/65 and then systematically studied by dozens of boreholes on the flanks of several smaller oil field domes near Carmopolis. Similar structures and also salt deposits also have off-shore continuations near Aracaju, the capital of Sergipe, as well as on the eastern side of the Atlantic Ocean in Africa.

2. General Geological Situation and Development of the Sergipe Saline Deposits

2.1 General Geological Situation and Tectonics

The Cretaceous basin of Sergipe and the corresponding basin of the Alagoas state in the north of Sergipe are situated at the most eastern edge of the Brazilian Shield between a latitude of approximately 10 and 11° south of the equator. The Brazilian saline formation of Aptian age finds a parallel in many features of the oil fields, as well as of the potassium salt and tachhydrite occurrences in the western coastal region of Central Africa from Gabon in the north to Angola in the south (cf. *Fernandes,* 1966; *Kroemmelbein* and *Wenger,* 1966). Both salt deposits are characteristically found in intimate connection with more or less adjacent oil field domes, most of them with dimensions of several km to several tenths of a km.

The initial phase of the separation of Africa and South America have been dated as late Jurassic (cf. *Kennedy*, 1965). However, tensional stresses in the earth's crust, which later produced the Atlantic Ocean, have been of influence from early the pre-Cambrian Age. They have been an essential factor, for example, in the formation of a special type of rich iron orebody from ordinary itabirites with much original chert material. It is well known that the old crystalline masses of the southern continents (i.e., Gondwana Continent) are generally in a phase of secular uplifting usually combined with tensional tectonics (cf. *Borchert*, 1961, 1967, 1968). The average uplifting rate of the Brazilian Shield may be in the order of at least 15—25 km over the last 1000 my, combined with local and more or less sporadic basin formations.

From deep-seated metamorphic masses of common itabiritic protoes and under the influence of temperatures of 400—600° C during the first pre-Cambrian tensional stress and uplifting stages, SiO_2-rich material may have been mobilized and may thus have been separated from the main mass of the iron oxides. In fact, oridinary itabirites in Brazil, Krivoi Rog, etc often demonstrate boudinage and cross-cutting structures opening fissure systems, and making way for the formation of pneumatolytic quartz aureoles which at distances of mostly hundreds of m, surround this special type of rich iron orebodies.

On the other hand, most of the rich iron orebodies were concentrated to their present content of 60—68% Fe by much younger lateritic weathering processes long after the uplift of the original cherty itabirites to the earth's surface.

Special continental drift processes, however, of much smaller dimensions within the Brazilian Shield itself, have also produced smaller blocks and basin areas and/or especially horst/graben structures. There are numerous transgressions of epicontinental seas in post-Cambrian times, often combined with restricted basin structures and the corresponding sedimentation of sapropelites followed by evaporites. Saline deposits are found in Brazil from the Pennsylvanian, Permian, Lower and Middle Cretaceous (cf. *Benavides,* 1968).

Besidesthe two principal potash sub-basins of the Aptian Age of Taquari-Vassouras and Santa Rosa de Lima near Carmópolis in Sergipe, other deeper basins occur, namely Piranhas, Timbó, Ilha das Flores, and Quiçama/Socorro/Aracaju, all of them containing intercalations of salt zones, besides the Mosqueiro Continental Shelf, with thick sections of soluble salts located at depths of over 2500 m (according to *Moacyr de Vasconcellos,* 1972, p. 41).

2.2 Evolution of the Sergipe Saline Deposits

a) Geological and Tectonic Conditions

In the region north of Aracaju three oil field anticlines are known so far, with rather irregular shape and dome-like structures, from E-SE to W-NW the domes of Carmópolis, Siririzinho, and Riachuelo (cf. Figs. 1 and 2).

The oil-bearing Carmópolis Series, together with the salinar Ibura Series, followed by the Oiteirinhos Series, belong to the Muribeca Formation of Lower Cretaceous and especially to the Aptian Age. The development of the three phases of the Muribeca Formation is schematically demonstrated in Figure 1.

The Carmópolis Series (or Carmópolis "Member") mainly consists of various sandstones and conglomerates. They often rest immediately on the pre-Cambrian crystalline basement. The subsequent evaporites of the Ibura Series will be discussed in detail later. They are covered in turn by foliated marine shales, marls and limestones of the Oiteirinhos Series.

Figure 2 shows the general geological situation, where more or less intensive tectonic movements already occurred before the sedimentation of the Middle Cretaceous Riachuelo Formation of the Albian Age. Similar weak tectonic events have caused the even more irregular regional distribution of the Upper Cretaceous Cotinguiba Formation (Santonian and Turonian) and the subsequent Piacabucu Formation of Lower Tertiary.

The marine epicontinental sedimentation was interrupted several times; it was accompanied by local erosion caused by further uplifting processes during the post-Ibura phase and was influenced by weak effects of salinar tectonics, tensional stress and, therefore, also by important subterranean leaching processes, mainly rainwaters descending along dilatation faults.

The last marine sedimentation during the early Tertiary age was follwed by a longer period of uplift and continental erosion covering all older formations with, however, mostly only thin beds of loose sands and pebbles of the Barreiras Formation (Pliocene, cf. Fig. 2).

It can be learned from the cross-sections of Figure 1 that on the dome-like structures of the Carmopolis and the Siririzonhos oil field, the chloridic salts of the Ibura Series have been completely leached away by descending atmospheric waters. The same happened on the Riachuelo oil field anticline situated still further to the west. In all these three areas only the two nearly insoluble main beds of anhydrite in the upper parts of the profiles have been more or less completely preserved.

It is important to note that the dome-like structures had already been partly developed during the pre–Oiteirinhos Age – cf. Figure 1 upper and lower part and Figure 3, which shows in cycles 8 and 9 the most important anhydritic beds AH 8 and A 9, as well as two anhydrite layers with shale intercalations in cycle 10 (AF 10).

However, as can also be seen in Figure 1, all of the most soluble chloridic salts have been leached away not only on top of the anticlines and the beginning slopes. There are also several places where erosional and secondary leaching processes by wandering solutions have occurred from below, forming "salt horses" (Vertaubungen) in the potash beds. Similar leaching effects and local salt horses caused by waters coming from above and below can be found in many special sections of the dozens of DNPM-boreholes which have cut the saline basins of Sergipe in different directions (W to E, SW to NE, SE to NW etc). It can often be clearly seen that the leaching effects are especially bound to faults, fissure and breccia zones. The percolating ground waters predominatly moved along unpermeable shales and bituminous clay intercalations over lateral distances of hundreds of m and more, whereas vertically penetrating solutions principally used fault zones, caused mainly by tensional stress in connection with the secular uplifting processes of the Brazilian Shield.

In the south of the Carmópolis anticline we find the saline sub-basin of Aguilhada, which has larger continuations especially eastwards into the Great Coastal Basin. In this area the saline formation of the Ibura Series has been studied with about 15 boreholes.

Fig. 1. Three phases of sedimentation of the Lower Cretaceous Muribeca Formation. (After a scheme ▶ of DNPM, 1970; for symbols cf. Fig. 3)

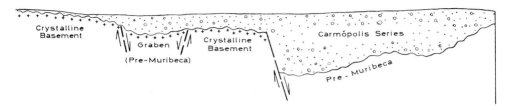

1 – 1   Phase of the Carmópolis Series in the Sergipe Basin

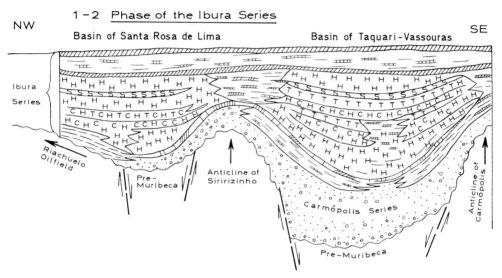

1 – 2   Phase of the Ibura Series

1 – 3   Phase of the Oiteirinhos Series

Fig. 1

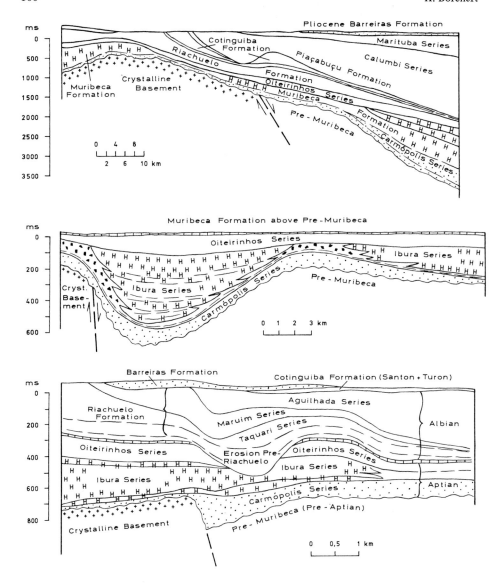

Fig. 2. Sedimentary sequence and tectonics of the Sergipe Basin. (After DNPM Report, 1970)

Fig. 3. General scheme of the 10 zones of evaporites in the Sergipe Basin. (After *J.C. Fonseca*, CPRM ▶
Report, 1972)

T T T T	Tachhydrite „T"	A A A A	Anhydrite „A"
C C C C	Carnallite „C"	═══	Foliated shale „F" and bitumenous clay
S S S S	Sylvinite „S"	⊥ ⊥ ⊥	Limestone „L" etc.
H H H H	Halite „H"	M1-M18	Marker

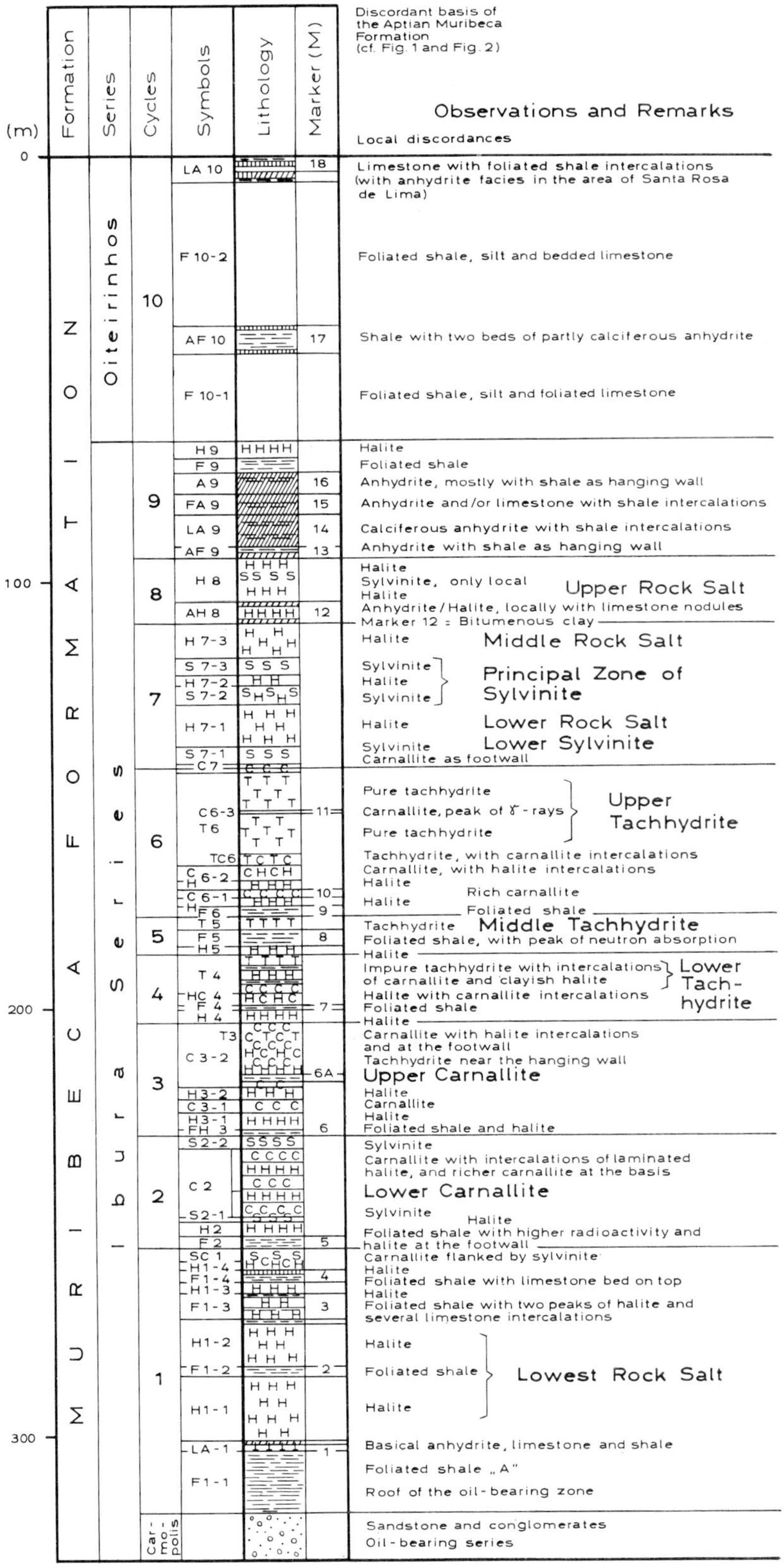

Fig. 3

	Formation	Series	Cycles	Symbols	Lithology	Marker (M)	Observations and Remarks

Discordant basis of the Aptian Muribeca Formation (cf. Fig. 1 and Fig. 2)

Observations and Remarks

Local discordances

(m)

0

Oiteirinhos	10	LA 10		18	Limestone with foliated shale intercalations (with anhydrite facies in the area of Santa Rosa de Lima)
		F 10-2			Foliated shale, silt and bedded limestone
		AF 10		17	Shale with two beds of partly calciferous anhydrite
		F 10-1			Foliated shale, silt and foliated limestone

NOITA (vertical letters on left margin within Formation area)

100

200

300

H 9	HHHH		Halite
F 9	===		Foliated shale
A 9		16	Anhydrite, mostly with shale as hanging wall
FA 9		15	Anhydrite and/or limestone with shale intercalations
LA 9		14	Calciferous anhydrite with shale intercalations
AF 9		13	Anhydrite with shale as hanging wall

Cycle 9

| H 8 | HHH / SSSS / HHH | | Halite / Sylvinite, only local   **Upper Rock Salt** / Halite |
| AH 8 | HHHH | 12 | Anhydrite/Halite, locally with limestone nodules — Marker 12 = Bitumenous clay |

Cycle 8

H 7-3	H H / H HH		Halite **Middle Rock Salt**
S 7-3	SSS		Sylvinite ⎤
H 7-2	HH		Halite } **Principal Zone of**
S 7-2	SHSHS		Sylvinite ⎦ **Sylvinite**
H 7-1	HHH / HHH		Halite **Lower Rock Salt**
S 7-1	SSS		Sylvinite **Lower Sylvinite**
C7	ccc		Carnallite as footwall

Cycle 7

C6-3	TTT/TTT	11	Pure tachhydrite ⎤ **Upper**
T6	TTT		Carnallite, peak of δ-rays } **Tachhydrite**
	TTT		Pure tachhydrite ⎦
TC6	TCTC		Tachhydrite, with carnallite intercalations
C 6-2	CHCH		Carnallite, with halite intercalations
H	HHHH		Halite
C 6-1	cccc	10	Rich carnallite
H	HHH	9	Halite
F6	===		Foliated shale

Cycle 6

T5	TTTT		Tachhydrite **Middle Tachhydrite**
F5	===	8	Foliated shale, with peak of neutron absorption
H5	HHH		Halite

Cycle 5

T 4	TTTT		Impure tachhydrite with intercalations ⎤ **Lower** of carnallite and clayish halite }
HC 4	CCCC		Halite with carnallite intercalations **Tach-**
F 4	HCHC	7	Foliated shale **hydrite**
H 4	HHHH		Halite

Cycle 4

T3	CCC/CTCT		Carnallite with halite intercalations and at the footwall
C 3-2	HCHC / HHHH	6A	Tachhydrite near the hanging wall **Upper Carnallite**
H3-2	HHH		Halite
C 3-1	CCC		Carnallite
H3-1	HHHH	6	Halite
FH 3			Foliated shale and halite

Cycle 3

S2-2	SSSS		Sylvinite
	CCCC / HHHH		Carnallite with intercalations of laminated halite, and richer carnallite at the basis
C 2	CCC / HHHH		**Lower Carnallite**
S2-1	CCCC		Sylvinite Halite
H2	HHHH		Foliated shale with higher radioactivity and halite at the footwall
F 2	===	5	

Cycle 2

SC 1	ScSS / CCHCH		Carnallite flanked by sylvinite
H 1-4			Halite
F 1-4		4	Foliated shale with limestone bed on top
H 1-3	HH		Halite
F 1-3	HH	3	Foliated shale with two peaks of halite and several limestone intercalations
H 1-2	HHH / HH / HHH		Halite
F 1-2	===	2	Foliated shale } **Lowest Rock Salt**
H 1-1	HHH / HH / HHH / HH		Halite
LA-1		1	Basical anhydrite, limestone and shale
F 1-1	===		Foliated shale „A"
			Roof of the oil-bearing zone

Cycle 1

| Car-mo-polis | o o o | | Sandstone and conglomerates / Oil-bearing series |

(Formation: MURIBECA; Series: Ibura; left vertical letters)

The best investigated potash deposits of Sergipe are

1. the Basin of Taquari — Vassouras (or Taquari — Jurema — Vassouras) situated north of the Carmópolis oil field and between the anticlines of Carmópolis and Siririzinho (cf. Fig. 1);
2. the Basin of Santa Rosa de Lima to the west of Siririzinho and north of the Riachuelo oil field anticline.

In these salinar sub-basins a *general and distinct zoning of mineral facies around the oil field structures can be observed.* Immediately bordering the anticlines, where the highly soluble chloridic salts have been completely leached away, the first salinar region begins, where only rock salt has been preserved. Further towards the centre of the synclines follow the potassium salt paragenesis of sylvinite and carnallitite. The extremely soluble tachhydrite masses are almost only found in the most interior, deepest parts of the basins.

At first sight the extension and distribution of sylvinite and carnallitite does not seem to be completely clear. In some cases the sylvinites occur somewhat nearer to the anticlinic structures and their slopes. In most of the regions — and characteristically more in the deeper parts of the basins — the carnallitic salts show greater extensions and thus also seem to be nearer to the slopes of the anticlines. Details of the lateral and vertical variations will be explained in the following pages.

b) The Sequence of Salt Deposition of the Ibura Series (Aptian Muribeca Formation)

Above the oil-bearing Carmópolis Series with mainly sands and conglomerates — the original source and enrichment of the oil is still somewhat problematic — the beginning of the Ibura Series is characterized by shaly sediments, cf. F 1 — 1 of Figure 3 (F = Portuguese Folhelho, foliated shale). The first salinar crystallization products LA 1 consist of limestone (L) and anhydrite (A) with intercalations of marl shale.

In a first unpublished report on the saline deposits of Sergipe, Prof. *G. Richter-Bernburg* (Hannover, BGR), divided the evaporation series with increasing depth according to the letters A, B, C, D, etc; later in 1968, *Hite* subdivided seven special sequences downwards from Cycle I to Cycle VII. Finally, on the basis of several dozens of systematically driven borings and several years of thorough investigations, the Brazilian geologist *Fonseca* (1972) separated ten cycles with prominent stratigraphic "Markers" and numbered them from 1 to 18. He was also able to constitute a general stratigraphic scheme (cf. Fig. 3, originally Anexo 5 of the unpublished report of *Moacyr de Vasconcellos*, 1972).

In the lower part of the alinar sequence the rather thin layers of mainly anhydrite LA 1 on top of some limestone with shale, and practically no dolomite or magnesite, permit the implication that important masses of these salinar phases of restricted basin evaporites usually present earlier must have been deposited in forebasins (cf. *Borchert*, 1940, 1959; *Borchert* and *Muir*, 1964). However, the special paleogeographic conditions and the systems of barrens and/or narrow sea channels are still completely unknown.

Cycle 1 already contains two rather thick halite beds separated by the shale F 1—2 which represents the important Marker M 2. The deepest halite beds of the Sergipe basins H 1—1 and H 1—2 can be characterized as "Lowest Rock Salt".

A sequence of two further halite layers with thicker shale intercalations then follows (Markers M 3 and M 4), and on top of them the first carnallitite layer can be found embedded between two sylvinite beds. There should be general agreement with the statement of *Moacyr de Vasconcellos* (1972) that the sylvinites of Sergipe are generally of secondary origin and not primary crystallization products from mother liquors of the original evaporation basin. This is also supported by the fact that the sylvinites mostly show red colours due to microscopic inclusions of hematite flakes which have been inherited from primary carnallite inclusions.

Carnallite is the principal and most widely spread primary potassium mineral and has here, as in many other places, been decomposed by wandering and impregnating ground waters, and by extraction of $MgCl_2$, leaving sylvite as residue. Besides the main mass of red sylvites there are also many fissure fillings of white sylvite, widespread in similar form in many potash basins all over the world. This happens especially where periodically occurring tensional stress is an important factor, particularly during more or less regional uplifting stages (cf. *Siemeister*, 1961/1969).

Similar leaching processes have occurred on a greater scale in Cycle 2 with Marker 5 as its basis. The main mass of this cycle consists of the rather thick carnallitite C 2 with intercalations of laminated halite beds. The most characteristic feature, however, is once more the fact that the carnallitite zone is flanked below and above by the sylvinite layers S 2–1 and S 2–2. Such a stratigraphic situation is most probably not of primary origin but is typical as a result of slowly percolating ground waters. This example can be considered as a real test case, since the potash salts are locked below and above by impermeable shales. Therefore, the wandering solutions and brines could move preferentially along the shale intercalations of Marker 5 and 6 in a lateral direction. The percolating and impregnating solutions could thus decompose mainly the upper and lower parts of the carnallitite, similarly to the technical process of sylvite production by the "cold decomposition" of carnallite.

A similar origin of secondary sylvite from primary carnallite has been described by many authors in very many potash formations of different geological ages and regions (cf. *Borchert* and *Muir*, 1964, p. 171 f.). The character of the same leaching effects by very slowly percolating ground waters in the Oligocene potash deposits of the Rhine Gragen have been well demonstrated in the corresponding Figure 17.3 (*Borchert* and *Muir*, 1964, p. 199).

In the next Cycle 3 (cf. Fig. 3) we find the first intercalations of tachhydrite T 3 together with the Upper Carnallite C 3–2. The widespread occurrence of tachhydrite in the Sergipe basins as well as in the potash deposits on the other side of the Atlantic Ocean in Gabon, Angola etc requires a special discussion in the following section. Here, it may be outlined only that cycles 3, 4, 5 and 6 with Markers 3–11 represent those central parts of the Sergipe basins, where the smallest secondary leaching effects by descending atmospheric waters have occurred (cf. also Fig. 1). Likewise the numerous vertical and lateral DNPM-sections of these cycles 3–6 further show practically no anhydrite but principally the predominance of carnallite (often together with the very soluble bischofite) and together with the extremely hygroscopic tachhydrite.

The most important sylvinite layers are found in Cycle 7 (S 7–1 as well as S 7–2 and S 7–3, which are separated from each other by a thin bed of the halite H 7–2). It is probably not by accident that we find in this cycle the purest thick rock salt beds H 7–1 and H 7–3, which can be indentified with the "Lower" and "Upper" Rock Salt of the Sergipe basins. Also in the German Zechstein and in many other saline formations we find the purest halite masses mostly in such regions, where re-solution and recrystallization processes have occurred in greater dimensions.

The Cycle 8 begins with thick anhydritic beds with halite intercalations containing Marker M 12 of bituminous clay which is distinguished by a high rate of neutron absorption. Probably organic and bituminous substances play an important role in the origin of tachhydrite. The upper part of the scheme of sequences and zones (cf. Fig. 3) shows once more a thicker halite layer H 8 together with a thin sylvinite bed which, however, seems to have no larger regional extension, probably on ccount of stronger secondary intracrustal leaching processes.

With Cycles 9 and 10 we have reached those parts of the sequence where several thicker layers of anhydrite represent the ending salinar stage. This phase can be characterized by new influx of sea water and by broader communication channels of the restricted basin with the open ocean. Probably these changes are in connection with subsiding effects and the further opening of the huge graben systen of the Atlantic Ocean. Combined with these events it is only natural that ordinary marine sediments such as limestone and shales with Markers M 13–M 18 prevail.

c) The Special Conditions of Tachhydrite Genesis

D'Ans (1961) has suggested several possible answers for the origin of concentrated $CaCl_2$ solutions in connection with oceanic salt deposits, none of which is satisfactory (cf. also Wardlaw, 1972, p. 156).

According to Braitsch (1971, p. 80) the lower temperature for the formation of tachhydrite or "tachyhydrite" ("Ca-carnallite" = $CaCl_2 \cdot 2 MgCl_2 \cdot 12 H_2O$) is about $22°$ C, but only at the extremely high concentration of 92.7 mol $CaCl_2/1000$ mol H_2O (approx. 450 g $CaCl_2/l$).

In fact, the not uncommon appearance of tachhydrite in paragenesis of carnallite and even bischofite — sometimes still with intercalations of thin anhydritic beds in the South Harz Potash district and elsewhere — can probably be related to reactions between intracrustal bitumen-impregnated brines of metamorphic origin resulting from deeper ground waters which decompoe carnallitic layers in the environment of calcium sulphate beds (cf. Borchert and Muir, 1964, p. 178).

Similar genetic conditions are most probably responsible for the development of very $FeCl_2$-rich brines which are necessary for the formation of the K-Fe^{2+} containing chloride minerals rinneite, zirklerite and douglasite. It seems probable that, for example, in the German Zechstein Formation, reactions may occur between ascending "oil field waters" with reducing character — originating from basal bituminous shales and dolomites of the Zechstein 1. Such deeper ground waters intermixed with Ca SO_4-solutions resulting from the metamorphism of gypsum into anhydrite, can ascend in anticlines or dome-like structures during tensional stress phases. They can then decompose carnallitic salts of the Stassfurt Potash Layer and react with Fe-containing solutions which have been leached by brines out of the neighbouring Gray Saliferous Clay (Grauer Salzton). Siemeister (1961) has discussed similar reactions of $FeCl_2$-rich brines in connection with rinneite paragenesis.

Wardlaw (1972) tends to explain the two very thick layers of tachhydritic salts and other intercalations in the central parts of the Sergipe basins by "dry lake" stages. Thus, up to 100 m of tachhydrite should have crystallized during two phases (Lower and Upper Tachhydrite T 4 and T 6, cf. Fig. 3) from little more than 200 m of tachhydrite-saturated brine. Such solutions, extremely rich in $CaCl_2$, are thought to have resulted from special formation waters to be driven into grabens similar to the recent Red Sea and the Dead Sea in Palestine, flanked by complementary uplifted blocks which in the case of the Red Sea have reached a relief of up to 3000 m above sea level.

The influx of such formation waters discharging larger masses of $CaCl_2$-containing brines into the marine Aptian basin of Sergipe should be enforced by similar great differences of relief which, however, probably have never existed in connection with the Lower Cretaceous basin and their surrounding continental blocks in Sergipe, Brazil. A similar morphological situation (as in the environment of the recent Red Sea) for the other side of the Atlantic Ocean is at least just as problematic (cf. *Fernandes*, 1967; *Kroemmelbein* and *Wenger*, 1966).

Moreover, it should not be forgotten that 1000 g of normal sea water have to be concentraged to less than 7 g before bischofite begins to crystallize, and an equivalent volume reduction would also be necessary for the beginning of tachhydrite crystallization. The very high evaporation temperature and an extremely dry climate for the Cretaceous period in eastern Brazil and western Africa is also doubtful, and nevertheless would be indispensable for such "dry lake" evaporation processes.

Other arguments against a "dry lake" stage and in favour of a secondary metamorphic formation of the Sergipe (and African) tachhydrite layers will be given later: such as breccia textures of neighbouring bituminous clays and foliated shales (Blätterschiefer), collapse structures in different dimensions, many salt horses (Vertaubungen), all of which favour subrosion processes, leaching and intracrustal reaction effects.

3. The Principal Causes of the Facies Variations in the Sergipe Basins

The facts and explanations of the foregoing chapters may have shown that practically the same sequence of sediment and salt crystallization products — especially with the "backbones" of Markers 1—18 of the Ibura Series (cf. Fig. 3) — are found in the three better-investigated sub-basins of Santa Rosa de Lima, Taquari—Jurema—Vassouras and Aguilhada. There can, therefore, be no doubt that the same or at least very similar evaporation conditions and salt sequences were primarily realized over distances of approximately 40 x 30 km^2 and probably even much more. Geological conditions similar to those in the Sergipe basins are also encountered in the adjoining Cretaceous area of the state Alagoas in the north of Sergipe and north of the San Francisco river.

However, there are many facies variations of the salt and potash layers in dimensions of km and often also much smaller extensions (up to more or less abrupt changes of mineral facies) in each of the special Sergipe sub-basins. Once more the question arises about the essentially primary or secondary character of these principal variations of salt paragenesis. It is, in fact, important to be able to distinguish which of these facies differences are due to primary causes — such as palaeogeography, basin morphology, synsedimentary subsiding, basin streams, temperature gradients etc (cf. *Borchert* and *Muir*,

1964) — and which of them are mainly due to secondary leaching and replacement processes in connection with descending atmospheric waters or with circulating and "impregnating" deeper ground waters, brines, and "oil field waters" or (in other cases) even with ascending hydrothermal solutions of magmatic origin.

In view of the vertical sedimentation and precipitation sequence it can easily be seen that *it does not represent one continuous evaporation series of a completely "closed basin" type.* The bromine content does not show a perfectly systematic concentration in the salt crystallization products moving up from below. Thus, new influxes of sea water may have occurred at several periods. Furthermore, great masses of the earlier crystallization products (dolomite, magnesite, anhydrite, etc) must have been precipitated already in forebasins. Nevertheless, however, there can be no doubt that in the Sergipe-Alagoas and neighbouring basins (as in western Central Africa) we have to deal with a marine source of the salt material and with systems of more or less restricted basins connected with each other and with the ocean by barrens or by narrow and more or less shallow sea channels.

The principal lateral facies variations in the Sergipe subbasins are clearly dependent on secondary replacement processes, and especially on leaching effects in connection with younger tensional stress tectonics (horst/graben structures, fault systems etc). The main changes of salt and potash facies show immediate relations to the anticlines and dome-like structures of the three oil fields of Carmópolis, Siririzinho and Riachuelo (cf. Fig. 1). The more easily soluble chloridic salts have been leached away on top of the anticlines, and there can be hardly any doubt that descending rein waters are the main reason for their complete disappearance, with conservation of the only slightly soluble anhydrite only (cf. Fig. 1, middle and lower part).

Likewise, the *lateral sequence of the beginning halite facies and the subsequent sylvinite zones and further carnallitite can be adequately explained only by slowly progressing dissolution and decomposition processes along the slopes of the anticlines.*

The remarkable continuity of the salt and sediment sequences with Markers 1—18 (cf. Fig. 3) in many boreholes on the slopes of the three anticlines of Carmópolis, Siririzinho and Riachuelo proves that previously *only one basin existed with rather homogenous geological conditions.* Similar relations probably exist in the environmental synclines and anticlines and also in the Alagoas Salinar Basin in the north of Sergipe.

The leaching effects, however, are mainly influenced by faults which in general correspond to the secular uplifting processes of the Brazilian Shield. These tectonics sometimes affected only broader or smaller regions forming local basins, some of them with the character of epicontinental seas with the facies of restricted basins and with corresponding evaporite deposits. In other cases they formed basins of larger dimensions, such as the Pennsylvanian Middle Amazonas Basin or the Permian basin of Pedro do Fogo (cf. *Benavides*, 1968, p. 254), with mainly gypsum and anhydrite in the Maranhão Basin in the states of Maranhão, Piaui, and the north-eastern part of Goias.

The most effective tensional stress tectonics and dilatation processes occurred during Jurassic-Cretaceous time, when the Old Shield of western Africa separated from the Brazilian Continent. Tensional stress especially must also have occurred after the deposition of the Ibura Salt Series, beginning before the still Aptian Oiteirinhos Member, but continuing during upper Cretaceous and earlier Tertiary epochs (cf. Fig. 2).

In the Sergipe sub-basins the principal movements of the mainly descending leaching solutions should have occurred along the more or less impermeable bituminous clays, shales and markers. The main vertical wandering paths, however, might have proceeded along the faults of the tensional stress tectonics. This is not only true for the regions of the more horst-like structures of the oil field anticlines themselves (middle and south-eastern part of Fig. 1 above and below) but also for the western region within the sub-basin of Santa Rosa de Lima, where for instance the facies of more or less pure rock salt masses in the lower section of the Ibura Series changes to more carnallitic potash salts to the north-west. In this area the corresponding leaching processes must have taken place in connection with deeper ground water ascending by artesian hydropressure from below.

Similar conditions are also representative of some deeper salt horses in the middle of the Taquari-Vassouras sub-basin. Several sections of the systematically arranged bore-hole series across the different saliferous sub-basins show similar effects of leaching along tensional fracture zones "from below", but the principal corresponding leaching processes have proceeded, probably in connection with descending rain waters along the slopes and impermeable beds of the oil field anticlines.

The secondary character of the most important facies variations, and the *zonal arrangement of the different salt facies,* with only rock salt on the highest parts of the slopes, then sylvinite and finally carnallite (and tachhydrite) in the central parts of the synclines, is also proved by the fact that the isopachs of rock salt, of sylvinite and/or carnallitic salts etc in many cases do not run parallel to the boundaries of the outcrops of the Ibura Salt Formation. On the contrary, the contours of sylvinite, carnallitite etc are often obliquely or even more or less vertically cut by the corresponding isopachs. Such relations can be expected only in connection with leaching effects along fissure and fault zones.

In contrast to this concept of predominantly secondary facies variations, many authors, among them *Moacyr de Vasconcellos* (1972) tend to atrribute these — as a more essential consequential contribution — on the one hand to primary sedimentation conditions (synsedimentary subsiding etc) and on the other to secondary tectonics, such as "halite-kinetic energy". Similar effects are considered to be the reason not only for the present stratigraphic-structural regional framing of the sub-basins, but also for more local changes of thickness of the different salt mineral masses and even for special facies variations, notwithstanding the fact that collapse structures have often been observed and described.

Detailed knowledge and correct interpretation of the primary or secondary character of the different types of facies variations are not only important for scientific reasons, but also of high interest for further optimal projects of exploration, mining technology and economy.

4. Potash and Salt Reserves

4.1 Potassium Resources

With special law No. 61 157 (August 17th, 1967) the Federal Government of Brazil constituted a National Reserve Area of 425 km^2 for carnallite, sylvinite and rock salt in
Sergipe, limited by the parallels 10°35'12" and 10°45' south latitude and the meridians
36°55' and 37°15'15" west longitude. Within this area the Lower and Upper Sylvinite
layers S 7–1 and S 7–2 together with S 7–3 (cf. Fig. 3) are those of highest economic
value.

The mineable Lower Sylvinite layer S 7–1 of the *basin of Taquari – Vassouras* extends within the line representing the 3-m isopach continuously over a distance of
4500 m. It has – though there may be local salt horses – an average thickness of 4.71 m,
corresponding to 75 million t of sylvinite with an average content of 20.14% K$_2$O and
2.34% insoluble residue.

The Principal Sylvinite layer (S 7–2 and 7–3) can be followed in the basin of
Taquari – Vassouras for a distance of more than 12 km. Its lower section is especially
persistant. However, the Lower Sylvinite S 7–1 has an immediately underlayer of carnallite followed by thick masses of pure tachhydrite (cf. Fig. 3). This is very unfavourable for conventional mining, and even more for solution mining. For the Principal
Sylvinite layers also, the probable existence of salt horses must be taken into consideration. The spacing of the investigating bore holes is between 1 and 3 km, leaving room
for many unknown irregularities.

In the sub-basin of Taquari – Vassouras, a mineable area of 30.06 km^2 with an average thickness of 6 m has been calculated. The resources in this area comprise 350 million t of good sylvinite at an average content of 24.0% K$_2$O.

In the *basin of Santa Rosa de Lima* the corresponding figures within the 3-m isopach
and above 900 m depth (but here only for the Principal Sylvinite layer S 7–2 and
S 7–3) show an average thickness of 4.2 m with 100 million t of sylvinite, having an
average content of 25.47% K$_2$O and only 1.3% insoluble residue [all datas from the
1970 report of the CPRM as Department of the Ministry of Mines and Energy (MME)
of Brazil].

4.2 The Resources of Rock Salt, Carnallitic and Tachhydritic Rocks in the National Reserve Area and 29 Adjacent Areas

In the original National Reserve Area of 425 km^2, the rock salt reserves have been
roughly calculated for the Lower, Middle and Upper Rock Salt (probably H 7–1,
H 7–3 and H 8; cf. Fig. 3). Summarizing the thicknesses at an average of 35 m the
"resources in place" are 525,000 t of rock salt.

The Lower and Upper Carnallite C 2 and C 3–2 in the basin of Taquari – Vassouras
– the average contents in the different boreholes show variations between 30.86% and
49.5% carnallite and thicknesses between 17 and 51 m – is calculated to be 6060 million t of carnallitic rock.

On the basis of the great thickness of the Upper Tachhydrite T 6, underlying the syl-vinite S 7–1 (cf. Fig. 3) a total mass of 400 million t of tachhydritic rock has been obtained.

"As the areas where carnallite and tachhydrite occur are close to each other it may be verified that there is a 'reserve in place' only in the Taquari – Vassouras area of 0.9 billion (German: milliard) t of metallic magnesium" (*Carneiro de Almeido*, CPRM report of May 1970).

During the following years, from 1970–1972, 29 areas adjacent to the National Reserve Area were also investigated by systematic borings, 12 areas A to L in the north, Nos. 1–13 in the east and M to P in the south-east of the original 425-km^2 concession. However, in this larger territory there are only small quantities of more sylvinite but even greater masses of halite, carnallitic and tachhydritic salts. This fact tends to suggest that in the 29 adjacent areas there have been less intensive leaching processes enriching (veredeln) the original carnallitites into sylvinites. Pure carnallite $KCl \cdot MgCl_2 \cdot 6 H_2O$ has contents of only 14.0% potassium and 9.7% magnesium, whereas pure sylvite con-tains 52.4% K.

The measured and indicated reserves of carnallitite in the layers C 2 and C 3–2 (cf. Fig. 3) have been calculated to 2.132 billion t. Including the inferred reserves the sum is 5540 billion t of carnallite corresponding to 469.62 million t K_2O.

a) Total of the Calculated Reserves of the Sergipe Salt Deposits

The summation of the halite quantities in the Lower and Upper Zones of Rock Salt in the 29 adjacent areas – in addition to the resources of the original National Reserve Area of 425 km^2 – has been calculated to be 5089 million t of rock salt.

A greatly condensed summary of the numerous data on reserves in the Sergipe terri-tory (National Reserve Area and the 29 adjacent areas) as per the CPRM reports of 1970 and 1972 is given in the following table:

Area	Sylvinite 10^6 t	Carnallitic rocks 10^6 t	Tachhydritic rocks 10^6 t	Rock salt 10^6 t
1. Basin of Taquari–Vassouras				
a) Lower layer S 7–1 with 20.14% K_2O	75	6060	400	525
b) Principal layer S 7–2 and 3 with 25.47% K_2O	350			
2. Basin of Santa Rosa de Lima				
Pricipal layer S 7–2 and 3 with 25.47% K_2O	100	–	–	–
3. 29 areas adjacent to the National Reserve Area of 425 km^2	–	5540	–	>5089
Total	525	11,600	400	>5600

On the basis of the measured reserves of 525 million t of sylvinite (with 130×10^6 t of K_2O in the basins of Taquari – Vassouras and Santa Rosa de Lima) and a prospective yearly production of 500,000 t of potash salts with 60% K_2O, the reserves would be, roughly calculated, sufficient for at least many decades, if not for some centuries.

Acknowledgments. Thanks to the kindness of the Companhia de Pesquisa de Recursos Minerais (CPRM) and the Departemento Nacional da Produção Mineral (DNPM) I was able to study several voluminous reports with numerous maps, sections, tables of mineralogical and chemical analysis etc, geophysical borehole investigations, coloured photos of the cores of many salt zones, macroscopic and microscopic work of core textures, geochemical investigations or bromine content and trace elements etc: the first report appeared on May 19, 1970, by *Carneiro de Almeido*, Chief Executive of the Potassium Project, and the second report on July 1972 by *Moacyr de Vasconcellos* as Technical Director of Operations. With Dr. *Rainald Heinisch, Krupp Industrie- und Stahlbau*, the interesting salt region was visited and also the well-preserved cores of many thousand m of saline formation were studied near Carmópolis, with the leading geologist *José Cornélio Fonseca* as a guide in the field; he has also studied in detail a series of borehole sections and constructed the original scheme shown in Figure 3. I am further much indebted to Dr. *João Neiva de Figueireido, Marcelo de Barros Oliveira* (Petrobras) as well as to Dr. *K. Wiesner, H.-P. Drescher* and *G. Haubold (Krupp Industrie- und Stahlbau)* for their helpful assistance and fruitful discussions. Last but not least, I have to thank Dr.-Ing. *C. Siemeister* (Technical University Clausthal) for critical reading of the manuscript.

References

Almeida, Sandoval Carneiro de: Final report on the Potassium Project in Sergipe of the Companhia de Pesquisa de Recursos minerais (CPRM) and the Departemento Nacional da Producao Mineral (DNPM). May 19th, unpubl. (1970)

Benavides, V.: Saline deposits of South America. Geol. Soc. Am. Spec. Paper 88, 249–290 (1969)

Borchert, H.: Die Vertaubungen der Salzlagerstätten und ihre Ursachen. Part I, Z. Kali **27**, 97, (1933). Part II, Z. Kali **28**, 290, **29**, 1 (1934/35). Part III, Z. Kali **35**, 33 (1941)

Borchert, H.: Die Salzlagerstätten des deutschen Zechsteins, ein Beitrag zur Entstehung ozeaner Salzablagerungen. Berlin: Arch. f. Lagerstättenforsch., Heft 67, 1940, 196 p.

Borchert, H.: Ozeane Salzlagerstätten – Grundzüge der Entstehung und Metamorphose ozeaner Salzlagerstätten sowie des Gebirgsverhaltens von Salzgesteinsmassen. Berlin: Borntraeger 1959,237 p.

Borchert, H.: Geosynklinale Lagerstätten, was dazu gehört und was nicht dazu gehört, sowie deren Beziehungen zu Magmatismus und Geotektonik. Freiberger Forsch.-H. C 79, 7–61 (1960)

Borchert, H.: Zusammenhänge zwischen Lagerstättenbildung, Magmatismus und Geotektonik. Geol. Rdsch. **50**, 131–165 (1961)

Borchert, H.: Vulkanismus und oberer Erdmantel in ihrer Beziehung zum äußeren Erdkern und zur Geotektonik. Bol. Geofis. Teor. Applic. **9**, 35, 194–213 (1967)

Borchert, H.: Der Wert gesteins- und lagerstättengenetischer Forschung für die Geologie und Rohstoffnutzung. Ber. Deut. Ges. Geol. Wiss., B Mineral. Lagerstättenforsch. **13**, 1, 65–116 (1968)

Borchert, H., Muir, R.O.: Salt Deposits – The Origin, Metamorphism and Deformation of Evaporites. London–Princeton–New York–Toronto: Van Nostrand Ltd. 1964, 338 p.

Braitsch, O.: Salt Deposits – Their Origin and Composition. Berlin–Heidelberg–New York: Springer 1971, 297 p.

D'Ans, J.: Über die Bildungsmöglichkeiten des Tachhydrits in Kalilagerstätten. Kali Steinsalz **3**, 119–125 (1961)

Dunham, K.C.: Mineralization by deep formation waters: a review. Trans. Inst. Min. Metall. (Sect.B), 79, 127–136 (1970)

Fernandes, G.A.: Analogia des vacias saliferas de Sergipe, Gabão, Congo e Angola. Bol. Tecnico PETROBRAS 9, 349–365 (1966)

Fernandes, G.A.: A geologia de subsuperfície indica o novo parâmetro para futuras descobertas de oleo em Sergipe. PETROBRAS, Coletânea de Relat. Exploração, Rio de Janeiro 2 (4), 111–135 (1967)

Fonseca, J.C.: General scheme of the 10 zones of evaporites in the Sergipe Basins (cf. Fig. 3). In: Vasconcellos, unpubl. (1972)

Harbort, E.: Zur Frage nach der Entstehung gewisser devonischer Roteisenerzlagerstätten. N. Jb. Mineral. etc., Sect. A, 1, 179–192 (1903)

Hite, R.J.: Recommendations and observations pertaining to Projeto Potássio, State of Sergipe, Brazil. U.S. Geol. Surv., unpubl. (1968)

Höll, R.: Die Scheelitlagerstätte Felbertal und der Vergleich mit anderen Scheelitvorkommen in den Ostalpen. Habil.-Schr. München: Bayr. Akad. Wiss. 1975, 114 p.

Kennedy, W.Q.: The influence of basement structure in the evolution of the coastal (Mesozoic and Tertiary) basins of Africa. In: Salt Basins Around Africa. London: Inst. of Petroleum 1965, pp. 17–40

Klockmann, F.: Montangeologische Reiseskizzen. Z. prakt. Geol. 8, 265–275 (1900)

Kroemmelbein, K., Wenger, R.: Sur quelques analogies remarquables dans les microfaunes Crétacées du Gabon et du Brésil Oriental (Bahia et Sergipe), p. 193–196, In: Reyre, D. (ed.), Sedimentary Basins of the African Coasts. Paris: Assoc. African Geol. Surveys 1966, 304 p.

Maucher, A.: Zur alpinen Metallogenese in den bayrischen Kalkalpen zwischen Loisach und Salzach. Tschermaks Mitt., 3. Ser., 4, 454–463 (1954)

Maucher, A.: Die Antimon-Wolfram-Quecksilber-Formation und ihre Beziehungen zu Magmatismus und Geotektonik. Freiberger Forsch.-H. C 86, 173–188 (1965)

Oftedahl, C.: A Theory of Exhalative-Sedimentary Ores. Geol. Fören. Förh., Stockholm 80, 1–19 (1958)

Richter, A.J.: Geologia do "horst" que separa as Bacias do Reconcavo – Tucano e Alagoas – Sergipe. PETROBRAS, Coletânea de Relat.Técn., Rio de Janeiro 2 (4), 167–177 (1967)

Schaller, H.: Revisão estratigráfica da Bacia de Sergipe – Alagoas. PETROBRAS, Bol. Técnico 12 (1), 21–86 (1969)

Schmidt, W.: Festigkeit und Verfestigung von Steinsalz. Z. angew. Mineral. 1, 1–21 (1939)

Schneiderhöhn, H.: Die Erzlagerstätten der Erde. Vol. II, Die Pegmatite. Stuttgart: Fischer 1961, 720 p.

Schulz, O.: Die Pb-Zn-Vererzung der Raibler Schichten im Bergbau Bleiberg–Kreuth (Grube Max) als Beispiel submariner Lagerstättenbildung. Carinthia 2 (22), 1–93 (1960)

Shearman, D.J.: Origin of Marine Evaporites by Diagenesis. Trans. Inst. Mineral. Metall. (Sect. B) 75, 208–215 (1966)

Shearman, D.J., Fuller, J.G.: Anhydritic Diagenesis, Calcitization, and Organic Laminites, Winnipegosis Formation, Middle Devonian, Saskatchevan. Bull. Can. Petrol. Geol. 17, 496–525 (1969)

Siemeister, G.: Primärparagenese und Metamorphose des Ronnenberglagers nach Untersuchungen im Grubenfeld Salzdetfurth. Dissert. Clausthal 1961; Beih. Geol. Jb., B.f.B., Heft 62, Hannover 1969, 122 p.

Teixeira, A.A., Saldanha, L.A.R., Maia, A., Fernandes, G.A.: Bacia salifera Aptiana de Sergipe/Alagoas ocorrencias de sais soluveis. PETROBRAS, Bol. Tecnico 11, 221–230 (1969)

Vasconcellos, F. Moacyr de: Report on Mineral Exploration of Potassium Salt – Rock Salt in the area adjacent to the National Reserve Area, CPRM, State of Sergipe. July 1972, unpubl. (1972)

Wardlaw, N.C.: Unusual marine evaporites with salts of calcium and magnesium chloride in Cretaceous Basins of Sergipe, Brazil. Econ. Geol. 67, 156–168 (1972)

Precambrian Deposits

Timing Aspects of the Manganese Deposits
of the Northern Cape Province (South Africa)

P.G. SÖHNGE, Stellenbosch

With 1 Figure

Contents

Summary

The origin and timing of the manganese deposits of Griqualand West are briefly reviewed within the framework of the new South African stratigraphic nomenclature. Though strata-bound, the ores of Postmasburg type formed intermittently along a major unconformity over a time-span of 2000 my. The stratiform ores of the Kuruman field relate to a very limited period of deposition dated at ca. 2200 my when biological evolution greatly influenced chemical sedimentation in the exogenic cycle.

1. Introduction

The manganese deposits of Postmasburg and Kuruman in the northern Cape Province are essentially strata-bound. Several milliard t of various types of ore are located in an early Proterozoic bedded sequence about 1000 m in thickness – a comparatively limited section in the Precambrian stratigraphy of South Africa.

The literature of the past 50 years may be found confusing, as there have been major changes in the subdivision and naming of stratigraphic sequences, including adjustments with respect to the relative age of formations, unconformities, and the tectonic history of the area.

To clarify the setting, a composite diagrammatic section is presented, bringing into single view the manganese deposits of both the Postmasburg and Kuruman fields. The two parts of the section are geographically 100 km apart, but structurally in correct relative position. Unconformable contacts are given special emphasis inasmuch as they designate intervals in the geological history when manganese may have been mobilized and redeposited. For detailed descriptions of the geology the reader is referred to the writings of *de Villiers* (1960), *Boardman* (1964), *de Villiers* (1970) and the beautifully illustrated review by *Button* (1976). The writer has been privileged to visit the mines in Griqualand West on numerous occasions and is greatly indebted to their personnel for informative discussions.

Reference is made mainly to the 140-km manganese belt between Postmasburg and Black Rock. Further ore occurrences of the same type are known over a total strike length of 600 km northward and eastward through Botswana into western Transvaal. Taking up the arguments expressed by *Maucher* (1974), an appraisal is made of the time-bound or time-dependent character of these giant manganese deposits.

2. Outline of Stratigraphy

The new lithostratigraphic nomenclature recommended by the South African Committee for Stratigraphy has been briefly stated in a paper by *Kent* and *Hugo* (1976) at the International Geological Congress held in Sydney. For the area of present concern the subdivisions are presented in Table 1.

Table 1. General stratigraphy

	Group	Formation	Age (my)
	Kalahari		60
	Karoo	Dwyka Tillite	300
Griqualand West Supergroup	Olifantshoek	Volop Quartzite Hartley Andesite Lucknow Quartzite	2070
	Postmasburg	Voëlwater Jasper Ongeluk Andesite Makganyene Diamictite Gamagara Shale	2224 2263
	Griquatown	Koegas Mudstone/jaspilite Asbesheuwels Ironstone	
	Campbell	Ghaapplato Dolomite Schmidtsdrif Shale-dolomite Vryburg Siltstone	
	Ventersdorp		2600
	Undifferentiated basement		

As the dominant lithology of the sequence is reflected in the formation names, attention is drawn only to those features relevant to manganese deposition.

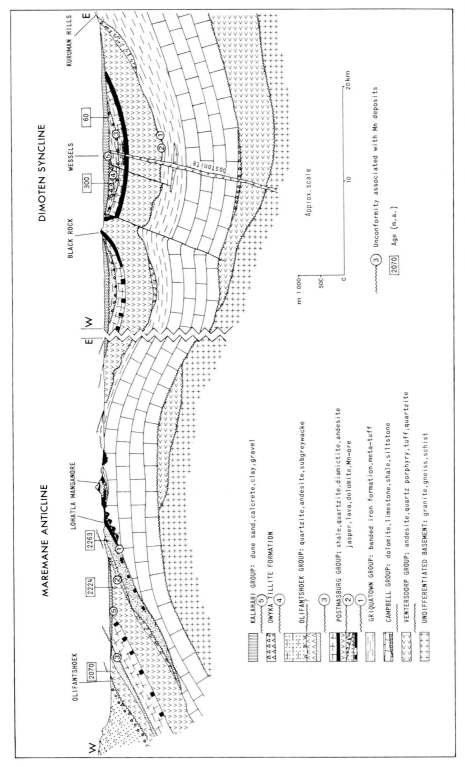

Fig. 1. Composite diagrammatic section illustrating stratigraphic relationships of the manganese deposits of Postmasburg and Kuruman. (For explanation see text)

The stromatolite-rich Ghaapplato Dolomite is a shallow water (bio-)chemical sedimentary formation carrying on the average 0.5% Mn. The banded iron formation and jaspilites of the overlying Griquatown Group contain only 0.3% Mn and include countless bands of stilpnomelane rock that have been shown to represent altered volcanic tuff. In the western Transvaal (Zeerust) the contact between the equivalents of the Griquatown and Campbell Groups is discordant. No such relationship has been reported from Griqualand West.

The Postmasburg Group starts with sedimentary breccia ("Blinkklip") at the base of the Gamagara Formation. The breccia, crowded with fragments of banded ironstone and chert, fills hollows up to 10 m deep in the dolomite and also caps erosional relics of the resistant Griquatown Group. Most ore exposures reveal slump synclines of Gamagara Shale between massive ribs of dolomite, but in some the overlying shales rest undisturbed upon the steep-walled breccia bodies. Thus the base of the Postmasburg Group is an erosional unconformity (marked 1 in Fig. 1). The shale has been dated at 2263 my (*Button*, 1976). Patches of diamictite, thought to be of fluvioglacial origin, together with extensive flows of andesite, cover the Gamagara Shale and lap over eastward on to the Griquatown. The contact is reported to be unconformable, though the magnitude of the hiatus was limited considering that the age of the Ongeluk is 2224 my (marked 2 in Fig. 1). Intercalated units of banded jasper appear in both Formations. Upon the lava follows an alternating succession of banded ironstone, red jasper, manganiferous-ferruginous-siliceous carbonate rock and layers of black manganese ore; several interbeds of tuff and lava have also been identified. Lenses of agglomerate in the Ongeluk and Voëlwater Formations are so thoroughly carbonated that the original textures of lava inclusions have virtually faded out.

Two unconformities are associated with the Olifantshoek Group. At the base Lucknow Quartzite with minor units of shale and lava rests on an eroded surface of Voëlwater Jasper (marked 3 in Fig. 1). Again, above the Hartley Andesite (age 2070 my) a great thickness of purple to brown hematite-bearing Volop Quartzite follows with apparent concordance, transgressing across its floor eastward to rest on the Voëlwater in the Black Rock area. Some authors consider this contact to represent the greater of the two unconformities (marked ? in Fig. 1). If Hartley volcanism correlates approximately with the Loskop Formation of the Transvaal Supergroup and Volop sedimentation equates with the early Waterberg dated about 1900 my, then a lengthy period of erosion may have intervened, coinciding with the first appearance of red clastic sediments. Perhaps this hiatus is not present in Griqualand West. Manganese ore pebbles of Hotazel type occur in the basal conglomerate of the Volop Formation.

Though appearing insignificant on the diagrammatic section (Fig. 1) the Dwyka Tillite marks the end of a 1000 my era of erosion. It is unlikely that Nama sediments ever extended east beyond the Langeberg fold range, so that the unconformity forming the floor of the Dwyka (marked 4 in Fig. 1) represents a paleosurface of great importance to supergene ore deposition. A glacial palaeovalley up to 800 m deep has been proved by drilling south of Black Rock (*Smit*, 1972). The time gap between Karoo sedimentation and the deposition of Kalahari beds implies another 100 my exposure to weathering; this unconformity (marked 5 in Fig. 1) merges into the present land-surface.

3. Sedimentary-Tectonic Setting

The Transvaal basin was an enlargement of the preceding depression upon the Kaapvaal craton in which the volcanic rocks and sediments of the Ventersdorp Group had accumulated. Gentle arching has since divided the basin into two structural entities, the western of which contains the strata of the Griqualand West Supergroup. A comparatively uniform shallow marine environment prevailed while the chemical sediments of the Ghaapplato and Griquatown Groups were deposited. The succeeding Postmasburg Group accumulated in a more varied environment than the corresponding Pretoria sequence of the Transvaal, as reflected in diverse lithofacies and formation thicknesses. Along the western and north-western margin of the craton, unstable conditions caused recurring interruptions in sedimentation and differences in depth of the sea. Thus the Gamagara sediments appear to have formed in a small basin reaching from Sishen to Postmasburg and separated from a larger depression farther south. They are not known to occur to the north in Griqualand West. The succeeding Makganyene Diamictite and Ongeluk Andesite fill ill-defined paleovalleys, the lavas spreading out over the entire Transvaal basin. Subsurface exploration has demonstrated that the variable development of manganese ore beds in the Voëlwater Formation is related to several well-defined local depressions strung out northwards; these were protected against clastic sedimentation not by depth, but by distance from the Limpopo Belt provenance. *Button* (1976) draws attention to the gradual deepening of the epicratonic sea from north-east to south-west, the manganese deposits appearing where the limestone facies takes over from shale.

Following general uplift and erosion, a new basin configuration evolved along the north-west periphery of the Kaapvaal craton in which the formations of the Olifantshoek Group accumulated.

There is evidence that deeper marine volcanism occurred in the far south (Marydale) and west (Vioolsdrif) when Hartley lavas erupted in the shallow water of a long, marginal trough reaching north and east as far as the Soutpansberg. The further development of this basin in Waterberg time (= Volop) has been discussed by *Jansen* (1975) and does not bear directly on manganese deposition.

4. History of Ore Deposition

4.1 Early Geochemical Concentration

The Ventersdorp volcanic sequence (dated at ca. 2600 my) is regarded as the starting point of manganese concentration in the north-western sector of the Kaapvaal craton. Through long exposure and peneplanation a level surface was fashioned that became the floor of the Transvaal basin. Until sealed off by the carbonate sediment, the bottom mud and underlying lavas could supply abundant Mn to the water. The major source of later additions must have been the deeply weathered surrounding land. The manganese was taken up in the limestone-dolomite precipitate at a fairly regular rate for the duration of chemical sedimentation of the Ghaapplato Group. Subsequent volcanism of the Griquatown Group appears to have been unimportant as a contributor of manganese to the province.

4.2 The Gamagara Period

The siliceous breccias and paleokarst surface at the base of the Gamagara tell of vigorous physical and chemical erosion. The contact is no longer regarded as a tectonic thrust. Several hundred m of banded iron formation were removed from the Maremane rise to expose dolomite and to activate water circulation deeper underground. Probably no large manganese oxide deposits were formed while free drainage prevailed down the westward slope.

As soon as Gamagara sediments were being laid down, the flow of ground-water was impeded, artesian conditions set in, and Mn dissolved from the dolomite could begin to precipitate in the breccias as well as replace directly overlying aluminous shale. Even with a comparatively O_2-poor atmosphere the formation of manganese oxides in this milieu seems acceptable, considering the very shallow depth and nearshore position along the eastern edge of the Gamagara basin. This early cycle of deposition finds support in the metamorphosed character of the ore, consisting mainly of braunite and bixbyite, with which specularite, diaspore and ephesite are associated; these minerals probably formed or recrystallized when "bostonite" dykes and sills were intruded. At Sishen and Beeshoek the more ferruginous shale resting on banded iron formation became enriched with hematite to form laminated replacement iron ore.

The concentration of manganese accomplished during this depositional period is difficult to assess, as repeated later effects have been imprinted upon the rocks and ores. The first of these weathering cycles occurred when the land was exposed and eroded just before Makganyene/Ongeluk time. Little is known about the influence of this unconformity on ore deposition, but certainly the area of dolomite outcrop must have been enlarged, yielding additional manganese by dissolution.

4.3 The Voëlwater Period

Two stratiform bodies of black manganese ore appear in the banded jasper sequence from 30 m above the Ongeluk lava upward. At the established mines the lower ore bed is 3–25 m thick and the upper (where present) 4–6 m. The beds are separated by 5–22 m of banded ironstone. Braunite and cryptomelane are the dominant ore minerals, with hausmannite, bixbyite, and jacobsite plentiful in places. Even where massive, the ore has lenticular to laminated structure with elongate "oolites" of calcite and manganiferous carbonate. The upper and lower contacts are essentially parallel to the general stratification. The ore is considered to have originated as a chemical sediment of "oolitic carbonate in a matrix of gelatinous manganese and iron hydroxide, manganese and iron carbonate and in places hydrous silicates of magnesium and iron" (de Villiers, 1970). Its average grade ranged from 20 to 30% Mn.

Whence came the manganese? It could not have been supplied to the depositional basin by weathering of the Ghaapplato Dolomite, buried under thick sequences of the Griquatown and Ongeluk, nor by general leaching of the flooring andesite, which shows no impoverishment.

The influx of manganese-bearing thermal waters accompanying volcanic episodes during the deposition of the Voëlwater Jasper is a much more likely alternative. Block-faulting had probably already started. There is evidence of hydrothermal alteration of the tuff and agglomerate units and it can be argued that the banded jasper as well as the carbonate beds transitional into manganese ore represent chemical sediments derived from a volcanic source. Such a view would accord with the general conclusion drawn by *Hewett* (1966) for most stratified deposits of manganese. It does not, however, indicate why the world's largest ore field should be associated with this particular episode in a well-established type of sedimentary-volcanic sequence.

The writer proposes that the deciding factor lay in biological activity. Proliferation of iron-oxidizing bacteria contributed to world-wide deposition of banded iron formation in the time-span 2300–2000 my. Conceivably, this may have brought about such impoverishment of dissolved iron in the shallow seas that microorganisms resorted to using manganese to rid themselves of oxygen. The changed chemical environment may have stimulated a population explosion of microbes of *Metallogenium personatum* type that use mainly Mn^{2+} as energy source and deposit Mn^{3+} as oxide. In experiments using manganese-iron-rich mud *Gabe* et al. (1975) found that a stratified microzonal profile was produced at the oxidation-reduction interface where iron did not precipitate unless specific iron-oxidizing microbes were also present. Bacteria furthermore helped to dissolve manganese and iron from the mud in the reducing zone and thus provided for continuous accumulation at the interface. The model can be adapted to fit the realm of a long bottom pool in the shallow sea, stretching from Olifantshoek northward round the arc through Botswana to the Gopane mine in western Transvaal. Variations in water depth and the position of centres of volcanic-fumarolic activity along the trough probably determined the locales of maximum microbial productivity. Such a setting can explain the occurrence of oxidic ore lenses in the Kuruman field in positions that have not since been accessible to oxidizing vadose water.

4.4 Later Paleolandscape Cycles

Transport of manganese and iron by meteoric water has augmented, modified and depleted the ore deposits for 2000 my. Uplift and erosion following immediately on the Voëlwater enabled the manganiferous limestones to become enriched. Considering the long interval of 150 my between Ongeluk and Hartley volcanism, it is likely that such supergene changes were well advanced when Lucknow sedimentation started, and that they continued during the hiatus following the Hartley. Ore outcrops eventually yielded fragments of manganese oxide to the basal sediments of the Volop Formation, after which they vanished from sight until this cover had again been removed by later prolonged erosion and peneplanation.

The flat land surface of the Ghaap plateau may date back to the interval between the 1000 my metamorphic event and the start of Nama sedimentation; it had certainly been fully fashioned by the time Carboniferous glaciation set in. The pre-Dwyka valley, that used to be the main drainage feature between Mamatwan and Black Rock, was apparently carved in Volop shale and quartzite just west of a more resistant outcrop of Voëlwater Jasper. The relief was again locally favourable to migration and reconcentration of manganese and iron, guided along fractures and formational contacts. In the Kuruman field this accounts for not only ferruginous zones of steep as well as flat attitude, so clearly displayed in the Wessels mine, but also important manganese enrichment. In the Postmasburg field the main deposition of later descendant manganese ore probably occurred in Nama times and continued through the Phanerozoic until the present.

Thus, interrupted cycles of mobilization and redeposition by circulating ground water led step by step to the final features as seen today.

5. Time-Bound Manganese Deposition

In retrospect it is clear that the manganese ores of Postmasburg evolved from a dolomite source through meteoric processes repeated many times over some 2000 my. Though strata-bound, they do not portray a unique event in geological history.

The vast Kuruman ore beds, on the other hand, represent a single episode of extreme manganese deposition at ca. 2200 my. Its duration, considering the thickness of litho-logical units between the Ongeluk (2224 my) and Hartley (2070 my) and allowing for the intervening unconformity, could have been of the order of 10–20 my. *Button* (1976) draws attention to the east Hamersley basin of Australia where manganiferous Noreena Shale overlies the Woongarra Volcanics (2000 my) and the lithological frame-work broadly resembles that of the Kuruman field in the Transvaal basin. One may be tempted to take this as evidence for intercontinental reach of time-bound manganese deposition, but there is an age-gap of some 200 my.

The Kuruman manganese deposits are unique not for ore type, but for their total magnitude, which is ascribed to the impact of biological evolution in a favourable physi-cal realm at a critical time in the earth's history.

References

Beukes, N.J.: Precambrian iron formations of Southern Africa. Econ. Geol. **68**, 960–1004 (1973)
Boardman, L.G.: Further geological data on the Postmasburg and Kuruman manganese ore deposits, northern Cape Province. In: The Geology of Some Ore Deposits in Southern Africa, Vol. II. Haughton, S.H. (ed.). Johannesburg: Geol. Soc. S. Afr. 1964, pp. 415–440
Button, A.: Transvaal and Hamersley Basins – review of basin development and mineral deposits. Miner. Sci. Eng. **8**, 262–293 (1976)
Gabe, D.R., Troshanov, E.P., Sherman, E.E.: The formation of manganese-iron layers in mud as a biogenic process (1965). In: Benchmark Papers in Geology, Vol. XVIII: Geochemistry of Iron. Lepp, H. (ed.). Dowden: Hutchinson and Ross Inc. 1975
Hewett, D.F.: Stratified deposits of the oxides and carbonates of manganese. Econ. Geol. **61**, 431–461 (1966)
Jansen, H.: Precambrian basins on the Transvaal craton and their sedimentological and structural features. Geol. Soc. S. Afr. Trans. **78**, 25–34 (1975)
Kent, L.E., Hugo, P.J.: Aspects of the revised South African stratigraphic classification and a pro-posal for the chronostratigraphic subdivision of the Precambrian. Unpubl. paper delivered at 25th Intern. Geol. Congr., Sydney, 1976
Maucher, A.: Zeitgebundene Erzlagerstätten. Geol. Rdsch. **63**, 263–275 (1974)
Smit, P.J.: The Karoo System in the Kalahari of the northern Cape Province. Annals Geol. Surv. S. Afr. **9**, 79–81 (1972)
Villiers, J. de: The manganese deposits of the Union of South Africa. Geol. Surv. S. Afr. Handbook 2. 1960
Villiers, P.R. de: The geology and mineralogy of the Kalahari manganese-field north of Sishen, Cape Province. Geol. Surv. S. Afr. Mem. 59 (1970)

Time- and Strata-Bound Features
of the Michigan Copper Deposits (USA)

G.C. AMSTUTZ, Heidelberg

With 13 Figures

Contents

Summary

The Lake Superior Copper deposits in the Keweenaw lavas may be explained by a uniform single origin. The congruence analysis in the field, in hand specimens and with the microscope, as well as paragenetic and chemical analyses led to the conclusion that the copper is a comagmatic (contemporaneous) constituent accumulated in hydromagmatic (deuteric) phases of the Keweenaw basalt magmas. The copper was carried up, in and with the lavas and crystallized in and with the deuteric phases of these lavas. Some parts of these phases escaped into fractures and into overlaying later flows or sediments (mostly porous conglomerates). The hydromagmatic (deuteric) phases show a distinct crystallization sequence in which the native copper occupies a distinct place. The high degree of congruence existing between the compositional variations and the primary lava features (flow lines, layering, breccia generations, dyklets, etc) rules out a later overall redistribution of elements (or minerals) for most of the copper sections. Burial metamorphism may have changed some of the originally basaltic material, but did not upset the stability of the deuteric phases, which are also of a low temperature – low pressure origin.

1. Introduction

The problem of the genesis of the Lake Superior copper deposits in the Keweenaw lavas (Fig. 1) has been discussed for over 100 years. *Pumpelly* provided some of the most stimulating discussions in papers published between 1869 and 1873. A summary of the essential work on the area was provided by *Lindgren* (1933, p. 517–526), *Cornwall* (1951a, b) and *White* (1956). The theories which resulted from the many papers published over the past 120 years are summarized in Figure 2.

The present paper is based on field work done in 1957 while working for the Bear Creek Mining Company. The permission to use the data for various publications (indicated in the References) is here again warmly acknowledged. It is also based on laboratory

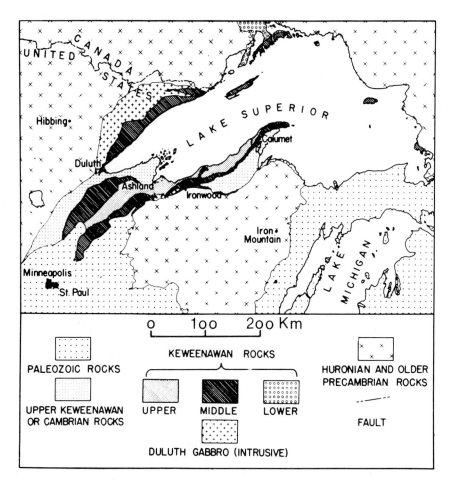

Fig. 1. Generalized geologic map of the Lake Superior region. Modified after *White* (1956)

work done in Missouri and in Heidelberg over the past 10 years, in part with Dr. *Dorai-Babu,* the late Mr. *Kellner,* Dr. *Ottemann* and Mr. *Krouzek.* Their cooperation is also acknowledged on this occasion.

A first re-interpretation was presented in 1958 (*Amstutz,* 1958a) which to a certain extent corresponded to *Cornwall*'s interpretations. (The abstract of the 1958 paper as well as other additional material is incorporated in the conclusions.)

2. Observations on the Relationship between the Textures and the Distribution of Copper and Its Associated Minerals

The textures observed in the Lake Superior lavas and in the intercalated sediments can be classified into five geometric types, according to Figure 3. The scale of these geometric patterns obviously varies from some m to mm and even μ (in thin or polished sections).

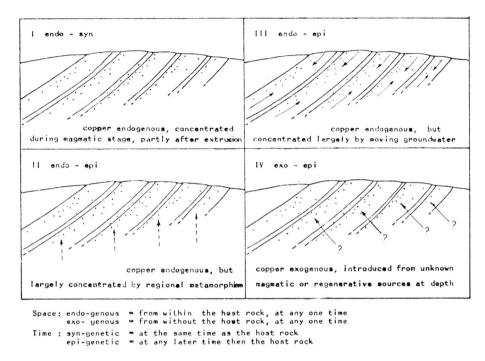

Space : endo-genous = from within the host rock, at any one time
        exo- genous   = from without the host rock, at any one time
Time : syn-genetic   = at the same time as the host rock
        epi-genetic   = at any later time then the host rock

Fig. 2. Four major theories on the genesis of the Lake Superior copper deposits (on lavas)

Essential for a genetic discussion is now the spatial distribution of the copper in these fabric or texture types, and, likewise, the distribution of the minerals associated with the copper. The latter are listed in Table 1 in which the "molecular fraction" is given to facilitate compositional calculations with the formulas. It is obvious that most or all of these "gangue minerals" commonly associated with the native copper are of a hydro-thermal or deuteric origin; they are therefore termed "hydrous minerals".

Figure 4 offers a schematic drawing with the five basic patterns of distribution of these hydrous minerals in the lavas and the interbedded conglomerates.

An investigation into the relationship between textural (fabric) patterns of the lavas as displayed in Figure 3 and the distribution of hydrous (or also dry) minerals resulted in Figures 5 and 6. The copper is perfectly "cobatic" with the hydrous minerals and not with the textures. Some textures are naturally bound to occur frequently with more abundant hydrous minerals, but not always and not all of them. In other words, it is essential to emphasize that the hydrous minerals may or may not be present in the fabric types I to V. One can thus find just as much or as little copper and hydrous minerals in the massive or tight types I and IV as in the amygdaloidal types and the frag-mented (breccia) portions, provided the mineralogical composition is the same.

In exploration for new native copper occurrences the relationship displayed by Fig-ures 5 and 6 is vital. Beyond this useful guide in exploration. an attempt was made in the 1957 statistical field study with many hundred paragenetic observations, to detect

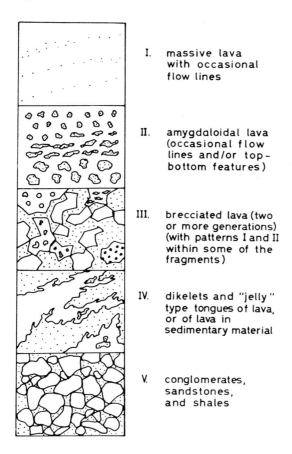

I. massive lava with occasional flow lines

II. amygdaloidal lava (occasional flow lines and/or top-bottom features)

III. brecciated lava (two or more generations) (with patterns I and II within some of the fragments)

IV. dikelets and "jelly" type tongues of lava, or of lava in sedimentary material

V. conglomerates, sandstones, and shales

Fig. 3. The five basic fabric types in the copper bearing Keweenaw lavas and conglomerates

an additional "affinity" of native copper with one or more of the hydrous minerals. The result can be summarized as follows: a relationship is seen only locally, and the "affinity" to one phase, for example with prehnite, appears to be typical for individual flows or portions of flows. A general rule could not be observed.

On the other hand, the paragenetic position of these minerals and of copper appears to be the same in most samples. The quantitative study (counts of sequences) confirmed thus the sequences reported previously by other observers: copper is usually earlier than most of the calcite and prehnite, and some types of chlorite and epidote, but essentially contemporaneous with, or later than, other types of chlorite and epidote.

Silver is (as worked out in a term paper by the late Mr. *Kellner*, in connection with microprobe work of Dr. *Ottemann*) as a rule contemperaneous with the earliest 10 to 30% of copper crystallization. This is best seen in the amygdules or amygdales filled with, or rimmed by copper. Here the silver blebs are found along or close to the rim. Crystallization of the fillings of amygdules and amygdales procedes as a rule from the rim towards the centre. In quite a number of amygdules or amygdales, epidote or other minerals lines the rim, and the inside is either empty or filled partly or completely with calcite.

Table 1

"Dry" Minerals

Andesine – Oligoclase

Augite $\begin{cases} \text{Diopside} \\ \text{Hedenbergite} \end{cases}$ $\begin{aligned} & 1/4\ (2SiO_2 \cdot MgO \cdot CaO) \\ & 1/4\ (2SiO_2 \cdot FeO \cdot CaO) \end{aligned}$

Olivin $\begin{cases} 1/3\ (SiO_2 \cdot 2FeO) \\ 1/3\ (SiO_2 \cdot 2MgO) \end{cases}$

Sphene $1/3\ (SiO_2 \cdot TiO_2 \cdot CaO)$

"Wet" Minerals

Albite	$1/10\ (6SiO_2 \cdot Al_2O_3 \cdot Na_2O)$
Microcline	$1/10\ (6SiO_2 \cdot Al_2O_3 \cdot K_2O)$
Chlorite	$1/5\ (SiO_2 \cdot Al_2O_3 \cdot 2\ (Mg, Fe)\ O \cdot 2H_2O)$
Serpentine	$1/5\ (2SiO_2 \cdot 3\ (Mg, Fe)\ O \cdot 2H_2O)$
Epidote (Ep. = Pistacite)	$1/16\ (6SiO_2 \cdot Fe_2O_3 \cdot 2Al_2O_3 \cdot 4CaO \cdot H_2O)$
Zoisite	$1/16\ (6SiO_2 \cdot 3Al_2O_3 \cdot 4CaO \cdot H_2O)$
Pumpellyite	$1/8\ (4\ (Si, Al)\ O_2 \cdot (Al, Fe, Mg)\ AlO_3 \cdot 2CaO \cdot H_2O)$
one possibility:	$1/16\ (6SiO_2 \cdot Fe_2O_3 \cdot 3Al_2O_3 \cdot 4CaO \cdot 2H_2O)$
Prehnite	$1/7\ (3SiO_2 \cdot Al_2O_3 \cdot 2CaO \cdot H_2O)$
Laumontite	$1/7\ (4SiO_2 \cdot Al_2O_3 \cdot (Ca, Na)\ O_{1-2} \cdot 4H_2O)$
Leucoxene	$1/3\ (SiO_2 \cdot TiO_2 \cdot CaO)$
Calcite	$1\ (CaO \cdot CO_2)$
Quartz	$1\ (SiO_2)$

At this time the discussion of the origin of amygdule-filling cannot be taken up again. The extensive treatment of this topic by *Rosenbusch, Mügge* and others showed clearly that each case has to be considered separately. The excellent detailed study of *Walger* (unpubl. thesis) under the direction of *Tröger* left no doubt either that a late magmatic, deuteric filling is certainly possible and must apply in cases where one is dealing with closed systems. The present paper supports this critical, differentiating approach to the problem with detailed observations on the degree of *congruence* in regard to the most common textural (fabric) types in volcanic rocks.

In short, the criteria of congruence or noncongruence developed for the genesis of the native copper deposits discussed here apply equally well to the genesis of any other mineral phase in the volcanic rocks or, as a matter of fact, in any other rock.

Illustrations of a few examples of features displaying clear congruence, will serve *pars pro toto* to show the principle. Figure 7 is a rather common fluidal type of lava from La Salle 2 Kearsarge Lode. The amygdules connected with the whitish-greenish matrix material (a mixture of albite and quartz) are filled at least partly with this "internal sediment". The schlieren texture of the scoriacious lava is paralleled by the congruent schlieren type distribution of the composition of the amygdule minerals.

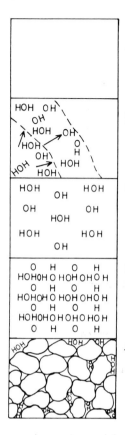

A Lava with practically no "hydrous minerals" present. Perfectly dry "orthodox" basalt.

B Lava with large scale alteration gradients, partly or completely altered and/or even replaced "dry," normal minerals. Thus partly or completely altered A.

C Disseminated hydrous minerals (such as chlorite) in a tight rock mass. No large scale alteration gradient visible (large scale=larger than unit cell of homogeneous rock). (Textures essentially primary or deuteric.)

D Hydrous minerals build up matrix or main rock mass. Textures primary. (Epidotites, pumpellyites, carbonatites, albitites, etc.)

E Hydrous minerals cement partly or completely foreign material of basically different composition (pebbles, sands, shales, etc.).

Fig. 4. The five basis mineralogic rock types in and around the copper bearing Keweenaw lavas

Figure 8a displays a similar congruence of two layers of amygdaloidal lava, each with a different shape of amygdules and a different composition. A later filling is hard or impossible to defend and would require many more assumptions than a comagmatic crystallization. The sample was collected in the Laurium Mine, Upper Peninsula.

Figure 8b corresponds basically to 8a, but the overlaying rock is a fine siltstone which was deposited on top of the scoriacious surface. The sample also displays a neat horizontal zoning (parallel to the flow surface) with dominating simple-shaped calcite amygdules on top and finer, more complicated epidote-filled or epidote-lined amygdules below. The sample was taken at the Centennial No. 3 Mine (label upside down). The same zoning was pictured with a different sample in Figure 9 by *Amstutz* (1963).

Figure 9 corresponds to Figure 2 of the paper on *Space, Time and Symmetry in Zoning* (*Amstutz*, 1963) and shows three lava generations. Each one has a different copper content, as follows (analyzed by *E. Krouzek*):

Generation 1: 0.009% Cu; Generation 2: 0.050% Cu; Generation 3: 0.100% Cu. Generation 3 is hard to sample and surely must contain often up to 3% Cu.

These results correspond to the megascopic observations on about ten similar samples cut from the same block at the City entrance of Houghton: the first generation has no visible copper and is dense, with only about 5–10% of hydrous minerals. Generation two has about 20–40% of chlorite spots and amygdules, with a few points of Cu within

Fig. 5. Comparison of copper
content and fabric types

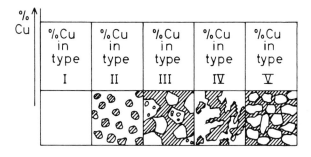

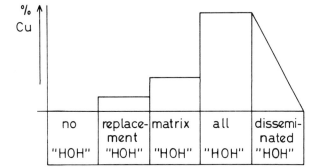

Fig. 6. The relationship between "hydrous" minerals
and copper

the chlorite blebs. Generation one was brecciated by generation two, which is, consequently, the matrix of generation one. The diklet cutting both these earlier generations
has still another mineralogical composition with much calcite and epidote in the large
and numerous amygdales. These also contain native copper blebs with a diameter of

Fig. 7. Common fluidal type of lava from La Salle 2 Kearsarge Lode

Fig. 8 a and b. Congruence of two layers of amygdules with different shapes and different composition
In 8b, the fine siltstone on top fills the open amygdules (label upside down)

1 to 10 mm. Again, this degree of congruence between textures and compositions does
not allow an epigenetic diffusion origin for either the copper or the hydrous minerals.

The conventional idea was based on the apparent analogy between certain properties
of the copper deposits in our lavas and in the nature of hydrothermal vein deposits. The
hydrous minerals in our lavas are in fact basically the same as those of hydrothermal
veins. This, however, may only prove that similar conditions existed. It does not explain
whether these conditions were epigenetically imposed on the rock or whether they are
syngenetic and existed during the formation of the rock, i.e., the deuteric phase.

A hydrothermal mineral assemblage is, in itself, no proof for or against syn- or epigenetic origin. We need further criteria and these are, logically, of a geometric, textural

Fig. 9. Three lava generations, each one with a different copper content (see text). *Points* mark the last lava generation which cuts the two former ones as a diklet

nature. There exists the possibility of a syngenetic-hydrothermal *intramagmatic* origin. This origin has been given the name deuteric. There is no field evidence and no physico-chemical reason why the hydrous or volatile phases of a magma should always separate completely from the parent magma. Logically seen, gradual transitions in the course of the accumulation of volatiles to the formation of hydrothermal fluids must exist. These transitions are present in the form of hydrous portions of magmas, or what we may thus call hydromagmas. Finally, hydromagmatic or deuteric rocks of basaltic composition have for over sixty years been called spilites. The Keweenaw lavas are an outstanding example of basaltic flows with associated spilitic facies.

Consequently, on the basis of physicochemical consideration and the observations on perfect congruence, the conclusion can not be drawn that our copper deposits are epigenetic, i.e., introduced after the formation of the lavas — this conclusion is untenable in the light of the observations outlined above. What other alternative does then exist? It is the possibility of a syngenetic-hydrothermal *intramagmatic* origin. And it should be emphasized again that there is no field and no physicochemical reason why the hydrous or volatile phases of a magma should always depart from the parent magma. They remain at least partly in the silica melt and thus form hydromagmas.

Following are a few pertinent arguments against an epigenetic theory and for a hydro-magmatic nature:

1. A diffusing fluid needs pathways. Such pathways could only be the intergranular film and obviously none of the megafabrics such as the amygdules, the fragmentals, etc because many light rocks have just as much copper as the amygdaloidal ones and the fragmentals. However, gradients suggesting diffusion are virtually absent as discussed above. Also, if we assume diffusion along the intergranular film, we again reach no con-

clusion, because then the fine-grained rocks should be the ones richest in copper because their intergranular surface is greatest. However, we find just as much copper in coarse conglomerates and in coarse-grained amygdules and matrices of fragmentals.

2. Also, epigenetic diffusion is, as know from nature, experiments and from theoretical consideration (*Amstutz*, 1967) always crosscutting and not congruent, if the phenomenon is considered statistically (and not in an insignificant local spot). Congruent mimetic replacement is non-existent statistically, on any scale.

3. If one assumes an introduction of all or most of the hydrous minerals later on, epigenetically, one cannot explain the mineralogic difference between closely neighboring rock portions and fragments. The dyke of later, more hydrous lava with copper cutting earlier, less hydrous lava without copper is a clear illustration of the genesis (Fig. 9). The mineralogic difference between the amygdules of individual fragments in a fragmental lava, and also between these amygdules and the matrix is a striking simple proof of the age of these minerals. Any epigenetic introduction into open spaces or coreplacement would homogenize the composition.

4. Furthermore, any epigenetic idea must assume that some replacement has occurred. However, of 100 phase boundary relationships associated with Cu only about five to ten show possible replacement nature. And even in these few cases replacement may have taken place only along rims and cannot be used *pars pro toto*.

5. What we find are perfectly normal primary crystallization textures with crystalliuation generations instead of partial and selective replacements. Amygdular fillings can often be shown to be closed-system, last-generation crystallizations, not only in basalts the world over, but also in the Keweenaw lavas (e.g. Figure 10). The only difference

Fig. 10. Microphoto of amygdule with fluidal congruent pattern surrounding it. Uppermost "Greenstone Flow", N of Calumet. Inside of amygdules are hydrothermal (deuteric!) minerals

in the hydrous portions of the Keweenaw lavas and other hydrous lavas is that the range of the hydrous or hydrothermal conditions often extends beyond the restricted space of an amygdule. It reaches into the rock so that the major minerals may crystallize in the hydrothermal field. Therefore, there is good justification for calling this process a hydromagmatic crystallization as shown in Figure 11.

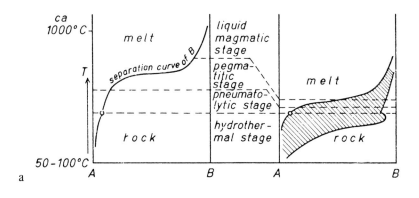

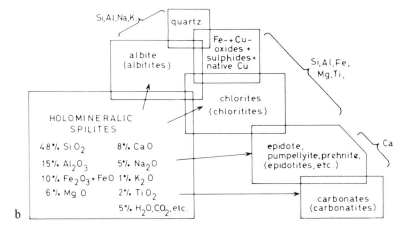

Fig. 11. (a) Temperature-composition diagram for (*left*) normal magmatic (*P. Niggli*) and for hydro-magmatic crystallization (*right*). B presents the non-volatile, A the volatile phases. (b) Differentiation scheme of spilitic rocks (*Amstutz*, 1974, Fig. 7)

Occasionally some of the minerals had formed before the temperature had dropped into the hydrothermal field and/or before an appreciable amount of volatile material had accumulated and lowered the crystallization temperature. In these cases we observe deuteric or auto-hydrothermal alterations. These alterations can again not be taken *pars pro toto* as proof of epigenetically introduced hydrothermal fluids.

6. Any scientific explanation is more logical if less assumptions have to be made. This is the case for a syngenetic origin of the Keweenaw copper deposits and of many other ore deposits and rocks.

7. Instead of replacement or epigenetic fracture filling we find thus a gradation from normal "dry" or "semi-dry" rocks, i.e., of normal basalts, to completely hydrous or hydrothermal material such as epidotites, albitites, pumpellyitites, extrusive carbonatites etc, and from these a gradual change-over to the adjacent sediments by way of jelly-type textures which illustrate the state of the material at the time of formation (see Fig. 7). These jelly, chert, jasperoid, or jaspilitic rocks are very common in similar rocks in many parts of the world (cf. the spilite volume, *Amstutz*, 1973).

8. Hydrothermal alterations as a rule preserve or reduce the original grain size of the altered rocks. For example, the silifications, kaolinizations, saussuritizations etc, all usually reduce the original average grain size. In our case the opposite is observed: the more "altered", the bigger the grain size, and in may cases the latest mineral assemblages even reach pegmatitic grain sizes, for example in the excellent and numerous museum specimens of calcite, quartz, zeolites, epidote, native copper etc from this area.

The term "alteration" is thus inadequate. We do not have alterations except in the local, marginal and restricted cases of mineralogic type B. Instead we find no other way of explaining the paragenesis of constantly increasing grain size than by a syngenetic intramagmatic crystallization from a hydrothermal rest fluid. The copper is then just one of the phases which crystallized from this rest fluid which happens to have occurred in quantities which could lead to the formation of commercial ore deposits. In addition to these, there are many more criteria and observation which can be mentioned and explained. All of them speak against an epigenetic origin and strengthen the syngenetic comagmatic explanation.

As I pointed out, it was recognized a long time ago that the amygdular crystallization in the basalts the world over is often hydrothermal. We obtain, in normal basalts, which may be called "dry", an accumulation of volatiles, such as water, CO_2, etc. If there is a slight surplus of a major element available from the main rock crystallization a hydrothermal crystallization of quartz, epidote, prehnite, zeolites, calcite etc in these amygdules may take place. As *Cornwall* (1951a, b, 1956) has shown, many of these amygdules also contain traces of native copper. This is in normal basalts.

During a study of spilitic rocks extending from 1945 to 1974 it became evident that apparently portions of the Keweenaw basaltic lavas and other spilitic rock provinces (very many are listed in the volume on *Spilites and Spilitic Rocks, Amstutz*, 1974) crystallized under hydrous conditions. Table 2 shows what types of mineral form under such hydrous conditions. On the left side is the normal "dry" mineral assemblage. On the right side is a list of the crystallization under hydrous conditions.

The main differences are these:

Ca can no longer enter the plagioclase structure. It remains in solution until epidote, or calcite are formed. Furthermore, the anhydrous ferrosilicates are unstable and do not form or, if formed, decompose. Chlorite and/or iron oxide are usually formed instead. The mineralogy becomes thus more complex. The large amount of water and other volatiles present during the crystallization leads, in addition, to another property of the crystallizing magma, a property which is most important for the understanding of the geometry and mineralogy of Michigan copper rocks, as well as any other similar rock types. It leads to the high mobility of the rest fluids during crystallization. This mobility is much higher in such a hydrous system than in a "dry" system, and, consequently, the rest fluid has a higher tendency for separating from the obviously slow crystallizing system.

Table 2. Hydrous or spilitic crystallization of a basalt magma takes place according to the following scheme (a spilite is but a basalt which formed under hydrous conditions)

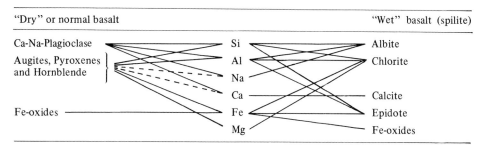

This is why we find so many mono- or oligo-mineralic rocks in the Keweenaw area and in spilite provinces in general. Figure 11b displays this tendency and the resulting occurrence of special rocks in congruent spaces, such as epidotites, pumpellyitites, chloritites, etc occurring for example as breccia fragments in a totally different matrix.

It is thus perfectly normal that the tendency to form more or less monomineralic differentiation products is strong. One often gets the impression of a pegmatitic or micro-pegmatitic nature. The texture of most of these hydrous magmatic rocks is basically the same as those of hydrothermal veins: rapidly changing from mono- to poly- and back to mono-mineralic.

There is thus a logical transition in nature between "dry melts" – the majority of the basalts – and "wet melts". In the case of wet melts, the melts which are so to say still "pregnant with hydrothermal solutions", we could speak about hydromagmas because, after all, we see that the bulk of the rock was influenced by hydrous, i.e., hydrothermal conditions. On the classical separation diagram of *P. Niggli* the hydrous crystallization of a magma looks thus like that shown in Figure 11a. The pegmatitic field is narrow or almost absent, and the bulk crystallization takes place in the hydrothermal range (cf. also *Amstutz*, 1968a, b).

Based on my investigations since 1945 on hydromagmatic rocks and ore deposits and on field work around Lake Superior, I have reached thus the following conclusions with regard to the Lake Superior coppers in the lavas.

The Lake Superior copper deposits are best explained by a uniform single origin. The field evidence and the microscopic, paragenetic statistical and geochemical analyses lead to the conclusion that the copper is a normal co-magmatic syngenetic constituent accumulated in hydromagmatic phases of the Lake Superior basalt magmas.

First, the copper was carried up in and with the lavas and stayed in the hydromagmatic and hydrothermal portions of the lavas or escaped into the sediments and fractures. After the lavas ceased to extrude, the volatile fractions still continued to leak from the same magma chambers as hydrothermal fluids, most of which reached the surface. (This conclusion is essentially identical with my abstract of a lecture given in 1958a.)

This uniform single cycle of ore genesis which formed the Lake Superior copper province is illustrated in Figure 12 and compared with the pan-epigenetic theory. The arrows in the lower drawing represent a consecutive outpouring of lavas ("flood-basalts")

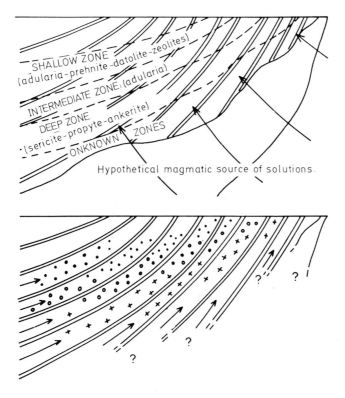

Fig. 12. Uniform single cycle of ore genesis which formed the Lake Superior copper province. (For explanation, see text)

with hydromagmatic portions. It is identical with drawing I in Figure 2 and represents a syn-endo-origin for the copper.

   Naturally, processes of load or burial metamorphism may have affected the rocks of the area. The paper by *Smith* (1974) has its justification for the normal basaltic portions of the Keweenaw province. It is unable to explain, however, the congruent features in the hydromagmatic portions here described, and which contain the major deposits of native copper in the lavas of the Lake Superior Province.

3. Discussions and Conclusions

The copper-bearing lavas of the Lake Superior copper district extend almost all around Lake Superior. These lavas are marked in solid black on Figure 1.

   The stratigraphy, petrology, mineralogy and chemistry of these lavas of Middle Keweenaw age is well known and little needs to be added to the observations (cf. *Cornwall*, 1951a, b; *Niggli*, 1952)

Marked discrepancies exist only with regard to the genetic interpretation of the observations. Figure 2 is a schematic summary of the four major genetic interpretations of the Michigan copper deposits in lavas and the associated hydrothermal or deuteric minerals, including the native Cu.

The *first theory* proposes that the copper was brought up in and with the lavas and solidified during the cooling of the lavas, essentially in the form observed today. Only minor rearrangements are assumed by this theory, as having taken place through epigenetic processes, as for example burial metamorphism.

The *second theory* assumes also that the copper was always contained within the lavas, but defends that the concentration was produced by regional metamorphism.

The *third theory* proposes that the copper was always contained in the lavas, but that the concentration was produced epigenetically by circulating ground water.

The *fourth theory* proposes an entirely foreign and epigenetic nature of the copper. It assumes the introduction of the copper from unknown magmatic or regenerative sources at depth.

Excellent mineralogic and textural evidence was offered for the first theory by *Cornwall* in the two papers published in 1951. In 1958 statistical evidence on mineral distribution was offered by *Amstutz* which also supports this first mode of formation. The present paper adds the geometric and paragenetic observations and some new (and probably more consistent) steps of interpretation, which strengthen the first theory.

It is recognized in general that the Middle Keweenaw rocks can be classified roughly into five texture or fabric groups shown in Figure 3.

In studying lavas or any other type of rock we may focus our attention on the composition *or* on the texture. In the present investigation the compositional or mineralogical gradients ΔM were compared systematically with textural or *geometric* gradients ΔG.

Figure 4 lists seven features, F_1 to F_7, which are probably recognized by anyone familiar with lavas as representing typical primary features. It is important to remember that they are truly primary geometric patterns and could not be produced by any secondary process. Still more such patterns could be listed and they yield the same result. The space occupied by these spatial patterns may be called ΔS_1, ΔS_2, and so on.

The compositional, i.e., the *m*ineralogical gradients or slopes within these *s*paces are marked as ΔM_1, ΔM_2, etc., whereas the textural or *geometric gradients are marked as ΔG_1, ΔG_2, and so on.

Basically or theoretically one may have many different types of relationship between the geometric and the compositional gradients. One may find a completely haphazard or random relationship, or a certain degree of coincidence or interdependence. The distribution of chlorite, epidote and native copper, for example, would not have to coincide with the space of a primary pattern ΔS_x.

The result of over 15.000 statistically recorded observations on this interrelationship between compositional and geometric gradients revealed, however, a very high degree of coincidence. Since this coincidence is a geometric or space property of these rocks, we may apply the mathematical term congruence or congruency (Fig. 13).

Congruence is defined as the property to be "geometrically superposable so as to be coincident throughout" (*Webster*).

Observations				Interpretations
Primary lava features (F)	Space of features (S)	Mineralogic composition (M)	Texture or "geometry" (G)	Primary T–P slopes or gradients
F_1 flow lines and layers in lava	$\Delta S_1 \cong$	$\Delta M_1 \cong$	ΔG_1	
F_2 lava dikelets in lava (2 ore more genera-tions)	$\Delta S_2 \cong$	$\Delta M_2 \cong$	ΔG_2	
F_3 breccia pieces and matrices (2 or more ge-nerations)	$\Delta S_3 \cong$	$\Delta M_3 \cong$	ΔG_3	Coincidence or con-gruence of space changes (i. e. geome-tric changes, i. e. of size, shape, texture, structure, symme-try, etc.) and of com-positional, i. e. mi-neralogical changes $$\frac{(\Delta M, \Delta G)}{\Delta s} = \frac{f(\Delta p, \Delta t)}{\Delta s}$$
F_4 amygdales and amygdules (sizes and shapes)	$\Delta S_4 \cong$	$\Delta M_4 \cong$	ΔG_4	
F_5 micro-fabric of lava in general	$\Delta S_5 \cong$	$\Delta M_5 \cong$	ΔG_5	
F_6 micro-fabric around amyg-dales and amygdules	$\Delta S_6 \cong$	$\Delta M_6 \cong$	ΔG_6	
F_7 gravity or geo-petal features at all scales	$\Delta S_7 \cong$	$\Delta M_7 \cong$	ΔG_7	

Fig. 13. Congruence derivations between mineralogic distribution spaces ΔM and primary textural spaces ΔG. (From *Amstutz*, 1966b)

A high degree of congruence was thus observed between the compositional gradients and the gradients of the geometric features which are considered to be primary. This congruence has been illustrated with various figures.

It is inconceivable how this high degree of congruence between such clearly magmatic or primary features as flow breccias and lava diklets and their composition could be caused, or left undisturbed, by any secondary process.

Even the composition of the sedimentary features in the interfragmental spaces coin-cides with that of the amygdales which are connected with them by channels or frac-tures. The amygdales which are closed off show a different composition. If there had been epigenetic soaking through the amygdaloidal layers, the existing congruence just described would most certainly have been destroyed and the observed perfect geometric coincidence between composition and primary geometric features obliterated.

On the whole even the shape of the amygdales changes with the composition as already mentioned. Epigenetic gradients can hardly be made responsible for this high degree of congruence, because we would end up with the meaningless conclusion that the various minerals had a preference for certain specific shapes of amygdales.

The perfect congruence does not apply only to the megascopic scale. Unter the microscope, primary microtextures are seen to change congruently with the mineralogic composition and this congruence is more dominating on the whole than, and certainly predates, the cross-cutting features of burial metamorphism or other later changes. Also, there is very often a change of the microtexture towards the amygdules. It is possible to recognize a compositional relationship between the amygdules and the microscopic zone immediately surrounding it, as also studied in detail by *Walger* (unpubl. thesis) in lavas of the Pfalz, SW Germany.

Thus, the only logical parameters responsible for the close congruence relationship between composition and shape of the amygdales and all the other primary geometric features mentioned appear to be primary composition-temperature-pressure slopes.

Some previous papers defend the existence of regional non-congruent zoning lines produced by one of the epigenetic mineralization processes pictured in Figure 2. According to the evidence presented today, these zoning lines are connecting patches of similar mineralogical composition which are, however, unrelated in time. Moreover they are also unconnected in space, since their occurrence is patchy, which is natural for the distribution of primary volatile-rich fractions in a lava.

In conclusion, we are unable to visualize any epigenetic, secondary temperature-pressure-slopes and compositional gradients which could explain the majority of relationships just described. It appears therefore logical to conclude that the copper is of primary deuteric origin, as also the majority of the other deuteric minerals.

Consequently, the conclusion reached in 1958 (A*mstutz,* 1958a) was confirmed by the additional work done on these and similar rocks of an essentially spilitic composition (*Amstutz*, 1974). In the light of syngenetic ore-forming processes which have become known in the past 20 years the exhalative sedimentary hypothesis proposed for the White Pine copper deposit may, on the other hand, no longer be valid. The copper content of the None-Such Shale is probably of a diagenetic age and its source may well be exhalative and/or purely erosional in origin.

The extensive work done on the geometric distribution of deuteric minerals (including native Cu) in this area also cast great doubt on the recent theories on Ocean water circulation in oceanic basalt piles as a means of metal enriching process. Such a simplistic mechanism ignores the many volcanic petrographic textural features, some of which were described in this paper and also those pictured and analized by *Cornwall* (1951a,b). It is much more probable that the heat-flow-deficiency over the oceanic ridges are due to the heat loss through expansion and cooling and, specifically, through the expansion of the gases in the amygdales and other voids (fissures, breccia spaces, surfaces). Moreover, the geochemical balance and the alteration geometry rule out a sweeping alteration and migration process. But this wider topic will be discussed in a forthcoming paper.

References

Amstutz, G.C.: The genesis of the Lake Superior copper deposits. Inst. Lake Superior Copper Geol. Program, p. 25 (abstract) (1958a)

Amstutz, G.C.: Spilitic rocks and mineral deposits. Bull. Missouri School of Mines., Tech. Ser. **96**, 11 p. (1958b)

Amstutz, G.C.: Syngenetic zoning in ore deposits. Proc. Geol. Assoc. Canada **11**, 95–113 (Fig. 7 on zoning in Michigan lavas) (1959a)

Amstutz, G.C.: Syngenese und Epigenese in Petrographie und Lagerstättenkunde. Schweiz. Mineral. Petrogr. Mitt. **39**, 1–84 (1959b)

Amstutz, G.C.: Spilites and spilitic rocks. In: The Poldervaart Treatise on Rocks of Basaltic Composition. Hess, H.H., Poldervaart, A. (eds.). New York: Wiley Interscience 1968, pp. 737–753

Amstutz, G.C.: Space, time and symmetry in zoning. Symp. Problems of Postmagmatic Ore Deposition (Prague), Vol. I, 1963, pp. 33–37

Amstutz, G.C.: Some comments on the genesis of ores. Symp. Problems of Postmagmatic Ore Deposition (Prague), Vol. II, 1965, 147–150.

Amstutz, G.C.: Die Kupfererzlagerstätten in den Laven vom Oberen See: Deutung neuer Beobachtungen. Clausthal: GDMB, 1966a, pp. 67–74

Amstutz, G.C.: La symmétrie comme critère génétique en géochimie et en gîtologie. Bull. Suisse Mineral. Petrol. **46**, 329–335 (1966b)

Amstutz, G.C. (ed.): Spilites and Spilitic Rocks. IUGS Ser. A, No. 4. Berlin–Heidelberg–New York: Springer 1974, 482 p.

Amstutz, G.C.: The logic of relations in ore genesis. Proc. 15th Inter-University Geol. Congress, 1967, Leicester, England, pp. 13–26

Broderick, T.M.: The origin of the Michigan copper deposits. Econ. Geol. **47**, 215–220 (1952)

Broderick, T.M.: Copper deposits of the Lake Superior region. Econ. Geol. **51**, 285–287 (1956)

Cornwall, H.R.: Differentiation in magmas of the Keweenawan Series. J. Geol. **59**, 151–172 (1951a)

Cornwall, H.R.: Differentiation in lavas of the Keweenawan Series and the origin of the copper deposits of Michigan. Bull. Geol. Soc. Am. **62**, 159–202 (1951b)

Cornwall, H.R.: A summary on ideas of the origin of native copper deposits. Econ. Geol. **51**, 615–631 (1956)

Jenney, C.P.: The Coppermine River Area, Northwest Territories, Canada. Proc. Geol. Assoc. Can. **6**, 11–26 (1954)

Leith, C.K., Lund, R.J., Leith, A.: Pre-Cambrian rocks of the Lake Superior region. U.S. geol. Survey Prof. Paper **184**, 34 p.

Lindgren, W.: Mineral Deposits. 4th ed. New York: McGraw-Hill 1933, 930 p.

Moffit, F.H., Capps, S.R.: Geology and mineral resources of the Nizira district, Alaska. USGS Bull. **448**, I–III (Amygdaloidal copper, 79–83) (1911)

Niggli, P.: The chemistry of the Keweenawan lavas. J. Sci., Bowen Vol. 381–412 (1952)

Pumpelly, R.: Copper District. Geol. Survey of Michigan; Upper Peninsula (1869–1873), I, Part 2, New York.

Pumpelly, R.: The paragenesis and derivation of copper and its associates on Lake Superior. Am. J. Sci., 3rd Ser. **2**, 188–198, 243–258, 347–355 (1874)

Rickard, T.A. (ed.): Ore deposits. Discussion in: Ore Deposits, A Discussion. Eng. Mining J., New York, May 1903, 1–17; 56–60 (1903)

Smith, R.E.: The production of spilitic lithologies by bural metamorphism of flood basalts from the Canadian Keweenawan, Lake Superior. In: Spilites and Spilitic Rocks. Amstutz, G.C. (ed). Berlin–Heidelberg–New York: Springer 1974, pp. 403–416

Walger, E.: Über die postmagmatischen Umwandlungserscheinungen an den Melaphyren des Pfälzer Berglandes. Unpubl. thesis, Freiburg (Germany), 98 p.

White, W.S.: Regional Structural Setting of the Michigan Native Copper District. Michigan: College of Mining and Technology Press, Institute on Lake Superior Geology, 1956, pp. 3–16

Strata-Bound Scheelite Deposits
in the Precambrian Basement of San Luis (Argentina)

M.K. DE BRODTKORB and A. BRODTKORB, Buenos Aires

With 4 Figures

Contents

Summary

It is proposed that the studied deposits of Central Argentina lie in tactized rocks produced by regional metamorphism and the wolfram is singenetic with the marine sediments. Therefore, both sediments and singenetic mineralization have suffered regional metamorphism. The scheelite belts of the San Luis and Córdoba Provinces are of late Precambrian age (580–620 my) and the geotectonic setting is characterized by their location in a metamorphic mobile belt marginal to the Brazilian shield.

1. Introduction

Scheelite deposits occurring in metamorphic rocks have not, until recently, been extensively studied. The difficulties have been numerous, for example, the Precambrian and Eopaleozoic rocks involved, the doubts about the origin of tungsten, its paragenetic location and its distribution in the geochemical cycle.

The different views presented in the last decade on the new genetic interpretations of scheelite deposits in Precambrian and Eopaleozoic metamorphic rocks (*Maucher*, 1965, 1972. 1974; *Maucher* and *Höll*, 1968; *Höll*, 1971; *Höll* et al., 1972; etc) led the authors to review different deposits in the San Luis and Córdoba Provinces of Argentina (*Brodtkorb* and *Brodtkorb*, 1975).

The deposits of Sierrita de Yulto, the El Morro mining district, were studied, and Los Reventones, Pampa de Tamboreo and La Florida were shortly examined. Data from the references were gathered from other ores from San Luis and Córdoba, such as Sierra de la Estanzuela, Cerro Aspero, Pampa de Olaen and Sierra de Altautina (*Angelelli,* 1950; *Angelelli* et al., 1970; see Fig. 1).

All deposits are included in the crystalline basement rocks of the San Luis and Córdoba Provinces, where the main rock types consist of schists and gneisses, with NNE-SSW to NNW-SSE strikes and easterly dips.

Most scheelite deposits occur within carbonatic metamorphic belts, and the latter, associated with amphibolites, lie conformably within the basement. Quartz veins with scanty scheelite and/or wolframite cut the regional or local structures. In the literature the so-called "tactites" are considered to be of metasomatic origin, and quartz veins with scheelite and/or wolframite, of hypothermal origin. In the present study, the strata-bound mineralization is proposed to be of singenetic origin, and the tungsten content of pegmatites and quartz veins of later formation due to remobilization.

The tungsten deposits of San Luis and Córdoba Provinces have been the object of intensive exploitation during the two World Wars. In 1912 the quartz veins were already being exploited and the carbonatic beds since 1918.

Actually, only a modest exploitation is being undertaken in shaft 5 of the El Morro mining district. Here (Fig. 1: 1, 2 and 3) from 1952 until 1962, 300,000 t of mineral were extracted; 200,000 t of these were selected manually and after treatment in table classifiers produced 1000 t of a 65% concentrate in WO_3.

Since 1962 mining activity has been notably reduced.

2. Geological Setting

The Sierras Pampeanas represent a morphostructural province of Argentina. It consists of block mountains of crystalline rocks, surrounded by flat, wide depressions filled up mostly by Quaternary sediments. This province comprises a considerable part of central and north-western Argentina.

According to *Gordillo* and *Lencinas* (1971) the Sierras of Córdoba and San Luis are made up of crystalline basement with an extremely complex geologic history. The first event of this history was the deposition, within a geosincline, of a sequence of pelitic strata, with local facies of psamitic, marly, or limy character, deposited in one or several basins. The tectonic-sedimentary character and its sequence have been masked by later deformation, magmatic intrusions, and metamorphism.

Post-metamorphic granitic batholitic intrusions, with aplitic and pegmatitic byproducts, comprise 25 to 30% of the basement area.

The geomorphologic picture was completed with the uplift of the Sierras Pampeanas during pre-Permian or neo-Carboniferous times and this brought as a sequel an important erosional degradation. A peneplain formed over the postmetamorphic granites, and over it a sedimentary Paleozoic couverture was deposited. The poor record of these rocks indicates that during this era the area was a positive geomorphic element.

The Cenozoic tectonic reactivation accentuated existing paleostructures and formed new ones. The western areas were affected by trachyandesitic vulcanism associated with these orogenies.

The age problem of the crystalline basement of the Sierras Pampeanas has not been solved. To the numerous regional, stratigraphic, petrologic, etc contributions, geochro-

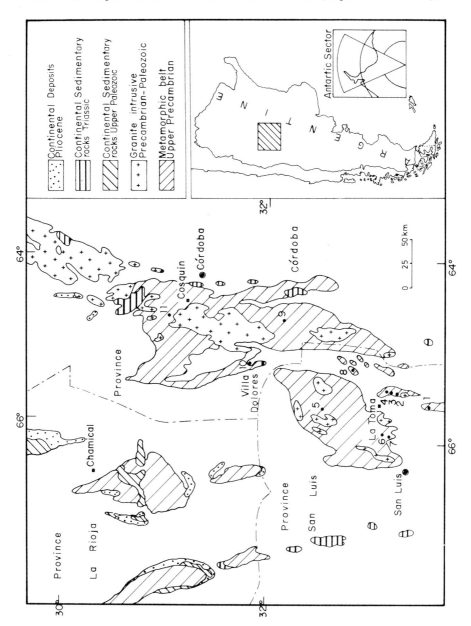

Fig. 1. Regional geology of the area

Deposits studied	Deposits overviewed	Other deposits
1: Sierrita de Yulto	*5:* Los Reventones	*8:* Sierra de la Estanzuela
2: Mina Loma Blanca	*6:* Pampa de Tamboreo	*9:* Cerro Aspero
3: Mina Morro No. 1	*7:* La Florida	*10:* Sierra de Altautina
4: San Antonio		*11:* Pampa de Olaen

nologic ones have been added in recent years. According to *Linares* and *Latorre* (1975) and *Linares* and *Turner* (1975), the metamorphic rocks correspond to the late Precambrian (580–620 my), and the granitic rocks belong to diverse cycles, the first of which would be located in the Precambrian-Cambrian boundary (570–600 my), with three other peaks of activity during the Paleozoic.

The differentiation in the field of the amphibolites has not been completed but both para- and ortho-amphibolites have been recognized. Future petrologic and geochronologic studies will help to solve this problem.

The metamorphic process of the Sierras Pampeanas can be correlated to the Brazilian Cycle of Almeida, but *Cingolani* and *Varela* (1975) have dated rocks within the Sierras Pampeanas of 900–1300 my that could represent remnants correlated with the Uruaçuano Cycle of Brazil.

The orogenic belts of the Brazilian Cycle are generally peripheric with respect to the shield.

3. Regional Geology in the Area of the Deposits

In the area of interest the crystalline basement is characterized by wide concordant belts of quartz-oligoclase-biotite schists and gneisses with intercalations of carbonatic beds up to 1.50 m thick, and different types of amphibolic rocks. The strikes are predominantly N-S and the dip is 40 to 60° east.

Clayton (1971) places these rocks in the amphibole facies of regional metamorphism, based on the presence of oligoclase in the schists and gneisses, and on the hornblende-plagioclase (andesine) association developed in the amphibolites.

Near the scheelite-bearing district of El Morro and Sierrita del Yulto there is a relatively large intrusive body, the San José del Morro granite; pegmatite and quartz veins are profusely distributed throughout the area.

3.1 Deposits Studied

The type-lithology of some characteristic deposits is described below.

3.1.1 El Morro Mining District

a) **El Morro Mine.** The crystalline rocks of the region are characterized by quartz-oligoclase-biotite schists interbedded with two carbonatic scheelite-bearing layers (see Fig. 2). Belts of amphibolites and biotite schists are found between the metamorphic limestone. The quartz-oligoclase-biotite schists on either side of the carbonatic beds show a maximum thickness of 10 cm of epidote at the contact with the latter beds. The carbonatic layers ("tactite") are somewhat lenticular and are 0.2 to 0.8 m thick. These are composed of calcite, tremolite, epidote, phlogopite, garnet, quartz, fluorite, scheelite, pyrrhotite, pyrite, sphalerite, chalcopyrite, magnetite, molybdenite and gold.

The presence of discordant quartz veins and pegmatites produced local enrichments in scheelite of some economic importance.

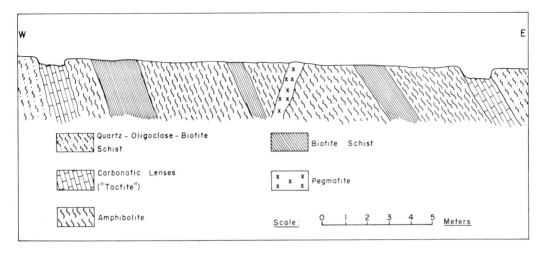

Fig. 2. Schematic sketch of the geology of the Morro 1 deposit

b) Loma Blanca Deposit. This is characterized by four carbonatic mineralized beds, 0.3–0.4 m thick, showing longitudinal and cross-sectional variation in composition. Gray, markedly foliated calci-chloritic marbles are the most conspicuous rocks. These lenses of primary depositional origin wedge out between quartz-oligoclase-biotite schists and amphibolic-type rocks. The country rocks consist of schists and gneisses of quartz-oligoclase-biotite composition (Fig. 3). The following minerals are present: calcite, antinolite, tremolite, epidote, phlogopite, fluorite, apatite, scheelite, beryl, pyrite, sphalerite, chalcopyrite and gold.

c) San Antonio Prospect. Mineralizations north of the El Morro Mining District, San Antonio and Los Halcones, belong to a relatively unknown belt several km long with a N-S trend. The scheelite mineralization is again included in the tactized carbonatic rock and is concordant with the schists and gneisses characteristic of this area.

3.1.2 Deposits of Sierrita del Yulto

Scheelite mineralization is found within rocks with a greater degree of migmatization. The lithologic sequence is similar to that of other deposits: quartz-oligoclase-biotite schists and gneisses, thin epidotized beds, variably tactized mineralized limestones, amphibolites, and again quartz-oligoclase-biotite schists and gneisses. In this area are found a great number of pegmatite-like bodies associated to the carbonatic beds, injected preferably at the contact between these beds and the schists and amphibolites.

3.2 Overview of Deposits

The following deposits were briefly visited in the field in order to have a more complete view of the scheelite mineralization of the San Luis Province.

In Los Reventones, a tactized carbonatic rock lies conformably to the biotitic schists of the region. The tactite consists of calcite, epidote, garnet and scheelite.

Fig. 3. Loma Blanca deposit, San Luis Province. Mine working. Illustrated in this picture is a mineralized lens (*t*) in contact with the quartz-oligoclase-biotite schist (*s*)

In Pampa de Tamboreo several mineralized and exploited trenches were observed, parallel to the biotitic schist and gneisses.

The La Florida Mine has a complex mineralization, partly formed by scheelitic impregnations in biotitic schists.

3.3 Other Deposits

From the references we can refer to the Pampa de Olaen deposits, Córdoba Province. The mineralized bodies are located between limestone beds, and schists and gneisses. They contain scheelite, epidote, garnet, calcite, quartz, fluorite, pyrite, chalcopyrite and sphalerite.

The Sierra de Altautina prospects conform a 22-km long belt to the N and NE of Villa Dolores and consist of micaschists, amphibolites and limestone bodies parallel to the schists. The mineralization is associated either with the limestone or with the micaschists and is composed of calcite, scheelite, garnet, actinolite, fluorite, pyrite and chalcopyrite.

In Sierra de la Estanzuela, the scheelite mineralized belts are found in dolomitic limestones interbedded with schists and associated with epidote and garnet.

Finally the wolfram district of Cerro Aspero with an area of 5 by 8 km, has diverse types of deposit. They correspond partly to beds concordant with regional schists with the following mineralogy: scheelite, pyrite, chalcopyrite, sphalerite, molybdenite, bismuthinite, fluorite and beryl.

Fig. 4 a and b. El Morro deposit, San Luis Province. (a) Normal light; (b) UV light. This sample's cut is approximately perpendicular to the original stratification. It consists of a very rich scheelite mineralization that is located in the contact of the tactized limestone with the quartz-oligoclase-biotite schist. Metamorphic mobilization has produced larger scheelite porphyroblasts

4. Geologic Evidence of the Scheelitic Belt of the Province of San Luis

The following conclusions can be drawn from the study of the regional distribution of the deposits and of the country rocks.

    1. The most important mineralization is found in tactized limestones that are concordant with the regional metamorphic rocks. Occasionally the tactite is banded, reflecting a relatively well-preserved primary sedimentary texture (Fig. 4).

2. The mineralization lies within lenticular bodies up to 100 m in length, 2 m in thickness and 100 m in depth (the extent of the mineralization in the direction of the dip is unknown). There are several mineralized beds.

3. The mineralogic composition of the tactized rock is very uniform: calcite, epidote, actinolite, phlogopite, garnet, associated with scheelite, fluorite, beryl, pyrite, pyrrhotite, phalerite, chalcopyrite and magnetite.

4. Biotitic schists locally present an economically important scheelitic impregnation.

5. Amphibolites are related to the carbonatic rocks. Some preliminary studies indicate an igneous origin for them.

6. There are no intrusive bodies in the vicinity of the deposits; thus the "tactites" are not contact metamorphic rocks, but a type of regional skarn (*Clayton*, 1971).

7. Pegmatites may or may not be associated with the deposits. If present, they are small and generally lack wolfram. The quartz veins occasionally bear scheelite and/or wolframite, specially when in contact with the tactized limestones.

8. Deposits with paragenesis of wolfram and Au, Bi and Mo as minor elements are known at a regional scale.

9. There are several beds within each deposit and the deposits themselves are repeated towards the west, separated by tens of km.

5. Genetic Interpretation

It is impossible to discuss within this commemorative volume all Professor *Maucher* and his followers' ideas on the strata-bound time-bound scheelite deposits. We will refer only to those that are related to the present study.

In 1965, *Maucher* published his first results on deposits of the Sb, W, Hg paragenesis and its relationship to magmatism and geotectonics. In more recent studies (*Cardoso*, 1973) characteristic minor elements (Mo, Bi, Au, Ag, Sn and Be) of the same paragenesis were determined.

According to these new ideas, the mineralization of these deposits is contemporaneous to the formation of the country rocks and is localized in a specific facies of a certain lithologic sequence. The scheelite-bearing metamorphic rocks are very similar to those described as tactites, but they do not have a contact metamorphic origin, as they lie far from intrusive bodies. A volcanic origin is postulated for the tungsten, derived from submarine volcanic processes that took place during the deposition of the marine sediments.

If we compare the geologic evidence of our deposits with the numerous regional, petrologic and geochemical world-wide data (*Angermeier*, 1964; *Höll*, 1966; *Höll* and *Maucher*, 1967; *So*, 1968; *Urban*, 1971; *Burchard*, 1972; *Lahusen*, 1972; *Skaarup*, 1974) it is clear that our deposits correspond to the strata-bound mineralizations of the Sb-W(Mo)-Hg paragenetic association, and that it is singenetic with sedimentation. Like most of the deposits found world-wid, they lie within limestones, concordant with other regional metamorphic rocks. The bodies are generally lenticular and the repetition of mineralized beds is common. The mineralization in the carbonatic rock is very uniform, both with respect to the silicate minerals and to the singenetically originated cations.

References

Angelelli, V.: Recursos minerales de la República Argentina. I. Yacimientos metalíferos. Rev. Inst. Nac. Inv. Cs. Nat. Serie Geol. 2, Bs. As. Argentina (1950)

Angelelli, V., Fernandez Lima, J.C., Herrera, A., Aristarain, L.: Descripción del mapa metalogenético de la Republica Argentina. Dir. Nac. Geol. Min. Bs. As. Argentina. Anales XV (1970)

Angermeier, H.O.: Die Antimonit-Scheelit-Lagerstätten des Gerrei (Südostsardinien, Italien) und ihr geologischer Rahmen. Dissert. Univ. München (1964)

Brodtkorb, M.K. de, Brodtkorb, A.: Especulaciones sobre un origen singenético-sedimentario de la scheelita presente en las metamorfitas del NE de la Provincia de San Luis. VI Cong. Geol. Argentino 2 (1975)

Burchard, U.: Geologische Untersuchungen zur Genese der Scheelit-Lagerstätte King Island, Tasmanien. Dissert. Univ. München (1972)

Cardoso, J.: Zeit- und Schichtgebundenheit des Berylliums im Paläozoikum der Ostalpen und dessen Beziehung zu Wolfram (Scheelit). Dissert. Univ. München (1973)

Cingolani, C.A., Varela, R.: Geocronología Rb-Sr de rocas igneas y metamórficas de las Sierras Chica y Grande de Córdoba, Rep. Argentina. II. Cong. Iberoamericano de Geol. Ec. 1, 1–35 (1975)

Clayton, R.: Estudio petrológico de algunos yacimientos de tungsteno en la Sierra del Morro y la Sierrita del Yulto, prov. de San Luis, Argentina. Contr. Cient. y Tecn. Univ. del Estado, No. 2, Chile (1971)

Gordillo, C.E., Lencinas, L.A.: Sierras Pampeanas de Córdoba y San Luis. In: Geología Regional Argentina. Leanza, A. (ed.). Acad. Nac. Cs. en Córdoba, 1972, pp. 1–39

Höll, R.: Genese und Altersstellung von Vorkommen der Sb-W-Hg Formation in der Türkei und auf Chios, Griechenland. Bayer. Akad. Wissen. math. nat. Kl. Abh. n. F. 127 (1966)

Höll, R.: Scheelitvorkommen in Österreich. Erzmetall 24, 273–282 (1971)

Höll, R., Maucher, A.: Genese und Alter der Scheelit-Magnesit-Lagerstätte Tux. Sitz. Bayer. Akad. Wiss. math. nat. Kl. pp. 1–11 (1967)

Höll, R., Maucher, A., Westenberger, H.: Synsedimentary-diagenetic ore fabrics in the strata- and timebound scheelite deposits of Kleinarltal und Felbertal in the Eastern Alps. Mineralium Depositum 7 (2), 217–226 (1972)

Lahusen, L.: Schicht- und zeitgebundene Antimonit-Scheelit-Vorkommen und Zinnober-Vererzungen in Kärnten und Osttirol, Österreich. Mineralium Deposita 7 (1), 31–60 (1972)

Linares, E., Latorre, C.O.: Edades radimetricas obtenidas para las rocas graníticas y metamórficas de las Sierras Pampeanas de las provincias de Córdoba, San Luis y Santiago del Estero y su significación geológica. 6th Cong. Geol. Arg. 1 (1975)

Linares, E., Turner, J.C.: Comarcas con rocas de edad precámbrica, sobre la base de dataciones radimetricas. 6th Cong. Geol. Arg. 1 (1975)

Maucher, A.: Die Sb-W-Hg-Formation und ihre Beziehungen zu Magmatismus und Geotektonik. Freib. Forsch. C 186, Leipzig, 173–188 (1965)

Maucher, A.: Time and stratabound ore deposits and the evolution of the earth. 24th IGC, Sect. 4, 83–87 (1972)

Maucher, A.: Zeitgebundene Erzlagerstätten. Geol. Rundsch. 63 (1), 263–275 (1974)

Maucher, A., Höll, R.: Die Bedeutung geochemischer-stratigraphischer Bezugshorizonte für die Altersstellung der Antimonit-Lagerstätten von Schlaining im Burgenland, Österreich. Mineralium Deposita 3 (3) 272–285 (1968)

Skaarup, P.: Stratabound scheelite mineralisation in skarns and gneisses from the Bindal area, Northern Norway. Mineralium Deposita 9 (4), 299–308 (1974)

So, C.S.: Die Scheelit-Lagerstätte Sangdong. Dissert. Univ. München (1968)

Urban, H.: Zur Kenntnis der schichtgebundenen Wolfram-Molybdän-Vererzung in Orsdalen (Rogaland), Norwegen. Mineralium Deposita 6 (3), 177–195 (1971)

Paleozoic Deposits

Contribution to a New Genetical Concept on the Bolivian Tin Province

H.-J. SCHNEIDER, Berlin (West), and B. LEHMANN, Berlin (West)

With 7 Figures

Contents

Summary

The Bolivian tin province is composed of four different groups of tin deposits distinguished by geo-tectonic setting, paragenesis, and geometric shape. There are Precambrian tin-bearing granites of the Brazilian Shield, Silurian meta-sedimentary strata-bound "manto" deposits, and two cycles of magmatic regeneration in Early Mesozoic (vein deposits) and Cenozoic time (tin porphyries). Heredity of tin derived from the Precambrian metallotect can be traced by exogene processes (paleo-placers) in Silurian, deep-seated magmatism in Mesozoic, and subvolcanic intrusions in Tertiary. The new evidence of Silurian synsedimentary cassiterite enrichment provides the missing link for a conclusive model of the metallogenetic development in the Bolivian crustal segment, thus proving the formation of different types of time- and strata-bound tin deposits in accordance with the geotectonic development of the continent.

1. Preface

Stimulated by modern concepts on global ore belts, the exceptional Bolivian tin province has over the last decade once again become the subject of genetical discussions (e.g., *Sillitoe*, 1976). According to the pattern of the South American tin belts traced out by *Schuiling* (1967), the Bolivian tin province is situated on a point of intersection of global importance (Fig. 1). This evidence was to lead to imagining the heredity of tin within the same crustal segment during the whole history of the earth. With respect to the prevailing close relationship of tin and acid magmatism, the regeneration of tin has been observed chiefly in connection with magmatic events, as pointed out, for example, by *Stoll* (1964a, b, 1976).

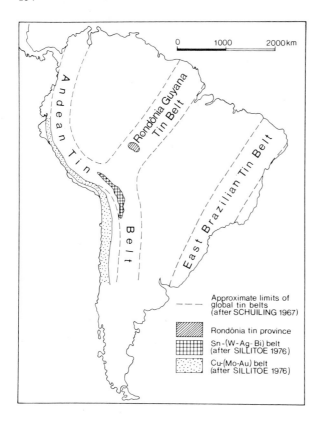

Fig. 1. Tin belts and tin provinces in South America. (According to *Schuiling*, 1967, and *Sillitoe*, 1976)

On scanning critically the geological development of the tin province discussed here, one finds that a gap exists between the Precambrian metallotect and the Early Mesozoic tin feeding magmatic intrusions of the Andean belt. However, if we take into account the record of different "generations" of ore deposits in the order of the geological age of their host rocks, a chronological sequence of metallogenetic stages becomes traceable (*Schneider*, 1975). This concept, discussed below, may be regarded as a suggestion of an exceptional model for time- and strata-bound ore deposits, intercalated into a geological sequence of a period of over 1000 my within the Bolivian crustal segment.

Basic field survey has been sponsored by DAAD in connection with a Universities Cooperation Programme of the Bolivian and German governments, the following lab investigations have been partly supported by the German governmental research programme (NTS 60) which is gratefully acknowledged.

2. Introduction: Development of a Problem

The classic model of a sole Mid-Tertiary epoch of mineralization in the Andean tin province of Bolivia which explained the paragenetic variety of the Sn-Ag-W-Bi-Sb-Pb-Zn ore deposits by one cycle of magmatic differentiation (*Ahlfeld*, 1937) has had to be corrected decisively since the first radiometric datings of some of the magmatic rocks of the metallogenic province. Based on the preliminary K-Ar data of *Evernden* (1961) a significant bi-phased distribution of the magmatic events and the connected mineralization has been pointed out by *Schneider-Scherbina* (1962):

1. 199–180 my Early Mesozoic deep-seated magmatism;
2. 19–7.8 my Late Tertiary magmatism at shallow depth.

More extensive geochronological studies by *Robertson* (1974) and *Clark* et al. (1976) on the whole confirm these two major magmatic events.

The granitic intrusions of the Cordillera Real occurred in the Upper Triassic (about 210–200 my) perhaps persisting into the Lower Jurassic (up to 180 my).

The smaller plutons of the Illimani, Quimsa Cruz and Santa Vera Cruz districts (Fig. 2), situated between the Cordillera Real batholiths to the north and the subvolcanic province of Central Bolivia to the south, represent an intermediate magmatic episode, corresponding to Upper Oligocene–Lower Miocene (28–22 my). The few available data from the subvolcanic province of Central and Southern Bolivia indicate a Mio-Pliocene age (Llallagua: 8.6–9.4 my; *Evernden,* 1961).

The temporal classification of the igneous rocks and their associated ore deposits reveals a close relationship to the geometric shape and environment of emplacement of these deposits (Fig. 2).

1. Vein deposits are bound: (a) to Mesozoic intrusive bodies of the Cordillera Real (Fabulosa, Milluni, Chacaltaya, Chojlla Mines etc.), or (b) to Tertiary intrusive bodies of the Cordilleras Quimsa Cruz and Santa Vera Cruz (Viloco, Caracoles, probably Colquiri and also Chicote Grande Mines).

2. Breccia pipes and dykes, stocks with strong hydrothermal alteration and stockworks of multidirectional veinlets carrying Sn mineralization are found within, or are associated to, Tertiary high-level subvolcanic intrusives and are here referred to as "tin porphyries" (*Sillitoe* et al., 1975), e.g., Oruro, San Pablo, Llallagua, Potosi, Chorolque etc.

As a function of the geotectonic position *Ahlfeld* and *Schneider-Scherbina* (1964), *Ahlfeld* (1967) and *Turneaure* (1971) could accentuate a regional zoning of the metallogenic province by the prevailing Sn-W paragenesis of the northern Bolivian sector (batholithic mineralization) and the Sn-Ag paragenesis of Central and Southern Andean Bolivia (subvolcanic mineralization).

The discovery and first investigations of the Precambrian tin province of Rondônia in the borderland between Bolivia and Brazil provided a new impulse for discussion on genetical relations to the "classical" Andean tin province (*Ljunggren*, 1964; *Grabert*, 1966).

The tin deposits of Rondônia are associated with Precambrian granites of the Brazilian Shield (*Priem* et al., 1966; *Kloosterman*, 1967). Recent investigations (*Botello* et al., 1972) have shown that the extension of the Brazilian Shield area under a thin Quaternary cover goes far more to the west than thought before, and the Precambrian Basement is now thought to have intensely influenced the geotectonic development of the Andes, as proved by the deflection of the East Andean Cordillera at Cochabamba–Santa Cruz or the paleogeographical significance of the "Arequipa Massif" (see below).

Magmatic remobilization of tin and wolfram has been postulated by *Stoll* (1964a, b, 1976) in the crustal segment of Argentina and Bolivia over five metallogenetic provinces of different ages from the Precambrian Sn-W deposits of the Pampean Ranges of Northern Argentina up to the latest wood-tin occurrences in Pliocene volcanic rocks of Southern Bolivia. This idea of a close interrelationship between a first Precambrian metallotect

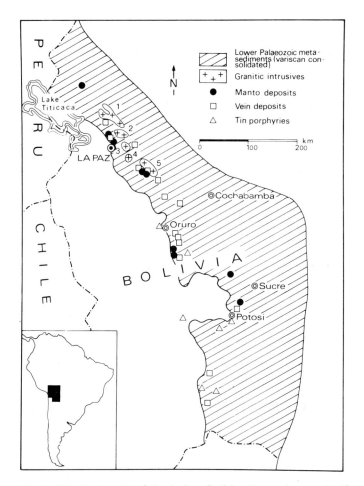

Fig. 2. Main Sn deposits of the Andean Bolivian tin province as classified by genetic features with special regard to manto deposits. Mesozoic granitic intrusions: *1*: Zongo-Yani; *2*: Tiquimani-Huayna Potosí; *3*: Taquesi-Mururata. Tertiary granitic intrusions: *4*: Illimani; *5*: Quimsa Cruz and Santa Vera Cruz

and various later magmatic mineralization epochs has been stressed by *Schuiling* (1967) explaining the position of the Andean Bolivian tin province at the intersection of a circumpacific tin belt and a Precambrian E—W oriented tin belt by polycyclic orogene reactivation of geochemical preconcentrations of tin (Fig. 1).

While *Stoll* (1964a, b) and *Schuiling* (1967) consider magmatogene mechanisms of tin remobilization by anatexis of tin granites only, *Ljunggren* (1964, p. 435) points out the possibility of heredity of tin by exogene erosion processes feeding geosynclinal sedimentation: " . . . tin included in the hydrolyzates of the marine sediments". Improving this concept *Fleischer* and *Routhier* (1970, p. 90) suppose a preconcentration of tin in the Andean Paleozoic sediments and refer to a tempting "étagement temporel" of the host rocks of tin ore deposits.

These most interesting ideas did not attract general attention, as plain field evidence was apparently lacking. Exactly this evidence was indicated by *Ahlfeld* and *Schneider-*

Scherbina (1964) for those tin deposits which have been described as "mantos". The typical strata-bound "manto" tin deposits have been interpreted until now, however, almost without exception as high-temperature replacement bodies in association with igneous rocks.

Because of this interpretation the problem arising from the enormous time interval between the formation of Precambrian and Mesozoic/Cenozoic tin deposits has been ignored. In addition, the geotectonic development of the continent during Paleozoic and Mesozoic epochs has been neglected, considering the Mesozoic and Tertiary magmatic events as isolated processes. By taking the Paleozoic formation of the mantos into account, this gap in metallogenetic development could be filled.

3. The Manto Deposits

About 20 km north of La Paz, in the foothills of the Cordillera Real, two of the most striking examples of the manto type Sn-deposits are situated: the Kellhuani and the Huallatani mines (Figs. 3, 4). On a regional scale both are bound to the lower member of about 500 m thickness of the Silurian "Catavi Formation" (*Ahlfeld* and *Branisa*, 1960), which consists of a characteristic series of alternating layers of quartzites and black shales (*Lehmann*, in prep.). In the field this member is characterized not only by its typical lithology but also by hundreds of small mines, adits etc., which can be traced easily over a small stripe of at least 20 km of strike.

On a local scale these mines are bound to the quartzitic layers representing about one third of this member of the Catavi Formation. The quartzitic layers range in thickness from some cm up to 30 m and are mineralized heterogenously. The mineralization of the different meta-sandstone layers ranges from economically valuable contents down to trace amounts of cassiterite.

From approximately 15 bigger quartzitic layers, more than 2 m thick, Kellhuani Mine produces only from four to five selected horizons and Huallatani Mine is limited to one single quartzite layer of about 6–8 m thick. The exceptional strata-bound Sn mineralization of the Huallatani area has been mapped in detail by *Schultze-Westrum* (1966), showing the distribution of ancient mining acivities on the surface (Fig. 3). Present mining activities of the Huallatani Mine are levelled only at the lowermost manto with underground operations of some 1000 m of galleries.

Some km south-east of the Chacaltaya, an observatory and ski resort at 5200 m above sea level. the Kellhuani Mine displays an outstanding congruency of meta-sandstone layers and the Sn mineralization (Fig. 4). Kellhuani carries a tremendous Sn potential and may become one of the biggest mining operations in Bolivia. There are at least four to five quartzitic layers of a total thickness of 40 m, extending over more than 1000 m with an average grade of about 1% Sn. This proved potential may multiply with further exploration. Present-day operations are still small-scale and restrained by dressing problems, but large open-pit production has been scheduled for the next years.

The manto-type Sn deposits can be followed with the strike of the Silurian series over 250 km distance up to the Poopo region, south of Oruro. In a stratigraphic position somewhat lower than the northern deposits, the Poopo mines are restricted to the Silur-

Fig. 3. Huallatani Mine. Sketch map of outcropping Sn-bearing mantos and the younger, disconformable gold-quartz vein. (Generalized after *Schultze-Westrum*, 1966)

Fig. 4. Kellhuani Mine – a typical manto deposit. Strata-bound Sn mineralization of some of the Silurian quartzites within the Catavi Formation: partial view from W in direction of striking of the sequence. Former mining activity (adits!) indicating the intensity of Sn mineralization of the mantos no. M4a, M4b, M5, M6, and M7. (According to local mining nomenclature)

ian Llallagua Formation, also consisting of alternating sandstones/quartzites and black shales (Fig. 5). Besides the structural control of the mineralization by a large reverse fault, there is a significant lithologic control by some sandstone/quartzite layers distinctly exemplified by the Candelaria and Mujicani Mines. At present the mining activities of the region are small-scale and at intervals completely paralyzed. The necessary change-over from rich lode mining to low-grade ore production is rendered difficult by lack of geological concepts and available funds.

In a preliminary survey by *Schneider* (1975), these deposits have already been out-lined, and it has been stated that they represent the "missing link" within the sequence of heredity of tin of the Andean crustal segment of Bolivia: there is increasing evidence that these deposits represent a Lower Paleozoic sedimentary tin enrichment as part of the crustal cycle (paleo-placers). The existence of such paleo-placers as preparatory exogene phases of tin enrichment and their causality for subsequent intrusions of "tin granites" has been repeatedly considered.

From the Erzgebirge/Germany *Baumann* and *Weinhold* (1963) described Proterozoic cassiterite-bearing "felsite horizons" which are classified as a first, stratiform-syngenetic tin mineralization of the Erzgebirge (*Bolduan*, 1972).

From the Billiton district/Indonesia *Adam* (1960, p. 410) indicated "the vast majority of the workable tin-ore lodes" in the Upper Paleozoic meta-sediments as "bed veins" and pointed out the distance of several km to the nearest outcrops of granitic rocks. Moreover from Western Tasmania/ Australia *Solomon* and *Green* (1976) stressed the lense-shaped stratiform emplacement of tin min-eralization in a Lower Paleozoic meta-volcano-sedimentary complex. For the genesis of the Leeuw-port and Nieuwpoort Mines /S. Africa *Garnett* (1967, p. 144) takes into consideration that the Bush-veld granite "would have acted only as an agent of transportation and conversion" by assimilation of "tin-bearing sediments".

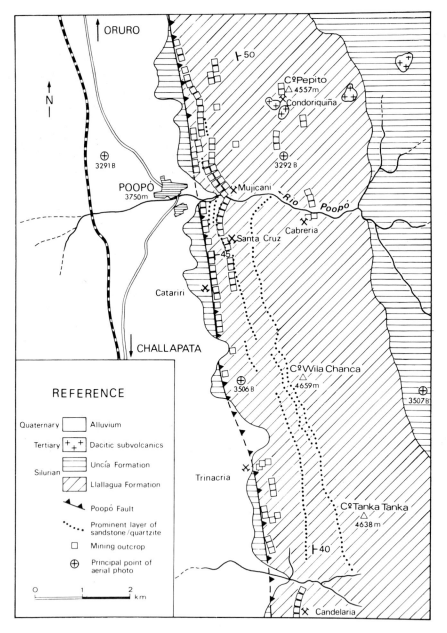

Fig. 5. Geological situation and mining activities of the Poopó region indicating the striking outcrops of some mineralized mantos. (Airphoto interpretation)

A survey of the literature on tin deposits yields many hints of sedimentary features, but the well-known prevailing textbook opinion has usually been able to neutralize these contrary observations by auxiliary constructions such as "wall rock control", "replacement textures", etc. A consequent remodelling of the genetical conceptions for these tin provinces is still lacking. For the Bolivian tin province we shall try to present a conclusive genetical model from the previous results of our current studies.

What characterizes the manto deposits, and how do they differ from the Sn-bearing veins?

3.1 Geometry

a) Congruency on a regional scale: control of Sn mineralization by the Silurian Catavi Formation in the La Paz area and by the Silurian Llallagua Formation in the Poopo area.

b) Congruency on a local scale: lithologic control of Sn mineralization by the meta-sandstone layers of the Catavi or Llallagua Formation. Additional to the lithological control there must be another factor operative, distinguishing between mineralized and "barren" quartzites of the same lithology. As these two types of quartzite are on the whole identical and are found closely sandwiched together, this second factor must lie in the history of sedimentation of this sequence (oscillation of the shoreline, change in source rock, etc.). An interpretation of this phenomenon by wall rock control, which would have had to act selectively only in some of the quartzitic layers over distances of some km, seems very doubtful, especially as there are veins cutting the whole sequence as at Huallatani (Fig. 3), Milluni, or in the Chacaltaya area. There can also be added the Chojlla Mine, where the vein system of more than 500 m of mining depth cuts through thick alternating quartzitic and argillaceous layers without traceable mineralization inside the quartzites.

3.2 Mineralogy

The manto deposits of the La Paz area are very simple in paragenesis. Nearly all the ore material consists of fine-grained cassiterite, tourmaline and abundant pyrite and hematite. This is in contrast to the complex mineralogy of the vein deposits of this region, which consist of sphalerite, galena, cassiterite, pyrite, stannite, fluorite, wolframite, chalcopyrite, siderite, bismuthinite, etc.

Thin sections of the mineralized quartzites show distinct metamorphic recrystallization textures. Equigranular quartz mosaics with mutual grain boundaries and strong interlocking testify to intense grain growth, often destroying the sedimentary features. In many cases abundant zircon grains of rounded shape indicate relics of old ss-planes. Green tourmaline appears as an apparently clastic component enriched in some layers and with concordant bedding. Cassiterite is found as fine disseminations or in small fissures on the mm and cm scale. Concordant cassiterite concentrations seem to be rare and raise the question of the mode of a possible cassiterite mobilization under special geochemical conditions. The broadly accepted inertness of cassiterite and the known difficulty of transportation of tin compounds under regular natural conditions is puzzling both in a synsedimentary and in a magmatogene genetical concept.

Certainly chemical processes are involved which we still cannot estimate. From the Monte Blanco Mine (Cordillera Quimsa Cruz) *Rumbold* (1909) already pointed out the strata-bound nature of the ore deposit and described the cassiterite mineralization in the laminae of the quartzite schist host rock. As this noteworthy observation may have fallen into oblivion, we will reprint here his original drawing (Fig. 6).

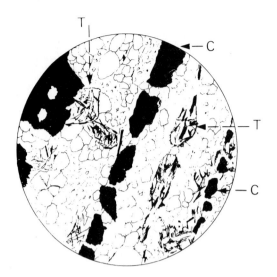

Fig. 6. "Mineralized quartzite schist ore from Monte Blanco Mine. Groundmass = coarse quartzite with tourmaline and cassiterite in the laminae. *C*: Cassiterite; *T*: Tourmaline. Crossed Nic. x 28". (Drawing and legend reprinted from *Rumbold*, 1909, Fig. 51)

The paragenesis of the Poopo Mines with abundant pyrite, sphalerite, galena, stannite, cassiterite and the famous cylindrite closely resembles the common massive sulphide deposits. Striking similarities can be seen to the Pb-Zn-Cu-Sn Sullivan Mine/Canada, where a synsedimentary origin seems probable (*Mulligan*, 1975). Perhaps the prevailing sulphide paragenesis of Poopo is controlled by paleogeographic environment (euxinic facies?).

Summarizing the old and new observations, the conspicuous existence of old Paleozoic strata-bound tin deposits can be accepted, and it may be assumed that they represent real paleo-placers of tin. Their formation, belonging to an old Paleozoic geosynclinal development, fits in close accordance to the modern concept of the Paleozoic sequence of the High Andean area.

4. Geological Setting of the Paleozoic

In the past 15 years, the knowledge of the stratigraphy and sedimentology of the Paleozoic of Andean Bolivia has improved considerably, and the compilation of Bolivian geology by *Ahlfeld* and *Branisa* (1960) has been revised in some basic features. According to *Schlatter* and *Nederlof* (1966), the Lower Paleozoic sequence comprises more than 10,000 m of clastic meta-sediments. The sedimentary sequence ıs of remarkable monotony. In general there are only facial changes, ranging between clay, silt, and medium-grained sandstone fractions. Coarse clastic rocks are nearly absent. There are no limestone occurrences of any importance and Lower Paleozoic volcanism seems of very restricted extent.

In part this may be due to still unsatisfactory field investigations, but reflects also the miogeosynclinal character of the Lower Paleozoic sedimentation basin. The enormous thickness of this sedimentation cycle from Lower Ordovician to Lower Devonian indicates large, intense erosion processes.

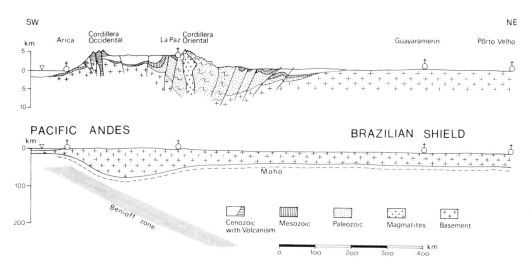

Fig. 7. Schematic cross section through the Andes and adjacent part of the Brazilian Shield; Arica – Pôrto Velho. (Compiled after *Ahlfeld*, 1970; *James*, 1971; *Botello* et al., 1972; *Audebaud* et al., 1973; *Stauder*, 1973)

Sedimentological studies of the Lower Devonian of Bolivia by *Isaacson* (1975) could reveal a westerly source area in the present Altiplano and Western Cordillera region. The homogeneity of the Paleozoic sequence permits generalizing this result also for the Silurian. Additional evidence for the postulated "Arequipa Massif" has been provided by drilling data in the Altiplano area south of Lake Titicaca, which found the Precambrian Basement at 2745 m below the surface of 4000 m above sea level, Paleozoic strata lacking (*Lehmann*, 1978). These results and various Precambrian outcrops in Southern Peru, Northern Argentina, and at the Pacific coast around Arequipa indicate the intracratonic position of the Lower Paleozoic sedimentation basin (Fig. 7).

Investigations by *Mégard* et al. (1971), *Martinez* et al. (1972) and *Audebaud* et al. (1973) established a Hercynian orogeny with culminations in the uppermost Devonian and in the Permian.

This Paleozoic orogeny affected the area of the present Eastern Cordillera by ample folding, faulting, and regional metamorphism of greenschist facies. A synorogene magmatism has been detected by field studies of *Bard* et al. (1974) in the Zongo Valley/Cordillera Real, still to be checked by radiometric dating.

Doubtless the majority of the granitic intrusions are of posttectonic type as, for example, the Huayna Potosi batholith or the Chacaltaya granite porphyry. Their associated Sn mineralization builds up ac vein fillings, cutting thick sedimentary complexes nearly without lithologic control and being younger than the tin mineralization of the mantos. Thus the Paleozoic tin mantos of the Cordillera Real are the predecessors of the Mesozoic tin veins, which are further succeeded by a next generation of tin deposits associated with Mid-Tertiary intrusives followed in Mio-Pliocene by the culmination of the tin porphyries.

5. Geochemical Indications

Although the Bolivian Sn-Ag province has been subject to intense geological and mineral-ogical studies for a long time, geochemical investigations have only recently been under-taken.

First systematic analyses for trace elements in cassiterites were performed by *Fesser* (1968) and *Wolf* and *Espozo* (1972) by emission spectrography. Although based on conventional genetical concepts on the Bolivian metallogenic province, they nevertheless already outlined some interesting trends in the trace-element distribution of different cassiterites. *Fesser* (1968, Fig. 1) demonstrated a remarkable distribution of Nb and Ga in the Bolivian cassiterites: there is a systematic decrease in Nb content from the north-ern to the southern tin ore deposits with Ga showing a counter trend. This regional pat-tern can be correlated with the temporal sequences: prevailing Mesozoic ore deposits to the north and Tertiary ones to the south.

Because emission spectrography is of restricted usefulness for trace element analysis due to its relatively high detection limits for elements especially desirable in cassiterite geochemistry, in our Berlin team we are coming to use spark source mass spectrometry, recording a broad range of elements. First results of these still current studies (*Schneider* et al., 1977) show a distribution pattern of some important trace elements in the cassite-rites which is of startling congruency with the temporal relationship of four different groups of tin deposits (see Sect. 6). Thereby it must be noted that until now only about 50 analyses of mineral samples from various tin ore deposits have been completed. From these studies ist may be summarized briefly:

According to the Nb/Ta vs. Nb/Sn ratios the different cassiterites show a distinct fractionation pattern. There is a well-defined field representing the Precambrian cassite-rites from Rondônia and another related to Tertiary tin mineralization. Contrary to this, the cassiterites from Mesozoic vein deposits, e.g., the Chojlla Mine, are characterized by a distinct scattering centered at the Precambrian field. The cassiterites analyzed to the present from manto deposits, e.g., the Kellhuani Mine, show a broad scattering between the Precambrian and the Tertiary groupings. Repeated mobilization processes seem to have caused this shifting of Nb/Ta ratios.

Analogous trends can also be recognized for other trace elements. There is a decrease in Zr content of the cassiterites from the Precambrian to Tertiary age, whilst the W con-tent increases from Precambrian to Tertiary ore assemblages.

In addition, the rare-earth-element contents of fluorspars associated with some tin deposits are under investigation by neutron-activation analysis under the research scheme of *Möller* et al. (1976). First results exhibit a large scattering of the indicative data, advocating repeated remobilization of different fluorspar generations caused by the various magmatic activities.

These fractionation trends, as summarized above, may be considered as an additional indication of a geochemical and genetical consanguinity of the different tin mineraliza-tions of the Bolivian metallogenic province.

6. The Heredity of Tin During Earth's History

Surveying the Bolivian tin province from the point of view of geotectonic position and age of host rocks of the different types of ore deposits, we can summarize four main groups. This new grouping is also correlated by the distinct leading paragenesis and geometrical shape of the ore deposits:

Group 1. Precambrian tin granites (greisen) and tin-bearing pegmatites occur in the Rondônia region of the Brazilian Shield. They are distinguished by their paragenetic association with Nb- and Ta-bearing minerals in general common to Precambrian tin belts.

Group 2. Between the thick Lower Paleozoic sequence of clastic sediments there are strata-bound tin mineralizations in Silurian quartzites (Catavi and Llallagua Formation), the so-called mantos. Their paragenetic monotony in the northern part of the tin province of the Cordillera Real (quartz, tourmaline, cassiterite, zircon and Fe minerals) points to a reworked fossil-placer origin. The complex sulphidic paragenesis of the southern part may be regarded as due to special environmental conditions (euxinic facies).

Group 3. In close connection with Mesozoic intrusive bodies there are the typical disconformable vein deposits of the Cordillera Real (Fabulosa, Milluni, Chojlla mines, etc.). They display a paragenetic variety with additional ore minerals such as e.g., galena, sphalerite, chalcopyrite, wolframite, scheelite, stannite, and bismuthinite.

Group 4. Late Tertiary intrusive and extrusive magmatism, predominating in the southern tin province, exhibits a last and intense tin mineralization. Magmatic activity can be classified according to the intrusive level: plutonic granite bodies occur in the Cordillera Quimsa Cruz with their typical hydrothermal vein swarms, and more to the south, the subvolcanic tin porphyries which are characterized by their famous silver abundance. Various stages of mineralization here gave rise to a complex mineralogy with strong telescoping phenomena.

A characteristic feature, common to the last two magmatogene epochs in Mesozoic and Tertiary times, is their geological setting within the Lower Paleozoic sedimentary complex. There are changing magmatic environments from the northern plutonic subprovince to the southern subvolcanic tin province accompanied by disconformable vein deposits to the north and disseminated and stockwork types of mineralization to the south. There are, however, strata-bound manto deposits in the northern as well as in the southern province, thus providing evidence of their pre-magmatic age.

By admitting an independent Lower Paleozoic genesis for the manto deposits we arrive at a coherent and conclusive metallogenic succession which fits without constraint the geological development of this crustal segment: in Lower Paleozoic time the uplifted regions of the Precambrian tin belt are eroded; clastic debris is accumulated in the sedimentation basin of the Hercynian geosyncline. The source rock area lies to the west in the present Altiplano and Western Cordillera region and represents the westerly continuation of the Guyana-Rondônia tin belt (*Schuiling*, 1967; see Fig. 1).

Repeated reworking of the nearshore sand bars results in an excellent classification of the quartz detritus. "Uprooting" of some components by physical and chemical agents and accumulation of other, more resistant, minerals such as zircon, tourmaline, and cassiterite leads to high compositional maturity and qualifies some of the sand bars as tin placers.

During Hercynian orogeny, parts of the Paleozoic sediments are subject to metablastic transformation, metatexis, and partial anatexis (*Wolf*, 1973; *Bard* et al., 1974). Granitic magmas are generated and the conversed tin content of the Paleozoic source rock is enriched in magmatic differentiates feeding the Mesozoic vein deposits.

A geotectonic-magmatic reactivation takes place in the Tertiary age, caused by plate tectonic movements at the Pacific margin. Heat transfer gives rise to quartzlatitic to rhyodacitic magmas of strong differentiation, fed by the same crustal source as in the Mesozoic period. This last mobilization of tin is accompanied by outstanding quantities of silver as well as other base metals. The origin of these elements may be correlated to deep-seated crustal sources (hybridization), but will not be further discussed here.

In any case, it seems evident that the Bolivian tin belt shows no genetical relationship to the Coastal Cordilleran copper belt, which may be related to subduction of the oceanic crust (*Sillitoe*, 1976; see Figs. 1 and 7). In addition there is no indication of a circumpacific tin belt as postulated by *Radkewitsch* (1972). The "Andean tin belt" can be traced by Sn ore deposits only at the one section where the Precambrian Guyana-Rondônia tin belt intersects the Andean mountain range. Beyond the extent of the Bolivian tin province there are only minor tin occurrences of mineralogical interest in the Cordilleran chains. Thus the Bolivian crustal segment may be regarded as an exceptional model for the heredity of tin by various stages of geotectonic-magmatogene and exogene-sedimentary reactivation during 1000 my of earth's history, closely correlated to the geological development.

References

Adam, J.W.H.: On the geology of the primary tin-ore deposits in the sedimentary formation of Billiton. Geol. Mijnbouw **39**, 405–426 (1960)

Ahlfeld, F.: Über das Alter der zinnbringenden Magmengesteine Boliviens. Zbl. Miner. Geol. Paläont. Abt. A (Miner. Petr.), 34–38 (1937)

Ahlfeld, F.: Metallogenetic epochs and provinces of Bolivia. Mineralium Deposita **2**, 291–311 (1967)

Ahlfeld, F.: Zur Tektonik des andinen Bolivien. Geol. Rdsch. **59** (3), 1124–1140 (1970)

Ahlfeld, F., Branisa, L.: Geologia de Bolivia. La Paz: Inst. Bol. Petrol. Don Bosco 1960, 245 p.

Ahlfeld, F., Schneider-Scherbina, A.: Los Yacimientos Minerales y de Hidrocarburos de Bolivia. Dept. Nac. Geol. Bol. **5** (Especial), 388 p., La Paz (1964)

Audebaud, E., Capdevila, R., Dalmayrac, B., Debelmas, J., Laubacher, G., Lefèvre, C., Marocco, R., Martinez, C., Mattauer, M., Mégard, F., Paredes, J., Tomasi, P.: Les traits géologiques essentiels des Andes Centrales (Pérou-Bolivie). Rev. Géogr. Phys. Géol. Dynam. 2nd ser. **15** (1–2), 73–114 (1973)

Bard, J.P., Botello, R., Martinez, C., Subieta, T.: Relations entre tectonique, métamorphisme et mise en place d'un granite eohercynien à deux micas dans la Cordillère Real de Bolivie (Massif de Zongo-Yani). Cah. ORSTOM, sér. Géol. **6** (1), 3–18 (1976)

Baumann, L., Weinhold, G.: Zum Neuaufschluß des sog. "Felsithorizontes" von Halsbrücke. Z. angew. Geol. **9** (7), 338–345 (1963)

Bolduan, H.: Die Zinnmineralisation im Erzgebirge; Typen und Verteilung. Geologie **21** (6), 677–692 (1972)

Botello, R., Martinez, C., Subieta, T., Tomasi, P.: La carte tectonique de Bolivie. Maquette à l'échelle de 1/5.000.000. Cah. ORSTOM, sér. Géol. **4** (2), 149–152 (1972)

Clark, A.H., Farrar, E., Caelles, J.C., Hajnes, S.J., Lortie, R.B., McBride, S.L., Quirt, G.S., Robertson, R.C.R., Zentilli, M.: Longitudinal variations in the metallogenic evolution of the Central Andes: a progress report. Geol. Assoc. Canada Spec. Pap. **14**, 23–58 (1976)

Evernden, J.: Edades absolutas de algunas rocas de Bolivia. Noticiero Soc. Geo. Boliv. **2**, 3–4 (1961)

Fésser, H.: Spurenelemente in bolivianischen Zinnsteinen. Geol. Jb. **85**, 605–610 (1968)

Fleischer, R., Routhier, P.: Quelques grands thèmes de la géologie du Brésil. Miscellanées géologiques et métallogéniques sur le Planalto. Sci. Terre **15** (1), 45–102 (1970)

Garnett, R.H.T.: The underground pursuit and development of tin lodes. In: Fox, W. (ed.): A Technical Conference on Tin, Vol. I. London: Intern. Tin Council 1967, pp. 137–200

Grabert, H.: Die neue Zinnerz-Provinz Rondonia am Oberen Amazonas im brasilianisch-bolivianischen Grenzland. Erzmetall **19** (8), 398–399 (1966)

Isaacson, P.E.: Evidence for a western extracontinental land source during the Devonian period in the Central Andes. Bull. Geol. Soc. Am. **86**, 39–46 (1975)

James, D.E.: Plate tectonic model for the evolution of the Central Andes. Bull. Geol. Soc. Am. **82**, 3325–3346 (1971)

Kloosterman, J.B.: A tin province of the Nigerian type in Southern Amazonia. In: Fox, W. (ed.): A Technical Conference on Tin, Vol. 2. London: Intern. Tin Council 1967, pp. 383–398

Lehmann, B.: A Precambrian core sample from the Altiplano/Bolivia. Geol. Rdsch. **67** (1), in print (1978)

Ljunggren, P.: The tin deposits of Rondonia, Brazil, as compared with the Bolivian tin mineralization. Geol. Fören. Förhandl. **85** (4) 431–435 (1964)

Martinez, C., Tomasi, P., Dalmayrac, B., Laubacher, G., Marocco, R.: Caractères généraux des orogènes Précambriens, Hercyniens et Andins au Pérou et en Bolivie. 24th IGC, Montreal, 1972, Sec. 1, pp. 136–146

Megard, F., Dalmayrac, B., Laubacher, G., Marocco, R., Martinez, C., Paredes, P.J., Tomasi, P.: La chaîne hercynienne au Pérou et en Bolivie; premiers résultats. Cah. ORSTOM, sér. Géol. **3** (1), 5–44 (1971)

Möller, P., Parekh, P.P., Schneider, H.-J.: The application of Tb/Ca–Tb/La abundance ratios to problems of fluorspar genesis. Mineralium Deposita **11**, 111–116 (1976)

Mulligan, R.: Geology of Canadian tin occurrences. Geol. Surv. Canada, Econ. Geol. Report **28**, 155 pp. (1975)

Priem, H.N.A., Boelrijk, N.A.I.M., Hebeda, E.H., Verschure, R.H., Bon, E.H.: Isotopic age of tin granites in Rondonia, N.W. Brazil. Geol. Mijnbouw **45**, 191–192 (1966)

Radkewitsch, E.A.: Die metallogenetische Zonalität des Pazifischen Erzgürtels (Review by W. Beyer). Z. angew. Geol. **18** (11), 497–500 (1972)

Robertson, R.C.R.: Notas sobre el método del K/Ar de datación de rocas e interpretación de edades obtenidas hasta ahora. Serv. Geol. Boliv., Proy. Plutonismo, Inf. Prel. **2**, 11 pp. (1974)

Rumbold, W.R.: The origin of the Bolivian tin deposits. Econ. Geol. **4**, 321–364 (1909)

Schlatter, L.E., Nederlof, M.H.: Bosquejo de la geología y paleogeografía de Bolivia. Serv. Geol. Boliv., Bol. **8**, 49 pp. (1966)

Schneider, H.-J.: Ein neues genetisches Konzept für die sogenannte bolivianische Zinn-Silber-Provinz. Pap. present. at Intern. Clausthal-Kolloqu. SGA, GDMB a. COM, Clausthal-Zellerfeld, 1975, 1 p.

Schneider, H.-J., Dulski, P., Luck, L., Möller, P., Villalpando, A.: Correlation of trace elements distribution and geotectonical position of cassiterites in Bolivian ore deposits. Mineralium Deposita, in print (1978)

Schneider-Scherbina, A.: Über metallogenetische Epochen Boliviens und den hybriden Charakter der sogenannten Zinn-Silber-Formation. Geol. Jb. **81**, 157–170 (1962)

Schuiling, R.D.: Tin belts on the continents around the Atlantic Ocean. Econ. Geol. **62**, 540–550 (1967)

Schultze-Westrum, H.H.: Mina Huallatani. Mapa Geológico 1:2.000. Unpubl. report, Fabulosa Mines Cons., Milluni (1966)

Sillitoe, R.H.: Andean mineralization: A model for the metallogeny of convergent plate margins. Geol. Assoc. Can. Spec. Paper **14**, 59–100 (1976)

Sillitoe, R.H., Halls, C., Grant, J.N.: Porphyry tin deposits in Bolivia. Econ. Geol. **70**, 913–927 (1975)

Solomon, M., Green, G.R.: Ore deposits of Western Tasmania. 25th Intern. Geol. Congr. Excursion Guide 31 AC, Sydney, 1976, 49 pp.

Stauder, W.: Mechanism and spatial distribution of Chilean earthquakes with relation to subduction of the oceanic plate. J. Geophys. Res. **78**, 5033–5061 (1973)

Stoll, W.C.: Sn-W-Bi provinces and epochs in Argentina and Bolivia and their genetic interrelationship. Geol. Soc. Am., Abstr. Pap. submitt. Meeting Miami Beach. Fl., Nov. 1964, Annual Meet., 1964a, pp. 197–198

Stoll, W.C.: Metallogenetic belts, centers, and epochs in Argentina and Chile. Econ. Geol. **59**, 126–135 (1964b)

Stoll, W.C.: Provincias metalogénicas en Argentina, Bolivia y Chile: Aspectos de una teoria evolutiva de la Metalogenia. Bol. R. Soc. Española Hist. Nat. (Geol.) 74, 171–189 (1976)

Turneaure, F.S.: The Bolivian tin-silver province. Econ. Geol. 66 (2), 215–225 (1971)

Wolf, D., Espozo, E.: Zur Geochemie bolivianischer Kassiterite. Z. angew. Geol. 18 (10), 459–468 (1972)

Wolf, M.: Zum Magmatismus der Cordillera de Potosi in Bolivien. Leipzig: Freiberger Forschungsh. C 275, 1973, 174 pp.

Early Paleozoic Ore Deposits of the Sb-W-Hg Formation in the Eastern Alps and Their Genetic Interpretation

R. HÖLL, München

With 3 Figures

Contents

Summary

A widespread, strata- and time-bound metal deposition with W, Sb, and Hg provides an example of metallogeny at a convergent plate margin in the Eastern Alps probably from Upper Ordovician to Silurian Time. The ore formation is genetically, temporally, and spatially related to a submarine, basic, in part also ultramafic and acid volcanism and associated igneous-hydrothermal activities. These volcanics are similar to volcanic rocks of modern marginal basins. The arrangement of the different ore assemblages seems to reflect the different ore supplies and specific features of magmatism and facies in probably parallel troughs and uplifts as a function of the depth and angle of a north-dipping Benioff zone. The cinnabar occurrences were formed near the trench. Stibnite, (scheelite-)stibnite, and scheelite-arsenopyrite deposits were produced more to the north. Still further north, "pure" scheelite ores were deposited together with carbonate rocks. An ubiquitous scheelite mineralization with a complex paragenesis, especially W, Mo, Bi, Cu, Au, Ag, within an unusually thick volcanic rock series, resulted from the furthest down-dip ore generation above the Benioff zone.

1. Introduction

In 1965, *Maucher* presented a paper on the *Sb-W-Hg formation*, interpreting a voluminous literature on stibnite, cinnabar and tungsten deposits, and utilizing new data on a
Sardegnan scheelite-stibnite deposit (*Angermeier*, 1964) and on many cinnabar, stibnite,
and tungsten deposits in Turkey (*Höll*, 1966). The main ideas and conclusions of this
paper are:

1. The paragenesis of tungsten minerals, especially scheelite, with stibnite or stibnite-
cinnabar is not coincidental, but isogenetic. The former Sb-Hg formation must be extended to the Sb-W-Hg formation.

2. Many of the Sb-W-Hg formation deposits are strata-bound, often in black schists,
and genetically related to submarine volcanism.

3. World-wide, many of these ore deposits are connected with Early Paleozoic, especially Silurian, host rocks, and genetically related to a contemporaneous volcanism.
Such ore deposits are scarce, or perhaps even lacking in the Precambrian.

4. The distribution on the earth is not homogeneous. Early Paleozoic ore deposits are
restricted to geosutures which indicate the margins of the continental masses at that
time. These ore deposits are distributed within two belts, a Circum-Pacific and an Eura-
sian-Mediterranean belt.

5. These belts of Early Paleozoic age and those of the "young" Sb and Hg deposits of
Cretaceous-Tertiary age coincide. Many, if not all of these young ore deposits are con-
nected with events of regeneration, caused by syntexis or anatexis of the old, Early
Paleozoic Sb-Hg supply.

Our investigations of ore deposits of the Sb-W-Hg formation in the Eastern Alps
began in autumn 1965 during a private stay in Carinthia. The cinnabar deposits Eisenkappel, Stockenboi, Hohes Kohr and Rottrasten showed characteristics similar to those of
many well-known Turkish cinnabar deposits, e.g., Kalecik and Karareis, which I studied
in 1964. These Turkish cinnabar deposits were interpreted by me, for the first time, as
strata-bound and genetically related to an Early Paleozoic submarine volcanism and a
hydrothermal supply (*Höll*, 1966). For Kalecik, the same genetic interpretation is submitted by *Sözen* (1973, and this volume). The investigated cinnabar deposits in the
Eastern Alps were also regarded as strata-bound and genetically connected with probable Early Paleozoic volcanics. This interpretation was incompatible with other genetic
explanations at that time (*Höll*, 1970b).

Further investigations, carried out as an assistant of Prof. *Maucher*, were concerned
with the Kreuzeckgruppe. This is a mountain chain, where some stibnite deposits (Lessnig, Radlberg, Gurserkammer, Edengang, Rabant and Johannisstollen) are within or in
the immediate vicinity of probable Early Paleozoic submarine basic metavolcanic layers
and associated with a newly discovered rare scheelite mineralization. Moreover, research
was made in the stibnite deposit Schlaining (Burgenland) (*Maucher* and *Höll*, 1968), and
in the scheelite-magnesite deposit Tux (*Höll* and *Maucher*, 1967). In Tux, the scheelite
mineralization was also considered as strata-bound and genetically related to underlying
basic metavolcanics. Fossils, found for the first time in the overlying sparry magnesite
bed, provided an Upper Silurian to Lower Devonian age. These genetic interpretations
of the scheelite-stibnite deposits in the Kreuzeckgruppe, as well as the ore deposit Tux,
were in contradiction to the then accepted theories. The scheelite mineralization of the

ore deposit Tux, and of other ore deposits of the Eastern Alps, had been considered as of Alpidian or Variscan origin and genetically related to granites or their metamorphic representatives (*Wenger*, 1964; *Petrascheck*, 1966; *Vohryzka*, 1968).

A review of the scheelite literature for the Eastern Alps up to 1965 reveals 12 occurrences, without the Tux deposit, discovered by chance, not by systematic research. These occurrences are situated in the area of the Hohe Tauern, seven in the eastern section and five in the middle (*Höll*, 1970a, 1975). Those in the eastern section lie in different rock series, and it seems likely that they are of different origin. Scheelite discoveries in the middle Hohe Tauern are remarkably restricted to outcrops of the so-called "Habachserie". This series was described as a metamorphic, most probably Early Paleozoic, rock series with abundant basic and calc-alkaline metavolcanics (*Frasl*, 1958). This statement, that several scheelite occurrences are apparently restricted to a submarine, Early Paleozoic rock series, rich in different volcanics, was exactly suited to our conclusions about the strata- and time-bound scheelite mineralizations in the investigated areas. A delimitation of favourable areas for a scheelite prospect in the Eastern Alps appeared possible to me, taking this genetic interpretation as well as the proved Paleozoic age of the ore deposit Tux, and the newly discovered scheelite mineralizations of the Kreuzeckgruppe as a basis. Furthermore, the Habachserie was regarded as the most favourable rock series for new scheelite discoveries. This idea was published by *Höll* and *Maucher* in January 1967. We wrote that the scheelite occurrences in the Habachserie between Krimmler Achental, Stubachtal and Granatspitze are of the greatest interest and we would start work there. The research began in summer 1967 with a short prospecting course for scheelite by the Institut für allgemeine und angewandte Geologie und Mineralogie of the University of Munich, followed by my own scheelite prospecting, which lasted several years. In retrospect, we can report that this idea proved very successful. Many new scheelite occurrences in the Habachserie, especially the large scheelite deposit Felbertal, and in other areas of the Eastern Alps have been discovered (*Höll*, 1970a, 1971, 1975; *Maucher*, 1972).

2. General Geology

In the Eastern Alps, Precambrian rocks have not yet been clearly identified. It is still a fundamental problem to trace pre-Variscan metamorphic rocks because of Variscan and/or Alpidian overprinting. *Schönlaub* and *Daurer* (1977) propose a pre-Upper Ordovician, i.e., "pre-Sardinian", rock genesis and metamorphism in some parts of the Eastern Alps.

Early Paleozoic rock series within the Eastern Alps indicate a locally very mobile sedimentary basin, with elongated parallel troughs and uplifts, lasting from Upper Ordovician to Devonian/Carboniferous time. The volcanic activity may indicate an island arc evolution, with individual trough systems and intervening ridges as in modern examples. Other detectable features of this model are submarine to subaerial stages. There is no evidence, however, for spreading centres or "zebra" stripes of oceanic crust. The original position of the different rock series within this basin, from north to south, corresponds to the different tectonic units of the Eastern Alps, formed by the Alpidian nappe transport. These tectonic units are, from top to bottom: Upper Austro-Alpine unit, middle Austro-Alpine unit, lower Austro-Alpine unit, and Penninic zone. As presently accepted, The Early Paleozoic rock series of the Penninic zone, e.g., the Habachserie, are part of the northernmost region of this old basin. The contemporary rock series of the upper Austro-Alpine unit (particularly in the Northern Graywacke zone and in the Gurktal nappe) represent the southernmost deposition known to date.

Basic, and to a minor extent, ultrabasic submarine volcanics followed the sedimentation of the oldest, argillaceous to arenaceous sediments, which were probably deposited in the whole area of this geosynclinal basin. This basic volcanism began in the southernmost trough at the latest in the Caradocian stage, and the main activity ended at about the close of this stage. Subsequent basic volcanic deposits exist in local Silurian and Devonian strata. Uncertainty concerning the age of the basic metavolcanics extends to rock series within other parts of this old basin (e.g., the Habachserie, with 3000 m outcrop of basic metavolcanics).

In the Northern Graywacke zone a widespread, at least regionally subaerial acid volcanism (ignimbrites) probably occurred during the Caradocian stage ("Blasseneck-Porphyroid"). However, layers of acid metavolcanics can also be found in other Early Paleozoic rock series, such as in the Habachserie, and are perhaps of the same age (*Frasl*, 1958; *Frasl* and *Frank*, 1966; *Höll*, 1975). Silurian acid volcanics are, however, also present. At the time of the Ordovician-Silurian transition, the southernmost trough was uplifted and its development interrupted without creating fold mountains (*Mostler*, 1970). In other regions of the basin, an abrupt structural change is as yet unproved. It seems possible that these northern features gradually evolved into the Variscan geosyncline.

The final consolidation of the crust was completed later through the Variscan orogeny. The "central gneisses" of the Tauern window are derived, to a great extent, from Variscan (Permian) plutonites. Permian acid volcanic rocks in the Southern Alps ("Bozen Quartzporphyry") and in Vorarlberg are perhaps consanguineous.

The ore deposits of the Sb-W-Hg formation in the Eastern Alps are, excepting a few deposits, of two genetic types (*Höll*, 1975; *Höll* and *Maucher*, 1976):

1. Strata- and time-bound Early Paleozoic deposits in metavolcanics and metasediments with locally intense metamorphic mobilization. Most ore deposits belong to this genetic type, but the different ore parageneses are not equally dispersed in the main tectonic units of the Eastern Alps. They show strict dependencies (*Höll*, 1971, 1975; *Höll* and *Maucher*, 1976): thus, cinnabar mineralizations occur in the upper Austro-Alpine unit, with the exception of a deposit in the middle Austro-Alpine unit. Scheelite-stibnite and scheelite-arsenopyrite deposits are in several locations within the middle Austro-Alpine unit. Pure scheelite mineralizations, mainly associated with carbonate host rocks, occur only in the lower Austro-Alpine unit. In the Penninic zone scheelite outcrops are very abundant, above all within thick volcanic sequences (e.g., scheelite deposit Felbertal); they contain Mo, Bi, Be, Cu, Au, Ag, and other elements. This distribution is not only dependent upon the tectonic movements, especially the Alpidian nappe transport, which displaced the rock series to their present geographical arrangement, but is linked to specific features of facies and magmatism, originally dependent upon the former paleogeographic boundaries of the troughs and uplifts in the Early Paleozoic sedimentary basin (Fig. 3).

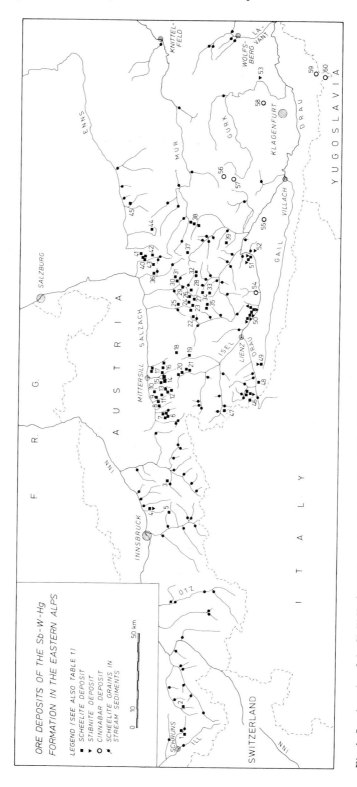

Fig. 1. Ore deposits of the Sb-W-Hg formation in the Eastern Alps (*Höll*, 1975)

2. Quartz-rich scheelite-bearing veins and impregnations in the central gneisses of the Hohe Tauern. This ore mineralization is explained as a result of Variscan palingenic regeneration and partial anatectic granitization of the primary, scheelite-bearing Early Paleozoic strata.

Some deposits certainly do not belong to these two genetic types: the strata-bound (cinnabar-)stibnite deposit Schlaining (Burgenland), in the easternmost part of the Alps, is genetically connected with submarine Cretaceous metabasites (*Maucher* and *Höll*, 1968; *Schönlaub*, 1973a; *Höll* and *Maucher*, 1976). The strata-bound Triassic cinnabar deposit Vellacher Kotschna in the southernmost edge of Austria is within the Southern Alps, just outside the Eastern Alps.

The origin of a few depleted or insufficiently exposed outcrops is still uncertain. Pegmatitic-pneumatolytic and contact-pneumatolytic tungsten deposits are reported (*Vohryzka*, 1968; *Frisch*, 1972: scheelite occurrence Lappgraben; Fig. 1, No. 47), but not yet ascertained. Yet, such deposits may exist in the Eastern Alps, especially in connection with Variscan intrusions of sialic-palingenic magmas.

3. Scheelite Deposit Felbertal

This ore deposit is the major scheelite deposit of the Eastern Alps and one of the largest tungsten deposits on the earth. It is about 9 km to the south of the town of Mittersill (Fig. 1, No. 14) and has been investigated by two galleries and 52 core drillings (with a total of 6400 m drilled).

This scheelite deposit and its geological environment show the following characteristics:

1. The ore mineralization lies exclusively within the Habachserie, the principal rock series of the "Lower schist cover" (Untere Tauernschieferhülle) of the middle part of the Hohe Tauern. The age of the Habachserie is unconfirmed by fossils, but is commonly considered Early Paleozoic.

2. This ore deposit was metamorphosed by the Alpidian orogeny (Tauernkristallisation) during the Tertiary (Oligocene). This barrovian-type metamorphism took place in the area of the ore deposit under conditions of the "low-stage of metamorphism, isograd (17–20) + hornblende", using the classification of *Winkler* (1970).

3. The Habachserie can be divided into three units (Fig. 2):a) Habach phyllites (Habachphyllite), b) volcanic rock sequence (Eruptivgesteinsfolge), c) basal schist sequence (Basisschieferfolge).

4. The basal schist sequence is predominantly composed of dark schists, originating from arenaceous-argillaceous sediments. It shows an unusual thickness of more than 400 m in the Felbertal (valley), whereas elsewhere it is less than 100 m. Volcanic intercalations are precursors to the volcanic rock sequence. They are remarkably accumulated in the Felbertal. Traces of scheelite ($\leqslant 240$ ppm WO_3), connected with these intercalations, have been found at four small places in the Felbertal.

5. The volcanic rock sequence overlies the basal schist sequence with a sharp boundary, caused by an abrupt ending of the arenaceous-argillaceous sedimentation and the sudden intensification of the submarine volcanic activity. The volcanics are up to

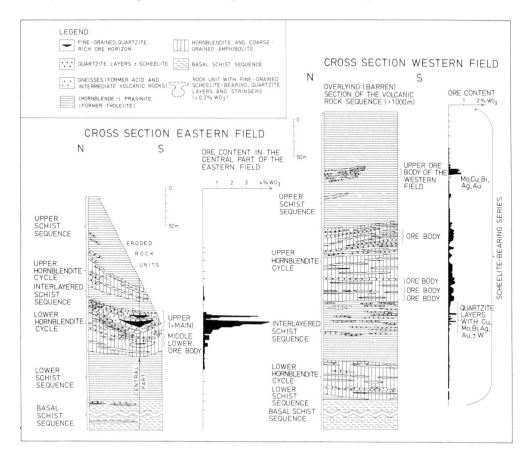

Fig. 2. Schematic cross sections through the scheelite-bearing series of the eastern and western field of the ore deposit Felbertal (*Höll*, 1973)

1500 m thick in the Felbertal, and further west and east they thicken to at least 3000 m.

6. The lowest, 300-m or more, wide section of this volcanic rock sequence is the "scheelite-bearing series" (Scheelit-führende Serie). Workable ores are found only within this section.

The sole plane of the scheelite-bearing series is sharp and coincides with that of the volcanic rock sequence. However, the scheelite-bearing series undergoes a gradual transition into the overlying, fairly uniform and barren part of the volcanic rock sequence (primarily tholeiite), through a slow, irregular decrease in the tungsten content.

7. The scheelite-bearing series shows intensive, alternate bedding and interfingering. It contains ultramafic, tholeiitic, intermediate and acid metavolcanic rocks, as well as reassorted material, and precipitations of hydrothermal supplies.

The original ore mineralization is strata-bound. The ore supply, the ore deposition, and the diagenetic ore alteration, were apparently contemporaneous and conformable with the volcano-sedimentary host rocks.

8. Dark, chromium oxide-green hornblendites and coarse-grained amphibolites are very characteristic for the scheelite-bearing series of the ore deposit Felbertal and other scheelite deposits in the Habachserie. These Cr-rich rock types are characterized by high Ca and high Mg as well as low Al and low alkali metal components in comparison with tholeiites. They show striking similarities to "komatiitic" volcanics, which are found in Archean greenstone belts and are regarded as of primitive mantle origin.

The hornblendites and coarse-grained amphibolites are concentrated within two successive rock units, thus making feasible the following division of the scheelite-bearing series (from top to bottom; Fig. 2):

a) Upper schist sequence,
b) upper hornblendite cycle,
c) interlayered schist sequence,
d) lower hornblendite cycle,
e) lower schist sequence.

The formation of both hornblendite cycles is connected with two critical pressure releases causing a rapid ascent of ultramafic magma successions. The hornblendite and coarse-grained amphibolite layers are particularly related to submarine lava flows, while most bands of hornblendite schists are derived from volcanic tuffs. This ascent of ultramafic volcanics was overlapped by other volcanics (tholeiit-basaltic, intermediate, and acid) and associated by the most intensive hydrothermal solutions within the whole scheelite-bearing series.

9. The upper schist sequence, the interlayered schist sequence and the lower schist sequence consists mainly of former tholeiitic basalts, altered to (hornblende-)prasinites (viz. hornblende-chlorite-epidote-albite-oligoclase assemblages).

10. The scheelite deposit Felbertal contains two ore fields, an "eastern field" and a "western field".

Two adjoining special troughs, probably along a prominent volcanic trend are regarded as initial troughs for both ore fields. A paleogeographic high, active from the time of formation of the basal schist sequence until the upper hornblendite cycle, forms the boundary between both troughs. At present, this high lies beneath the Felbertal valley floor. It caused an extensive independency of each trough. However, regional events, such as the formation of both hornblendite cycles, are common in both fields.

11. The estimated length of the elongated eastern field is somewhat more than 2.5 km and the mineable zone is hardly more than 100 m wide in the well-probed easternmost portion of the field.

The western field has been explored only in its eastern part, and the width of the mineable ore zone cannot be calculated exactly because of imbrication, but is much broader than that of the eastern field. After evaluating the extent of the scheelite mineralization and host rocks in the neighbouring Hollersbachtal (about 6 km to the west of the Felbertal), it seems possible that the western field is longer than the eastern.

12. The lower schist sequence and the lower hornblendite cycle are thicker and more significant in the eastern field, whereas the interlayered schist sequence, the upper hornblendite cycle, and the upper schist sequence are especially thick and important in the western field.

13. Strict temporal, spatial and genetic relationships between the formation of the hornblendite cycles, the metal supplies, and the most important phases of ore mineralization are confirmed.

The presently proven, mineable ore bodies of the eastern field are linked with the lower hornblendite cycle. Those of the western field are associated with the upper hornblendite cycle and with layers in the middle section of the upper schist sequence. Sparse scheelite mineralization can be found in all types of rocks of the scheelite-bearing series of both fields. In the western field, the ore bodies are, to a great extent, distinguished from the surrounding rocks only by their enriched scheelite content ($\geqslant 0.3$ WO_3).

14. The ore minerals of the ore deposit Felbertal are: scheelite, powellite, wolframite, tungstenite-molybdenite solid solutions, molybdenite, pyrrhotite, pyrite, marcasite, pentlandite, sphalerite, galena, tetrahedrite, bornite, cobaltite, arsenopyrite, native Au, native Ag, native Bi, bismuthinite, galenobismuthite-cosalite, emplectite, stibnite, beryl, cassiterite, columbite, ilmenite, hematite, magnetite, chromite, apatite, fluorite, barite.

Some of these ore minerals are rare. Statistical tests showed significant differences in the ore mineralization between the eastern and the western field. Although most of the ore minerals have been found in both fields, the eastern field contains essentially pure and high-grade scheelite ores over large areas. In the western field, however, concomitant ore minerals occur frequently, particularly in quartzite lenses.

The following elements are significantly enriched in at least some layers of the scheelite-bearing series: W, Mo, Cu, Bi, As, Sb, Au, Ag, Be, Sn, Ta, Nb, Se, Te, Li, Cs, Ni, Co, Pb, Zn, Cd, Hg, P.

15. The tungsten supply occurred apparently to a large extent as Si-W-heteropoly acids (with minor amounts of P, Sb, As and other elements). The presence of $Na_2 WO_4$-bearing solutions and of W-halogenids seems possible.

16. Three generations of scheelite are known:

a) The primary generation represents a syngenetic-diagenetic, submarine ore formation. The ore-forming processes are related to complex interactions between igneous-hydrothermal activities, and the precipitation and sedimentation processes in the submarine environment.

The ore mineralization within non-quartzitic host rocks occurred on the sea floor, alternating with tuffs and reworked volcanic material, as well as within original lava layers.

All quartzite ores are derived from chemical sediments (ore-bearing silica gel). There are two quartzite ore types divergent in their ore fabrics and ore contents, through differences in the original silica gels, the associated ore mineralization, diagenetic alteration and metamorphic change. The first quartzite ore type with only fine-grained scheelite is usually of high quality. It is restricted to some sequences of the lower hornblendite cycle of the eastern field. The second, mostly with low WO_3 content, can be found in all parts of the scheelite deposit Felbertal, except those sequences of the first type.

b) In the second generation, recrystallization during Alpidian regional metamorphism produced porphyroblasts of scheelite. Their WO_3 content can be deduced from the old syngenetic-diagenetic ore mineralization. This generation is common in the western field and in many outcrops of the Habachserie outside the ore deposit Felbertal.

c) Scheelite crystals in fissures (third generation) are very rare within the ore deposit Felbertal, but common in some other mineralized regions of the Hohe Tauern. This fissure mineralization took place during a phase of tensional tectonism caused by the uplift of the Habachserie. It followed the main period of tectonic activity associated with the Alpidian regional metamorphism. These usually Mo-free scheelite crystals must have formed at slightly lower temperatures than the $CaMoO_4$-bearing scheelite porphyroblasts of the second generation. However, there is no evidence to indicate that fissure mineralizations occurred during a late, hydrothermal event.

17. About 90% of the mineable ores of the eastern field are part of the first generation. The fine-grained scheelite (diameter of grains $\leqslant$ 0.35 mm) is likewise enclosed in fine-grained quartzite with mosaic texture (diameter of grains $\leqslant$ 1 mm). This ore type is linked to two major quartzite "rich ore horizons" and to a great number of thin (up to some cm thick), quartzite bands and stringers.

These rich ore horizons, bands, and stringers are interpreted as former scheelite-bearing silica gel precipitation upon the sea floor as well as within the sediments just below the sea floor.

Both rich ore horizons show the best preserved syngenetic-diagenetic fabrics, exhibiting rhythmic sedimentation and submarine gliding (*Höll* et al., 1972, Figs. 6, 7, 8).

18. The scheelite-bearing quartzite bands and stringers, and both rich ore horizons are part of the middle to the uppermost section of the lower hornblendite cycle in the eastern field. There, three superimposed, elongated ore bodies are definable, if we take a workable limit of 2 m and a cut-off grade of 0.3% WO_3.

The lower and the middle ore body consist of several joint ore lenses.

The upper ore body (= main ore body) probably forms a continuous mineable ore mineralization along the extent of the elongated eastern field trough. It is up to 25 m thick along its central axis and contains both rich ore horizons (Fig. 2).

19. These three ore bodies reflect the configuration of a single thermal fissure fault with igneous-hydrothermal springs in the central area of the former trough. There, hydrothermal solutions flowed out on the sea bottom, while at the same time swells extended on the sides of this trough as the main centres of tholeiit-basaltic and ultramafic volcanic activity. The lenses of the lower and middle ore body are attributed to isolated spots of hydrothermal metal inflows along this thermal fissure fault and to ore deposition in the immediate surroundings. Such metal-rich inflows and very favourable conditions for precipitation of scheelite and silica gel must have existed near this thermal fissure fault during the formation of the main ore body. Thus, the lower rich ore horizon is an up to 100-m wide, elongated lense. Its average content is below 2% WO_3 at marginal positions, increasing proportionally to 3.5% WO_3 in the thicker lense centre. It swells abruptly to 8 m and shows intensive submarine gliding fabrics at several localities, considered as the most intensive emergent outlets of hydrothermal solutions along this thermal fissure fault. Along the elongated trend direction, the main ore body achieves its greatest thickness and highest average content of WO_3 (Fig. 2).

A slightly basic, physicochemical environment is thought especially favourable for the formation of these high-grade, quartzite, scheelite ores. Near the outflow source, the precipitation of $CaMoO_4$-bearing scheelite was favoured over the precipitation of silica gel and other ore minerals.

Furthermore, mostly scheelite-poor or barren, partially Na-rich, partially K-rich, acid volcanic tuffs (metamorphically altered to fine-grained gneisses) were temporarily and spatially bound to these hydrothermal activity (Fig. 2).

20. The axis of the elongated main ore body of the eastern field is almost parallel to the eastern slope of the valley. At the surface this ore body is broken into large boulders, caused by a post-glacial collapse. These boulders have slid down the slope for short distances, and formed two, some 100-m long, dumps of ore-bearing screes above the continuous solid ore-bearing rock piles. The eastern dump comprises 500,000 t of high quality ore.

21. In the western field the scheelite is concentrated above all in various nonquartzitic rocks. It shows an evident trend to coarse-grained recrystallization.

The broad, former trough of the western field comprises an up to 140 m thick section of the upper hornblendite cycle. This ore-bearing section was split up tectonically by the Alpidian orogeny into an upper and a lower ore-bearing wedge.

This upper hornblendite cycle does not contain a continuous ore enrichment throughout and within both wedges. However, rock sequences containing $0.2-0.25\%$ WO_3 are decameters thick, and within both wedges, lense-shaped ore bodies can be delimited with a 0.3% WO_3 cut-off grade. Many erratic, short-term hydrothermal springs along a wide thermal fissure fault were active throughout the broad trough of the western field.

In the upper schist sequence of the upper wedge, the eastern part of an elongated ore body has been investigated. This ore body is up to 10 m thick. It lies in metavolcanics, which can be related to former submarine layers of andesitic composition. The Mo-rich, coarse-grained scheelite mineralization is associated with scheelite-poor quartzite lenses.

On the other hand, some quartzite lenses within the upper two-thirds of the interlayered schist sequence of the lower ore-bearing wedge, are characterized by their remarkably enriched and unusual ore content. Thus, in a particular core drilling (named 3 H) 23 quartzite lenses, some up to several decameters thick, are found, with variable contents of WO_3 and particularly Bi (to 1.2%), Mo (to 0.7%), Cu (to 3.3%), Au (to 0.5 ppm) and Ag (to 120 ppm).

22. The ore reserves of the ore deposit Felbertal are 2,245,000 t with 0.75% WO_3, using a cut-off grade of 0.3% WO_3 (eastern field 1,550,000 t, western field 695,000 t). The most important or "main ore body" of the eastern field has 1,480,000 t of ores with 0.88% WO_3 (inclusive both dumps of ore-bearing screes) and is being mined at present.

In the eastern field there is still a large, undrilled area, whose reserves are not available for the total ore calculation.

In the western field only a small region has been explored exactly. Large quantities of ore are suspected in areas to the west and southwest.

4. Scheelite Occurrences Within the Habachserie (Excepting the Scheelite Deposit Felbertal)

A widespread, sparse scheelite mineralization occurs in the volcanic rock sequence of the Habachserie outside the scheelite deposit Felbertal (40 km in E-W and up to 25 km in N-S direction).

Scheelite outcrops are known just from the westernmost extensions of the Habachserie, stretching along the Äußeres Söllnkar on the eastern slope of the Krimmler Achental (Fig. 1, No. 7)[1]. This scheelite mineralization in a northern wedge of the Habach-serie can be traced about 5 km in the ENE direction to the valley floor of the Obersulz-bachtal. Scheelite crystals are also long known from the famous epidote occurrence Knappenwand on the eastern slope of the Untersulzbachtal in the further continuation of this wedge to the NE (No. 8).

Poor scheelite outcrops can also be traced in a second, almost parallel, elongated wedge of the Habachserie, about 2 km to the south of the northern wedge, separated by central gneiss masses (No. 6). There, scheelite grains have been discovered north of Humbachkarkopf and SW of Seebachsee. A section through the scheelite-bearing series is well exposed on the steep slope of the Ochsenwinkelgraben, a western tributary of the Obersulzbachtal, at about 1400–1610 m a.s.l. and at the way about 200 m north of the bridge across the Obersulzbach.

An additional, sparse, widespread scheelite mineralization was found in the lower Habachtal just on the western slope east of Feschwand, and across the steep eastern slope north of Fazenwand (No. 9). A continuation of this scheelite mineralization may extend to the scheelite-bearing $PbS-ZnS-CaF_2$ deposit Reichertleiten Alm – Scharrn Alm – Achsel Alm (Hollersbachtal) below overthrusted central gneisses (No. 10).

Another scheelite location is the famous abandoned emerald mine in the Leckbach-scharte within the upper part of the Habachtal (No. 11). A further scheelite-bearing area lies on the western slope of the uppermost Habachtal below the Steinkarl and on the eastern slope to the south of Noitroi (No. 12). Scheelite outcrops are discovered inter-mittently across this eastern slope up to Knoflachlahner and Blessachkopf, as well as at the glaciated Seescharte and north of the Kratzenberg.

The eastern continuation within the steep western slope of the Hollersbachtal is not yet known. However, scheelite is found infrequently at its base (No. 13). Scattered scheelite-bearing localities in two wedges also lie on the steep western slope of the Fel-bertal, tectonically overlying the western field of the scheelite deposit (No. 15). From there, scheelite outcrops stretch several km south to Eastern Tyrol.

A scarce scheelite mineralization, mostly covered by debris, has also been found in Wilhelmsdorfer Graben and at Lannbach Asten SE of the town of Mittersill (No. 17).

A sporadic scheelite mineralization can also be detected in the scheelite-bearing part of the Habachserie, ringing and dipping away from the central gneisses of the Granat-spitz arch. This scheelite-bearing rock sequence circles at the surface with a diameter of about 18 km. The scheelite deposit Felbertal lies at its NW-side. In the continuation of the eastern field, the ore content diminishes rapidly to the east and falls below 0.3% WO_3 at the western slope of the Amertal, at less than 2100 m a.s.l. From there, only very poor scheelite outcrops have been found in the remaining parts of the Amertal, in the Guggernbachtal and on the western slope of the Stubachtal (No. 16). Long-known scheelite localities on the eastern side of the Granatspitz arch are in the area Oberes Rifflkees (Totenlöcher) – Totenkopf – Unteres Rifflkees (No. 18). Further SW, schee-lite crystals have been found in the Laperwitztal and lower Fruschnitztal (No. 19). Well-exposed scheelite occurrences are found between the old and modern Felbertauern road between Landeckgraben and Petersgraben (No. 20). Several scheelite outcrops stretch 3–6 km more to the NNW (No. 21). The poor ore mineralization is best exposed south

---

[1] Numbers in parentheses in the text refer to points on Figure 1.

of the Felbertauern tunnel about 1.5 km along the Felbertauern road. Further localities are in the Meßelinggraben, in the Dabergraben, and in the pipeline tunnel as well as in the region of the Grüner See and Schwarzer See. From there, we can trace the scheelite mineralization to the western slope wedges of the Felbertal.

The scheelite occurrences within the Habachserie show the following characteristics:

They lie within the lower, up to several 100-m thick section of the volcanic rock sequence. The scheelite-bearing series is composed predominantly of metabasites, and it shows more lithological facies changes than the upper part of the volcanic rock sequence. However, there is no sharp boundary between the scheelite-bearing series and the superincumbent rocks. The scheelite mineralization ceases within uniform metabasite sequences. Areas with an enriched scheelite content are usually marked by the presence of characteristic dark, chromium-oxide green hornblendites and/or coarse-grained amphibolites as in the ore deposit Felbertal. The most important differences between this ore deposit and all other Habachserie scheelite occurrences is related to the concentration of the metal supply combined with the igneous-hydrothermal activity. Quartzite layers and stringers with scheelite and/or sulphide minerals are only rare outside the ore deposit Felbertal. However, these scheelite occurrences may be enriched by the elements W, Mo, Cu, Bi, Sb, As, Be, Au, Ag, Sn, Ni, Co, Zn, Pb, F, Ba, P. The identified ore minerals correspond with those of the ore deposit Felbertal. Most scheelite occurrences show only a simple association with scheelite, pyrite, pyrrhotite, chalcopyrite and molybdenite. Pyrite usually exceeds pyrrhotite here, in contrast to the ore deposit Felbertal. The scheelite mineralization is usually fine-grained and Mo-bearing. Scheelite porphyroblasts, with or without Mo, are mostly concentrated in small WO_3-rich layers or stringers, but are apparently also dependent on the host rocks and the degree of metamorphism. Thus, biotite-rich layers are occasionally distinguished by large scheelite crystals.

An unusual ore deposit lies in the area Reichertleiten Alm – Scharrn Alm – Achsel Alm on the western slope of the Hollersbachtal (No. 10). This ore deposit displays local pecularities, but also evident conformities with the scheelite deposit Felbertal, which is about 7 km to the SE. The ore mineralization comprises galena, Cd-rich sphalerite, fluorspar, and subordinate scheelite, chalcopyrite, pyrite, pyrrhotite, and arsenopyrite. The galena-sphalerite-fluorspar ore body is up to some m thick and originally stratabound within the scheelite-bearing series, but tectonically disrupted (*Höll*, 1970a, 1975; *Kreis* and *Unger*, 1971). More to the north, at the locality "Bärenbad", pyrrhotite, sphalerite, arsenopyrite, stannite, cosalite, and gudmundite are reported (*Ramdohr*, 1960, p. 1027). Moreover, I found chalcopyrite, galena, native Bi and bismuthinite; traces of Au and Ag are also present.

Maucher and *Höll* (1968) have pointed to the paragenesis W-Be, which may be widespread in the Eastern Alps. Meanwhile this statement has been repeatedly confirmed (*Höll*, 1970a, 1975; *Meixner*, 1971, 1976; *Lahusen*, 1972; *Cardoso*, 1973; *Niedermayr* and *Kontrus*, 1974; *Niedermayr* et al., 1976). This paragenesis has been ascertained above all in the Habachserie and in the eastern section of the Hohe Tauern, but also in the scheelite-stibnite deposits of the Kreuzeckgruppe and in the Thurntaler Quarzphyllitserie. A prominent example is the old emerald mine of the Leckbachscharte (Habachtal; No. 11), where scheelite findings (*Kontrus*, 1953) are reconfirmed, and stibnite and small Au contents identified for the first time (*Höll*, 1975). However, the maximum enrichment of Be may be in the western field of the scheelite deposit Felbertal.

In the Penninic zone of the western Hohe Tauern, successful scheelite prospecting results have also been obtained (*Vohryzka*, 1968; *Höll*, 1970a, 1971). Meanwhile a poor scheelite mineralization has been found in the tunnels of the Zemm hydro-electric plant (Dr. *O. Thiele*, pers. comm.), but no details are available.

Two km north of the village Großarl (No. 36), I found some samples with Mo-bearing scheelite grains within a rock sequence, which has been compared with the Habachserie and considered to belong to the Penninic zone (*Mostler*, 1963). This occurrence shows striking differences to the adjoining scheelite deposits within the Kleinarltal and the Lambachgraben (Großarltal) with regard to the Mo-content and the composition of the host rocks.

5. Scheelite Occurrences in the Penninic Zone of the Eastern Hohe Tauern (Rauristal — Sonnblickgruppe — Gasteinertal — Liesertal)

The easternmost scheelite mineralization of the middle part of the Hohe Tauern has been found on the eastern slope of the Stubachtal. There, Permomesozoic, barren rock series of the Glockner depression cover the Habachserie. Scheelite is, however, found over a large area east of this depression within the eastern part of the Hohe Tauern, from the Rauristal to the eastern end of the Tauern window. There, the scheelite occurrences lie within the imbricated Tauern schist cover, especially the "basal and central schist cover" (*Exner*, 1957, 1964, 1971) as well as in the central gneisses.

5.1 Scheelite Occurrences Within the Schist Cover

Scheelite mineralization lies in amphibolites and carbonate lenses of the "Gneislamelle 1" (*Exner*, 1962, 1964) on the southern slope of the Groß-Fleißtal NW of Gjaidtroghöhe, Carinthia (No. 22). Several scheelite occurrences are in the Hüttwinkeltal (Rauristal) in the area Erfurter Weg — Maschingraben — Wintergasse — Melcher Böden (No. 23). There, scheelite-bearing fissures, pockets, and quartz veinlets are within the basal and central schist cover and in the central gneiss. An eastward continuation can be found in the schist cover southwest of the Niedersachsenhütte and in Schlapperebental (No. 27). Scheelite surface outcrops and subsurface mineralizations are scattered in the Woisgenschieferserie in the Radhausberg area, especially in the Heilstollen (gallery) and on the dumps of the former gold mines (No. 28). Scheelite impregnations and fissure mineralization have also been discovered in quartzites of the upper Reitalpengraben (Großarltal), above all 200 m north of the Hetteg Alm (No. 31). The scheelite-bearing area at Elschekamm, Alte Hannoverhütte, and Plattenkogel, Carinthia (*Kontrus*, 1961) stretches to the uppermost Anlauftal (No. 32). Scheelite has also been found east of the Dr. Rudolf-Weißgerber-Hütte in the uppermost Mallnitztal, Carinthia (No. 34), in schists of the upper Kölnbreinkar at the Kalte Wandspitz in the uppermost Maltatal (No. 37), and at the eastern slope of the Hintereggengraben about 1 km north of the Rupp Alm (No. 39).

The first scheelite findings within the Eastern Alps were in the old gold-sulfide deposit Schellgaden (No. 38) (*Baumgartner* and *von Ettingshausen*, 1826). This ore deposit lies within a 30-km long row of similar ore deposits ("ore deposit type Schellgaden") stretching from the upper Murtal to the Radlgraben (*Friedrich*, 1953, 1968). This type of mineralization with scheelite was also found, as expected, in the highway tunnel through the Katschberg (*Strasser*, 1973). These metamorphic ore deposits are strata-bound within the upper part of the Mureckdecke (Mureck nappe), which is rich in amphibolites, hornblende gneisses, and migmatites (*Medwenitsch* et al., 1964). They exhibit conformable features, such as ore bodies up to several m thick, consisting of long parallel quartzite layers intercalated with schist lenses. The quartzite layers contain alternations of quartzite and a few-mm thick parallel bands and lenses with the following ore minerals: Pyrite, pyrrhotite, chalcopyrite, galena, sphalerite, scheelite, wolframite, stolzite, molybdenite, native Au, native Ag, altaite, tetradymite, sylvanite, arsenopyrite, cubanite, bornite, tetrahedrite, hematite (*Ramdohr*, 1952; *Friedrich*, 1953, 1968; *Meixner*, 1953). Pyrite is the prevailing ore mineral in all deposits. Scheelite is often concentrated in lenses together with the sulphide ore bands (especially in the abandoned mine Schulterbau) or within pockets (often in the old mine Stüblbau). These mineralizations in quartzite layers are very similar to quartzite layers with sulphide minerals, scheelite and native Au and Ag in the scheelite deposit Felbertal, especially within the interlayered schist sequence of the western field. In my opinion the ore formation must be related to submarine, cyclically alternating silica gel and ore mineral deposition connected with an igneous-hydrothermal activity. Associated basic and ultrabasic metavolcanics appear to be approximately contemporaneous and most probably Early Paleozoic in age. This interpretation contradicts former genetic explanations suggesting an Alpidian origin (*Petrascheck*, 1966; *Friedrich*, 1968; *Exner*, 1971). *Frisch* (1972) regarded these quartzite layers as epigenetic quartz veins, and their ore content as a remobilization product out of the surrounding amphibolites during the Variscan orogeny. However, this model conflicts with all the features of these ore deposits and is mere conjecture.

The strata-bound scheelite occurrences within the schist cover of the eastern Hohe Tauern, together with Mo, As, Sb, Bi, Au, and Ag mineralizations, correspond to the scheelite deposit Felbertal and other scheelite occurrences within the Habachserie. They are often intimately connected with basic metavolcanic layers. Syn- and posttectonic, coarse-grained recrystallization of the ore minerals is evident above all in high-grade metamorphic rock sequences.

5.2 Scheelite Occurrences Within the Central Gneisses

The central gneisses of the eastern Hohe Tauern are usually regarded as Variscan intrusions, which were later subjected to the Alpidian regional metamorphism. Their hybride character has been demonstrated especially by *Exner* (1953, 1957, 1964). Genetic evidence for the scheelite mineralization exists particularly in the hydro-electric plant gallery between Wurtental and Großes Zirknitztal (Nr. 35). There, the central gneiss masses of the Sonnblick contain finely dispersed scheelite impregnations, which must be interpreted as original components of the old intrusions. However, during the Alpidian metamorphism, scheelite was also mobilized and concentrated along foliation planes, or in pockets, fissures, and veinlets with or without quartz and carbonate minerals.

The scheelite mineralization of the central gneisses is considered as an ore content out of older, Early Paleozoic rock series in connection with the Variscan palingenic magma melting. Scheelite occurrences of this genetic type are presented in Figure 1 and Table 1, Nos. 24, 25, 26, 29, 30, 33, 35.

Table 1. List of mineral locations of the Sb-W-Hg formation in the Eastern Alps, presented in Figure 1

1. Sporadic scheelite mineralizations to the west of Fellimännle and in the valley NE of Furkla (within the Silbertal)
2. Scheelite deposit near Konstanzer Hütte
3. Scheelite-magnesite deposit Tux
4. Run down Cu-Sb mine Volderwildbad with scheelite
5. Scheelite deposit Navistal
6. Scheelite outcrops on the western slope of the Obersulzbachtal (within the southern wedge of the Habachserie)
7. Scheelite outcrops on the western slope of the Obersulzbachtal and on Äußeres Söllnkar (Krimmler Achental) (within the northern wedge of the Habachserie)
8. Sporadic scheelite crystals within the old epidote mine Knappenwand (Untersulzbachtal)
9. Scheelite outcrops on both slopes of the Habachtal to the north of the Fazenwand
10. Dispersed scheelite mineralizations in the Reichertleiten Alm − Scharrn Alm − Achsel Alm area (Hollersbachtal)
11. Scheelite in the abandoned emerald mine in the Leckbachscharte (Habachtal)
12. Several scheelite occurrences on the eastern slope of the uppermost Habachtal to the south of Noitroi including Knoflachlahner, Blessachkopf, and Kratzenberg, as well as on the western slope below Steinkarl
13. Sporadic scheelite mineralizations on the western slope of the Hollersbachtal (Schachern − Marchlecker Kar)
14. Scheelite deposit Felbertal (eastern and western field)
15. Scheelite outcrops in two wedges on the western slope of the Felbertal
16. Traces of scheelite from Amertal to Stubachtal
17. Scheelite mineralizations within the Wilhelmsdorfer Graben and at Lannbach Asten
18. Scheelite in the area Oberes Rifflkees (Totenlöcher) − Totenkopf − Unteres Rifflkees
19. Scheelite in Laperwitztal and Unteres Fruschnitztal
20. Several scheelite occurrences in the Tauerntal between Felbertauern tunnel and Landeckgraben
21. Several scheelite occurrences in the Tauerntal between Landeckgraben and Petersgraben
22. Scheelite outcrops on the southern slope of the Groß-Fleißtal NW of Gjaidtroghöhe
23. Several scheelite occurrences in the Hüttwinkeltal (Rauristal) from Erfurter Weg to Melcher Böden
24. Many scheelite-bearing veinlets in the accessible western part of the Imhof-Unterbau gallery, above all between gallery m 3300 and 3970
25. Scheelite in the valley below the Mitterasten Alm (Hut)
26. Some scheelite occurrences between Oberer Bockhart-See, Kleiner Silberpfennig, Schleierfallstollen (gallery), and Schöneck Alm
27. Scheelite SE of Niedersachsenhütte and within Schlapperebental
28. Scheelite surface outcrops and subsurface mineralization within the Radhausberg, especially in the Heilstollen (between gallery m 1340 and 1420)
29. Traces of scheelite in the quarry Hirschau SSW of Badgastein
30. Scheelite in Kötschachtal, especially the Döferlbach valley (1100−1400 m a.s.l.)
31. Scheelite localities in the upper Reitalpengraben (Großarltal) (above all 200 m north of Hetteg Alm)
32. Scheelite outcrops at Elschekamm, Alte Hannoverhütte, Plattenkogel, and uppermost Anlauftal
33. Scheelite in hydro-electric plant gallery between Großer Oschenik-See and Kleiner Oschenik-See
34. Scheelite occurrence east of Dr. Rudolf Weißgerber-Hütte in the uppermost Mallnitztal
35. Many scheelite-bearing localities in the hydro-electric plant gallery between Wurtental and Großes Zirknitztal
36. Scheelite about 2 km to the north of the village Großarl
37. Scheelite mineralization in upper Kölnbreinkar at Kalte Wandspitz
38. Scheelite-gold-sulphide deposits of the ore deposit type Schellgaden

Table 1 (continued)

39. Scheelite locality 1 km north of the Rupp Alm (Hut) at the eastern slope of Hintereggengraben
40. Scheelite deposit in the Kleinarltal to the south of the Seyfriedgraben near Griespalfen (1100–1250 m a.s.l.) and scheelite outcrop at the Wagrain-Mitterkleinarl road 50 m to the south of the Alpenrose inn
41. Scheelite mineralizations stretching from the Schneeleiten to the Fürbachgraben
42. Two adjacent scheelite occurrences in two brooklets (Stuhlgraben) (1490–1650 m a.s.l.), about 2 km east of the village Mitterkleinarl
43. Scheelite outcrops in Lambachgraben and Plojergraben
44. Traces of scheelite at the Seekar north of the village Obertauern
45. Scheelite occurrence Untertal near Schladming
46. Widespread scheelite and arsenopyrite mineralizations within the Turnthaler Quarzphyllitserie in the area Sillian – Innervillgraten – Außervillgraten – Panzendorf
47. Scheelite in Lappgraben
48. Traces of scheelite in Sägebach (1180 m a.s.l.) south of the village Panzendorf
49. Stibnite deposit Obertilliach with traces of scheelite
50. Scheelite-stibnite deposits Gomig, Marienstollen, Johannisstollen, Rabant, Edengang, Gurser-kammer, and Strieden at the southern slope of the Kreuzeckgruppe
51. Scheelite-stibnite deposits Lessnig, Radlberg, Tränkergraben, Pirkeben near Kleblach at the southeastern slope of the Kreuzeckgruppe
52. Stibnite deposit Guginock near Lend
53. Stibnite deposit Brückl
54. Cinnabar deposit Glatschach near Dellach
55. Cinnabar deposit Stockenboi
56. Cinnabar deposit Hohes Kohr
57. Cinnabar deposit Rottrasten
58. Cinnabar mineralization in Magdalensberg area
59. Cinnabar occurrence Eisenkappel
60. Cinnabar deposit Vellacher Kotschna

6. Scheelite-Magnesite Deposit Tux

As mentioned in the introduction, *Höll* and *Maucher* had published new results concerning genesis and age of this ore deposit (No. 3) by 1967. This deposit is intercalated between a rock sequence of the Innsbrucker Quarzphyllitserie, belonging to the lower Austro-Alpine unit. It consists of an up to 3-m thick scheelite-bearing horizon and an overlying dolomite – sparry magnesite bed. Conodonts within this bed provided an age from the Upper Ludlovian to Lower Emsium (*Höll* and *Maucher*, 1967). *Mostler* (1973) showed that this up to several tens of m thick carbonate bed starts in the high Middle Ludlovian and reaches as far as Lower Emsium. For the origin of the sparry magnesite, an organogenic Mg-preconcentration is postulated, and a Mg-metasomatism, Mg-concentration and sparry magnesite crystallization caused by the Variscan metamorphism are considered. Close temporal and spatial relations are proven between this carbonate bed and the scheelite-bearing horizon. Thus, this horizon may be of Lower or Middle Ludlovian age. It contains (carbonate-)quartz bands and graphitic schists, as well as stratiform laminas, lenses and impregnations of ore minerals, above all scheelite, pyrite and apatite. This strata-bound ore mineralization and the host rocks show the same tectonic deformation (ore type I, *Wenger*, 1964). Moreover, there are mobilized ores in quartz-scheelite veinlets, originated during metamorphism (ore type II) and some scheelite crystals in fissures and pockets (ore type III). The scheelite of all ore types shows a blue fluorescence and is characterized by a poverty of Mo. Further, usually rare minerals are: tungstenite, wolframite, hydrotungstite, stibnite, molybdenite, galena, sphalerite, bournonite, boulangerite, chalcopyrite, chalcocite, tetrahedrite, bindheimite, malachite and azurite (*Wenger*, 1964; *Höll* and *Maucher*, 1967). Geochemical investigations revealed at least traces of W, Sb, Cu, Pb, Zn, As, Mo, Be, Bi, Sn and Hg.

The metal supply for the tungsten mineralization and the associated ore minerals may be related to an igneous-hydrothermal activity probably connected with submarine basic volcanic layers in the footwall. The cessation of the stratiform ore mineralization, where the graphitic facies thins out, suggests strong environmental control.

Other scheelite occurrences within the Innsbrucker Quarzphyllitserie: scheelite-bearing samples have been found at the small, run down Cu-Sb mine Volderwildbad (No. 4; *Höll*, 1971). A strata-bound scheelite mineralization is also known in the Navistal (Nr. 5) within graphitic schists and carbonate layers, where conodonts provide a Silurian age (Dr. *Th. Bechstädt*, pers. comm.).

7. Scheelite Deposits in Kleinarltal and Adjacent Lambachgraben/Großarltal

These strata-bound scheelite deposits, which I discovered in 1968 (*Höll*, 1971a), are scattered within an area of 8 km in N-S and up to 6 km in E-W direction. They are confined to the Radstädter Quarzphyllitserie within the lower Austro-Alpine unit. Fossils, also within the scheelite-bearing carbonate rocks favour an Early Paleozoic, probably Silurian age (*Schönlaub*, 1973b, 1975). The ore-bearing rock sequence is locally underlain by submarine volcanic rocks (*Wagner*, 1972). It is intensely tectonically disturbed and separated in different ore deposits. It consists of carbonate rocks, especially iron-bearing dolomite, locally also of quartzite intercalations, quartzphyllites, and black and grey phyllites. This rock sequence is from several decimeters up to 20 m thick.

The main scheelite deposit is about 300 m long. It lies to the south of the Seyfriedgraben near "Griespalfen" at 1100–1250 m a.s.l. (No. 40). There, laminar and pocket-like scheelite concentrations (to 7% WO_3) adjoin barren host rocks. A poor scheelite mineralization is exposed at the road Wagrain–Mitterkleinarl, 50 m to the south of the Alpenrose inn. Scheelite outcrops stretching from the Schneeleiten to the Fürbachgraben (at 1520 m a.s.l.) are also indigent (No. 41). Two adjacent scheelite occurrences lie about 2 km east of the village Mitterkleinarl in two brooklets (Stuhlgraben) at 1490–1650 m a.s.l. (No. 42). Iron-bearing dolomite lenses with sporadic scheelite grains can be traced over a distance of 1.5 km in the uppermost Plojergraben. About 1 km further west, ore-bearing carbonate boulders are concentrated in the Lambachgraben, a tributary of the Großarltal, at 1680–1705 m a.s.l. (No. 43). A continuation of the scheelite-bearing strata from the Plojergraben to the Lambachgraben may exist below a debris cover.

All these scheelite deposits display conformable characteristics: they are strata-bound in an up to 20-m thick rock sequence, composed of carbonate rocks, above all iron-bearing dolomite, locally also of quartzite intercalations, quartz-phyllites and black and grey phyllites. Thick beds with massive, often siliceous dolomite are usually intersected by veinlets of quartz and calcite; the ore mineralization is distributed irregularly in thin fissures as well as in lenses and pockets with coarse scheelite grains. On the contrary, dark, bituminous, finely laminated carbonate layers are dominant in places, where the scheelite-bearing rock sequence is only up to 1 m thick. There, above all in the Stuhlgraben, the most excellent fabrics of synsedimentary-diagenetic scheelite mineralizations of all strata-bound scheelite deposits of the Eastern Alps are present and represent a quiescent-area facies. A distinct stratification caused by variable contents of calcite, dolomite, quartz, graphite and scheelite is indicated. The fine-grained scheelite lies strictly parallel to the layering of the host rocks and shows the same tectonic deformation. Laminas are displaced along minute faults as a result of subaqueous sliding and diagenetic compaction. In places, the scheelite forms small lenses, deposited in gentle depressions of an erosional disconformity (*Höll* et al., 1972, Figs. 2–5). Thin, stratiform ore enrichments can be found at the lower or upper boundary of some coarse-grained dolomite lenses.

Geochemical and ore microscopical investigations prove that the ore deposits of the Kleinarltal represent the purest scheelite mineralization in the Eastern Alps. The scheelite is only blue fluorescent and extremely Mo-poor. The associated ore minerals pyrite, pyrrhotite and arsenopyrite are rare and there are no indications for traces of Au, Ag, Bi, Be and Sn.

A comparison of the scheelite deposits of the Kleinarltal with the ore deposit Tux at a distance of about 120 km reveals striking similarities and common features: both lie within quartzphyllite series of the lower Austro-Alpine unit on the northern side of the Tauern window. In Tux, the scheelite-bearing black schist horizon is overlain by a dolomite – sparry magnesite bed and locally underlain by basic metavolcanics. In Kleinarltal, the scheelite deposits are also connected with unusual carbonate rocks and locally underlain by basic and acid metavolcanics.

Fossils confirmed the Silurian (Ludlovian) age of the Tux scheelite mineralization. Silurian age is also indicated for the scheelite deposits of the Kleinarltal. In both, the scheelite is extremely Mo-poor, and synsedimentary-diagenetic ore fabrics are identified.

8. Scheelite Occurrences in the Silvretta-Kristallin

In westernmost Austria, several scheelite-bearing areas have been delineated between the Swiss-Austrian border and the town of Landeck (Tyrol) in high-grade metamorphic rocks of the middle Austro-Alpine unit.

Scheelite was found in the valley northeast of Furkla and in an occurrence west of Fellimännle in the Silbertal (Vorarlberg; No. 1). Veinlets and pockets with scheelite lie within amphibolite and hornblende gneisses in a paragneiss series.

A scheelite deposit probably within the same staurolite-bearing paragneiss series is exposed just north of the Konstanzer Hütte at 2100–2200 m a.s.l. (No. 2). Other scheelite outcrops are in the Fasultal SSW of this hut and in the adjacent Rosannatal. The ore is concentrated within a sequence of amphibolite layers and hornblende gneisses. The scheelite reveals striking features of metamorphic migration, especially a strong tendency to form discordant fissure veins and pockets with coarse-grained recrystallization. The ores contain small amounts of Mo, Cu, Ag, Au, Bi, Sn, and additional Co and Ni.

9. Scheelite-Stibnite Deposits of the Kreuzeckgruppe

As mentioned in the introduction, these deposits were investigated by the author in 1966. The main results concerning the newly discovered scheelite mineralization and the genesis of the strata-bound ore deposits, connected with probable Early Paleozoic basic metavolcanic layers, have been published by *Höll* and *Maucher* (1967). Later on, *Lahusen* (1969, 1972) has confirmed and extended these results, which are in contradiction to all former genetic explanations, based on an Alpidian replacement.

These scheelite-stibnite deposits are concentrated in two districts. The Rabant district comprises the deposits Gomig, Marienstollen, Johannisstollen, Rabant, Edengang, Gurserkammer and Strieden (No. 50). It stretches about 7 km in E-W direction north of the villages Nikolsdorf and Oberdrauburg. The Lessnig-Radlberg district is about 30 km to the east near Kleblach and comprises the deposits Lessnig, Pirkeben, Tränkergraben, and Radlberg (No. 51).

These scheelite-stibnite deposits, except the metamorphic lithogene occurrences Gomig and Marienstollen, are strata-bound within an approximately 100-m thick sequence in the middle section of the Rabantserie (*Lahusen*, 1969, 1972). This series is built up mainly by phyllites, quartzphyllites, mica schists, and quartzites. In the Lessnig-Radlberg district there are also layers of carbonate with some magnesite and siderite. The ore-bearing sequence is characterized by several, concordant layers of metabasites, quartzite schists, mica schists, and the strata-bound ore mineralization.

Fossil debris, found at Lessnig, enables one to exclude a Precambrian age of the Rabantserie and indicates an Early Paleozoic origin (*Lahusen*, 1969). The metamorphism and the major tectonic deformations of the Rabantserie with intense block faulting are of pre-Alpidian (Variscan) age. The "Altkristallin" of the southern Kreuzeckgruppe, enclosing the Rabantserie, has been assigned to the middle Austro-Alpine unit by *Tollmann* (1963), and to the upper Austro-Alpine unit by other authors.

The parageneses of all strata-bound deposits are uniform, with stibnite the prevailing ore mineral; pyrite, marcasite and arsenopyrite are abundant, and scheelite is usually only subordinate. Pyrrhotite, chalcopyrite, sphalerite, galena, and jamesonite are present. Enhanced contents of Sb, As, W, Au, Cu, Pb, Zn and traces of Mo, Bi, Ag, Hg, and Be have been found in most deposits. In the Rabant district, stibnite-rich ores are concentrated within an up to 5-m thick graphite schist horizon, which is locally intercalated by basic metatuff bands. This main ore horizon is often tectonically disturbed or even mylonitized. In the ore deposit Lessnig the stibnite mineralization is concentrated within at least two major graphite schist horizons, and separated by layers of metavolcanics, quartzite lenses, and mica schists.

The originally stratiform ore mineralization comprises bands with parallel lenses, disseminations, and pockets with ore minerals. Stibnite rich ore bands interlaminated by graphite schists and thin bands of metavolcanics are abundant in the ore deposits Rabant, Gurserkammer, and Lessnig. These ore bands consist of finely laminated and fine-grained stibnite, locally associated by thin pyrite and arsenopyrite lenses and traces of scheelite.

The ore supply must be seen in spatial, temporal and genetic connexion with the submarine volcanism of the ore-bearing series and an associated hydrothermal activity. Veinlets and fissure veins with coarse-grained stibnite, scheelite, quartz, and carbonate minerals are also concentrated within the ore-bearing series. They are metamorphic lithogene products.

10. Scheelite and Arsenopyrite Deposits in the Thurntaler Quarzphyllitserie

In 1969 and 1970, I discovered scheelite deposits within the area bound by the villages Sillian – Innervillgraten – Außervillgraten – Panzendorf (No. 46) and a small outcrop at Sägebach (No. 48). Later on details were investigated by *Krol* (1974).

The Thurntaler Quarzphyllitserie is very similar to the Rabantserie of the Kreuzeck-gruppe and also incorporated in the middle Austro-Alpine unit. Its age is also considered Early Paleozoic and its metamorphism as pre-Alpidian (Variscan) (*Höll*, 1971; *Krol*, 1974; *Heinisch* and *Schmidt*, 1976). Its lower section is composed mainly of mica schists and quartz-phyllites. Basic and acid metavolcanic layers, as well as quartzites and graphite schists, are intercalated in the middle section. Quartz-phyllites, chlorite-phyllites, chlorite-epidote schists, and quartzite layers are prevailing in its upper part.

The scheelite mineralization is concentrated within the sequence of basic metavolcanics (mainly metatuffs and metatuffites), which thickens up to 200 m between the hills Gumriaul and Marchkinkele at the Austrian-Italian border. From there, the axes of these basic metavolcanic layers plunge to the ENE. The volcanic rock masses diminish in the northwestern and southeastern marginal positions, where the interdigitations of facies are most common. The maximum scheelite concentration is obviously conformable to these axes. Laminas, lenses, and impregnations of scheelite represent primary components of the submarine, basic volcanic host rocks and of intercalated phyllites. The same genetic interpretation is favoured for arsenopyrite mineralizations within these basic metavolcanics, often together with scheelite. Discordant veins, veinlets, and pockets with quartz, carbonate minerals, scheelite, and arsenopyrite are widely distributed and of lithogene origin.

Geochemical investigations of ore-bearing samples verified the enrichment of W, As, Au (to 8.5 ppm), Sb, Ag, as well as traces of Mo, Cu, Pb, Zn, and Hg.

Arsenopyrite and pyrite ores in small veins are also present in the upper section of the Thurntaler Quarzphyllitserie, especially around the Thurntaler hill.

A comparison of the ore deposits of the Sb-W-Hg formation within the Rabantserie and the Thurntaler Quarzphyllitserie shows stibnite is the foremost and arsenopyrite an ubiquitous, but subordinate ore mineral in the Kreuzeckgruppe deposits (excepting the cinnabar deposit Glatschach). However, the ratio As to Sb is undoubtedly shifted in favour of As in the Thurntaler Quarzphyllitserie.

Moreover, temporal, spatial, and genetic relationships exist with old known sulphide deposits within the Rabantserie and the Thurntaler Quarzphyllitserie. Thus, several Cu-bearing sulphide deposits are situated east of Innervillgraten – Panzendorf. They are strata-bound in the basic metavolcanic sequence of the Thurntaler Quarzphyllitserie, and locally they contain small amounts of scheelite. In addition, many strata-bound sulphide occurrences and deposits (e.g., Strieden) are in the neighbourhood of the Rabant district scheelite-stibnite deposits, probably also of contemporaneous origin within the same basic metavolcanic sequence.

11. Stibnite Deposit Obertilliach

This run-down mine is about 3 km NW of the village Obertilliach (No. 49). According to former reports, this ore deposit in the Gailtal-Kristallin is similar to the stibnite deposits in the southern Kreuzeckgruppe. Fine-grained stibnite is concentrated within a graphite schist horizon. Stibnite-bearing samples proved the presence of As, W, Hg, Cu, Pb, and Zn (*Lahusen*, 1969; *Höll*, 1975). A basic metavolcanic layer in the footwall contains traces of W and Be. With reservation, this ore deposit may be regarded as strata-bound within possibly Early Paleozoic host rocks.

12. Stibnite Deposit Guginock

This mine lies about 1.5 km east of the village of Lind and about 4 km from the schee-lite-stibnite deposit Lessnig. It is collapsed in large parts. Veinlets, lenses, and pockets, usually with long-columnar stibnite within a marble bed, and ore enrichments at the boundary to adjacent phyllites have been proved. The primary ores are possibly strata-bound, and genetic relationship with concomitant basic metavolcanics may exist. However, a strong syn- and posttectonic ore mobilization and recrystallization are evident.

13. Stibnite Deposit Brückl

This small, old mine and dumps were discovered by *Thiedig* in 1966. Surface outcrops of the ore mineralization are lacking. The host rocks are of Early Paleozoic (probably Silurian-Devonian) age. *Meixner* and *Thiedig* (1969) suggest that this deposit is genetically related to a Cretaceous-Tertiary fault zone. According to my own studies, the coarse-grained stibnite samples show characteristics known from many recrystallized stibnite ores within ore deposits of different origins. It seems possible to me that originally strata-bound ores within the Early Paleozoic rock sequence are the source of these recrystallized ores. Thus, a genetic conformity with other strata-bound Early Paleozoic ore deposits of the Sb-W-Hg formation may not be excluded.

14. Cinnabar Deposit Glatschach

This abandoned mine in the southern Kreuzeckgruppe is about 10 km east of the (scheelite-)stibnite deposits of the Rabant district and 20 km west of those of the Lessnig-Radlberg district (No. 54). It lies within a block, representing a continuation of the Rabantserie. The ore minerals are cinnabar, pyrite, arsenopyrite, marcasite, rare stibnite, chalcopyrite, galena and sphalerite (*Friedrich*, 1965). The major ore body is a pyritic graphite schist horizon interbedded with basic metavolcanics. Cinnabar-bearing pyrite-arsenopyrite bands alternate with graphitic quartzite layers and basic metatuff lenses, where impregnations of cinnabar are prevailing. This rock sequence represents a submarine ore formation connected with a basic volcanism and igneous-hydrothermal solutions similar to the rock series of the strata-bound (scheelite-)stibnite deposits to the west and to the east. Moreover, a basic metavolcanic layer in the hanging wall displays small concentrations of W, Sb, As, Hg, Cu, Pb, Zn (*Lahusen*, 1969), proving a further affinity with the neighbouring (scheelite-)stibnite deposits of the Rabantserie.

This deposit is the only known cinnabar deposit within the middle Austro-Alpine unit, the others are within the upper Austro-Alpine unit. Excepting the higher metamorphic grade, conformable characteristics of the ore mineralization are evident in these proved or probable Early Paleozoic cinnabar deposits.

15. Cinnabar Deposits Hohes Kohr and Rottrasten

These ore deposits are about 5 km apart (Nos. 56 and 57). The adit and old stopes of the deposit Hohes Kohr near Turracher Höhe are well preserved, the old trenches and galleries of Rottrasten near the village Ebene Reichenau are collapsed. Both deposits lie in similar rock sequences of the Eisenhut-Schieferserie, which belongs to the Gurktaldecke (nappe) within the upper Austro-Alpine unit. This series is composed of slates, and additional carbonate beds and volcanic layers. At Hohes Kohr, the ore mineralization is strata-bound within an up to 15-m thick sequence of basic tuffs and tuffites, as well as amygdaloidal rocks. At Rottrasten, graphite schists are also present. The host rocks and the ore mineralization of the deposit Hohes Kohr must be older than Wenlockian, probably of Upper Ordovician or Lower Silurian age (*Höll*, 1970b). The ore mineralization of the deposit Rottrasten may be contemporaneous. Both deposits contain only a poor Hg content: Hohes Kohr 0.05–0.11% Hg, Rottrasten still less (*Friedrich*, 1965). The ore minerals are cinnabar, pyrite, chalcopyrite, tetrahedrite, malachite, azurite, covellite, hematite and magnetite.

These ore deposits are not vein or metasomatic deposits of Alpidian origin, as regarded by *Friedrich* (1965) and *Petrascheck* (1966), but show characteristics of an originally synsedimentary-diagenetic ore deposition, including:

1. Fine-grained cinnabar (diameter usually less 0.2 mm) is impregnated within all volcanic rocks of the ore-bearing rock sequence, indicating it was a primary component.

2. Additional fine-grained cinnabar enrichments within amygdaloidal infillings (*Zirkl*, 1955; *Höll*, 1970b) in a layer at Hohes Kohr. It follows that these infillings (quartz, calcite and cinnabar) have a diagenetic origin.

3. Thin lenses with quartz, carbonate minerals, pyrite, and cinnabar, exist above all at the boundary of several layers, and at Rottrasten additionally within graphite schists. This ore mineralization may be seen in connection with a precipitation of submarine hydrothermal solutions during a pause of the volcanic deposition.

4. Cinnabar concentration, with quartz and calcite along foliation planes and within irregular pockets, is common in both deposits. This mineralization proves a syn- to posttectonic ore mobilization, migration and redeposition.

5. The posttectonic cinnabar mineralization in fissure veins and veinlets must also be considered as of lithogene origin.

Such features are also well known from many other strata-bound cinnabar deposits (*Maucher*, 1965, 1972, 1976; *Höll*, 1966, 1970b; *Maucher* and *Höll*, 1968).

16. Cinnabar Deposit Stockenboi

According to my own studies (*Maucher* and *Höll*, 1968; *Höll*, 1970b) and investigations by *Lahusen* (1969, 1972) and *Schulz* (1969) this ore deposit (No. 55) is also strata-bound within a rock series very similar to the proved Early Paleozoic Eisenhut-Schiefer-serie. The ore mineralization is confined to quartzite layers and adjoining schists.

17. Cinnabar Mineralization in the Magdalensberg Area

Names of places and brooklets containing the word "Zinnober" (cinnabar) indicate former discoveries of cinnabar mineralizations in the Magdalensberg area. However, no cinnabar could be found on the dumps in the last years. Yet geochemical investigations of some of my samples proved poor Hg concentrations in basic volcanic tuffs, which are most probably of Caradocian or Ashgillian age (Dr. *Riehl-Herwirsch*, pers. comm.). The host rock series is strikingly similar to those of the strata-bound cinnabar deposits Hohes Kohr and Rottrasten.

18. Cinnabar Occurrence Eisenkappel and Cinnabar Deposit Vellacher Kotschna

A small gallery at the Leppenberg (680 m a.s.l.), some hundred m SE of the village Eisenkappel (No. 59), intersects a cinnabar-bearing rock sequence, which has no distinct boundaries to the barren overlying and underlying rocks. Very fine-grained, rare cinnabar impregnations, representing a primary rock component, are within basic volcanic layers and tuffs in a thick volcanic pile of Upper Ordovician age. Small cinnabar enrichments, together with quartz and carbonate minerals, occur in fissure veins, veinlets, and pockets, concentrated by lithogene processes.

The cinnabar deposit Vellacher Kotschna (No. 60) lies in the southernmost part of Austria about 11 km SSW of the village Eisenkappel, south of the tectonic boundary between the Eastern and the Southern Alps. A gallery and outcrops of cinnabar ores demonstrate that the ore mineralization is strata-bound within an up to 20-m thick, dolomite bed of Triassic (Ladinian) age.

The small cinnabar deposit St. Ana at Loiblpass and the famous large cinnabar deposit Idrija (Yugoslavia), are also in the Southern Alps, strata-bound and of Triassic age (*Berce*, 1963; *Mlakar* and *Drovenik*, 1971, 1972).

19. Conclusions

1. The characteristics of the large, newly discovered, Early Paleozoic scheelite deposit Felbertal have been submitted in detail.

2. A widespread metal deposition with W, Sb, Hg, and associated elements Mo, Bi, Au, Ag, Cu, As, Be, Sn, Pb, Zn, F, P, Ni, Co, Te, Nb, Ta, Li, Cs in some scheelite and stibnite deposits, took place in the Eastern Alps during Early Paleozoic time.

This period of mineralization may have extended from Upper Ordovician (Caradocian-Ashgillian stage; probably most ore deposits) to Silurian time.

3. The ore formation is genetically, temporally and spatially related to a submarine, basic, in part also ultramafic and acid volcanism and associated igneous-hydrothermal activities.

The Early Paleozoic volcanics of the Eastern Alps are similar to volcanic rocks of modern marginal basins. This hypothesis is favoured by their chemical composition, the volume distribution, the bathymetric development and the stratigraphic succession (*Höll* et al., in press). Geochemical data of "immobile" elements (Ti, Zr, Y, P) of volcanic rocks are interpreted as indications for the existence of a contemporaneous subduction zone (*Loeschke*, 1975, 1976). The volcanic processes reflect an evolution from submarine to subaerial stages. There are domains of volcanism with interrupted activity and shifting eruption centres.

4. The major tectonic boundaries within the Eastern Alps are in part also sharp boundaries reflecting the distribution of the different, strata-bound, Early Paleozoic ore deposits of the Sb-W-Hg formation. These relationships allude to the different ore parageneses as well as specific features of magmatism, facies, and thickness of the ore-bearing host rocks. Thus, the known scheelite deposits in the Eastern Alps are restricted to the Penninic zone and the lower and middle Austro-Alpine unit.

The most intensive scheelite mineralization is within the Penninic zone, often accompanied by small concentrations of Mo, Cu, Bi, Au, Ag, Be, Sn (ore deposit Felbertal), usually in thick volcanic rock sequences.

In the lower Austro-Alpine unite, the scheelite mineralization is restricted to a few regions (e.g., ore deposits Tux and Kleinarltal), where extremely Mo-poor scheelite is present in some layers together with graphite schists and carbonate host rocks (dolomite – sparry magnesite in Tux, iron-bearing dolomite in Kleinarltal).

In the middle Austro-Alpine unit, scarce scheelite occurrences and several stibnite deposits with scheelite and arsenopyrite in graphite schists and metavolcanic layers (Rabant district and Lessnig-Radlberg district/Kreuzeckgruppe) as well as arsenopyrite and scheelite deposits with gold and traces of stibnite (ore deposits in the Thurntaler Quarzphyllitserie) are present. Moreover, small stibnite occurrences, a cinnabar deposit (Glatschach) and numerous sulphide deposits with Cu lie in Early Paleozoic rocks of this tectonic unit.

The upper Austro-Alpine unit is characterized by several strata-bound cinnabar deposits and sulphide deposits with Cu, Pb and Zn.

5. These different tectonic units were formed by the Alpidian orogeny. Their arrangement corresponds with the palinspastic position of the Early Paleozoic rock series within a marginal basin with probably parallel, individual trough systems and intervening uplifts.

The Early Paleozoic rock series of the upper Austro-Alpine unit represent the accepted southernmost depositions on palinspastic schemes. Those of the middle Austro-Alpine unit are sediments within regions more to the north, and those of the lower Austro-Alpine unit must have been still further to the north. The Early Paleozoic rock series of the Penninic zone with thick volcanic piles must have been part of the assumed northernmost regions (Fig. 3).

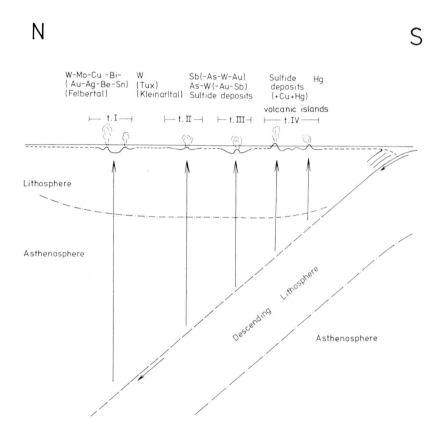

Fig. 3. Schematic diagram showing relative positions of emplacement of the different Early Paleozoic W, Sb, and Hg deposits (*t. I:* trough I, presently incorporated in Penninic zone; *t. II:* in lower Austro-Alpine unit; *t. III:* in middle Austro-Alpine unit; *t. IV:* in upper Austro-Alpine unit) (*Höll,* 1976)

Taking this palinspastic scheme as a basis, cinnabar was preferably deposited in volcano-sedimentary rocks of the southernmost trough. (Scheelite-)stibnite und scheelite-arsenopyrite mineralizations as well as sulphide deposits were formed more to the north. Still further north, "pure" scheelite ores were deposited together with carbonate rocks. The widespread strata-bound scheelite mineralization of the Penninic zone, enclosing the scheelite deposit Felbertal with a complex paragenesis, formed within an unusually thick volcanic rock sequence of the northernmost trough (Fig. 3).

6. Similar modern ore assemblages and ore belts are known from the Circum-Pacific area, where they are bound to subduction systems. Thus, the Early Paleozoic arrangement of the different ore assemblages of the Sb-W-Hg formation in the Eastern Alps seems to reflect the different ore supplies and specific features of magmatism and facies in different troughs and uplifts as a function of the depth and angle of a northdipping Benioff zone. The cinnabar deposits were formed near the trench, while the scheelite deposits of the Penninic zone resulted from the furthest down-dip ore generation above the Benioff zone.

7. This Early Paleozoic metallogenesis of the Eastern Alps is time-bound, but it did not occur within a geologically very short time period. This time-bound ore mineralization may imply all volcanic and igneous-hydrothermal ore supplies and ore-forming processes combined with a magma generation at a Benioff zone during a distinct period. This period may be separated by an inactivity of magma generation and ore concentration during times of diminished plate motions or drastic variations of this Benioff zone.

8. The source of the metals Hg, Sb, W and the associated typical elements remains a matter of debate. Three models are open to discussion, which are used for other ore deposits connected with subduction systems, such as porphyry deposits (*Clarke* et al., 1976; *Mitchell*, 1976; *Nielsen*, 1976; *Sillitoe*, 1976; *Sutherland Brown* and *Cathro*, 1976; *Swinden* and *Strong*, 1976):

a) Metals have been leached and concentrated from the host rocks within the crust of the upper plate through the influence of hydrothermal solutions generated by the intrusions of a hot magmatic body into cooler host rocks which are saturated with saline, connate water.

b) Partial melting of basaltic oceanic crust and metalliferous sediments, containing saline, connate waters, took place in the descending plate and produced a metal concentration. Magmas then evolved with the formation of metal-rich partial magmas and hydrothermal solutions during the process of ascent and intrusion or extrusion.

c) Metals have been concentrated by partial melting of peridotitic rocks of the upper mantle within the plate above the Benioff zone. This magma then evolved (as in b) with the formation of metal-rich partial magmas (e.g., the ore-bearing hornblendites of the scheelite deposit Felbertal) and of metal-rich hydrothermal solutions (e.g., the quartzite, rich ore horizons also of the scheelite deposit Felbertal) during the process of ascent and intrusion or extrusion. Model c) has been used in Figure 3.

9. Mercury with a high degree of mobility especially in reaction to volcanic heat formed the prevailing ore deposits near the trench. On the other hand, tungsten with significantly less mobility has a deep-seated origin. Thus, the belts with W (and Bi and Sn) are far from the trench. Sb is usually a connecting link between Hg and W (Fig. 3).

10. The statement that the strata-bound ore deposits of the Sb-W-Hg formation are remarkably restricted to Early Paleozoic rocks in the Alpine-Mediterranean areas does not mean that all strata-bound scheelite deposits on earth are only of Early Paleozoic age. Widespread, large-scale juvenile metal supplies with these elements took place during several periods of geological history. There are old strata-bound scheelite, stibnite, and cinnabar deposits connected with basic metavolcanics in Archean greenstone belts (*Cunningham* et al., 1973), and there are modern porphyry tungsten deposits in western Canada, related to a young subduction system (*Sutherland Brown* and *Cathro*, 1976).

11. The scheelite occurrences within the central gneiss masses of the eastern Hohe Tauern are considered as ore mineralizations derived from Early Paleozoic rock series in connexion with Variscan, sialic-palingenic meltings. The present features of these ore mineralizations are largely influenced by remobilizations during the Alpidian regional metamorphism.

Pegmatitic-pneumatolytic and contact-pneumatolytic tungsten deposits have not yet been proved with certainty in the Eastern Alps. However, such deposits may exist particularly in connexion with the Variscan intrusions of sialic-palingenic magmas.

References

Angermeier, H.-O.: Die Antimonit-Scheelit-Lagerstätten des Gerrei (Südostsardinien, Italien) und ihr geologischer Rahmen. Inaug.-Diss., 62 p., Univ. München (1964)

Baumgartner, A., Ettingshausen, A.v.: Mitteilungen zusammen mit Berichten von W. Haidinger, J. Lhotsky, J. Russegger, A. Schrötter und A. Wehrle. Z. Phys. Math. 1, Wien (1826)

Berce, B.: The formation of the ore-deposits in Slovenia. Rendic. Soc. Miner. Ital. 19, 25–40 (1963)

Cardoso, J.: Zeit- und Schichtgebundenheit des Berylliums im Paläozoikum der Ostalpen und dessen Beziehungen zu Wolfram (Scheelit). Inaug.-Diss., 45 p., Univ. München (1973)

Clarke, A.H., Caelles, J.C., Farrar, E., Haynes, S.J., Lortie, R.B., McBride, S.L., Quirt, S.G., Robertson, R.C.R., Zentilli, M.: Longitudinal variations in the metallogenic evolution of the Central Andes. In: Strong, D.F.: Metallogeny and Plate Tectonics. Geol. Ass. Can. Spec. Paper 14, 23–58 (1976)

Cunningham, W.B., Höll, R., Taupitz, K.C.: Two new tungsten bearing horizons in the Older Precambrium of Rhodesia. Mineralium Deposita 8, 200–203 (1973)

Exner, Ch.: Zum Zentralgneis-Problem der östlichen Hohen Tauern. Radex-Rundschau Jg. 1953 (7/8), 417–433 (1953)

Exner, Ch.: Erläuterungen zur Geologischen Karte der Umgebung von Gastein; Maßstab 1:50000, 168 p., Wien 1957

Exner, Ch.: Geologische Karte der Sonnblickgruppe; Maßstab 1:50000, Wien 1962

Exner, Ch.: Erläuterungen zur Geologischen Karte der Sonnblickgruppe; Maßstab 1:50000, 170 p., Wien 1964

Exner, Ch.: Geologie der peripheren Hafnergruppe (Hohe Tauern). Jb. Geol. B.-A. Wien 114, 1–119 (1971)

Frasl, G.: Zur Seriengliederung der Schieferhülle in den mittleren Hohen Tauern. Jb. Geol. B.-A. Wien 101, 323–472 (1958)

Frasl, G., Frank, W.: Einführung in die Geologie und Petrographie des Penninikums im Tauernfenster. Der Aufschluß, Spec. Issue 15, 30–58 (1966)

Friedrich, O.M.: Zur Erzlagerstättenkarte der Ostalpen. Eine Einführung zur Karte 1:500 000 der Erz- und einiger Minerallagerstätten. Radex-Rdsch. 1953, 371–407 (1953)

Friedrich, O.M.: Monographien Kärntner Lagerstätten. II. Die Quecksilberlagerstätten Kärntens. 3. Teilbericht und Schluß. Arch. Lagerstättenforsch. Ostalpen 3, 71–124 (1965)

Friedrich, O.M.: Die Vererzung der Ostalpen, gesehen als Glied des Gebirgsbaues. Arch. Lagerstättenforsch. Ostalpen 8, 136 p. (1968)

Frisch, W.: Scheelit-Lagerstätten in Zentral-Afrika und in den Ostalpen – ein genetischer Vergleich. Österr. Akad. Wiss., Anzeiger math.-naturw. Kl. 1972/14, 324–333 (1972)

Heinisch, H., Schmidt, K.: Zur kaledonischen Orogenese in den Ostalpen. Geol. Rdsch. 65, 459–482 (1976)

Höll, R.: Genese und Altersstellung von Vorkommen der Sb-W-Hg-Formation in der Türkei und auf Chios/Griechenland. Bayer. Akad. Wiss., Math.-naturw. Kl., Abh., N.F. 127, 118 p. (1966)

Höll, R.: Scheelitprospektion und Scheelitvorkommen im Bundesland Salzburg/Österreich. Chem. d. Erde 28, 185–203 (1970a)

Höll, R.: Die Zinnobervorkommen im Gebiet der Turracher Höhe (Nock-Gebiet/Österreich) und das Alter der Eisenhut-Schieferserie. N. Jb. Geol. Paläont. Mh. 1970, 201–224 (1970b)

Höll, R.: Scheelitvorkommen in Österreich. Erzmetall 24, 273–282 (1971)

Höll, R.: Die Scheelitlagerstätte Felbertal und der Vergleich mit anderen Scheelitvorkommen in den Ostalpen. Habilitationsschrift. Bayer. Akad. Wiss., Math.-naturw. Kl., Abh., N.F., 157A, 114 p., (1975)

Höll, R., Loeschke, J., Maucher, A., Schmidt, K.: Early Paleozoic Geodynamics in the Eastern and Southern Alps (in press)

Höll, R., Maucher, A.: Genese und Alter der Scheelit-Magnesit-Lagerstätte Tux. Sitzungsber. Bayer. Akad. Wiss., Math.-naturw. Kl. 1971/1, 1–11 (1967)

Höll, R., Maucher, A.: The strata-bound ore deposits in the Eastern Alps. Handbook of Strata-Bound and Stratiform Ore Deposits, Vol. V. Amsterdam: Elsevier 1976, p. 1–36

Höll, R., Maucher, A., Westenberger, H.: Synsedimentary-diagenetic ore fabrics in the strata- and time-bound scheelite deposits of the Kleinarltal and Felbertal in the Eastern Alps. Mineralium Deposita 7, 217–226 (1972)

Kontrus, K.: Vorlage neuer Mineralfunde aus dem Pinzgau. Tschermaks Miner. Petrogr. Mitt., 3. F. 3, 406–407 (1953)

Kontrus, K.: Neue Scheelit- und Datolith-Vorkommen in den Ostalpen. Tschermaks Miner. Petrogr. Mitt., 3. F. 7, 497–498 (1961)

Kreis, H.H., Unger, H.J.: Die Bleiglanz-Flußspat-Lagerstätte der Achsel- und Hinteren Flecktrog-Alm bei Hollersbach (Oberprinzgau/Salzburg). Arch. Lagerstättenforsch. Ostalpen 12, 3–53 (1971)

Krol, W.: Geologisch-lagerstättenkundliche Untersuchungen im Gebiet nördlich von Sillian (Osttirol/Österreich). Inaug.-Diss., 79 p., Univ. München (1974)

Lahusen, L.: Die schicht- und zeitgebundenen Antimonit-Scheelit-Vorkommen und Zinnobervererzungen in Kärnten und Osttirol, Österreich. Inaug.-Diss., 139 p., Univ. München (1969)

Lahusen, L.: Schicht- und zeitgebundene Antimonit-Scheelit-Vorkommen und Zinnobervererzungen in Kärnten und Osttirol/Österreich. Mineralium Deposita 7, 31–60 (1972)

Loeschke, J.: Spurenelement-Daten von paläozoischen Spiliten aus den Ostalpen und ihre Bedeutung für geotektonische Interpretationen. Geol. Rdsch. 64, 62–74 (1975)

Loeschke, J.: Kaledonischer geosynklinaler Vulkanismus im Vergleich mit rezentem Vulkanismus unterschiedlicher Krustenbereiche. Nachr. deutsch. Geol. Ges. 15, p. 27 (1976)

Maucher, A.: Die Antimon-Wolfram-Quecksilber-Formation und ihre Beziehungen zu Magmatismus und Geotektonik. Freiberger Forschungsh. C 186, 173–188 (1965)

Maucher, A.: Time- and Stratabound Ore Deposits and the Evolution of the Earth. 24th Intern. Geol. Congr. Montreal 1972, Sect. 4, 83–87 (1972)

Maucher, A.: The strata-bound cinnabar –stibnite–scheelite deposits (discussed with examples from the Mediterranean region). Handbook of Strata-Bound and Stratiform Ore Deposits, Vol. VII. Amsterdam: Elsevier 1976, p. 477–503

Maucher, A., Höll, R.: Die Bedeutung geochemisch-stratigraphischer Bezugshorizonte für die Altersstellung der Antimonitlagerstätte von Schlaining im Burgenland, Österreich. Mineralium Deposita 3, 272–285 (1968)

Medwenitsch, W., Schlager, W., Exner, Ch.: Ostalpenübersichtsexkursion. Mitt. Geol. Ges. Wien 57, 57–100 (1964)

Meixner, H.: Mineralogisches zu Friedrichs Lagerstättenkarte der Ostalpen. Radex Rdsch. 1953, 434–444 (1953)

Meixner, H.: Zur „Salzburg-Exkursion" der Österreichischen Mineralogischen Gesellschaft, Oktober 1971. Der Karinthin 65, 236–250 (1971)

Meixner, H.: Gadolinit und andere Berylliumminerale aus den Plattengneisbrüchen der Rauris (Salzburg), mit einer zusammenfassenden Übersicht über die alpinen Berylliumminerale. Der Aufschluß 27, 309–314 (1976)

Meixner, H., Thiedig, F.: Eine kleine Antimonitlagerstätte bei Brückl, Saualpe, Kärnten und ihre Minerale. Carinthia II, 79, 60–67 (1969)

Mitchell, A.H.G.: Tectonic settings for emplacement of subduction-related magmas and associated mineral deposits. In: Strong, D.F.: Metallogeny and Plate Tectonics. Geol. Ass. Can. Spec. Paper 14, 3–21 (1976)

Mlakar, I., Drovenik, M.: Structural and genetic particularities of the Idrija mercury ore deposit. Geol. Razpr. Porocila Ljubljana 14, 67–126 (1971)

Mlakar, I., Drovenik, M.: Geologie und Vererzung der Quecksilberlagerstätte Idrija. Proceedings of the Second International Symposium on the mineral deposits of the Alps, Bled, 1971. Geol. Razpr. Porocila Ljubljana 15, 47–62 (1972)

Mostler, H.: Geologie der Berge des vorderen Großarl- und Kleinarl-Tales. Unveröff. Diss., 96 p., Univ. Innsbruck (1963)

Mostler, H.: Struktureller Wandel und Ursachen der Faziesdifferenzierung an der Ordoviz/Silur-Grenze in der Nördlichen Grauwackenzone (Österreich). Festband Geol. Inst., 300-Jahr-Feier Univ. Innsbruck, 507–522 (1970)

Mostler, H.: Alter und Genese ostalpiner Spatmagnesite unter besonderer Berücksichtigung der Magnesitlagerstätten im Westabschnitt der Nördlichen Grauwackenzone (Tirol, Salzburg). Univ. Innsbruck 86 (Festschrift Heißel), 237–266 (1973)

Niedermayr, G., Kirchner, E., Koller, F., Vetters, W.: Über einige neue Mineralfunde aus den Hohen Tauern. Ann. Naturhist. Mus. Wien 80, 57–66 (1976)

Niedermayr, G., Kontrus, K.: Neue Funde von Phenakit, Bertrandit und Chrysoberyll aus Salzburg, Österreich und über die Verbreitung von Be-Mineralfundstellen in den Ostalpen. Ann. Naturhist. Mus. Wien 77, 7–13 (1974)

Nielsen, R.L.: Recent development in the study of porphyry copper geology: a review. In: Sutherland Brown, A.: Porphyry Deposits of the Canadian Cordillera. Can. Inst. Min. Metall. Spec. Vol. 15, 487–500 (1976)

Petrascheck, W.E.: Die zeitliche Gliederung der ostalpinen Metallogenese. Sitzungsber. Österr. Akad. Wiss., Math.-Naturw. Kl., Abt. I, 175, 57–74 (1966)

Ramdohr, P.: Einige neue Beobachtungen an Erzen aus den Ostalpen. Der Karinthin 17, 99–101 (1952)

Ramdohr, P.: Die Erzmineralien und ihre Verwachsungen. 3rd. ed. Berlin: Akademie-Verlag 1960, 1089 p.

Schönlaub, H.P.: Schwamm-Spiculae aus dem Rechnitzer Schiefergebirge und ihr stratigraphischer Wert. Jb. Geol. B.-A. Wien 116, 35–49 (1973a)

Schönlaub, H.P.: Conodontenstratigraphische Arbeiten im Altpaläozoikum. Verh. Geol. B.-A. 1973, A 81–83, Wien (1973b)

Schönlaub, H.P.: Zum Alter der Radstädter Quarzphyllite (Unterostalpin, Salzburg). Ann. Naturhist. Museum Wien 79, 47–55 (1975)

Schönlaub, H.P., Daurer, A.: Review of pre-Variscan events in the Eastern Alps. Geodynamics and Geotraverses around the Alps. Meeting 28.2.–1.3.77 in Salzburg, Abstracts, 47–49 (1977)

Schulz, O.: Schicht- und zeitgebundene paläozoische Zinnober-Vererzung in Stockenboi (Kärnten). Sitzungsber. Bayer. Akad. Wiss., Math.-naturw. Kl. 1968, 113–139 (1969)

Sillitoe, R.H.: Andean Mineralization: A model for the metallogeny of convergent plate margins. In: Strong, D.F.: Metallogeny and Plate Tectonics. Geol. Ass. Can. Spec. Paper 14, 59–100 (1976)

Sözen, A.: Geologische Untersuchungen zur Genese der Zinnober-Lagerstätte Kalecik/Karaburun (Türkei). Inaug.-Diss., 46 p., Univ. München (1973)

Strasser, A.: Vier neue Scheelit-Fundpunkte in Salzburg. Der Aufschluß 24, 61–62 (1973)

Sutherland Brown, A., Cathro, R.J.: A perspective of porphyry deposits. In: Sutherland Brown, A.: Porphyry Deposits of the Canadian Cordillera. Can. Inst. Min. Metall. Spec. Vol. 15, 7–16 (1976)

Swinden, H.S., Strong, D.F.: A comparison of plate tectonic models of metallogenesis in the Appalachians, the North American Cordillera, and the East Australian Paleozoic. In: Strong, D.F.: Metallogeny and Plate Tectonics. Geol. Ass. Can. Spec. Paper 14, 441–471 (1976)

Tollmann, A.: Ostalpensynthese. Wien: Deuticke 1963, 256 p.

Vohryzka, K.: Zur Scheelitprospektion in Österreich. Mitt. Ges. Geol. Bergbaustud. 18, 447–458 (1968)

Wagner, H.: Geologisch-lagerstättenkundliche Untersuchungen im Bereich des Kleinarltales, Salzburg, Österreich. Inaug.-Diss. 59 p., Univ. München (1972)

Wenger, H.: Die Scheelitlagerstätte Tux. Radex-Rdsch. 1964, 109–132 (1964)

Winkler, H.G.F.: Abolition of metamorphic facies, introduction of the four divisions of metamorphic stage, and of a classification based on isograde in common rocks. N. Jb. Miner. Mh. 5, 189–248 (1970)

Zirkl, E.: Bericht über geologisch-petrographische Aufnahmen in den Gurktaler Alpen. Verh. Geol. B.-A. Wien 1955, 85–89 (1955)

Genesis of the King Island (Tasmania) Scheelite Mine

U. BURCHARD, Toronto

Contents

Summary

The scheelite mine at Grassy close to the south-east coast of King Island, which itself is situated within the Tasman geosyncline, is the most important tungsten producer of Australia. In this area a mixed vulcano-sedimentary sequence (coarse conglomerates, dolomites, silty and tuffaceous beds overlain by pillow lavas and picrite basalts) represent, at least partly, the unaltered equivalent of the contact metamorphic Mine Series, located within the aureole of an intrusive granodiorite.

Contrary to the metasomatic replacement concept favoured by all previous authors, it is shown that the tungsten tenor of the orebodies was not epigenetically introduced, but was intrinsic to the various relevant beds of the Mine Series prior to metamorphism.

1. Introduction

The scheelite mine at Grassy, near the south-east coast of King Island, is Australia's largest individual tungsten producer[1]. The island is located in Bass Strait, halfway between the Australian mainland and Tasmania. In 1970, the author investigated the geology of the open pit and examined all available drill core. Peko-Wallsend Ltd., the sole owner of the mining operation, gave full access to all data and generously supported this study.

Since this study was completed, open pit mining has ceased and production commenced from two underground mines. The Dolphin orebody, part of which is situated underneath the ocean, is a faulted-off continuation of the ore zone exposed in the open

---

[1] Annual production within the last years ranged from 120,000 to 130,000 M.T.U. tungstic oxide.

pit. The Bold Head orebody was discovered 2.5 km north of the old mine site in a stratigraphically correlatable sequence.

This abstract deals exclusively with the geology of the open pit which is very similar to the geological environment of the Dolphin and Bold Head orebodies.

2. Regional Geology

King Island is situated within the structural province of the Tasman Geosyncline, which encompasses most of eastern Australia. Outcrops are generally restricted to coastal sections of the island, whereas most of the inland areas are covered by sand dunes. The general strike of sedimentary and metasedimentary rock formations is north-south, i.e., parallel to the long axis of the island; dips vary greatly.

The rocks of the western third of King Island are composed of a mixed metasedimentary sequence (quartzites, quartz-mica-, muscovite-andalusite- and garnet schists) intruded by pegmatites and foliated granodiorites. Rubidium-strontium datings of these west-coast intrusives yield an age of 750 my, corresponding to the Upper Proterozoic Adelaidian stage. The west-coast rocks are probably part of an old crystalline basement, unconformably overlain by a monotonous sequence of muscovite-, quartz-muscovite-, and muscovite-sericite schists, grading eastward into shales, siltstones and sandstones, which are most ubiquitous in the central part of King Island. It is suggested that the angular unconformity, separating rock types in the western and central portions of the island, is caused by the Penguin Orogeny which has been identified in northern Tasmania. Thus, a Lower Palaeozoic age is assigned to the rocks covering the central parts of the island, postdating the Penguin Orogeny with its peak of activity established at the Proterozoic-Cambrian boundary.

At the south-east coast of King Island, near "City of Melbourne" Bay, outcrops of a mixed volcano-sedimentary sequence are observed. The base consists of a coarse conglomerate with some carbonate matrix, followed by dolomite, silty and tuffaceous beds, which are overlain by pillow lavas and picrite basalts. Although field evidence is inconclusive, similar chemical composition indicates that at least parts of the volcano-sedimentary sequence described above are the unaltered equivalents of the contact metamorphic Mine Series in the aureole of the Grassy granodiorite intrusion. This unfoliated granodiorite is exposed along the coast line, south of the township of Grassy. More intrusive stocks occur along the east coast of King Island, all dating approximately 345 my. This uppermost Devonian age coincides with the final phases of the Tabberabberan Orogeny. Therefore, a Lower to Middle Palaeozoic age (pre-Carboniferous and probably post-Cambrian) brackets the east-coast sediments and the Mine Series.

3. Geology of the Mine

The scheelite mineralization of all three known orebodies at King Island is hosted in stratigraphically correlatable beds of the Mine Series which were originally clastic and carbonate sediments with interbedded volcanics. Within the aureoles of shallow granodioritic intrusions, the country rocks have been altered to calcsilicates and hornfelses under conditions of the upper hornblende-hornfels contact metamorphic facies (500–600° C). The open pit is situated on the southern flank of an easterly plunging anticline. Beds dip at an average of 50° south towards the granodioritic contact. There are several faults, the two most important ones cutting off the ore lenses at the eastern and western extremity of the open pit. Between the barren basal quartzites and the topmost metavolcanics, representing metamorphosed picrite basalts, the Mine Series consists of seven well-defined horizons with a total accumulated thickness of 150 m. These are described sequentially as follows:

3.1 Older Metavolcanics (10 m)

A dense, mafic rock with an acicular texture forms the base of the Mine Series. Relict mineral assemblage (olivine, feldspar, spinel, titanite in a grammatitic matrix) and ophitic texture suggest that the original rock was an olivine basalt. These older metavolcanics are devoid of scheelite mineralization.

3.2 Biotite-Pyroxene-Hornfels (30 m)

This sequence is characterized by finely laminated bands of varying chemical composition representing primary sedimentary bedding. Quartz-rich bands are attributed to interbedded arenaceous sediments. Beds composed almost entirely of oriented biotite flakes originated from marly pelites. Lenses rich in ferrosalitic clinopyroxene are commonly developed with some quartz, plagioclase and titanite. Bands consisting of an actionolite matrix with minor plagioclase phenocrysts are interpreted as metamorphosed tuffs and tuffites containing varying amounts of pyroclastic and sedimentary detritus. There are some sporadic lenses composed of an andradite-quartz assemblage containing minor scheelite mineralization and thus heralding the orebodies higher up in the stratigraphic column.

3.3 Banded Footwall Beds (21 m)

This member of the Mine Series is distinguished by a pronounced alteration of colourless, coarse-grained marble beds and dark-coloured bands of a biotite-pyroxene-amphibole assemblage. At the contact between the marbles and the silicate hornfelses there is always a development of a narrow grossularite seam. Wollastonite reaction veins are only present where the carbonate layers, surrounded by silicate beds, do not exceed several cm in thickness. In these exceptional cases, a spatially restricted dilution of CO_2 liberated from the carbonates and H_2O liberated from the silicates must have been possible. The resulting low mol fraction of CO_2 supported the formation of wollastonite at the temperatures realized during the metamorphic process. This dilution did not exist where carbonate beds have greater thickness, and consequently wollastonite was not in equilibrium.

Microscopically, these diffusion reaction bands show a zonal succession from calcite-wollatstonite-grossularite-pyroxene-actinolite-biotite to quartz.

Within the upper sections of the banded footwall beds some intercalations of dense andradite-clinopyroxene-quartz skarn occur. These can host scheelite mineralization that in places reaches economic ($> 0.5\%$ WO_3) grade. Chemical composition and relict sedimentary features indicate a primary mixed sedimentary sequence of carbonate rocks, marls and possibly some tuffitic beds.

3.4 C-Lens (18 m)

Most of the ore was derived from the C-lens, which is subdivided into a bottom orebody (4 m), a consistent barren marble marker bed (2.5 m), the top orebody (9 m), and a barren top biotite hornfels (2.5 m). The host rock of the bottom orebody that has an average grade of 0.74% WO_3 is a dense andradite-clinopyroxene-quartz skarn frequently banded by thin beds of barren clinopyroxene hornfels. Under the microscope, scheelite can be seen either enclosed by garnet or within a quartz matrix along the edges of andradite crystals.

A forsterite-phlogopite marble separates the bottom from the top orebody. This marble marker has a consistent thickness and can be traced along the entire strike length of the open pit.

The top orebody (average grade 1.05% WO_3) has a mineralogical composition very similar to the bottom orebody, although the host rock is more massive and lacks characteristic banding. A barren zone of 2 m occurs within the top orebody, showing alternating bands of grossularite and clinopyroxene zones.

C-lens is concluded by a top biotite-hornfels with a biotite-plagioclase-actinolite-quartz assemblage.

It is assumed that the marble bed and the top biotite hornfels resulted from a pre-metamorphic dolomitic limestone and from pelites. As for the andradite skarns, it is much more difficult to define the original rock type accurately. The skarns could be either the product of pure contact metamorphism of mixed carbonate-silicate rocks with suitable chemical overall composition, or they could be the product of replacement of carbonate rocks by infiltration metasomatism. In thin section there are no petrographical criteria to distinguish between the two hypotheses.

3.5 Metaconglomerate (12 m)

This horizon is well characterized by subrounded to rounded ovoids and angular fragments up to 30 cm in diameter embedded in pyroxene or actinolite matrix. These ovoids are composed predominantly of calcite, showing concentrical rection rims towards the matrix similar to the banded footwall beds. Other fragments consist of an assemblage of andradite-clinopyroxene-quartz with some scheelite. This assemblage is very similar to the garnet skarns of C-lens. The matrix is devoid of scheelite. The grade is generally too low to justify mining of the metaconglomerate. Angular fragments point to a close source, the composition of the psephites indicates redeposited carbonates probably derived from the banded footwall beds and from C-lens.

3.6 Hanging Wall Hornfels (30 m)

This is a uniform and barren hornfels originating from pelitic sediments. Main constituents are biotite, actinolite, quartz and plagioclase.

3.7 B-Lens (26–28 m)

B-lens is made up of a variety of beds with complex composition. It is dominated by three marble horizons and there are common intercalations of biotite, biotite-albite, biotite-actinolite and actinolite-phlogopite, as well as actinolite hornfels. Several thin bands of clinopyroxene-titanite and clinopyroxene-garnet-titanite hornfels occur at various levels throughout B-lens not imperatively associated with marble beds. It is only within these narrow clinopyroxene-rich beds that scheelite is concentrated up to grades of 8.5%. Consequently production from B-lens is on a selective basis.

B-lens represents a metamorphosed sequence of carbonates and clastic sediments. The presence of volcanics is evidenced by high Cr and Ni values in some of the actinolite hornfels.

4. Mineralization

Stratiform scheelite mineralization is finely disseminated as idiomorphic grains (0.2–0.3 cm) predominantly in an andradite-clinopyroxene-quartz host and to a lesser extent in clinopyroxene-andradite-plagioclase-titanite rocks. Chemical analyses of scheelite grains indicate less than 1% molybdenum; yellow scheelite fluorescence indicates that the molybdenum is contained in the mineral powellite ($CaMoO_4$).

Iron, nickel and titanium minerals like titanite, pyrrhotite, pentlandite are rather abundant; magnetite, ilmenite, pyrite and marcasite constitute accessories. Polished sections reveal molybdenite, traces of chalcopyrite and sphalerite and, very rarely, minute grains of native bismuth, bismuthinite, joseite and gold.

5. Ore Genesis

All previous authors regarded the King Island scheelite deposits as metasomatic replacement products of suitable carbonate beds. According to their theory, solutions would have infiltrated the carbonates, commencing at the granodioritic contact and migrating up dip or along fault zones. The solutions would have introduced WO_3 along with large quantities of SiO_2, Fe_2O_3, Al_2O_3 and MgO, and at the same time would have removed CaO and CO_2. Thus, most of the tungsten would have been precipitated within the carbonate beds, leaving remnants of unreplaced marble. Only small amounts of tungsten would have been precipitated either in the carbonate ovoids, of the metaconglomerate, or along quartz veins.

The observations and criteria listed below led the author to revise the above metasomatic concept for the genesis of the King Island scheelite deposit:

1. scheelite mineralization is almost exclusively stratiform and is restricted to several beds within the Mine Series;

2. the ore lenses and beds can be traced for several 100 m along strike and appear to decrease in thickness down dip approaching the intrusive contact. At the same time, the scheelite mineralization tends to diminish in grade;

3. mineralized beds are commonly associated with marble horizons, but also occur without any contact to carbonate beds. The bottom and top orebodies of C-lens are separated by a marble marker having a consistent thickness;

4. as shown in drill core, the various marbles of the Mine Series appear to be virtually unreplaced at the immediate granodiorite contact where maximum replacement would be expected. There is no evidence of any tungsten concentration gradient around the intrusive;

5. the host rocks are commonly characterized by relict bedding accentuated by rapid alterations of barren and mineralized beds, in places of identical mineralogical silicate composition. These aspects cannot be explained by selective replacements under metasomatic conditions;

6. within the metaconglomerate scheelite mineralization is restricted to the ovoids. It is hard to conceive how metasomatic solutions could have penetrated the dense silicate matrix and the impermeable grossularite reaction rims around the pods, without leaving traces in the barren matrix;

7. scheelite is the only significant ore mineral apart from minor molybdenite and traces of bismuth;

8. discordant, cloudy replacements occur very rarely and are spatially restricted. Very little scheelite is present in quartz veins cross-cutting stratiform ore lenses;

9. the various mineralized beds are cut by several faults, some of which pre-date the intrusion, or at least its aplitic apophyses. Detailed geochemical sampling of all faults and fractures revealed no significant tungsten or molybdenum values.

It is concluded from the criteria listed above that the tungsten tenor of the orebodies did not result from metasomatic replacement, since — among other arguments — none of the possible avenues for infiltrating solutions was used. Hence the tungsten content was not epigenetically introduced, but was intrinsic to the various relevant beds of the volcano-sedimentary Mine Series prior to metamorphism. The mineralization must be seen in context to basaltic volcanism which occurs at the base and the top, and also as thinner strata, throughout the entire sequence of the Mine Series.

The King Island scheelite deposit shows striking similarities (genesis, age) to other scheelite occurrences studied elsewhere by *A. Maucher* and co-workers.

References

Edwards, A.B., Baker, G., Callow, K.J.: Metamorphism and metasomatism at King Island scheelite mine. J. Geol. Soc. Aus. 3, 55–98 (1956)

Kinnane, N.R.: The geology of the contact metamorphic and metasomatic rocks at the King Island scheelite mine, Grassy King Island, Tasmania. Dept. Metallurgy and Min. Geol., Roy. Melbourne Inst. of Technology, 1968, 27 p.

Large, R.R.: Metasomatism and scheelite mineralization at Bold Head, King Island. Proc. Aust. Inst. Min. Met. 238, 31–45 (1971)

Maucher, A.: Die Antimon-Wolfram-Quecksilber Formation und ihre Beziehung zu Magmatismus und Geotektonik. Freib. Forsch.H. C 186, 173–188 (1965)

McDougall, I., Leggo, P.J.: Isotopic age determinations on granitic rocks from Tasmania. J. Geol. Soc. Austr. 12 (2), 295–332 (1965)

Scott, B.: The petrology of the volcanic rocks of South East King Island, Tasmania. Pap. Proc. Roy. Soc. Tas., for 1950, pp. 113–136 (1951)

So, C.S.: Die Scheelit Lagerstätte Sangdong. Inaug. Diss. Univ. München, 1968, 67 p.

Solomon, M.: Geology and mineralization of Tasmania. Eighth Commonwealth Min. Met. Congr. Geology of Australian ore deposits 1, 464–477 (1965)

Solomon, M.: The nature and possible origin of the pillow lavas and hyaloclastic breccias of King Island, Australia. Quart. J. Geol. Soc. London 124, 153–169 (1969)

Geological Investigations on the Genesis
of the Cinnabar Deposit of Kalecik/Karaburun (Turkey)

A. SÖZEN, Izmir-Bornova

With 4 Figures

Contents

Summary

The paper deals with the genesis of the strata mercury deposit of Kalecik/Karaburun (Turkey). First the different stratigraphic Paleozoic complexes in which the mineralizations are imbedded are described shortly and compared with similar sequences in Greece. Thus the ore-bearing series are explained as vulcano-sedimentary stratigraphic sequences where ore occurs strictly associated with tuffic layers limited to Upper Silurian–Lower Devonian age. The Kalecik deposits are assigned to the maximum of the time-bound Sb-W-Hg formation of early Paleozoic in the sense of *Maucher* (1965).

1. Introduction

The deposit of Kalecik is situated in the north-east corner of the peninsula of Karaburun which bars the bay of Izmir to the west. The deposit lies on a hill about 400 m above sea level, 3 km as the crow flies SSW of Karaburun. A mountain track, about 6 km long, connects the deposit with Karaburun. The highway distance from Izmir is 110 km.

The first geological mapping of the peninsula of Karaburun, at a scale of 1:100,000, was done by *Kalafatçioğlu* (1961) and *Akartuna* (1962). Both investigators succeeded in subdividing especially the Mesozoic series. The Paleozoic series, which are of importance for the deposit of Kalecik, were first studied in detail by *Höll* (1966). He assigned the Denizgiren Formation to the Ordovician, or Lower to Middle Gotlandian (= Silurian). The Kalecik Limestone is of Upper Silurian age according to his statement. The Kalecik Formation which, as he asserts, "overlies conformably, without hiatus" the Upper Silurian Kalecik Limestone and "intertongues with its upper layers" is "still to be assigned to the Upper Gotlandian". According to *Höll* the NNW-SSE striking Kalecik Limestone overlies the Denizgiren Formation with a true conformity. From this he infers a folding of Younger Caledonian age. *Brinkmann* et al. (1967) report for the first time a pelagic Triassic from Karaburun which probably extends at least from the Scythian to the Carnian.

In the survey of the geology of the island of Chios by *Besenecker* et al. (1968), they discriminate between an autochthonous and an allochthonous series: "The structure is based on several orogenetic phases. Caledonian folding is not entirely certain. The Variscan orogenesis created a fold structure striking NW to NNW. The vigorous Alpidic orogenesis has in most cases obliterated the older structures".

Lehnert-Thiel (1969) assigns the Kalecik Formation, which he maintains to be older than the Kalecik Limestone, to the Silurian-Ordovician with reservations. He claims that the Denizgiren Formation overlies the Kalecik Limestone with a slight unconformity. For the Denizgiren Formation, he assumes a Middle Devonian age. According to him the Variscan orogenesis is definitely traceable, but he doubts a Younger Caledonian folding.

2. Lithostratigraphy

2.1 Paleozoic

2.1.1 The Kalecik Formation

This complex, consisting mainly of sandstone, siltstone and slate, sporadic layers of greywacke and carbonatic rocks, as well as tuffs and tuffites, represents the oldest rock formation of the investigated area. It is characterized by a discordered mixture of shale, greywackes with graded-bedding, and breccia stemming from submarine sliding. As all findings show, they are turbidites deposited in a long, deep trough. During the investigations some units could be distinguished and mapped separately, though they are vigorously intertongued and tectonically deformed (Fig. 1).

1. **The Sandstone-(Silstone)-Slate-Breccia.** These rocks, quantitatively prevailing in the Kalecik Formation, were met down to a depth of 192.7 m at drillings, where their lower limit was not yet reached. Their main constitutents are sand, silt, clay and rock fragments. Ocassionally considerable quantitites of volcanic material (tuffs and tuffites) are involved. There are moreover fine lenticular carbonate layers alternating with layers of clays and fine conglomerates.

The main feature of this facies, which shows a black colour when fresh, is a curious subaquatic breccia structure. Within a clayey-carbonatic, sometimes silicified matrix, there occur markedly angular components as well as rounded, cylindrical, partly primarily laminar ones. In addition there are long, curved, generally displaced, up to several dm-thick lenses of sandstone, greywacke and occasionally limestones, which all indicate

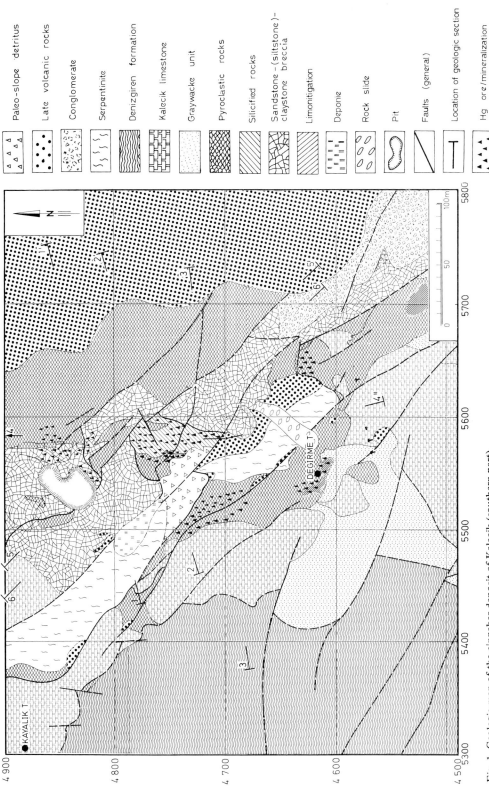

	Paleo-slope detritus
	Late volcanic rocks
	Conglomerate
	Serpentinite
	Denizgiren formation
	Kalecik limestone
	Graywacke unit
	Pyroclastic rocks
	Silicified rocks
	Sandstone - (siltstone) - claystone breccia
	Limonitigation
	Deponie
	Rock slide
	Pit
	Faults (general)
	Location of geologic section
	Hg ore/mineralization

Fig. 1. Geologic map of the cinnabar deposit of Kalecik (southern part)

formerly continuous layers. A further special attribute of this rock unit is the flysch-like
graded bedding. Apart from an upwards-diminishing size of the particles, not always dis-
cernible, there are again and again abundant silty and sandy components in the clay-
stones. Probably we have to assume several phases of sedimentation during the forming
of this rock sequence. The flysch-like graded stratification, with convolution and injection joints formed by the uprising compaction fluid and the formation of breccias
through submarine slides in an early diagenetic stage, shows that these sediments were
deposited rashly and were already reworked during consolidation. In cases of less
marked breccia formation, lcad casts, flute casts and groove casts can be observed
between the layers of slates and sandstones. The planes characterized by groove casts
are frequently tilted, sometimes overturned. Subsequent subaquatic slides are probably
responsible for this feature, as these planes are in several cases overlain unconformably
by groove casts with different orientation.

2. **Massive Light-Coloured Limestone Beds**. The thick-bedded limestones, interlayered
with thin-bedded ones, occur as intercalations in a complex of sandstone-(siltstone)-
slate-breccias. They are composed of a dense, massive white limestone; dolomitized
parts have a light yellow colour. We can almost certainly deduce that all occurrences of
limestone are stratigraphically intercalated with the breccia.

The ore-bearing highly silicified rocks (Çakmak) which outcrop to the south of the old
open mine show under the microscope that the allotriomorphous quartz mosaic has displaced the dolomitic carbonates. Therefore the Çakmak also is a silicified carbonate roock.
These massive, light-coloured limestone beds do not contain any macro- or micro-fossils.

3. **Pyroclastic Rocks**. These include various rocks which occur almost invariably
intertongued with each other and consist of volcanogenous material, with the exception
of some carbonatic lenses: siliceous tuffs and tuffites, tuffitic sandstones and siltstones,
agglomerates, fine-bedded limestones and marly detrital limestones. Of special interest
are tuffites with a conspicuous layer structure and with clayeous and breccious components. They are found as larger dislocated rock fragments within marly detrital limestones. The breccia components include, in addition to very small rock fragments, larger
scraps from the sandstone-(siltstone)-slate-breccia. Usually these pyroclastic rocks overlie the sandstone-(siltstone)-slate-breccia, and this prove to be of younger age within the
same sequence.

4. **Thin-Bedded Tuffites**. The tin-bedded tuffites consist of rhythmic bands of a
thickness within the range of 1 mm; they are gray, white, bluish-green, and brown. The
tuffites form an almost continuous lens, constant in its position throughout the area
of outcrop. Their mm-thick rhythmic layers, however, are folded paradiagenetically and
are partly broken. In their upper beds, they pass into silicified dolomitic parts, about
3 m thick, which form the transition to the Çakmak. The thin-bedded tuffites occur
only in connection with the "main ore formation". They seem to be in direct connection with the cinnabar ore contents of the overlying silicified rocks (Çakmak).

2.1.2 The Greywacke Formation

The Greywacke Formation is met at the borders of the turbidite trough, vigorously
intertongued with the Kalecik Formation. It consists of sandstones, slates, predominantly greywacke, subordinately limestone lenses with silexite nodules, conglomerates,
radiolarites and a few tuffs and porphyrite dikes.

2.1.3 The Kalecik Limestone

Owing to its content of fossils, however scarce, this limestone range, known as Kalecik Limestone in the literature (*Ktenas*, 1925 calls it Kajandjik Limestone; *Höll*, 1965; *Lehnert-Thiel*, 1969), forms the only stratigraphically fixable key horizon among the barren rocks of the surroundings.

The compact limestone is grey-coloured in tis lower, fossiliferous parts, otherwise white. It is generally massive, stratified only in some parts. Marly layers and small, some cm-thick, light-coloured red limestone laminae occur in the form of schliers. *Höll* (1966) quotes the following fossils from the Kalecik Limestone (determination by *P. Wellnhofer*):

Entelophyllum articulatum (Wahlenberg); *Spongophylloides* cf. *perfectus* (Wedekind); *Columnaria breviradiata* (Weissermel); *Thamnopora reticulate* minor (Weissermel); *"Heliolites n. sp."* (Weissermel 1939); *Platyorthis* cf. *cimes* (Kozlowski); *Leptaena rhomboidales* (Wilckens); *Camarotoechia nucula* (Sowerba); *Camarotoechis* aff. *bieniaszi* Kozlowski; *Camarotoechia* sp.; *"Rhynchonella" henrici* Barrande; *Calymene* sp.; Crinoidenreste.

Lehnert-Thiel (1969; det. *K.O. Felser*) adds the following:

Palaeofavosites sp.; *Entelophyllum* sp.; Crinoidenreste.

Prof. Dr. *H.W. Flügel*, Graz, informed me about the corals I collected as follows: "The material conveyed to me consists of the genus Favosites without exception. One specimen is of interest as it is connected with a commensal Asterosalpinx. Similar forms have been reported from the Lower and Middle Devonian". Based on the fossils quoted above, *Höll* (1966) assigns the Kalecik Limestone to the Upper Silurian (Ludlow). He compares it with the Agrelopos Limestone of Chios (Greece) and concludes, guided by the fossils: "The Kalecik Limestone and the Agrelopos Limestone are therefore not only similar in their lithological character and tectonical position, but they are Upper Gotlandian (= Upper Silurian) equivalents". However, from the Agrelopos *Roth* (1968) and *Herget* (1968) describe, in addition to the Upper Silurian fauna, also fossils with a range up to the Lower Devonian.

Just as the Agrelopos Limestone, the Kalecik Limestone contains fossils which could be assigned to the Lower or Middle Devonian. Therefore I would rather advocate an age of Upper Silurian – Lower Devonian for the Kalecik Limestone. The limestone rests on the Kalecik Formation and the Greywacke Formation. The contacts are generally not sharp, only at a few places it can be observed that the lower, most sandy limestones overlie the greywacke with a sharp contact.

2.1.4 The Denizgiren Formation

A formation consisting of well-bedded and folded clastic rocks, scarcely 2000 m thick (*Höll*, 1966), begins at the south-west of the limestone range. It is built up mainly of sandstone beds some cm to dm thick, furthermore greywacke, slates, conglomerates, silexites, lydites, a few thin beds of limestone and limestone breccias. Because of missing stratigraphical contacts, it is difficult to assert the stratigraphical position of the Denizgiren Formation. They always adjoin the Kalecik Limestone with a sharp tectonic unconformity, their contact with the Greywacke Formation, however, sometimes implies an interference of facies change.

2.2 Mesozoic

In the investigated area the Mesozoic consists of a conglomerate formation and of thick, massive limestones. As the Mesozoic rocks are of no importance for the Hg-deposits of Kalecik, they were not studied in detail.

2.3 Caenozoic

To the Caenozoic belong Upper Tertiary sediments rubble breccias and young volcanites (basalts and andesites). These rocks, likewise, are of no importance for the present ore formation.

3. Tectonics

Among the Paleozoic rocks of Kalecik it is possible to distinguish between two structural patterns. Whereas the Kalecik Formation and the Kalecik Limestone are scarcely folded, the Denizgiren Formation has been strongly affected by folding tectonics.

The regional extension of the Kalecik Formation, the Greywacke Formation, the Limestone ragen of Kalecik, and the Denizgiren Formation implies the assumption that the main strike of the shaping tectonics is NW-SE and may oscillate up to NNW-SSE.

The Kalecik Formation is interspersed with numerous dividing planes, usually with slickenside striae. These shear joints (hko-planes) hint at compression tootonics The beds of the Kalecik Formation have been deformed atectonically by subaquatic slides before the tectonic deformation. In the course of this event, their bedding structure disappeared. Tectonic compression folds do not occur anywhere in these beds. Shear folds are observed only locally. On the other hand, boudinages emerging especially from shear planes are abundant.

The tracing of the tectonic elements within the Greywacke Formation is very difficult due to the absence of clearly stratified members. Disturbances only conspicuous in the limestone lenses and through the porphyryte dykes connected with the normal faults, strike nearly always NW-SE.

Two directions of faults are predominant in the limestone range of Kalecik: the fractures striking NW-SE to NNW-SSE delimit the limestone range; they are displaced by younger fractures striking NNE-SSW to NE-SW.

In the Denizgiren Formation we are confronted with a very complicated pattern of folding. Along with the prevailing upright and inclined folds there are overturned and recumbent folds. Folds with heavily inclined axes are very frequent. In the fine structural survey, the points where the measured fold axes pierce the Schmidt net are dispersed fairly wide, but the north-west sector is occupied more densely. Paleozoic sequence of strata is characterized by facies intertonguing and by missing unconformities. We have to reckon with one single orogenetic phase. The axes measured and ascertained do not enable us to fix a main direction of folding in the paleozoic sequences.

However, relying on the prevailing NW-SE direction we can assume a compression in the NE-SW direction. The E-W and NNE-SSW directions of folding within this compression structure may be derived from the reimpression of a B $\perp$ B' (or B $\wedge$ B')-structure.

As the sediments of Upper Silurian — Lower Devonian age, as well as certainly younger strata, are folded together in this deformation, a Caledonian phase of folding can be excluded.

Legend for Figures 2 to 4

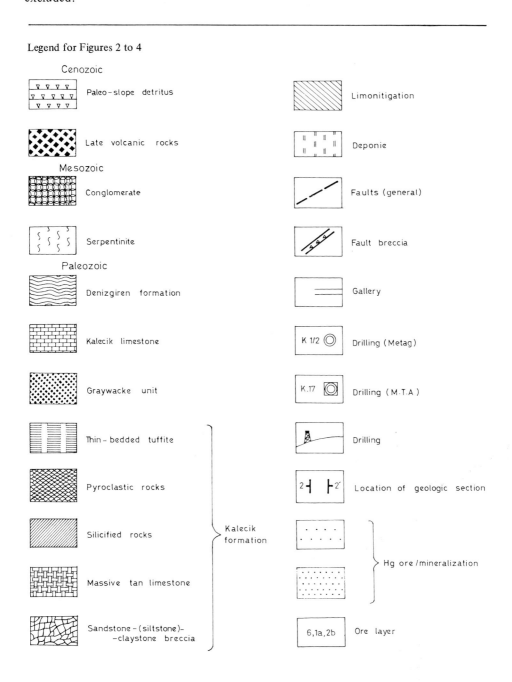

Cenozoic

Paleo-slope detritus

Limonitigation

Late volcanic rocks

Deponie

Mesozoic

Conglomerate

Faults (general)

Serpentinite

Fault breccia

Paleozoic

Denizgiren formation

Gallery

Kalecik limestone

K 1/2 ◎ Drilling (Metag)

Graywacke unit

K.17 ◙ Drilling (M·T·A·)

Thin-bedded tuffite

Drilling

Pyroclastic rocks

2⊣ ⊢2' Location of geologic section

Silicified rocks

Kalecik formation

Massive tan limestone

Hg ore/mineralization

Sandstone – (siltstone)–
       –claystone breccia

6,1a,2b Ore layer

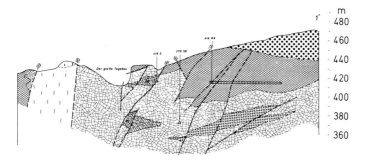

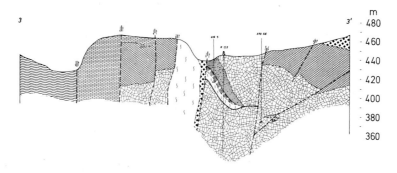

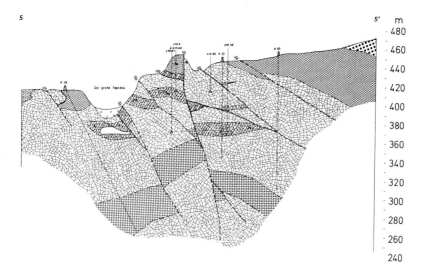

Fig. 2. Geologic sections

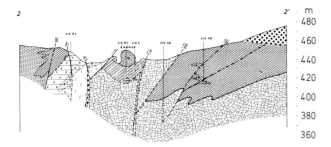

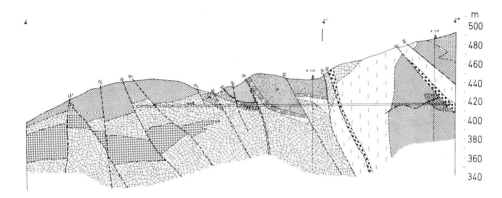

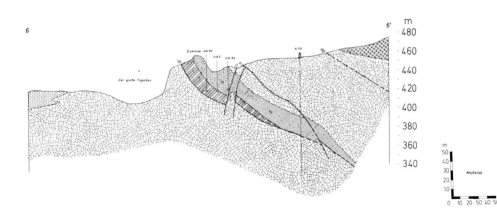

Fig. 2 (continued)

When comparing the above arguments with the geology of the island of Chios, where the main direction of the Variscan tectonics is NW to NNW (*Besenecker* et al., 1968), we can assign the Paleozoic deformations of our region also to the Variscan orogenesis. The Alpidic orogenesis has affected the area mainly through greater, deep-reaching joints. The old direction of NW to NNW was reactivated and prescribed the main lines of the joints. The great zone of fractures striking NW-SE south of the great open mine, with serpentinite slices and young volcanic dykes, demonstrates how deeply the Alpidic tensional tectonics penetrate the basement.

The age of the lenticularly sheared serpentinite bodies is unknown, but their emplacement is certainly prior to the Late Tertiary eruptions, for the serpentinites are pierced by the young volcanic dykes.

4. The Hg-Deposit of Kalecik

This deposit is confined to the Kalecik Formation and consists of two parts divided by the great fracture zone with serpentinite inclusions striking NW-SE. The north-eastern part of the main deposit can be traced for about 300 m in direction NNW-SSE: The south-western part of the secondary deposit, with scarce ore content, therefore economically insignificant, borders on the fracture zone in the south and extends for about 200 m in the direction NW-SE.

4.1 The Main Deposit

The main deposit is divided into two parts by a fracture line parallel to the major fracture zone striking NW-SE.

The Northern Part of the Main Deposit. A checkered structure marked by fractures striking mainly NW-SE, secondarily E-W to WSW-ENE, has subdivided the northern part of the main deposit into several blocks.

Based on our knowledge of stratigraphy, we can recognize these blocks as oroginally belonging together as parts of one single ore deposit (Fig. 2). The ore deposits are always strata-bound and show different types of ore formation depending on the different host rocks.

On the geological surface and subsurface maps, the ore deposits were numbered according to the host rocks, as well as to tectonic aspects (Figs. 3 and 4).

Deposits 2 and 3 are located in the sandstone-(siltstone)-slate-breccia. The cinnabar ore is mainly confined to the compact horizons with fine-grained components. These deep black pyritiferous parts are usually more or less silicified. Cinnabar is found in these rocks in the form of impregnation ores or widely dispersed schliers. Furthermore it appears in layers of some mm thickness alternating with pyrites, or, more rarely, in small discordant fissures. These fissures are almost constantly connected with the cinnabar-bearing layers. On the other hand, genuine joint deposits and vein deposits are entirely missing. The Variscan structure is certainly younger than the ores. We can frequently observe slickenside striae with cinnabar and pyrite on the Variscan-formed

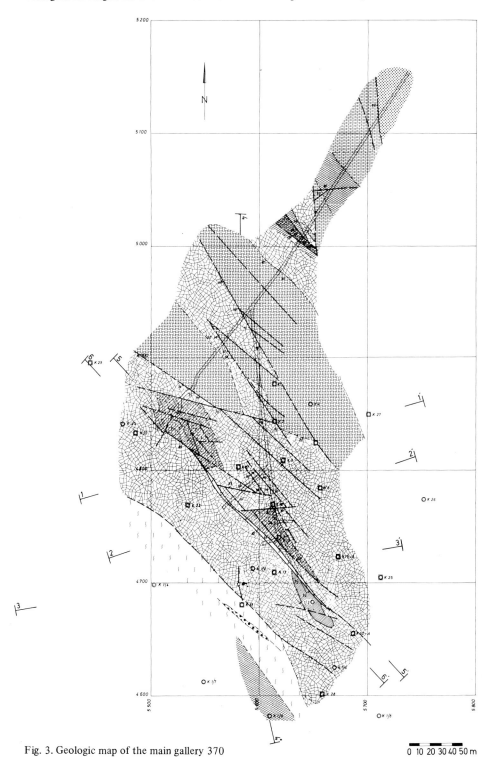

Fig. 3. Geologic map of the main gallery 370

0 10 20 30 40 50 m

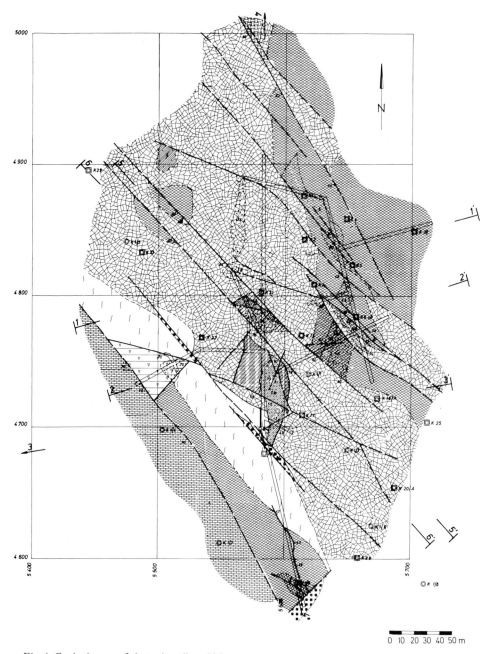

Fig. 4. Geologic map of the main gallery 416

hko-fault planes. There are close relations between the ore formation and the host rocks; the ores are affected by breccia formation which is also the characteristid feature of the country rocks. Frequently we find broken layers of pyrite and cinnabar which behave exactly like the other rock components. Furthermore we observe paradiagenetical drag and glide folds in the ore as well as in the country rocks. On the planes with groove marks we can sometimes observe dragged particles of cinnabar.

Under the microscope a drilling core showed extremely fine particles of cinnabar on quartz particles. These quartz particles show homoaxial circumferential growth layers which likewise enclose cinnabar particles. These textures are to a great degree similar to those described and illustrated by *Maucher* and *Saupe* (1967) from the cinnabar deposit of Almaden (Spain).

Deposit 4 represents the ore formation type in the marly detric limestones of the pyroclastic rock formation. In this deposit the cinnabar ore is met mainly as irregular ore enrichment from schliers and impregnations. However, in the silicified parts we also find thin beds of cinnabar and country rock. The thin beds are bent and displaced atectonically, as a result of subaquatic slides. Apart from these primary ores we sometimes observe rebedded cinnabar-bearing particles of the sandstone-(siltstone)-slate-breccia in the marly detric limestone. Deposits 5 and 6 are found in the siliceous tuffs and tuffites of the pyroclastic rocks displaying irregular shapes.

Characteristically, these deposits are widely distributed but poor in ore content.

The lenticularly outwedging tuffites contain sporadic, rarely enriched impregnations of cinnabar which alternate conformably. These tuffites are overlain by silicified carbonate rocks, which in their upper parts are markedly intertongued with the like-wise silicified pyroclastics. Here, the upper part of the first deposit contains abundant cinnabar, especially in a net-like fissure system of joints and fault planes. Apart from these unconformable ores, we again find bedded ores in the lower parts of deposit 1 where the silicified carbonates alternate with the sandstone-(siltstone)-slate-breccia.

4.2 The Secondary Deposit

The secondary deposit (deposit 8) represents the ore formation type in the thin-bedded carbonates of the pyroclastic rocks. This economically significant ore consists of diffuse inpregnations and sporadic, very small joint fillings.

5. Genesis of the Deposit

The genesis of the deposit of Kalecik can only be judged according to the mineral content, the host rocks, the texture and rock sequence. This occurrence of cinnabar ore is confined to a special facies, consisting of turbidites with the participation of tuffs and tuffites (Kalecik Formation). In this complex the cinnabar occurs only in certain horizons which contain either volcanic material (of acidic character) or which are marked by the immediate proximity of pyroclastics. The simple mineral paragenesis of the deposit consists of "cinnabar, metacinnabarite, pyrite, marcasite, arsenopyrite, vaesite and a mixed crystal of vaesite and illmanite" (*Höll*, 1966).

When describing the deposit, a predominantly stratified ore type and an unconformable type, occurring only locally, were described. The stratified type shows the following genetical characteristics:

a) constancy of stratigraphical level of the ore deposits; restriction to fractures; stratigraphical concordance;

b) alternation of cinnabar – pyrite with thin beds of country rock;

c) common paradiagenetical deformation of host rock and ore;

d) dragged particles of cinnabar on planes with groove casts (external deposition on the open sea bottom);

e) occurrence of cinnabar particles entrapped by quartz overgrowths (transport of cinnabar before diagenesis);

f) slickensides of cinnabar and pyrite on the polished shear planes of the compression tectonics (pretectonic attachment of ore);

g) absence of depth differences and of any hint of Tertiary ascendant supplies.

These characteristics of the ore-bearing tuffs and tuffites imply a confinement of the primary ores to the same strata and age of the ore mineralization and the volcanogenous rocks.

A further proof for synchronous sedimentation can be found in the resedimentation of cinnabar-bearing particles in the younger layers of the same rock sequence.

The unconformable ore type occurs in the great fracture zone striking NW-SE. It is found in the same rocks which are the host rocks of the synsedimentary-paradiagenetical ore type. The discordant ores are likewise accompanied by a strong silicification. Therefore the transporting solution contained abundant silica. The common occurrence of the unconformable ores with primary sedimentary ores and their absence outside the rocks bearing primary ores compels us to the conclusion that the ores in question are remobilized primary ores which have been replaced on tectonically prescribed tracks within the great fracture zone. The fact that this unconformable ore cannot be traced into lower or higher parts of the bedding sequence, i.e., the absence of genuine continuous veins, is especially in favour of this conclusion. The finer joint fillings confined to the fractures have been exposed to further mechanical stresses. Slickensides of cinnabar on the hko-shear planes and far going mylonitization of the ores suggest that this secondary discordant mineralization must have taken place in an early stage of the Variscan orogenesis. Young volcanic veins and updragged serpentinites do not contain ores anywhere. They always cut the ore body off sharply. They are therefore certainly younger than the stratified and the discordant ores and do not have any genetical relation to the ore transport. The ore-bearing "Kalecik Formation" is somewhat older than the Upper Silurian– Lower Devonian Kalecik Limestone. The Variscan tectonic is certainly postcrystalline in regard to the stratified ores. The age of these primary stratified ores has been fixed as at least Upper Silurian. The remobilized ore parts are of Variscan age. The cinnabar deposit of Kalecik, based on the above synsedimentary genesis, can be assigned to the maximum of the Sb-W-Hg Formation of early Paleozoic in the sense of *Maucher* (1965).

Acknowledgements. I am indebted to Professor Dr. Ing. *A. Maucher* for his encouraging support and to Dr. *H. Bremer* for the translation the manuscript into English. I thank Professor Dr. *H. Flügel* for the palaeontological determinations. This work has been supported financially by the Deutsche Forschungsgemeinschaft.

References

Akartuna, M.: On the geology of Izmir-Torbali-Seferihisar-Urla district. Bull. Min. Res. Inst. Turkey 59, 1–18 (1962)

Besenecker, H., Dürr, S., Herget, G., Jacebahagen, V., Kauffmann, G., Lüdtke, G., Roth, W., Tietze, K.W.: Geologie von Chios (Ägäis). Ein Überblick. Geologica et Palaeontologica 2, 121–150 (1968)

Brinkmann, R., Rendel, B., Trick, P.: Pelagische Trias bei Izmir. Sc. Rep. of the Fac. Sc. Univ. Ege 37, 1–3 (1967)

Herget, G., Roth, W.: Stratigraphie des Paläozoikums im Nordwest-Teil der Insel Chies (Ägäis). N. Jb. Geol. Paläont. Abh. 131, 46–71 (1968)

Höll, R.: Genese und Altersstellung von Vorkommen der Sb-W-Hg-Formation in der Türkei und auf Chios/Griechenland. Inaug. Diss., 149 p., Univ. München (1966)

Kalafatçioğlu, A.: A geological study on the Karaburun Peninsula. Bull. Min. Res. Inst. Turkey 56, 40–49 (1961)

Ktenas, K.A.: Contribution à l'étude géologique de la presqu'île d'Erythrée (Asie Mineure). Ann. Sci. Fac. Sci. A 1, 1–57 (1925)

Lehnert-Thiel, K.: Geologisch-lagerstättenkundliche Untersuchungen an dem Zinnober-Vorkommen Kalecik und dem Nordöstlichen Teil der Halbinsel Karaburun (Westl. Türkei). Bull. Min. Res. Explor. Ins, Turkey 72, 43–73 (1969)

Maucher, A.: Die Antimon-Wolfram-Quecksilber-Formation und ihre Beziehungen zu Magmatismus und Geotektonik. Freib. Forsch.-H. C 186, 173–188 (1965)

Maucher, A., Saupe, F.: Sedimentärer Pyrit aus der Zinnober-Lagerstätte Almadén (Provinz Ciudad Real, Spanien). Mineralium Deposita 2, 312–317 (1967)

The Strata-Bound Lead-Zinc Deposits
from Menderes Massif in Bayındır (West-Anatolia)

O. Ö. DORA, Izmir

With 10 Figures

Contents

Summary

North of Bayındır in the cover series of the Menderes Massif, Pb-Zn deposits occur conformable to the original stratification. The country rock of the ore layers is usually garnet-chlorite schist. The ore layers and country rocks exhibit conformable syn-sedimentary deformation and tectonic features. The ore layers form three separate horizons which crop out in the flanks and the core of a northward recumbent fold.

The Pb-Zn mineralizations are probably of early Ordovician age and their orogin seems to be comparable with Ordovician/Early Silurian Sb-Hg-W formations. As supported by microscopic and field evidences the Pb-Zn mineralizations can be related to contemporaneous submarine acidic volcanism.

1. Introduction

In the schists which constitute the cover series of the Menderes Massif lying north of Bayındır, the lead-zinc deposits occur in an area of 4–5 km^2 in stratiform bodies reaching to extensions of about 3 km (Fig. 1). The main layer of deposits on both sides of Ilıca Dere extends northward to the Kurudere area and becomes poorer in terms of lead-zinc content (Fig. 2). Deposits situated 100 km from Izmir are easily accessible by roads in good condition. The highest topographic expression in the surroundings of the deposits is about 900 m. The Macchias are the most predominant bushy cover of the area, typical for mediterranean regions.

The discovery of Bayındır lead-zinc deposits is not more than ten years old. In 1971, exploration studies began. After many drills in the area, in 1975 the deposits were found to be economically feasible; and have since been in production for a year.

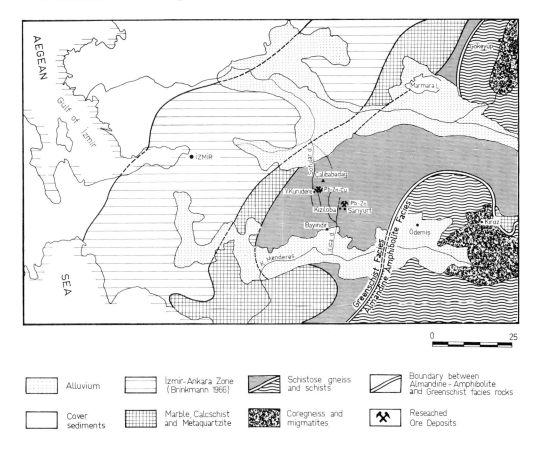

Fig. 1. Metamorphic units of the Menderes Massif

2. Geology and Petrology

In western Anatolia the Menderes Massif, composed of crystalline rocks, has an egg-like shape with a long axis of NNE trend (Fig. 1). The first detailed geological and petrographical investigation on the Massif was done by *Schuiling* (1958, 1962). To *Schuiling* the parental Cambrian and Precambrian sediments had undergone metamorphisms of Varistic and earlier phases. *Brinkmann* (1966, 1971) reports that metamorphism took place up to Lias. For augen gneisses, different parental rocks are suggested: *Graciansky* (1966) magmatites, *Izdar* (1971) both magmatites and sedimentites, *Başarır* (1970, 1975) and *Ayan* (1973) sedimentites. *Dora* (1972, 1975, 1976) regarded the gneisses and augen gneisses as having originated from sedimentary rocks and preferred the general use of the term "paragneiss".

The Menderes Massif can be divided, from base to top, into a generalized succession of augen gneisses, schists of variable lithologies and meta-carbonates. The gneisses include feldspars − large crystals of to 5 cm in size − quartz, biotite, muscovite and

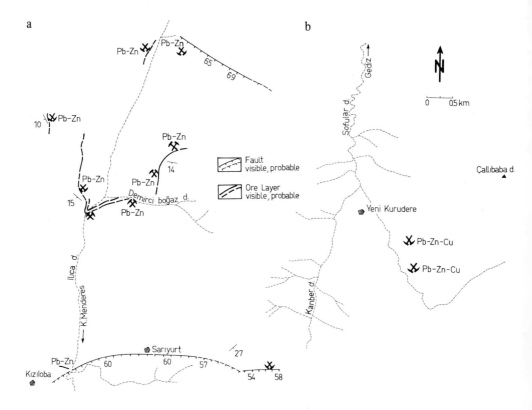

Fig. 2 a and b. Location map of (a) the Sarıyurt-Kızıloba and (b) Kurudere deposits

subordinate sillimanite and cordierite. The Plagioclases have a composition of up to 24% anorthite. The K feldspars have attained a Si/Al ordering to a transition boundary of monocline → tricline (*Dora*, 1976). The schists include mineral assemblages of kyanite, staurolite, garnet (almandine), biotite, muscovite, chlorite and chloritoid. Amphibolites and meta-volcanic rocks are local constitutents of the schists. The meta-carbonates are composed of marbles, calc-, talc- and graphite schists. In some places meta-quartzites overlie the meta-carbonates.

The oldest rocks of the studied area (Kurudere) are composed of almandine-weighted garnet- and staurolite schists. Biotite is represented by Fe-rich species, Muscovite is abundant. In the rocks with lepidoblastic texture, the schlierens of elongated quartz mosaic alternate with muscovite- and biotite-rich horizons. Large crystals of garnet (1–5 mm), have partially changed into biotite II + quartz + ore assemblage. Both garnet and staurolite are syndeformative products.

The high-grade metamorphites in the Sarıyurt region are garnet-chlorite schists. The schists, including additionally quartz, albite, biotite, clinozoisite, and calcite, are the country rocks of the lead-zinc layers. The rocks have probably resulted from the metamorphism of old acidic volcanics (lavas, tuffs and tuffites). In general the grains do not exhibit a particular arrangement to suggest a detrital origin. On the contrary, angular

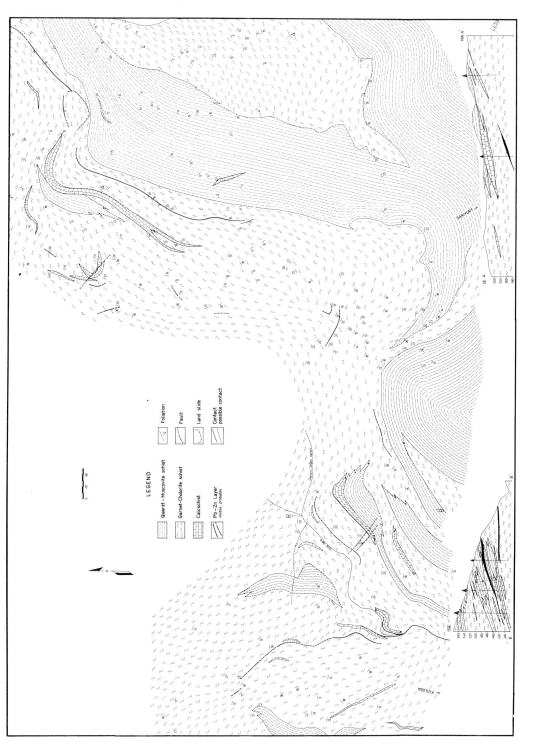

Fig. 3. Geologic map of the Pb-Zn deposits of the Bayındır/Sarıyurt-Kızıloba area. (After *H. Ozcan*, 1976)

fragments of quartz crystals show random distribution. The ore layers display parallel
stratification to the country rocks and, for short distances, conformable folding to the
latter (Fig. 4). Post-deformative chlorites cut the folds discordantly (Fig. 5). Around the
Kurudere, the ore layers are found in the same garnet-chlorite schists.

Fig. 4. The conformable ore layers in the closely folded country rocks, N X

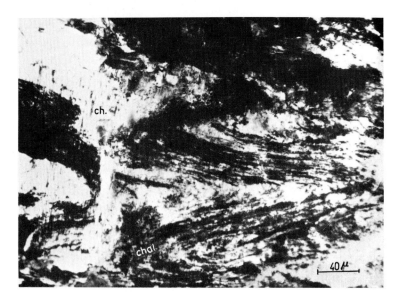

Fig. 5. Post-deformative chlorites disconformably cutting the minor folds, N //

In both Kurudere and Sarıyurt areas the marbles and calc schists overlie the garnet-chlorite schists. The calc schists include calcite, muscovite, and tremolite. Most particularly layers of pyrrhotite and chalcopyrite are observed in the black schists with graphite which probably include old volcanic ashes. The calc schists have been overlain by metaquartzites in Kurudere, and by quartz-muscovite schists in Sarıyurt. A generalized lithostratigraphic section is given in Figure 6.

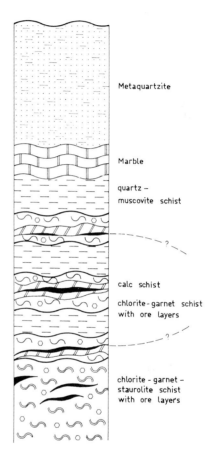

Metaquartzite

Marble

quartz – muscovite schist

calc schist

chlorite- garnet schist with ore layers

chlorite - garnet – staurolite schist with ore layers

Fig. 6. Idealized lithological section of the investigated area

The staurolite schist found in the studied ore, proves the metamorphic grade to have reached almandine-amphibolite facies. *Izdar* (1971), who investigated the 30 km east of the area, differentiated the rocks of the complete subfacies of the Barrovian type regional metamorphism reaching to migmatites. In the Sarıyurt and Kurudere areas, the complete subfacies has been restricted to staurolite-bearing and lower-grade subfacies.

3. Ore Deposits

In the Sarıyurt-Kızıloba area, as can be seen in the geologic map, there are three mineralization zones of 1–2 km length and 1–7 m thickness, interruptedly extending parallel
to the old S_1 planes of the garnet-chlorite schist or the contacts between calc schists and
garnet-chlorite schist. The ore layers extend parallel to each other, commencing in Sarıyurt Yayla with a trend to NE and changing across the Ilıca Dere to NW. The ore layers
display dips of 15–20° to SE and SW. The lead-zinc layers conforming to the bedding
of the country rock form a V whose apex is downstream in the Ilıca Dere (Fig. 3).

The ore layers in the Sarıyurt-Kızıloba area include galena and sphalerite as the main
ore minerals. From this ground, the layers with these minerals are the main zones of the
Bayındır lead-zinc deposit. Ore layers composed of fine-grained (0.1 mm) galena and
sphalerite alternate with country rocks. In all ore layers euhedral pyrites up to 1 cm in
size occur. Larger pyrites (up to 2 cm) have been observed in quartzite-rich portions of
the schists. Magnetite is subordinate in the paragenese. The fractures in the ore layers
and the conformable country rocks carry well-developed tetrahedrons of recrystallized
sphalerite (0.5 cm) and rhombohedrons of calcite and dolomite. Anglesite, serussite,
limonite and Mn-oxides are secondary products.

In the Sarıyurt-Kızıloba area, the upper part of the mineralization zone could be
laterally traced in part for 20–30 m. The most dominant mineral of this zone is galena.
The middle part with the exclusive presence of sphalerite constitutes the main mineralization zone. The middle zone is laterally traceable up to 600 m in distance without any
significant interruption. The lower zone crops out into discontinuous layers up to
10–20 m in extension. As in the upper layer, galena is the dominant ore species. In
some hand specimens the country rock and the ore layers display conformably recumbent isoclinal folds (Figs. 7 and 8). It is beyond doubt that larger examples of these
small structures exist in the field. Isoclinal folds in calc schists, up to sizes of 2–3 m are
frequently observed. It seems most probable that the ore layers of the Sarıyurt-Kızıloba

Fig. 7. The isoclinal folds of the
conforming ore layers and country rocks

Fig. 8. Isoclinal fold of the quartz-
vite schist. Disrupted thin layers of
extending along the bedding plane

ol. →

1.6 cm

area are situated on the flanks of an isoclinal anticline which is recumbent northward. Since the core of the anticline is the site of increase in the thicknesses of the constituent beds, it seems that the sphalerite-rich middle horizon is the exposed core of the anticline. The predominancy of galena in the lower and upper parts, and the lithologic similarities between these two parts, support this interpretation.

Under the microscope, inclusions of pyrrhotite in the anhedral crystal of galena or schlierens of pyrrhotite parallel to the bedding of the country rock have been recognized. In frequently patch-like chalcopyrite, indicating co-evel growth with galena, have been observed. The boundary between the contemporaneous galena and sphalerite is irregular. Sphalerite has randomly distributed inclusions of chalcopyrite which are oriented parallel to the crystallographic directions of the sphalerite. In part the galena has replaced the sphalerite. In the gangue columnar, crystals of rutile occur abundantly. Apart from the ore-bearing horizons, disseminated pyrite crystals have been observed in the uppermost part.

It was earlier pointed out that the ore layers occur in the garnet-chlorite and calc schists of same metamorphic grades in the Kurudere area. It seems that two occurrences at a distance of 7–8 km from each other, have been enclosed by similar country rocks. In Kurudere area the ore layers could be traced for 20–30 m. The ore layers crop out into two horizons, the upper one massive ore sphalerite 4–10 cm thick, and the lower one with mm-thick schlierens of chalcopyrite. Disseminated pyrite aggregates appear in both ore layers. The country rocks which carry schlierens of chalcopyrite include organic material such as graphite. The schlierens of galena form isoclinally recumbent small folds (Fig. 9), which are especially well developed in calc schists. The ore layers have either preserved their original bedding or have the same attitute with the S_2 schistosity in most places parallel to S_1 surfaces. Mineralizations in joints are relatively scarce (pyrite, marcasite etc). In every case, this type of occurrence in joints related to the main ore layers. It is suggested that the mineralization in joints has resulted from post-deformative crystallization following mobilization of the ore minerals in the main layers through metamorphism.

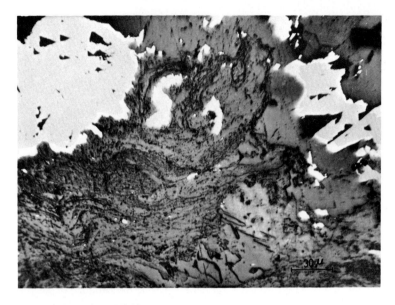

Fig. 9. Folded layers of galena, N //

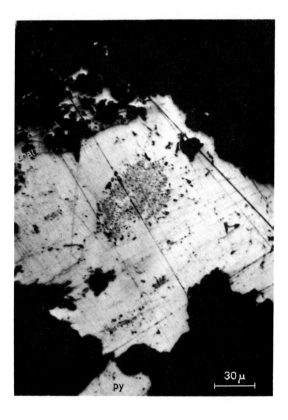

Fig. 10. Pyrite crystals formed from the old-generation ovate and prous colloform pyrite, N //

Under the microscope, flame-like streaks of pyrrhotite, schlierens of chalcopyrite, anhedral galena and patches of sphalerite are observed. The pyrites formed in the related paragenese are ovate and porous (Fig. 10). These features prove the pyrite to have formed from a colloform pyrite. Colloform marcasite accompanies colloform pyrite. Chalcopyrite appears to be the dominant mineral. Mineral composition of the gangue may include graphite.

4. Discussion

The lead-zinc mineralization in the Sarıyurt-Kızıloba and Kurudere areas, to the north of Bayındır, occurs in the schists which belong to the cover series of the Menderes Massif. The schists have suffered regionally a Barrovian type of greenschist facies of metamorphism. The ore-bearing layers of the schists include mineral assemblages marking the quartz-albite-epidote-almadine subfacies. The restriction of the ore minerals to the horizons rich in chlorite, graphite and garnet, or to the contacts between these rocks and the calc schists indicate the ore to be related virtually to the specific types of rock enclosed by parental calcareous and clayey sediments. These specific types of rock are most likely tuffs, tuffites and lavas of a submarine acidic volcanism. The presence of quartz fragments in the tuffs has already been discussed above. The flattened patches of quartz mosaics in the rocks, which are completely made up of muscovite and chlorite, support our inferences. The abrupt contacts of the patches of quartz to the surrounding material could be interpreted in favor of recrystallization of the old phenocrysts.

There are abundant criteria suggesting the metallic ions contributed by volcanic exhalations and syn-sedimentary precipitation. The lateral persistency of ore layers coinciding with stratification, the preserved glide structures of the ore bands, and conformable folding of the ore layer are the most reliable evidence for the syn-sedimentary origin. On the other hand, the ore layers either flat-lying or closely folded in the garnet-chlorite schist have been cut by the chlorites formed during the post-deformative phase of the metamorphism (Fig. 5). That is also an evidence supporting the contemporaneous formation of the ore. The spectacular erosional feature of an individual ore layer into 'V' rule in the Ilıca Dere valley, indicates the "relation of the lead-zinc deposits to time and stratification".

It was discussed earlier that apparently three separate ore layers occur in the lower and upper flanks, and in the core a northward recumbent fold. The anticline in the Sarıyurt area, the B_1-axis with NE trend, has been superposed by a B_2-axis of nearly N-S trend, and gained an attitude toward the NNW. As reflected by the hand specimens, the resultant gross structure is made up of northward recumbent folds (Fig. 7).

So far the Kurudere outcrops have not been mapped entirely and the dimensions have not been established strictly. These ore deposits occur in several layers in graphitic, chloritic and garnet-bearing host rocks as in Sarıyurt. The isoclinal folds of the ore and the host rock in Sarıyurt are also encountered in the Kurudere area. On the other hand, the same isoclinal folds are also found, especially in carbonate sections, in the area between Kurudere and Kızıloba. On this ground, the suggestion that the Kurudere deposits are the north-westward extension of those of Sarıyurt-Kızıloba, becomes greatly supported. If the traceable extension of the Sarıyurt-Kızıloba deposits for 3 km is taken into consideration, the reappearance of the ore related with folds conforms to the macro-structure of the region. Geologic exploration is still going on in the intermediate areas.

The exact age of the cover schists of the Menderes Massif has not yet been clearly established. The uppermost carbonate horizons range in age from Trias to Lias, based on fossil content (*Brinkmann*, 1976). The syn-sedimentary Sb-Hg-W formation has been recorded by *Höll* (1966) both in the nonmetamorphic Ordovician and Middle Silurian (?) detrital sediments and in their low-grade metamorphic equivalents on the Menderes Massif represented, mostly by graphite schists (*Höll*, 1966). The world-wide syn-sedimentary Sb-Hg-W deposits of the boundary of Ordovician-Silurian has been established by *Maucher* (1965). Lead-zinc deposits of Bayındır occur, as well as Sb-Hg-W mineralization, in the low-grade metamorphic rocks. All these country rocks which belong to Menderes Massif can be either slightly different in metamorphic grade or the same. The conclusion derived from the above relationships is that the Bayındır lead-zinc deposits are either just below or equivalent to Hg-Sb-bearing horizons. If all the deposits are considered as time-equivalent, and given a submarine volcanic origin, the Bayındır lead-zinc occurrences seem to have resulted in the most central part of the volcanic activity of the Ordovician/Silurian time interval. On the other hand, since lead-zinc mineralizations in the low-grade metamorphic cover series of the Menderes Massif are encountered in other localities (such is in Gümüldür), it is beyond doubt that widespread syn-sedimentary lead-zinc mineralization predated the Ordovician/Silurian, syn-sedimentary Hg-Sb mineralization. The strata- and time-bound formation of lead-zinc and minor copper occurrences has been recorded in the nearby massifs of western Anatolia (*Minčeva-Stafanova*, 1967).

Acknowledgements: I appreciate the help that came from the MTA Institute in Izmir in studying the unpublished report of the investigated area.

References

Ayan, M.: Gördes migmatites. Bull. Mineral. Res. Expl. Inst. Turk. **81**, 132–155 (1973)

Basarir, E.: Bafa Gölünün doğusunda kalan Menderes Masifi güney kanadının jeolojisi ve petrografisi. Rep. Fac. Sci. Ege Univ. **102**, 1–44 (1970)

Basarir, E.: Menderes Masifi güney kanadindaki metamorfik kayalarda görülen mineral transformasyonlari. Tübitak Congr. (1975)

Brinkmann, R.: Geotektonische Gliederung von Westanatolien. N. Jb. Geol. Paläontol. Mh. **10**, 603–618 (1966)

Brinkmann, R.: Das kristalline Grundgebirge von Anatolien. Geol. Rdsch. **60**, 886–899 (1971)

Brinkmann, R.: Geology of Turkey, 1st ed. Stuttgart: Ferdinand Enke 1976, p. 8

Dora, O.Ö.: Orthoklas-Mikroklin Transformation in Miagmatiten des Egrigöz-Massivs. Bull. Geol. Soc. Turk. **XV–2**, 131–152 (1972)

Dora, O.Ö.: The structural state of K-feldspars and their application as petrogenetic indicators in Menderes-Massif. Bull. Geol. Soc. Turk. **18–2**, 111–126 (1975)

Dora, O.Ö.: Die Feldspäte als petrogenetischer Indikator im Menderes-Massiv/Westanatolien. N. Jb. Miner. Abh. **3**, 289–310 (1976)

Graciansky, P. de: La Massif cristallin du Menderes (Taurus occidental Asie Mineure), un exemple possible de vieux socle granitique remobilisé. Rev. Géog. Phys. Géol. Dynamique **VIII/4**, 289–306 (1966)

Höll, R.: Genese und Altersstellung von Vorkommen der Sb-W-Hg-Formation in der Türkei und auf Chios/Griechenland. Bayer. Akad. Wiss. Abh. H. 127, 1–118 (1966)

Izdar, E.: Introduction to geology and metamorphism of the Menderes Massif of western Turkey. In: Geology and History of Turkey. Campbell, A.S. (ed.). Tripoli, Libya: Petrol. Explor. Soc. Libya 1971, pp. 495–500.

Maucher, A.: Die Antimon-Wolfram-Quecksilber-Formation und ihre Beziehungen zu Magmatismus und Geotektonik. Freiberger Forschsh. C 186, 173–188, Leipzig (1965)

Minčeva-Stefanova, J.: The genesis of the stratiform lead-zinc ore deposits of the "Sedmochislenitsi" type in Bulgaria. In: Genesis of Stratiform Lead-Zinc-Barite-Fluorite Deposits. Brown, J.S. (ed.). Lancaster: Penna 1967, pp. 147–168

Özcan, H.: Izmir-Bayındır-Ilıacadere Zn-Pb Madeni Raporu. MTA-Institute (unpublished, 1976)

Schuiling, R.D.: A zircon-study of an augen-gneiss in the Menderes-Massif. Bull. Mineral. Res. Expl. Inst. Turk. 51, 38–41 (1958)

Schuiling, R.D.: On petrology, age and structure of the Menderes Migmatite Complex (SW-Turkey). Bull. Mineral. Res. Expl. Inst. Turk. 58, 71–84 (1962)

Hydrothermal-Sedimentary Iron Ores Related to Submarine Volcanic Rises: the Teliuc-Ghelar Type as a Carbonatic Equivalent of the Lahn-Dill Type

H.G. KRÄUTNER, Bucuresti

With 14 Figures

Contents

Summary

In the metamorphic Devonian of the Poiana Ruscă Mts syngenetic iron carbonate ore deposits occur in a geological environment quite similar to the Lahn-Dill area. A model is presented according to which two main groups of hydrothermal-sedimentary iron ore deposits on submarine volcanic rises are distinguished: (1) Lahn-Dill-type deposits with mainly oxidic ores, formed on the culmination zone of the rise in shallow depths and oxygenated waters; (2) Teliuc-Ghelar-type deposits with mainly carbonatic ores, formed on the slope of the rise at greater depth where an acid aureole with high PCO_3 around the emission centres of the hydrotherms controls the local environment of the ore deposition. The available data point to the fact that these two types of iron ore are related to the evolution of intracratonic mobile zones.

1. Introduction

The iron oxide ores related to the diabasic volcanism are generally called Lahn-Dill ores, according to *Schneiderhöhn* (1955). The literature on mineralogy, petrography, geochemistry, paleostructural position as well as on the genetical problems of this type of ore, refers especially to classical regions in Central Europe, as the Renan massif, Harz, Eastern Thuringia and Eastern Sudets (Jesenik). Similar iron ore deposits appear also in other geological units with "basic initial magmatites". However, they have been rarely mentioned and described because of their small sizes and of the negligible amount of the ore reserves. It seems that this type of iron ore deposits is particularly characteristic in metallogenic provinces [1] related to the Paleozoic basic volcanism.

According to the literature referring to the deposits of the classical regions (*Kegel*, 1923; *Lippert*, 1951, 1953; *Knauer*, 1960; *Rösler*, 1960; *Gräbe*, 1962; *Dave*, 1963; *Steinike*, 1963; *Skacel*, 1964; *Bottke*, 1965; *Quade*, 1965; *Lutzens* and *Burchard*, 1972) or to the principal genetical problems of the ores (*Borchert* 1960, 1972; *Hentschel*, 1960) and especially the general surveys published by *Rösler* (1962) and *Quade* (1970), the main characteristics of the Lahn-Dill ores are as follows:

1. Syngenetic submarine deposition.

2. Close connection with the products of a basic "initial magmatism" (diabase-spilitic with keratophyres). The ores are disposed either at the top of the diabasic volcanic complexes (most frequently) or intercalated in the diabasic tuffs.

3. Stratiform concentrations with lenticular shape.

4. Variability of the facial development within the same ore body. On relatively small surfaces there occur primary ores represented mainly by siliceous ores, compact hematite ores, calcareous hematite ores (Flusseisenstein) and subordinately by iron-chlorite (chamosite) ores, iron carbonate ores, pyrite-melnicovite ores, as well as resedimented ores represented by hematite ores (Scheckenerze), epiclastic chlorite ores and schists with fragments of iron ores.

5. Simple and limited mineralogical and chemical composition: the principal elements are Fe, Si, Ca bound in hematite ($\pm$ magnetite), quartz, calcite and subordinately in iron-chlorite, iron carbonate, iron sulphide. In some places anthraxolite is present in the ore.

6. The ore contains about 40% Fe and forms deposits with an amount of the order of 5 million t iron ore, rarely exceeding 10 million t (*Quade*, 1970).

The Lahn-Dill ores therefore represent a well-described type of iron ore deposit, as a certain genetic process may be presumed. That is why we think that attempts to enlarge the concept of "the Lahn-Dill type" by using it also for other volcano-sedimentary ore deposits (e.g., *Rösler*, 1964) are both useless and confusing.

In a very similar geological environment there occur in the Middle Devonian of the Poiana Ruscă Mts (South Carpathians) stratiform and lens-shaped iron carbonate ores which locally pass laterally into magnetite and hematite ores of the Lahn-Dill type. These iron ores have undergone a Hercynian regional metamorphism in the greenschist facies.

Besides their mainly carbonatic character, the iron ores of the Poiana Ruscă Mts differ from the Lahn-Dill type sensu stricto, which is chiefly oxidic, not only by some geochemical peculiarities but also by a different palaeostructural position on the submarine volcanic rises. Therefore, we propose to use the denomination *Teliuc-Ghelar type* for this peculiar facial development of the hydrothermal-sedimentary iron ore deposits in the Poiana Ruscă Mts.

---

[1] In the acceptance of metallogenic units with ores from the same source, as employed on the Romanian metallogenic maps (*Rădulescu*, 1966; *Kräutner*, 1970).

2. Iron Ores of the Teliuc-Ghelar Type in the Poiana-Ruscă Mountains

In the southern part of the area with Paleozoic metamorphic rocks, syngenetic iron carbonate ore deposits occur, trending E-W over more than 50 km (Fig. 1; *Kräutner*, 1970). Some twenty groups or isolated deposits in the range of 0.5 to 20 million t of iron ore are already known, conformably intercalated in the Devonian crystalline schists. The mining in this region dates from the 1st century and is at present in an advanced stage, numerous deposits being worked.

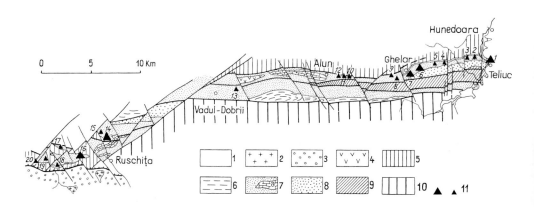

Fig. 1. Geology and iron ore deposits in the southern part of the metamorphic Paleozoic of the Poiana Ruscă Mts. *1:* Neogene sedimentary deposits; *2:* Paleocene granodiorites (Banatitic rocks); *3:* Upper Cretaceous of the Rusca Montană basin; *4–10:* Metamorphic rocks; *4:* Metaserpentinites; *5:* Upper Carboniferous (Padeş Series); *6:* Upper Devonian (Upper Ghelar Series); *7:* Middle Devonian (Lower Ghelar Series), Ruschiţa-Alun marbles; *8:* Lower Devonian (Govăjdia Series); *9:* Upper Ordovician-Silurian? (Bătrîna Series); *10:* Upper Precambrian metamorphites; *11:* Iron ore deposits. *Ore deposits: 1:* Teliuc; *2:* Pădurea Oraşului; *3:* Nicolson-Făgeţel; *4:* Mînăstirii; *5:* Filimon; *6:* Ghelar Central; *8:* Mihail; *9:* Cornet; *10:* Mătrăguna; *11:* Dragoş; *12:* Costiş; *13:* Vadul Dobrii; *14 and 15:* Pîrîul cu Raci (Ruschiţa); *16:* Dealul Boul; *17 and 18:*Valea Lupului; *19:* Afinar; *20:* Dealul Negrii

In the northern part of the Poiana Ruscă Mts, some small iron oxide deposits of the Lahn-Dill type are found in a similar stratigraphic position (*Kräutner*, 1970, p. 331).

2.1 Geological Setting

According to palynological data and Pb-Pb ages, the sedimentation of the sequence in which the iron ores are intercalated took place during the Paleozoic (*Kräutner* et al., 1973). Both the sedimentary rocks and the iron ores underwent a regional metamorphism of Barrovian type in the greenschist facies (*Kräutner* et al., 1976). The K-Ar ages date this metamorphism as 320 my ago (*Kräutner* et al., 1973), therefore in the Sudet phase of the Hercynian foldings.

The sequence of Paleozoic metamorphic rocks, known also as "Poiana Ruscă crystalline", has been subdivided into several lithostratigraphic units called, from the lower to the upper part, the Bătrîna Series, the Govăjdia Series, the Ghelar Series and the Padeş Series. Their lithological constitution, as well as their correspondence to the stratigraphic subdivisions, based on the palynological data and the Pb-Pb ages, are shown in Figure 2.

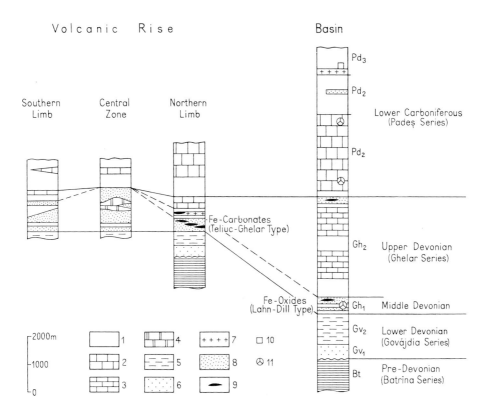

Fig. 2. Facies development of the Paleozoic in the Poiana Ruscă Mts. *1:* Sericite-chlorite schists; *2:* Hunedoara-Lucani dolomites and limestones; *3:* Dolomites and limestones from the Upper Ghelar Series; *4:* Ruschiţa-Alun marbles; *5:* Graphitic sericite-chlorite schists; *6:* Quartzitic sericite-chlorite schists; *7:* Quartz keratophyre metatuffs; *8:* Greenschists (basic metavolcanites, metatuffs and meta-epiclastites); *9:* Iron ore bodies; *10:* Pb-Pb age; *11:* Palynological data

The lower part of the known sequence – the Bătrîna Series – seems to correspond to the Upper Ordovician-Silurian. It is represented by a detrital formation with sericite-chlorite schists, graphitic schists, quartzites and basic metatuffs.

After a short exondation, due to the Caledonian movements, a detrital sedimentation begins in the Lower Devonian. The sequence, chiefly quartzitic at the lower part and graphitic at the upper part, corresponds to the Govăjdia Series and is characterized by facial uniformity in the whole area of the Poiana Ruscă massif.

A submarine basic volcanism starts in the Middle Devonian. Its products are widespread in the whole area and constitute a greenschist formation belonging to the lower part of the Ghelar Series.

In the Upper Devonian, the basic volcanism is followed by an uplifting of the southern region where the greenschist formation reached its maximum thickness. Thus a submarine volcanic rise appears in the south on which the sedimentation is very reduced and intermittent. Concomitantly, the northern area was sinking and a basin was formed. In this area detrital sediments were deposited, alternating with thick banks of limestones and dolomitic rocks, as well as some relatively thin layers of basic tuffs, the latter especially towards the end of the period. This sequence corresponds to the upper part of the Ghelar Series.

In the Lower Carboniferous, the two areas with different facial developments have been maintained. In the south, a pelito-psamitic sedimentation took place, while in the north the subsidence continued and important masses of dolomitic rocks and limestones (Hunedoara-Luncani dolomites and limestones) accumulated, corresponding to the lower part of the Padeş Series, and a thick pile of psamitopelitic rocks, corresponding to the upper part of the Padeş Series. Concomitantly to the latter, a basic volcanism appeared for a short time in the northern part, followed by a period of rhyolitic volcanism which developed both extrusively, forming some tuff layers, and intrusively by the emplacement of dykes. The latter are locally accompanied by pre-metamorphic hydrothermal alterations and Pb-Zn ores (*Kräutner*, 1963).

The iron ores are closely bound in time and space to the Middle Devonian basic metavolanites. The carbonate ores are associated with the greenschists from the southern part, while some oxide ores occur especially in the northern part of the Poana Ruscă Mts. The advanced exploration, particularly in the south, by mining and drillings, allows us to recognise that the distribution of the iron carbonate ores in this area seems to indicate a certain correlation with the morphology of the volcanic paleostructures.

As mentioned before, the Middle Devonian volcanic complex reaches maximum thicknesses of some 100 m in the south. It may be assumed that this zone of thickening of the greenschist formation corresponds to a submarine volcanic paleorise consisting of several eruption centres lined on a system of deep fractures. The volcanic products are mainly represented by basaltic (diabasic) and spilitic metatuffs. At the upper part of the volcanic formation and especially on the gentle slopes of the volcanic rise, there occurs an alternation of volcanic products with detrital-terrigenous schists. In this zone the mineral assemblage of the greenschists usually indicates a supply of terrigenous material in the premetamorphic rocks. The basic metatuffs pass therefore to metatuffitic rocks which often contain interlayered pelito-psamitic layers. All around the volcanic rise only metatuffitic basic rocks occur. These greenschists contain variable amounts of calcite, quartz, and sericite. They are considered as metamorphosed epiclastic products formed by the resedimentation of the primary basic tuffs and diabases, concomitantly with an intermittent supply of a pelito-psamitic fraction.

These facts indicate that the volcanic rise has been partly destroyed, probably even concomitantly with its building, but at the present metamorphic stage of the rocks no relicts of clastic structures have been preserved.

Intercalated in the greenschists of the central part of the rise or located on the top of the meta-volcanite pile, appear lenses of massive white marbles which pass laterally to bedded marbles and calc schists (the Ruschiţa-Alun marbles). It was considered that these massive marbles represent metamorphosed reefs, which have grown on the submarine rise during periods without volcanic activity and have been surrounded by perireefal detrital products. Subsequently, the reefal formations have been covered by the volcanic products of new extrusion phases.

The last eruptions had a rhyolitic and quartz-keratophyric character. Layers, 0.5–10 m thick, of white tuffs have been deposited. Laterally, these rocks pass also to tuffitic and epiclastic rocks.

An attempt to reconstitute the palaeostructure of the area at the end of the Middle Devonian (Fig. 3) shows that the iron carbonate ores are situated on the slopes of the submarine rise, more exactly in the zone where the diabasic tuffs and spilites pass gradually to epiclastic rocks, and interfingers with tuffites and terrigenous rocks. These deposits usually overlie the sequence of volcanic rocks but, in places, they are intercalated in the basic tuffs, tuffites and even in terrigenous rocks. Figure 3 indicates also that the big lenses of siderite and ankerite ore occur at a greater distance from the centre of the rise, while the smaller stratiform magnetite ore bodies are found closer to the culmination of the rise. The former always underlie the quartz-keratophyric metatuffs and correspond to the lower iron ore horizon. The latter are associated with, or disposed immediately over, the quartz-keratophyric metatuffs and constitute the upper iron ore horizon.

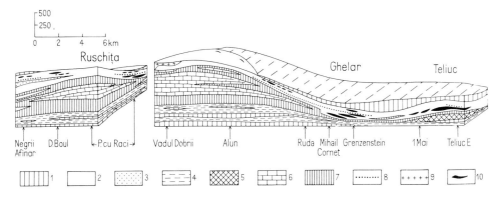

Fig. 3. The volcanic rise of the Teliuc-Ghelar-Ruschiţa region, at the beginning of the Upper Devonian (schematic reconstitution). *1:* Limestones and dolomites; *2:* Pelito-psamitic rocks; *3:* Quartzites; *4:* Pelito-psamitic rocks with organic matter; *5:* Dolomite; *6:* Reef limestones; *7:* Basic tuffs and epiclastites; *8:* Quartz keratophyre metatuffs; *9:* Serpentinites; *10:* Iron ore deposits

The iron carbonate ore bodies from Teliuc, Ghelar and Ruschiţa appear as thick lenses which occupy surfaces of about 0.5–0.8 km^2 and reach maximum thicknesses of 100–200 m. The ore lenses are concordantly intercalated in the crystalline schists and are folded together with them. In most of the cases, the ore bodies do not occur isolated but in groups of about two to five lenses. The ores of a group may be intercalated at the same stratigraphic level (e.g., at Teliuc) or they may be superposed, thus occupying a wider stratigraphic interval (e.g., at Ghelar) (for details see *Kräutner*, 1970, p. 330). The margins of the carbonate ore bodies with the surrounding rocks are usually sharp, however, small lateral interfingering may also be noticed.

The ore bodies generally present a complicated inner structure expressed by the numerous varieties of ores and rocks constituting them. The transitions between them are gradually or sharply limited, lateral passages and stratigraphic superpositions. A general tendency of stratification may be recognized in the distribution of these varieties of ferriferous rock. The present structure, as well as a reconstitution of the initial aspect of the main lenses belonging to the Teliuc and Ghelar deposits, is rendered in Figures 4 and 5.

At hand specimen size the ore structure is either massive or bedded (banded). The bedding results from the alternation of either layers with similar mineralogical but quantitatively different constitution, or bands with a different lithological constitution (e.g., quartzite-iron carbonate rock; siderite-parankerite, etc). Usually, the bedded rocks do not cross the deposit continuously over its whole extension.

Locally, especially in the marginal parts of the ore bodies relict structures of sedimentary breccias were found, which indicate the existence of some intraformational breccias in the ore. These rocks are represented, for example by siderite ore with inclusions of parankerite rocks or by parankerite rocks with fragments of siderite ore. Relict structures of pseudobreccias have been also found, especially in the parankerite rocks.

Microscopically, the structure and the texture of the ore show the characteristic features of metamorphic recrystallization. For the petrofabric and deformation see *Kräutner* (1970).

Looking at the form, the lithological constitution and the manner of the rock disposition inside the ore body, we can presume that the initial deposit was an accumulation of ferriferous carbonatic mud in which some irregular distributions of the component parts might be interpreted by submarine disturbances. The intercalations of crystalline schists in the ore body and the lateral passages of the ore to the surrounding rocks, although rare, indicate that the deposition of this mud took place simultaneously with the sedimentation of the volcanic and terrigenous material on the slope of the submarine volcanic rise. Taking into account that no volcanic or terrigenous detrital fraction has been found in the ore, we have to admit that the deposition of the ferriferous carbonatic mud took place during a relatively short time interval. The outflow of the thermal source was therefore high and probably temporarily interrupted when the intercalations of detritogenous rocks were deposited. The shape of the lenses, as it results from the attempts at reconstitution (Fig. 5), makes us think, at least as regards the thick lenses of Ghelar and Teliuc, of an accumulation in negative relief forms, because such "cones" of mud probably would not have had stability on plane or sloping surfaces. There are, however, deposits with stratiform concentrations (very flat lenses) for which such an interpretation is not necessary (e.g., The Vadul Dobrii deposit).

2.2 Petrographical, Mineralogical and Geochemical Characteristics

The iron carbonate ores from Teliuc, Ghelar, Vadul, Dobrii, and Ruschiţa are massive or bedded, mono- and polymineral rocks. The carbonates are especially associated with quartz and, subordinately, with magnetite. Silicates also occur frequently. They are represented, according to the intensity of the metamorphism, by ferro-stilpnomelane, thüringite, amosite, spessartine — in the chlorite zone — and by Fe-biotite (lepodomelane),

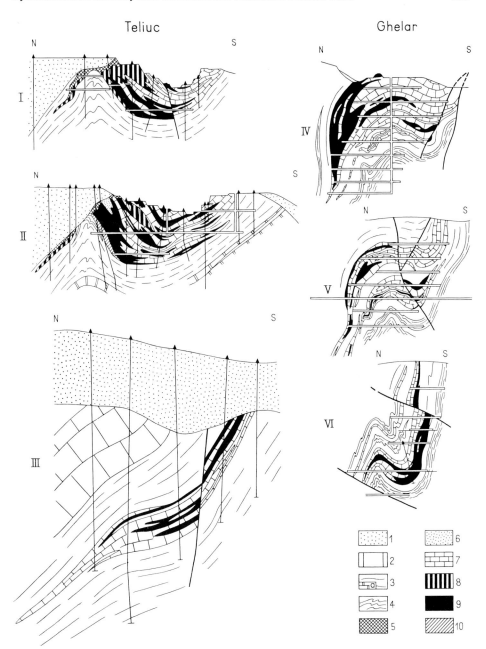

Fig. 4. Geological cross section through the main iron carbonate ore bodies from Teliuc and Ghelar. *1:* Neogene sedimentary cover; *2:* Limestones and dolomites; *3:* Sericite-chlorite schists and green-schists, *a:* dolomites; *4:* Graphitic sericite-chlorite schists; *5:* Limonite ore (superficial oxidation zone); *6:* Quartzites (Silica metagels); *7:* Parankerite rocks; *8:* Ankerite ore; *9:* Siderite ore; *10:* Hematite-magnetite ore

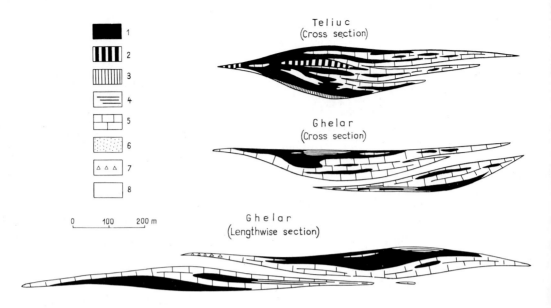

Fig. 5. Initial form of the main iron carbonate ore bodies from Teliuc and Ghelar (reconstitution). *1:* Siderite ore; *2:* Ankerite ore; *3:* Hematite-magnetite ore; *4:* Banded iron carbonate ore; *5:* Parankerite rocks; *6:* Quartzite (Silica metagel); *7:* Resedimented iron carbonate ore; *8:* Surrrounding pelito-psamitic and tuffitic rocks

cummingtonite-grünerite, almandine in the biotite and almandine zones. Albite, barium minerals such as celsian, barite and especially sulphides among which pyrite predominates, are also known, while pyrrhotine, mispickel, galena and sphalerite occur only sporadically. According to the manner of assemblage of these minerals with the various iron carbonates, numerous varieties of iron ores can be distinguished (*Kräutner*, 1964).

Considering only the carbonate components, the ores are mainly monomineralic. Consequently, one can distinguish sideritic, sideroplesitic, mangano-sideritic, ankeritic and parankeritic ore varieties. The composition of the carbonate components of these rocks is indicated in Figures 6 and 7. Rocks constituted of more than one iron carbonate mineral, i.e., with assemblages such as siderite + ankerite or siderite + parankerite are more rarely found. The composition of these ores falls into the field of unmiscibility for iron carbonates; therefore mineral assemblages with more than one carbonate phase seems to indicate a "resedimented" and "mixed" primary deposit. In fact, relict structures of sedimentary breccias have been found in these ores.

A characteristic of the ore deposits from Teliuc, Ghelar, Vadul Dobrii and Ruschiţa consists of a large amount of white or grey, massive and bedded parankerite rocks which, due to the low content in Fe (10—40 mol $FeCO_3$) cannot be recovered as ore. If fresh, these rocks are both macro- and microscopically very similar to dolomites.

Attention must also be given to some "glassy quartzites" which are associated with the iron carbonate ore. These rocks may be considered as recrystallized SiO_2 gels. Under the microscope they differ from the detrital quartzites by a specific structure due to an intimate intergrowth of small-sized irregular quartz crystals. There are continuous transitions from these quartzites to the iron ores by a gradual increase of the content in iron

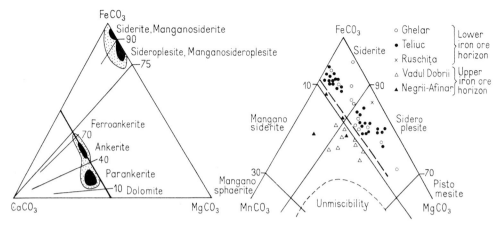

Fig. 6. Molecular composition of the iron carbonates (Teliuc, Ghelar, Vadul Dobrii and Ruschiţa: *high density black*)

Fig. 7. Molecular composition of the iron carbonates in the "siderite ore"

carbonates. Quartzitic rocks and quartzitic iron ore (with hematite instead of iron carbonates) are very characteristic of the Lahn-Dill deposits. The presence of such rocks in the ore deposit of the Poiana Ruscă Mts may be considered as another indication of some similitudes between the ores in the Teliuc-Ghelar region and those of the Lahn-Dill type. Relict gel structures, as found in the latter (*Cissarz*, 1924), could not be expected in the Poiana Ruscă massif due to the total recrystallization of the ores during the Hercynian metamorphism.

A general view of the chemical composition of the Teliuc and Ghelar iron carbonate ores is given in *Kräutner* and *Kräutner* (1963). A geochemical characteristic of the siderite ore consists of the ratios among Fe, Mn and SiO_2 (main values: 35% Fe:2,5% Mn:8% SiO_2). The SiO_2:Fe and SiO_2:Mn ratios of these rocks indicate a negative correlation.

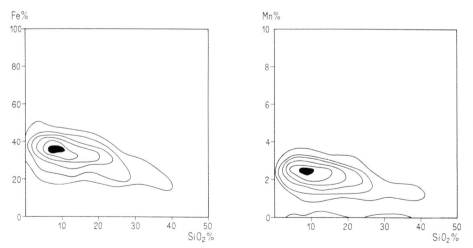

Fig. 8. Fe:SiO_2 and Mn:SiO_2 ratio in the "siderite ore" (Teliuc and Ghelar). Frequency isolines: 22 − 16 − 11 − 5 − 3 − 0.5%

As concerns the FeO:MnO ratio of the iron ores, higher values are to be found in the carbonate ores (4–10 in siderite-ankerite rocks; 4–20 in parankerite rocks) than in the oxide rocks (0–5 in magnetite and hematite rocks; Fig. 9). It should also be pointed out that the iron carbonates of the upper iron ore horizon are richer in MnO than those of the lower iron ore horizon (Fig. 7). Therefore, it seems that the Fe:Mn ratio was stratigraphically controlled.

In the rocks with over 15% SiO_2 the iron content of the carbonatic components decreases simultaneously with the increase of SiO_2 in the rock (Fig. 10). Thus, in the rocks rich in SiO_2, the ferriferous carbonate is always represented by sideroplesitic varieties, not by siderite.

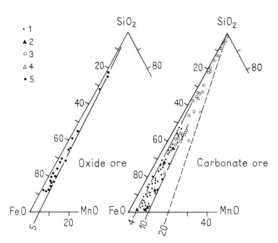

Fig. 9. FeO-MnO-SiO$_2$ ratio of the iron ores from Teliuc and Ghelar. *1:* Siderite ore; *2:* Ankerite ore; *3:* Parankerite rocks; *4:* Iron carbonate quartzites; *5:* Magnetite and hematite ore

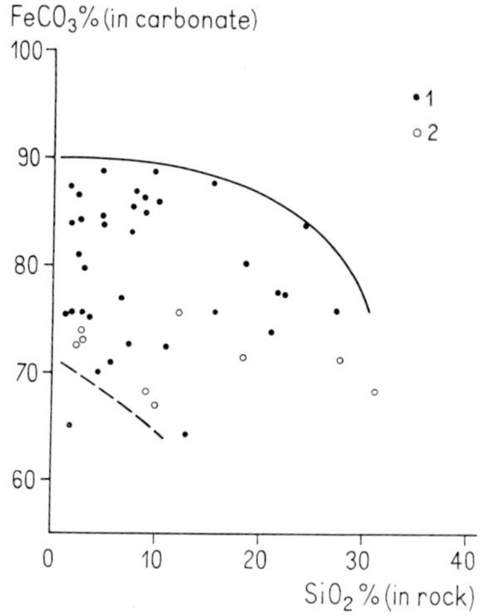

Fig. 10. Relationship between FeCO$_3$ (mol) in the carbonate and the SiO$_2$ content of the whole "siderite ore" from Teliuc, Ghelar and Vadul Dobrii. *1:* Lower iron ore horizon; *2:* Upper iron ore horizon

Geochemical profiles taken across the main ore deposits indicate that in the same ore body alternations of siderite and sideroplesite rocks occur with variable contents in SiO_2, as well as differences between the different ore lenses of the same deposit. Thus, the lenses of iron carbonate ore present a certain degree of nonhomogeneity.

3. Position of the Iron Ores on the Submarine Volcanic Rises

Figure 13 indicates schematically the position in the submarine volcanic structures of the deposits from Teliuc-Ghelar-Ruschiţa and of the Lahn-Dill type from the classical regions as it results, for the latter, from the data published by *Lippert* (1953), *Gräbe* (1962), *Dave* (1963), *Bottke* (1965), *Quade* (1965, 1970).

The *iron oxide ores of the Lahn-Dill type* are situated in the uppermost part of the submarine rise. The primary deposits are represented by siliceous hematite ores, compact hematite ores, calcareous hematite ores and chamosite ores with magnetite, in places. The deposition of the ore accumulations is admitted in an oxygen-rich medium, generally at shallow depths. The secondary deposits, formed by resedimentation of the primary ore, are found either on the slopes of the volcanic structures or in negative relief forms. Siderite and melnicovite-pyrite deposits (*Gräbe*, 1964; *Quade*, 1970) occur also in higher zones of the volcanic rise but over restricted areas. For these deposits a sedimentation in internal basins is assumed, in which the lack of oxygenated water circulation leads to reducing conditions.

The *iron carbonate ores of the Teliuc-Ghelar type* are situated on the lower parts of the slopes of the submarine volcanic rises, therefore at greater depth than the oxidic Lahn-Dill ores. The presence of magnetite in the basic metatuffs near the ore deposits and generally in the whole sequence indicates that weak oxidating conditions existed in this zone. As the mentioned depth is not favourable for a continuous oxygenation, the outflow of acid, CO_2-rich thermal solutions, modifies on reduced surfaces the environment of ore deposition (see Sect. 4.2). Therefore, the primary deposit is represented by siderite, ankerite, and parankerite ore with which local concentrations of silicates, oxides and iron sulphides are associated. Resedimented ores are rare and usually occur in the marginal zone of the iron carbonate ore bodies. It seems, however, that secondary redistributions of the material by submarine perturbations are more frequent.

4. Genesis Problems of the Iron Ore Deposits on Submarine Volcanic Rises

The hydrothermal-sedimentary iron ore deposits were much disputed especially in connection with the Lahn-Dill ores. General views on the genesis of these ores were given particularly by *Borchert* (1960), *Hentschel* (1960), *Rösler* (1962), *Harder* (1964a, b), *Bottke* (1965), and *Quade* (1970).

The main difficulty in a genetic interpretation concerns the source of the hydrothermal solutions, as well as the manner of iron transportation. At present, one may admit without doubt the deposition of the primary ore by a sedimentation supplied by a local source, facial variations on relatively reduced surfaces due to environmental control, a close association in space and time with a submarine basic volcanism and a certain position of the ore deposits on the relief of the volcanic structures.

In discussing the genesis of these iron ores one must take into account (1) the source of the ferriferous solutions; (2) the manner of the iron transport (constitution of the solutions); (3) the ore deposition; (4) resedimentation processes; (5) subsequent transformations due to the diagenesis and possibly to the regional metamorphism.

We shall later discuss some thoughts on the above points on the basis of new data offered by the iron carbonate ore deposits of the Poiana Ruscă massif.

4.1 On the Source of the Hydrotherms

To the older concept that assumes a hydrothermal source separated by magmatic differentiation (*Schneiderhöhn*, 1941; *Borchert*, 1957, 1960; *Cissarz*, 1957; *Quade*, 1970; *Lehmann*, 1972) a new idea has been opposed that states an extraction of the main elements from the volcanic rocks (*Hentschel*, 1960; *Rösler*, 1962). This extraction would be due to thermal solutions formed in water-rich nondiagenized basic tuffs, under the influence of the heating produced by basic intrusions in the subvolcanic domain. It is supposed that the thermal solutions would be enriched in Fe, Si, Ca by decomposition of the basic tuffs, and would carry these elements to the surface.

Certainly, both opinions may be considered only as hypotheses. Although the close association of the ore with the volcanic products could be in favour of a magmatic differentiation, it is hard to explain the particular composition of the solutions (Fe, Si, Ca ± Mg). The selective extraction of elements from volcanic or sedimentary rocks has been more and more frequently referred to, not only for the formation of hydrothermal sedimentary iron ores (*Harder*, 1964a, b; *Chukhrov*, 1973; *Müller* and *Förster*, 1973; *Puchelt* et al., 1973), but also for other elements. However, at present satisfactory explanations for the extremely selective extraction in each peculiar case (e.g., Pb, Pb-Zn, Cu, Fe-Si-Ca, etc) are generally lacking. It does not seem sufficient to assume only an adequate composition of the thermal waters or of the mineralogical and chemical composition of the rocks from which the elements are supposed to come. For instance, if we refer only to the iron ores we have to take into account that iron ore depositions, similar from the mineralogical and chemical point of view, are found in regions of various lithological constitution such as basic volcanites (Lahn-Dill), andesite-dacitic volcanites (Santorin), metamorphic rocks (Rusaia-East Carpathians). Iron ore accumulations in sediments from incipient zones of expansion (e.g., The Red Sea, the Lake Malawi), thus in a very different geological setting, can be also added to them.

Similar or apparently similar ore-forming processes seem therefore to take place under quite different geological conditions; this fact fits the concept of metal extraction by waters infiltrated from the surface. The experimental data on such processes are few and insufficient, both as regards the principal elements and the trace elements of the ores.

For the iron ores associated with submarine volcanic rises, considering also the peculiar situations of the deposits of the Carpathians, the following observations could be mentioned:

1. In the region of Teliuc-Ghelar, because of the lateral position of the ore deposits in the volcanic rise, the thickness of the underlaying basic rocks is very small (sometimes it is smaller than the thickness of the ore body). Therefore, the source of Fe, Si, Ca, Mg must be sought elsewhere. As in the Lahn-Dill region, intrusive masses of subvolcanic rocks which would have caused the heating and mobilization of the interstitial water in the basic tuffs are also missing in the Poiana Ruscă.

2. Ores very similar to the Lahn-Dill type appear in the East Carpathians (Russia), overlying Precambrian gneisses (*Kräutner*, 1967). In this case, the extraction of the iron from basic tuffs is out of question. To the same effect one may also mention that ores similar to the Teliuc-Ghelar ores, from the mineralogical and chemical point of view, appear in the sideritic veins of Siegerland and the Gemeride. Although the latter ores are a completely different type of iron carbonate ore, they show that probably solutions

somewhat akin to those of the Teliuc-Ghelar ores may be generated in a nonvolcanic environment. However, the occurrence of sulphides and barite in the Teliuc and Ghelar deposits supports the previously mentioned facts.

3. The stratigraphic distribution of the ores points to a change in time of the hydrothermal supply towards solutions enriched in Mn (see Sect. 2.2). However, small bodies of Mn ores are also found in classical regions of the Lahn-Dill ores, in Kellerwald (*Hummel*, 1923) and Harz (*Haage*, 1964); they also occur in a higher stratigraphic position, but are separated from the iron ores through a larger time interval than in the Poiana Ruscă massif.

The above data indicate that the rocks of the submarine volcanic structures cannot be considered as the only possible source for iron extraction. All the data fit better the presumption that iron and its associated elements originate in deeper zones than the rise made of basic volcanic rocks.

4.2 Hydrotherms and Ore Deposition

After the concept of the transport of Fe and Si as chloride gas (*Harbort*, 1903; *Schneiderhöhn*, 1941) had been rejected and a supply as solution had been admitted, the main controversy pointed to the constitution of these "hydrotherms". In fact, the idea of a transport as hydrolized irion silicate (*Cissarz*, 1957; *Rösler*, 1962) was opposed to the older concept of an ionic transport in bicarbonate solution (*Ahlburg*, 1917; *Kegel*, 1923). Finally, *Harder* (1964) and *Quade* (1970) argued that the carbonate or bicarbonate solution could be considered as the most probable form for the transport of iron.

New arguments in favour of the transport of iron as bicarbonate or carbonate may be found in the distribution of carbonates and oxides in the iron ore deposits of Teliuc.

Around the volcanic rise from Teliuc-Ghelar-Ruschiţa, a weak alkaline and oxidating environment may be admitted. This fact is indicated by the limestone and calc schist levels as well as by the presence of iron as magnetite in the schists of low metamorphic stage, where the oxidating degree of iron has not been changed by the Hercynian regional metamorphism. In the terrigenous or volcanogenous metamorphic rocks there occur locally stratiform disseminations of magnetite (10–20% Mt in the rock) which probably represents deposits from small hydrothermal sources with a reduced outflow, so that the iron oxides have been deposited simultaneously with the detrital or tuffitic material.

The deposition of iron carbonate ores in this oxidating and alkaline environment points to local changes of the physicochemical conditions. As indicated by *Harper* (1964), such local changes could be determined by hydrothermal solutions that have a long period of activity and high outflow. Some indication about such a process seems to have been preserved in the ore bodies of Teliuc.

A schematic reconstitution of these deposits (Fig. 11) shows that the carbonate ore is surrounded to the south (therefore towards the culmination of the rise) by an oxide facies represented by magnetitic rocks with low contents of quartz and ankerite, to which iron silicates and barite are, in places, associated. In the transition zone from the oxide to the carbonate facies, one can notice an intimate interfingering between the two types of ores. In fact, the magnetite layer is gradually divided into several tin bands

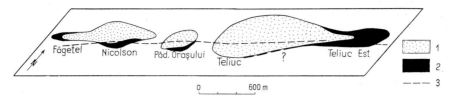

Fig. 11. Iron carbonate and iron oxide distribution in the ore bodies from Teliuc (schematic recon-stitution). *1:* Iron carbonate; *2:* Iron oxide; *3:* Actual erosion level

(some cm thick) which then disappear on variable distances in the iron carbonate ore (Fig. 12B).

Hematitic and magnetitic rocks quite similar to the Lahn-Dill ores occur also at the lower part of the main ore body (Fig. 12). This oxide ore, with thicknesses of about 10 m, occurs only on a small surface (about 100 x 150 m) and is situated below the zone of maximum thickness of the carbonatic ore body, where, as a matter of fact, the silica metagels also appear. Therefore, it can be presumed that the source of the solutions supplying the ore deposition was situated in this area. The transition from the oxide to the carbonate ore, overlying the former, takes place gradually (sometimes with recurrences) according to the stratigraphic column indicated in Figure 12A. The passing from hematitic to carbonate rocks by intermediary terms with magnetite points to a gradual decrease of the iron oxidation during the deposition.

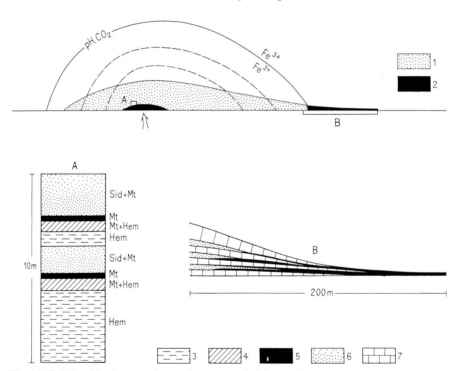

Fig. 12. Relationships between iron carbonate ore and iron oxide ore in the main ore deposit from Teliuc. *1:* Iron carbonate; *2:* Iron oxide; *3:* Hematite; *4:* Hematite + magnetite; *5:* Magnetite; *6:* Siderite with magnetite; *7:* Parankerite

This distribution of the iron ores is in disagreement with the models which admit variations in time of the iron oxidation in the hydrotherms (e.g., *Rösler*, 1962), as, in the first-mentioned case, the deposition of iron oxides and iron carbonates takes place concomitantly. Taking into account the oxidating and alkaline character of the environment, the situation can be better explained by a supply of acid carbonate or bicarbonate solutions which, on account of a high outflow over a long period, gradually changed pH and P_{CO_2} around the place of their penetration in the submarine environment.

As mentioned before, a high outflow rate has to be admitted for the hydrotherms as there are no detritic components in the constitution of the iron carbonate ore, although the formation of the deposits took place simultaneously with the sedimentation of the surrounding rocks. This fact points to an abundant deposition during a short time interval.

Parankerites represent one of the peculiarities of the iron ore deposits of the Teliuc-Ghelar type. They occupy significant volumes of iron carbonate bodies and constitute one of the main elements which contribute to the bedded structure of the deposits. The close connection of the parankerites with the iron ores excludes their formation outside the influence of the hydrotherms. Therefore, the problem arises whether Ca and Mg of these rocks also come from the hydrotherms or represent elements deposited from the seawater environment.

Considering that:

1. there is no direct connection between the bodies of iron carbonate rocks and the limestones or dolomites intercalated in the crystalline schists because these limestones and dolomites occur either at different stratigraphic levels or have been deposited simultaneously with the iron ores but in other areas of sedimentation (in the basin situated north of the volcanic rise);

2. the volume of parankerites in the deposits is very important whereas the detritic components are missing, although the formation of these rocks has taken place simultaneously with the deposition of the surrounding sedimentary rocks; it is more likely that the hydrotherms constitute the main source for Ca and Mg. Moreover, it is known that in hydrothermal iron carbonate veins (e.g., Siegerland, Gemeride), that is in formations where an external supply is out of the question, Mg and Ca terms occur frequently in the carbonate phase. Certainly, it is not out of the question that a part of Ca and Mg might come from the marine environment on account of their preferential solubility in the acid aureole of the hydrotherms. The calcareous basic tuffites as well as the beds of limestones and dolomites around the ore deposits indicate, as a matter of fact, the existence of a sedimentation of carbonatic rocks during the ore formation.

As concerns the stratified character of the ore deposits, it is hard to admit that the hydrotherms changed their composition continuously and thus caused an alternative deposition of siderite, parankerite, and ankerite. Therefore it is more reasonable to suppose that in the acid aureole, formed around the emission centre, Ca and Mg remained in the solution for a longer time, while Fe precipitated preferentially, When the rate of the outflow was diminished or the supply was temporarily stopped, pH became higher and, thus, Ca and Mg were precipitated as carbonate.

Therefore, the alternation of periods with abundant supply and periods of stagnation or flatness of the hydrotherms could explain the bedding of the iron carbonate deposits. The intercalations of schists in the ore body confirm the intermittent or pulsatory character of the supply. According to this model, the siderite ore and the parankerite rocks would result from a differentiation of Fe, Mn, Mg, Ca, carried concomitantly in the hydrotherms, due to the environmental control of the ore deposition.

Silica gels as well as iron-silica gels have been deposited simultaneously with the formation of the iron carbonate ore. Such products are characteristic of the Lahn-Dill ores but they have also been described in some recent formations (*Harder*, 1964) in connection with iron depositions in other geological environments. In the ore deposits of the Poiana Ruscǎ massif the silica gels seem to have been deposited towards the end of the pulsation phases of the hydrotherms. In these cases they are usually represented by fine-bedded varieties constituting an alternance of iron quartzites and pure quartzite bands. This alternance is hard to explain by a continuous change of the composition of the hydrotherms. It is easier to accept that from the same supply iron was preferentially deposited at first in an acid environment, whereas silica came to precipitation later when the environment was nearly alkaline. Therefore, the pulsatory character of the hydrotherms may be considered as the main cause of the mentioned bedding.

4.3 Resedimentation of the Ores

Unlike the Lahn-Dill ores, relict structures indicating processes of resedimentation in the ore bodies of Ghelar and Teliuc are not frequent. This fact cannot be explained only by the metamorphism, as such structures have been preserved in the carbonatic rocks which are not affected by the metamorphic schistuosity. Breccia structures occur only in very restricted areas near the primary ore. In fact, the greater depth of the ore deposition and even the primary carbonate mud could not favour an epiclastic redeposition. In exchange, frequent interruptions of the bedding, irregular contours and other aspects point to submarine disturbandes in the initial mud accumulation.

5. Conclusions

5.1 A Model

Bicarbonate thermal solutions enriched in Fe, Ca ($\pm$ Mg, Mn) and Si rise from the depth. They reach the sea floor in the area of the submarine volcanic rises formed of basic volcanites. Due to the new conditions of T, P, Eh and pH, they deposit their charge in elements, and accumulations of iron ores are formed. According to the position of the thermal sources in the volcanic structures and to the depth where the deposition takes place, two main types of iron ore deposits may be distinguished.

Iron oxide deposits of the Lahn-Dill type formed when the hydrotherms flow out at a shallow depth in a well-aerated, oxidating environment. Under these conditions the solution loses CO_2 and Fe is oxidated. This usually takes place in the culminant zones of the volcanic rises. The relief of these rises may cause various forms of ore deposits: (1) primary deposits of hematite, quartz-hematite or silicate ore in the high zones, (2) primary deposits of carbonate and rarely sulphide ore in depressionary zones with reducing environment, (3) secondary deposits of mainly hematite-calcite ore formed by the resedimentation of the primary ore on the slopes of the volcanic rises.

Iron carbonate deposits of the Teliuc-Ghelar type formed when the hydrotherms flowed at a greater depth, in a less aerated and oxidating environment. This takes place towards the lower part of the slopes of volcanic structures, where some negative relief forms may favour the accumulation of important quantities of ores.

If, at the beginning, the outflow is low, oxidic deposits may probably have formed in a first stage (e.g., Teliuc). Due to the persistance of the hydrotherms or to the increase of their flow, the acidity and the partial CO_2 pressure of the environment around the source increases continuously. Thus, in zones with hydrothermal emission on the sea floor, acid aureoles with high P_{CO_2} are formed (Figs. 12 and 13). According to the total dissolved carbonate and the changes of P_{CO_2} and pH in these aureoles, the stability field of siderite expands, and, therefore, iron carbonate ore may be deposited. Thus the local increase of P_{CO_2} and pH can cause the change of the iron deposition from hematite and magnetite to siderite, without any modification in the oxidation conditions of the environment (*King*, 1958; *Garrels*, 1960). In this way we can explain also the gradual increase of the Fe^{2+}/Fe^{3+} ratio in the lower part of the ore body from Teliuc (see Sect. 4.2; Fig. 12A).

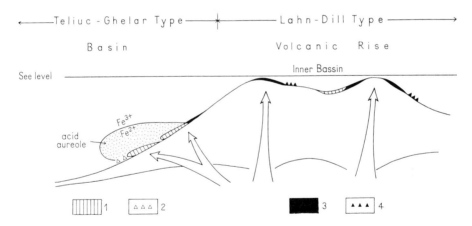

Fig. 13. Hydrothermal sedimentary iron ores on submarine volcanic rises (a model). *1:* primary iron carbonate ore; *2:* Resedimented iron carbonate ore; *3:* Primary iron oxide ore; *4:* Resedimented iron oxide ore

If the hydrothermal flow is high from the beginning, it may be assumed that the deposition of the siderite ore occurs directly without an intermediary oxidic stage (e.g., Ghelar).

If a part of the iron is carried outside the P_{CO_2} and pH limits which control the $FeCO_3$ stability, the iron oxide is deposited instead of the iron carbonate (Fig. 12). The relationships between the synchronous oxide and carbonate ores from Teliuc (see Sect. 4.2; Fig. 12B) indicate that the limit between the two facies has continuously modified its position. The extension and the shape of the aureoles around the hydrothermal sources are therefore continuously subjected to variations. These changes are probably due to the modification in time of the hydrothermal flow or of the composition (concentration) of the ascending solutions as well as to some external factors, such as marine streams or other disturbances.

If the hydrotherms temporarily take on a sulphide character, the increase of P_s in the aureole expands the stability field of FeS_2 and iron precipitates mainly as sulphide (e.g., Ghelar, Ruschiţa).

5.2 Common and Distinctive Features Between the Teliuc-Ghelar Ores and the Lahn-Dill Ores

According to the genetic model presented before, the following *common features* can be distinguished:

1. Syngenetic strata-bound iron ore deposits with stratigraphic and lithostratigraphic control.

2. Close connection with submarine volcanic rises mainly constituted of epiclastic rocks of basaltic and diabase-spilitic composition. Tuffs of quartz keratophyres occur subordinately.

3. Stratiform character of the ore.

4. Facial variations on short distances within the ore bodies, e.g., massive and bedded primary ore;sharp or gradual transitions among oxide, silicate, carbonate and sulphide ores.

5. Lateral transitions from primary ore to resedimented ores.

6. Close connection between iron ore and silica gels; mixtures and gradual passings between these two extreme components.

7. Main constituent elements: Fe, Si, Ca ($\pm$ Mg).

8. Local presence of organic substance (antraxolite, graphite).

9. Ore bodies of 0.5–10 million t iron ore.

Distinctive features are related to:

1. Mineralogical composition: TG^2 – mainly carbonatic, LD^3 – mainly oxidic.

2. Petrographical constitution: TG – mainly siderite, ankerite, parankerite ore more or less quartzitic; subordinately associated with magnetite rocks, rarely silicates and iron sulphides. LD – mainly hematite ore, quartz-hematite ore, calcareous hematite ore, subordinately associated with silicate rocks, more rarely with carbonates or sulphides.

3. Geochemical characteristics: TG – higher Mn:Fe ratio, relatively high Mg content. LD – lower Mn:Fe ratio. low Mg content.

4. Form of the ore bodies: TG – mainly thick (to 150–200 m) lenses. LD – mainly flat, stratiform lenses.

5. Paleostructural position: TG – in the marginal zone of the rise, on the lower part of the slope. LD – in the central zone of the rise, on culmination as well as in the local internal basins.

6. Depth and environment of deposition: TG – greater depth, calm (unagitated) and low oxidating environment with acid aureoles around the emission centres of the hydrotherms. LD – shallow depth, strongly oxidating, well-aerated and agitated environment, locally reducing conditions in closed basins.

---

[2] Teliuc-Ghelar type.
[3] Lahn-Dill type.

7. Resedimentation of the ore: TG — epiclastic ores are rare, syndepositional disturbances in the mud occur more frequently. LD — epiclastic ores occur frequently.

5.3 Relationships to Major Structural Elements of the Crust

Most of the available data about the Lahn-Dill deposits refer to the iron ores associated
with the Devonian basic volcanism in Europe, generally defined according to Stille as
"initial volcanism". This magmatism is well developed in the central part of Europe and
extends also to the south-western part of the British Isles and to the pre-Alpine basement of the Carpathian chain (*Kräutner*, 1970). The distribution of this magmatism, as
well as of the associated iron ores, make us assign them to four principal alignments
(Rhenohercynian, Saxothüringian, Silesian and Carpathian) which crossed Hercynian
Europe in approximately parallel positions. Figure 14 indicates that in the external
alignments the volcanic activity started earlier (Middle Devonian) than in the internal
zone (Upper Devonian).

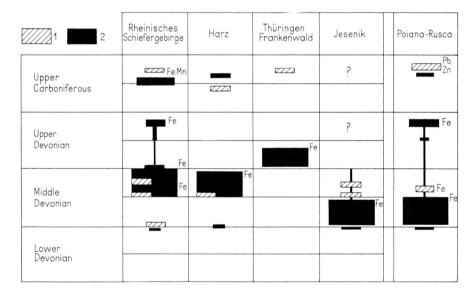

Fig. 14. Stratigraphic distribution of the Paleozoic basic volcanites and the associated hydrothermal
sedimentary iron ores (Fe) in Central Europe and the Carpathians. *1:* Quartz keratophyres; *2:* Basalts,
diabases, spilites

In the regions mentioned there are no proofs of any fossil oceanic crusts; on the contrary, the basaltic volcanites overlie older Paleozoic detrital deposits which transgres-
sively overlie a Precambrian basement. Therefore, the basic volcanism was active in the
mobile zones situated in cratonic areas and was probably supplied by magmas raised on
fracture zones from great depths. It seems that the iron ores of the Lahn-Dill and Teliuc-
Ghelar types (in the restricted acceptance used in this paper) are connected with the
evolution of the intracratonic mobile zones.

References

Ahlburg, J.: Über die Eisenerze und Eisenmanganerze des Lahngebietes und ihrer Beziehungen zu Eruptivgesteinen. Z. prakt. Geol. 25, 29–38, 49–56 (1917)

Borchert, H.: Der initiale Magmatismus und die zugehörigen Lagerstätten. N. Jb. Mineral., Abh. 91/1–3, 541–572 (1957)

Borchert, H.: Geosynklinale Lagerstätten, was dazu gehört und was nicht dazu gehört, sowie deren Beziehung zur Geotektonik und Magmatismus. Freiburger Forsch.-H. C 79, 7–61 (1960)

Borchert, H.: Zur Bildung marin-sedimentärer Eisen- und Manganerze in Verknüpfung mit spilitischen und keratophyrisch-weilburgitischen Gesteinsassoziationen. Mineralium Deposita 7/1, 18–24 (1972)

Bottke, H.: Die exhalativ-sedimentären devonischen Roteisensteinlagerstätten des Ostsauerlandes. Beih. geol. Jb. 63, 1–147 (1965)

Chukhrov, F.V.: On the genesis problem of thermal sedimentary iron ore deposits. Mineralium Deposita 8/2, 138–147 (1973)

Cissarz, A.: Mineralogisch-mikroskopische Untersuchungen der Erze und Nebengesteine des Roteisensteinlagers der Grube Maria bei Braunfeld a.d. Lahn. Mitt. Kaiser-Wilhelm-Inst. Eisenforsch. 5, 109–125 (1924)

Cissarz, A.: Lagerstätten des Geosynklinalvulkanismus in den Dinariden und ihre Bedeutung für die geosynklinale Lagerstättenbildung. N. Jb. Miner. Abh. 91, 485–540 (1957)

Dave, A.S.: Paragenetischer und geochemischer Aufbau der Eisenerzlagerstätte Braunesumpf bei Hüttenrode im Harz. Freiburger Forsch.-H. C 146, 1–110 (1963)

Garrels, R.M.: Mineral equilibra at low temperature and pressure. Harper's Geoscience Series. New York: Harper & Brothers 1960

Gräbe, R.: Beziehungen zwischen der tektonischen und Faziellen Entwicklung des Oberdevons und Unterkarbons sowie zur Genese der Eisenerze vom Lahn-Dill Typus am NW-Rand des Bergaer Sattels (Thüringisches Schiefergebirge). Freiburger Forsch.-H. C 140, 1–83 (1962)

Gräbe, R.: Über sulfidische Äquivalente der oberdevonischen Roteisenerze am NW-Rand des Bergaer Sattels (Ostthüringen). Ber. Geol. Ges. DDR 9/4–5, 527–537 (1964)

Haage, R.: Beitrag zur Genese des Kieselschiefer-Mangankieselvorkommens im Schärenholz bei Elbingerode (Harz). Ber. Geol. Ges. DDR 9/4–5, 567–580 (1964)

Harbort, E.: Zur Frage der Entstehung gewisser devonischer Roteisenerzlagerstätten. N. Jb. Miner. etc. I, 179–192 (1903)

Harder, H.: Können Eisensäuerlinge die Genese der Lahn-Dill-Erze erklären? Beitr. Miner. Petr. 9/5, 379–422 (1964a)

Harder, H.: Untersuchungen rezenter vulkanischer Eisenausscheidung zur Erklärung der Erze vom Lahn-Dill Typus. Ber. Geol. Ges. DDR 9/4–5, 470–474 (1964b)

Hentschel, H.: Zur Frage der Bildung der Eisenerze vom Lahn-Dill-Typ. Freiburger Forsch.-H. C 79, 82–105 (1960)

Hummel, K.: Über Manganerze im Kulm des Kellerwaldes. Z. prakt. Geol. 31, 89–93 (1923)

Kegel, W.: Zur Kenntnis der devonischen Eisenerzlager in der südlichen Lahnmulde. Z. prakt. Geol. 31/1–6, 20–29, 36–41 (1923)

King, N.H.: The environmental control of sedimentary iron minerals. Econ. Geol. 53/2, 123–140 (1958)

Knauer, E.: Quantitativ-mineralogisch-petrographische Untersuchungen an den mitteldevonischen Roteisenerzen vom Büchenberg bei Elbingerode im Harz. Beih. Geol. Jg. 9, 29 (1960)

Kräutner, H.G.: Die Zink-Blei-Lagerstätte von Muncelul Mic (Poiana Rusca). Assoc. Geol. Carp. Balk., V Congr. 2, 97–114 (1963)

Kräutner, H.G.: Gisements de fer dans la partie centrale est du massif Poiana Rusca. D.S. Com. Geol. 49/1, 345–360 (1964)

Kräutner, H.G.: Les gisements de fer de Rusaia et de Iacobeni. D.S. Com. Geol. 52/2, 5–29 (1967)

Kräutner, H.G.: Die hercynische Geosynklinalerzbildung und ihre Beziehungen zur der hercynischen Metallogenese Mitteleuropas. Mineralium Deposita 5/4, 323–344 (1970)

Kräutner, H.G., Kräutner, F.: Chemismus der Eisenerzlagerstätten des östlichen Zentralteils der Poiana Rusca. Rev. Géol. Géogr. Adac. RPR 7/1, 121–146 (1963)

Kräutner, H.G., Mureşan, M., Iliescu, V., Mînzatu, S., Vîjdea, E., Tănăsescu, A., Ioncică, M., Andăr, A., Anastase, S.: Le Dévonien-Carbonifère inférieur épimétamorphique de Poiana Rusca. D.S. Inst. Geol. 59/4, 5–63, Bucureşti (1973)

Kräutner, H.G., Sassi, F.P., Zirpoli, G., Zulian, T.: Barrovian-type Hercynian metamorphism from the Poiana Ruscă Massif (South Carpathians). N. Jb. Miner. Mh. 10, 446–455 (1976)

Lehmann, E.: On the source of the iron in the Lahn ore deposits. Mineralium Deposita 7/3, 247– 270 (1972)

Lippert, H.J.: Zur Gesteins- und Lagerstättenbildung in Roteisensteingruben des östlichen Dill-Gebietes. Abh. Senckenberg. Naturforsch. Ges. 485, 1–30 (1951)

Lippert, H.J.: Das Roteisenstein-Grenzlager von der Wende Mittel-Oberdevon in der Dill-Mulde. Z. deut. geol. Ges. 104/2, 260–276 (1953)

Lutzens, H., Burchardt, I.: Metallogenetische Untersuchungen an mitteldevonischen oxidischen Eisenerzen des Elbingeröder Komplexes (Harz). Z. angew. Geol. 18/11, 481–491 (1972)

Müller, G., Förster, U.: Recent iron ore formation in Lake Malawi, Africa. Mineralium Deposita 8/3, 278–290 (1973)

Puchelt, H., Schock, H.H., Schroll, E., Hanert, H.: Rezente marine Eisenerze auf Santorin, Griechenland. Geol. Rdsch. 62/3, 786–812 (1973)

Quade, H.: Zur paläogeographischen Entwicklung des Mittel- und Oberdevons im Bereich der Lagerstätte Eisenfeld. Notizbl. hess. L.-Amt Bodenforsch. 93, 207–228 (1965)

Quade, H.: Der Bildungsraum und die genetische Problematik der vulkanosedimentären Eisenerze. Clausthaler Hefte 9, 27–65 (1970)

Rădulescu, D.: Einige Bemerkungen über den Begriff „metallogenische Provinz". Ber. deut. Ges. geol. Wiss. B (Miner. Lagerst.) 11/4, 461–466 (1966)

Rösler, H.J.: Zur Petrographie, Geochemie und Genese der Magmatite und Lagerstätten des Oberdevons und Unterkarbons in Ostthüringen. Freiburger Forsch.-H. C 92, 1–275 (1960)

Rösler, H.J.: Zur Entstehung der oberdevonischen Eisenerze vom Typ Lahn-Dill in Ostthüringen. Freiburger Forsch.-H. C 138, 1–79 (1962)

Rösler, H.J.: Genetische Probleme der Erze des sogenannten erweiterten Lahn-Dill-Typus. Ber. Geol. Ges. DDR 9, 445–454 (1964)

Schneiderhöhn, H.: Lehrbuch der Erzlagerstättenkunde I. Jena: Fischer 1941

Schneiderhöhn, H.: Erzlagerstätten. Kurzvorlesungen. Stuttgart: Fischer 1955

Skacel, J.: Die Eisenerzlagerstätten des mährisch-schlesischen Devons. Ber. Geol. Ges. DDR 9/4–5, 487–506 (1964)

Steinike, K.: Quantitativ-mineralogische Untersuchungen an den Eisenerzen vom Typus Lahn-Dill aus Pörmitz bei Schleiz (Ostthüringen). Freiburger Forsch.-H. C 142, 1–123 (1963)

The Strata-Bound Magnesite Deposit of Eugui-Asturreta in the Spanish Pyrenees

W.E. PETRASCHECK, M. KRALIK, and A. RANZENBACHER, Leoben

With 3 Figures

Summary

A detailed lithological survey of the Asturreta magnesite deposit, carried out for practical reasons, led to the result that a correlation of the individual strata of magnesite and dolomite was not possible even for short distances. This phenomenon, unusual in marine-sedimentary deposits, can perhaps be explained by postdiagenetic and metamorphic mobilization processes. A study of the layered structure and microchemical analyses indicates a sedimentary genesis in a subsalinary environment. This is confirmed by the observation of some gypsum pseudomorphoses. Most convincing is the geological evidence, gained by reconstructing the original basin of sedimentation: the geometrical flattening of the folded carbonate rock series discloses a clear coincidence of greatest thickness and magnesite enrichment. The first sedimentary theory of *de Llarena* (1950) thus gains strong support.

The world-wide discussion between advocates of an epigenetic-hydrothermal vs. a syngenetic-sedimentary genesis of the crystalline magnesite occurrences did not leave out the strata-bound and stratiform deposit of Asturreta near Eugui in Spain. The scientific discoverer of this important deposit, Professor *Gómez de Llarena* was convinced from the beginning of its sedimentary origin. He derived his opinion from the spectacular stratification of the magnesite and the interbedded dolomite, visible on the wall of the quarry as well on the hand specimens, and from the strict connection of all the magnesite occurrences in the district to a well-defined carbonate horizon of carboniferous (Namurien) age within the Paleozoic sequence. This carbonate rock horizon is folded to simple anticlines and synclines and extends over 15 km in North-South direction and 8 km in East-West direction (see the excellent map by *Pilger*, 1959, Fig. 1).

The opposite view was defended by *Destombes* (1956) who laid stress on replacement features of magnesite in dolomite and particularly on the fact that magnesite deposits were also found in other stratigraphic horizons in the neighbourhood: in the Lower Devonian at Urepel near the French-Spanish border, and in the Middle Triassic near Puerto de Velate. Quite analogous arguments have been used by the respective partisans of both theories in the frequent discussions on the magnesite in the Austrian Alps, the Slovak Carpathes and the French Pyrenees near Canigou.

Over the last ten years metasomatic features have lost their distinctive genetic importance as evidence for "hydrothermal-epigenetic" formation of an ore. It was particularly Professor *A. Maucher* who, together with his collaborators, emphasized the existence of diagenetic replacement which has to be included into the synsedimentary ore-forming processes. Peculiarities of the sedimentary environment of a deposit became more

Fig. 1. Geological sketch of
the Quinto-Real-Massif with
the dolomite-magnesite belt.
(Simplified after *Mohr* and
Pilger, 1964)

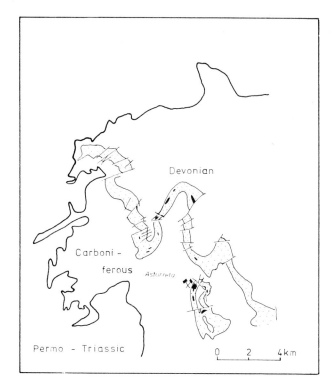

important as genetic criteria — although in the case of magnesite the different authors
referred to rather different facies conditions, sometimes without considering the inferred
magnesite facies within a corresponding palaeogeographic frame; also very different
sources of magnesium have been assumed.

On the other hand, the occurrence of magnesite in various stratigraphic horizons
within the same region, such as in Spain or in the Eastern Alps, of a kind of "repeated
strata-bound character", makes a synsedimentary conception more difficult. Metallogenetic heredity in this case is more a word than an explanation.

In 1972, one of the present authors (W.E.P.), was invited by the Mining Company
Magnesitas Navarras to make a detailed study of the mine Asturreta and the other magnesite indications, for mainly practical reasons; namely the exploration and reserve
evaluation in the quarries and the prospecting of the whole district. This work was
carried out together with *Kralik* (1977) and *Ranzenbacher* (1975); both of them spent
several months in the last years in Eugui for this purpose. The present author made frequent visits to the area. We all enjoyed most valuable help and support by the company,
particularly by the directors Mr. *Belzunce* and Mr. *Weiss* and the mine surveyer Mr. *Maza.*

The continuous exchange of views with these gentlemen was an essential stimulation
for gaining our results. In addition, one of the present authors (M.K.) is very much ob-
liged for the numerous suggestions he received in the Mineralogisch-Petrographisches
Institut in Heidelberg.

The first practical — and also genetical — problem which we considered as basic for
any planning of exploitation was the correlation of the magnesite strata in the strike

direction with regard to their lithological appearance and their technological quality. We tried to distinguish seven different types of magnesite — different in texture, colour, and dolomite content — and plotted them on a plane of the mine (scale 1:1000). A careful sampling along lines perpendicular to the strike, in the different levels, gave a surprising and unpleasant result: no quality correlation exists from line to line. None of the magnesite types, not even the dolomite layers, could be followed from one level to the next. There is a difference of the CaO and SiO_2 content from North to South, but, as the southern levels have a greater altitude, this difference could also be a function of depth. Proposed drilling and core sampling proved that the latter interpretation was correct, which means that the quality improves in the direction of the inclination. However, this again was an ambiguous result, because the beds dip in a westerly direction. Thus, the primary change could have been lateral in the East-West direction. A comparison of the magnesite in the Asturreta quarry, which is situated on the western flank of the anticline, with that of Rafael exposures in the eastern flank, has shown after some drilling and trenching that the minable ore layers are thinner in Rafael; the magnesite thins out towards the East (Fig. 2).

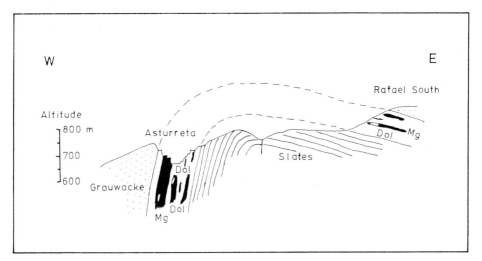

Fig. 2. The anticline of Asturreta

These observations and conclusions led to a systematic drilling programme. The cores were recorded according to the new lithological and technical classification and the programme was adapted to the current results. In this way, the ore reserves could be augmented remarkably (A.R.).

The absence of continuity of the individual layers is not usual in sedimentary marin deposits. For example, the Devonian limestone in Dornap or the Devonian dolomite near Bingen, both in Rheinland-Westfalen, are characterized by a good correlation of the strata within the quarries (*Gotthardt,* 1974; *Bottke,* 1974). Especially the SiO_2 content is a reliable parameter for correlation and key beds can be followed through the quarries.

This is not the case, even with gray, platy dolomite beds in Asturreta. Undisturbed gypsum deposits also show a regular sequence of strata. We attempt to interpret this irregularity by posterior diagenetic or even epigenetic activity of migrating solutions. We agree with the general statement of *A. Maucher* that such solutions must not be hydrothermal-magmatic, but that they can be simply reheated connate or surface water, and that mineral-forming processes in this context should be called "hydratrogenic".

The well-known ribbon texture of our magnesite, called "Asturreta type", is derived from the stratification of the primary sediment, combined with the effect of recrystallization under slight metamorphic conditions. The magnesite crystals have an elongated shape, their c-axes being arranged along the bedding planes. They grow in bipolaric direction perpendicular to the bedding from a thin black layer of clayish material. The white crystals growing from both sides meet along a suture line, where fine organic matter, clay, and pyrite were concentrated.

It is probable that these sediments belong to the group of seasonal-rhythmites, analogous to the recent deposits in algal marshes and in sabkha-environments. The rhythmic layering was later underlined by the recrystallization.

Quémeneur (1974) already suspected original gypsum in the sediment, according to some observations he made. It was possible, indeed, to identify some clear pseudomorphoses of gypsum in chert, partly in immediate contact with magnesite (M.K.). This is good evidence for a subsaline environment, supposed for some alpine magnesites by *W. Siegl* for many years.

A detailed investigation of trace elements has shown that these are very scarce in the magnesite, a fact which speaks against a provenance from a submarine basic magmatism. In spite of careful attention, we could not find any volcanic (diabasic) tuffs in the carboniferous sequence.

C-O-isotope determinations, carried out by Professor *J. Hoefs* in the Geochemical Institute of the University of Göttingen showed a comparatively high content of heavy carbon, indicating again an evaporitic formation. These investigations, for which our sincere thanks are expressed to Professor *Hoefs*, will be discussed in detail in a special paper by *Kralik*.

A decisive criterion for sedimentary deposits can be found by reconstructing the original area of sedimentation. For this purpose, the whole carbonate rock series, folded during the Hercynian orogenesis (Fig. 1) was geometrically flattened. Based on a detailed lithological survey of about 20 sections of the series, their original thickness as well as the distribution of high-quality magnesite (less than 6% CaO) could be established. The isopach map demonstrates a clear correlation between a zone of greatest thickness of the carbonate series (240–260 m) and the presence of magnesite enrichment (Fig. 3). Magnesite mixed with dolomite occurs in the areas with 140–240 m thickness, whereas the marginal parts of the basin consist of dolomite only. This correlation manifests *Quémeneur*'s idea of a magnesite-bearing trough.

Thus, the model of a "restricted basin" fits very well with the palaeogeographic situation in the Eugui magnesite district. The mechanism of the Mg concentration may have been similar to that described by *Guillou* (1970) for the Cambrian magnesite of Pacios in Northern Spain (Province Lugo).

However, we must also take into account all those phenomena which are characteristic of an epigenetic mineralization: the spatial limitation of rich magnesite to a few

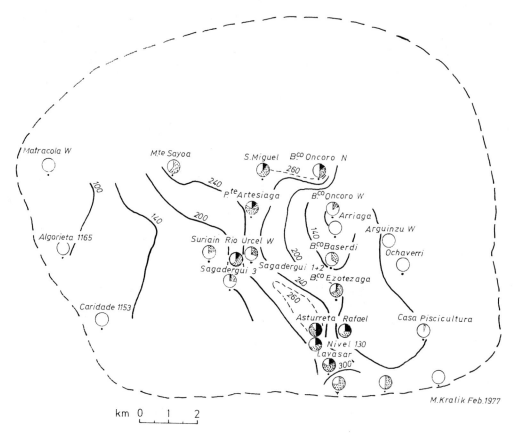

Fig. 3. Roughly reconstructed isopache-map of the carbonate formation with limits extrapolated from the visible extension (approx. 180 km²). *Black section:* magnesite (<11% CaO); *dotted:* mixed magnesite + dolomite; *white:* dolomite

places only; the regular occurrence of E-W-directed faults, filled with dolomite and some sulphides, even partly extending beyond the carbonate rock series. the big dolomite crystals disseminated in the magnesite and resembling the products of the so-called "redolomitization" in alpine magnesite deposits, declared as metasomatic. A younger metamorphism can perhaps be responsible for this.

In conclusion, after a critical reconsideration of all these pro and contra arguments, we give credit to *Gómez de Llarena* who first and persistently defended sedimentary interpretation of the carboniferous magnesite of Asturreta. We do not yet know how to fit the Devonian and the Triassic magnesite deposits of the neighbourhood into this picture [1].

---

[1]   A detailed publication of the magnesite of Eugui by *M. Kralik* is getting prepared.

References

Bottke, H.: Die devonischen Dolomite der Waldalgesheimer Mulde bei Bingen/Rhein als Lagerstätte der Feuerfestindustrie. Schr. Ges. deut. Metallhütten- u. Bergleute 25, 59–88 (1974)

Destombes, J.P.: Magnésites (Giobertites) des Pyrénées Occidentales. Bull. Soc. Géol. France (6) VI, 461–476 (1956)

Gómez de Llarena, J.: La magnesita de Eugui (Navarra). Bol. R. Soc. Hist. Nat. 48, 67–70 (1950)

Gotthardt, R.: Massenkalkvorkommen des Rheinischen Schiefergebirges, ihr geologischer Bau, ihr Abbau und ihre Verwendung. Schr. Ges. deut. Metallhütten- u. Bergleute 25 (1974)

Guillou, J.J.: Les magnésites cambriennes de Pacios (Province de Lugo-Espagne). Leur environement paléogéographique. Bull. Bur. Rech. Géol. Min. (IV) 3, 5–20 (1970)

Kralik, M.: Verbreitung und Genese der Magnesitlagerstätte Asturreta/Eugui (span. Westpyrenäen). Unpublished Diss. Univ. Wien (in preparation, 1977)

Pilger, A.: Zur Genese der Magnesite in den Westpyrenäen. Z. deut. geol. Ges. 111, 198–208 (1959)

Pilger, A.: Die Entwicklung von Oberdevon, Unterkarbon und Namur im Gebiet des von Eugui in den spanischen Pyrenäen. N. Jb. Geol. Paläont. Abh. 142, 44–58 (1973)

Quémeneur, J.M.: Les gisement de magnésite du Pays Basque: Cadre géologique et sédimentologique; genèse de la magnésite en milieu sédimentaire. Unpublished Diss. Univ. Paris VI

Ranzenbacher, A.: Erkenntnis der Verteilung der Magnesitqualitäten in der Lagerstätte von Eugui (Spanische Pyrenäen) und Ableitung eines Bohrprogrammes und einer Reserveberechnung. Unpublished Diss. Montanist. Hochsch. Leoben (1975)

Sedimentary Magnesite Fabrics Within the Sparry Magnesite Deposit Hochfilzen (Tyrol)

O. SCHULZ and F. VAVTAR, Innsbruck

With 6 Figures

Contents

Summary

In the fine-spathic magnesite deposit of Bürglkopf near Hochfilzen (northern zone of greywackes, Tyrol) a matter-concordant, locally limited, apparently lense-shaped intercalated layer of magnesite, hematite and sericite has been discovered. The sediment, which is at most 40 cm thick and has a striking red-white lamination caused by varying hematite and sericite portions, exhibits characteristic sedimentary appositional fabrics, e.g., space-rhythmic lamination, cross bedding, syndiagenetic superficial folds and slides. The polar surface reliefs which had originated in several mutually independent partial layers, were each covered by undisturbed laminated sedimentation. The disovery of colloform relic fabrics in magnesite laminae is interpreted to be probably a primary internal apposition of a Mg-hydrocarbonate in a solutional void. After the syndiagenetic deformation, accretional crystallization leading to fine-grained magnesite is thought to have occurred. The findings suggest a generally synsedimentary to diagenetic origin of the magnesite deposit.

1. Introduction

Paleozoic sedimentary genesis for alpine sparry magnesites has been assumed by *Leitmeier* (1917, 1951, 1953), *Rosza* (1925), *Rohn* (1950), *Llarena* (1953), *Leitmeier* and *Siegl* (1954), *Siegl* (1955, 1969), *Höll* and *Maucher* (1967), recently also by *Siegl* and *Felser* (1973). Quite different explanations have also been offered. A pioneering study was accomplished by *Höll* and *Maucher* (1967) using the example of the scheelite-magnesite deposit Tux, which made successful scheelite prospecting in Austria possible, following the discovery of the scheelite deposit Felbertal (*Höll*, 1975).

The proof of sedimentary fabrics has not been possible up to now because of the appearance of the magnesite as giant spar ("sparry magnesite"). Even in the relatively fine crystalline magnesite bed at the Bürglkopf-Weißenstein near Hochfilzen the search for primary sedimentary fabrics has not been successful − except for one case (*Vavtar*, 1974, 1976), which, as a rarity, deserves special attention and will be described and discussed here.

2. Findings

The magnesite aggregates which prevail in alpine sparry magnesite deposits are, in their present form, typical growth fabrics with corresponding lattice orientations (*Clar*, 1931, 1954). Tectonite orientation has, however, also been shown to exist in fine-crystalline aggregates of the deposits Entachen-Alm (Salzburg) by *Ladurner* (1965) and Hochfilzen by *Vavtar* (1974, 1976). These magnesite deposits are associated with Upper-Silurian to Lower-Devonian dolomite rocks of the northern zone of greywackes.

Along the southwest side of Bürglkopf an matter-concordant lens-shaped intercalation of a red-white laminated rock (locally $s = 105°\ 80°$ S) is embedded in the magnesite bed of the district Rettenwand, which is about 20 m thick and orientated E-W 80° S. The original planar extent of this sediment can no longer be surveyed because of manifold tectonic shearing (overthrust 45° 33° SE, transverse thrust 180° 80° E, R ± O), and due to mining of the deposit. The preserved remnants with a thickness of at most 40 cm can be traced vertically (along the dip) for only 2.60 m. After some further working of 1 m, this layer continues into the mountain along the strike, the exposure, however, looks more like a lens-shaped intercalation. The thinning out in the base surface and the apparent nonoccurrence in other levels of the open-cast mine confirm this impression. The striking red-white lamination is caused by interbedding of magnesite and sericite with varying hematite content.

Everywhere a sharp boundary exists towards the decimicron- to centimicron-sized granular, massive sparry magnesites typical of the Bürglkopf deposit. The sharp boundary contours are due mainly to the generous admixture of sericite, which in the under- and overlying strata facilitated sliding even locally for the tectonic formation of the intercalation. An example is the 5-cm thick red hematite-sericite bed in the underlying stratum (Fig. 1). The boundary to the over- and underlying beds seems to be wavy, and to have partly primary causes, but also appears to have originated partly during the tectonic formation. While mining about 1 m, it was striking to see that the series of strata shows great variation of detail, in the thickness of the individual laminated strata as well as in the fabric. In part the lamination is preserved undistorted, but severe distortion do exist. From the distribution of carbonate and clay in the lamination, the impression arises that the carbonate sedimentation happened continuously, whereas the sedimentation of the clay and iron component fluctuated and was even interrupted.

On the sericite-trimmed bedding planes (the stratification ss is identical to the schistosity s), plication axes may be found which are inclined 30° towards the SW, and which correspond to axes prevailing in the overall area of the magnesite mine. In addition to this, downright intensive continual deformation of the material by plication and sliding can be seen in part of the lamination (Fig. 2). This lineation, too, is evident as an axial formational element on the corresponding bedding planes. Both axes intersect at an angle of 15°. Ruptural deformations occur only as a minor feature.

Fig. 1. The magnesite-hematite-sericite intercalation inclined steeply along with the magnesite deposit. Rettenwand mine, Bürglkopf

Fig. 2. Magnesite-hematite-sericite laminated layers, exhibiting in parts polarity of the zones of formation, and folded, sheared and overthrust laminae. Large sections. Magnification is different for the upper and lower pictures. Length of marked-off scale = 75 mm

3. Microscopic Pattern

The difference in colour of the laminae on the mm scale is caused by changing mineral supply. Perfectly white layers consist of magnesite. Dolomite and calcite can occasionally be found in s-discordant joints and microfissures, which otherwise are healed principally with pigmentless magnesite. As a minor feature redolomitization may also be found. The intensively red layers consist of a felt of sericite pigmented with hematite. These microscopic findings in reflected and transmitted light were confirmed by X-ray diffraction analyses. Rutile also has its minor share as a pigment: in isometric grains (0.008–0.02 mm) and pillars (0.05 x 0.005 mm). The proportion of iron in the magnesite fine layers causes varying bright-red colours.

The iron minerals, though, are restricted mostly to the sericite layers. The change of material is strongly layered, but abruptly rather than continuously.

The magnesite fabric is fine-grained, individual crystals being xenomorphic. Within the individual layers, grains are mostly of similar size, the sericite-containing layers showing finer-grained magnesite spatite (often 0.07–0.3 mm) than the pure magnesite layers (most frequent diameters between 0.3 and 0.9 mm) — incidentally a striking and well-known phenomenon in carbonate-clay sediments!

The sericite spangles (0.03 x 0.003 mm) and tabular hematite crystals are orientated into the bedding planes. Here a discontinuity in grain size is remarkable. While the mica layers normally contain a larger share of hematite spangles with dimensions of 0.005 to 0.03 mm x 0.0004 to 0.006 mm, coarser aggregates with dimensions of 0.2 x 0.03 mm are also represented as an exception. Isometric grains and aggregates of hematite also occur. Exceptions to the above-mentioned arrangement of components are hematite-crystallizates (up to 0.3 mm spangle length) orientated in joints disformal to the sedimentary bedding. While the fine hematite spangles are orientated in the bedding plane (also with respect to their shape), the larger spangles and aggregates in most cases do not show this constant orientation within layers. This is why the larger hematite slabs are considered as secondary, younger products of crystallization.

4. Particularities of the Fabric

The thickness of the beds fluctuates extraordinarily from fractions of a mm to about 40 mm. Still larger accumulation of sericite in the substratum could also be caused tectonically. A certain space rhythm may be observed in various layers. Taking a closer look at the interbedding with its marked colouring, polar structure of many magnesite fine layers is evident, especially so within the thicker magnesite beds in the level, even lower boundary, and in the wavy, bumpy, upper boundary. Worthy of special attention are hemispherically shaped magnesite fine layers, where the curvatures are sometimes accentuated by laminated built-in hematite-sericite pigment (Fig. 3). Such shapes are not caused by deformation, but are rather a primary fabric with colloform growth of various zones of the structure. The polar structure in the strata is, moreover, often accentuated by the gradual and retarded discontinuation of the hematite-sericite supply in the respective overlying section of the ferrous mica laminae. This results in pigmentation

Fig. 3. Part of thin section, nicols ‖, with laminated intercalations of magnesite (varying greys and white), hematite and sericite (black). *Centre of picture:* Colloform fabrics with polarity of the zones of formation. Original size of section 7 x 10.2 mm

Fig. 4. Part of thin section, nicols ‖, with parallel magnesite-hematite-sericite laminated layers and a folded lamina. Thicker magnesite layers show larger grain size than thinner ones. Original size of section 7 x 10.2 mm

and contamination of the overlying lower parts of the magnesite deposits, causing in turn smaller-grained magnesite aggregates in contrast to the coarse-grained nature of pure magnesite sections. The crystalline structure of the magnesite laminae will be referred to later on. The other case, that is gradual setting-in of hematite spangles in the underlying parts of mica layers, is rare.

The carbonate-mica strata series offers other striking and manifold examples of crooked fabrics. The attentive observer will notice immediately that some of the thin magnesite layers of the primary inhomogeneous parallel fabric show monoclinic-symmetrical transformations by axial folding and gliding apart on the scale of cm (Fig. 4). These transformations include flexural gliding with overthrown folds, shear folds and flow folds. Anisotropy of the strength of the various layers was evidently effective during the mechanical deformation: in the hematite-sericite layers of high componental mobility, flowing apart with thinning out of the layers dominates, while the details of fold formation in the magnesite layers are developed beautifully. In some folded magnesite layers, for example, transport of matter towards the crest of the fold is obvious, demonstrating the plasticity of these parts during the folding. Complete flowing apart of various overthrown folds in the bends and shearing of magnesite beds with overthrowing may, however, also be observed.

All these phenomena of folding and flowing apart are reminiscent of subaqueous sliding of plastic sediments. Independent deformations occur in each of several laminae one over another: between folded laminae others remained intact. For dating and genetic interpretation of the deformation we have the useful observation that the relief which was caused on top by foliation is marked by geopetal fillings with mica-hematite-pelite. The succession with inhomogeneous parallel fabric is resumed only from there on up. This means that the deformations were restricted to the regions of sediment build-up, i.e., to "free" spaces. Pursuing these deformations further, the consideration and analysis of the grain fabric becomes necessary. As was mentioned in an earlier section, the magnesite laminae consist of more or less fine-sparry grain aggregates, the deformed laminae and those with conserved colloidal relics not being different with respect to the magnesite crystallization.

5. Deformation / Recrystallization

First the continual deformation will be considered, which has caused the cm foliation of several magnesite laminae. The present xenomorphic spatite fabric is of postdeformational origin. No matter how strongly the magnesite laminae were deformed by foliation and sliding apart, the presently observable magnesite grain fabric is quite uniform regarding grain size and shape. It is only the thickness or thinness of the layer and its degree of the folds as well as their hollows show undeformed grain fabric, independently of rotational position and of the degree of deformation.

The postdeformative (with ragard to the plication) crystallization of magnesite is also exemplarily confined by numerous internal fabrics in the magnesite grains. As an example, sericite-hematite contaminations, brought in, enclosed and deformed within the folds, have been preserved as internal sediments within the undisturbed magnesite aggregate (Fig. 5): exemplary cases of precrystalline deformation. Of course fine, layered sericites and hematites which may eventually have been contained in undisturbed magnesite beds, have been incorporated in the fine-grained magnesite as *si*. This is

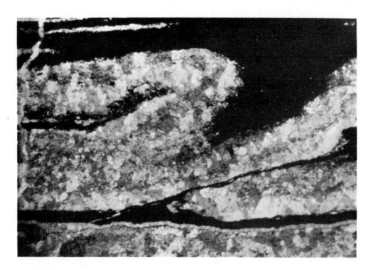

Fig. 5. Part of thin section (nicols ||). Precrystalline deformation of a magnesite fine layer: the *s*-parallel hematite pigment, bent along with the flexural fold, has been surrounded by magnesite grain growth, and is contained in the latter as "*si*". Original size of section 7 x 10.2 mm

admittedly true only for scattered contaminations, whereas in the case of clearly delineated hematite-sericite skins the magnesite grains do not normally extend beyond these disconnected layered magnesite subfabrics and mica-hematite subfabrics.

There is, however, one more remarkable deformation of the total succession within individual sections, in this case typically tectonic. A weak flexural deformation of the magnesite-mica-hematite intercalation on the scale of dm to half a m did not correspond in symmetry to the above-mentioned precrystalline deformation, which had been restricted to a few laminated layers. Two genetically entirely different processes have therefore acted in superposition.

The β-axes and the B-axes of this post-crystalline stress, imposed on the sericite layers, coincide with the axis which is known in the area of the deposit. Syntectonic shearings may also be observed. Admittedly the identification of some of these coarsely wavy deformations as syndiagenetic or tectonic is difficult if not impossible, unless unequivocal evidence for one or the other genesis is found.

Microscopic observation of the tectonically deformed layers reveals undulatory extinction of some magnesite grains. Statistical analysis of the grain fabric using the universal stage yields evidence of intragranular strain of the carbonate subfabric in a dm fold with 125° diverging limbs in the form of a weak, but distinct tectonite orientation. To eliminate errors caused by the effect of incision, three mutually orthogonal planes of section from the crest of the fold were examined, i.e., *ac, bc* and *ab*. In all three cases (Fig. 6 a–c) the typically split girdle of axes (*Sander*, 1950; *Ladurner*, 1965) shows clearly, caused by statistical accumulation of magnesite *c*-axes around the fold coordinate *b*, or in a distance of about 26° on both sides of the *ac*-plane, respectively. This tectonite orientation has been found in the magnesite deposit Hochfilzen, as well as in the wallrock which accompanies the dolomite (*Vavtar*, 1974, 1976). Due to lack

Diagram D1:   Diagram D2:

Diagram D3:

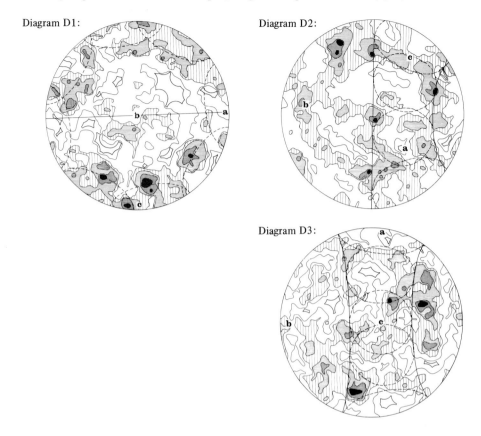

Fig. 6. (a) *Diagram D1:* Laminated magnesite, Hochfilzen, $\perp b$ (*ac*-section) 528 *c*-axes; $> 2.5 - 2$ $- 1.5 - 1 - 0.5 - 0\%$, abs. 0%. Peripheral ring with maxima, which are located on minor circles (r $\sim 26°$). The centre field of the diagram is not entirely free of axes, as should be expected from pure B-tectonite. This accumulation of axes around b is also detectable in the other two diagrams, hence not a random feature. (b) *Diagram D2:* Laminated magnesite, Hochfilzen, $\sim \perp a$ ($\sim bc$-section) 704 *c*-axes: $> 2.5 - 2 - 1.5 - 1 - 0.5 - 0\%$, abs. 0%. Split *ac*-girdle. The maxima of this broad girdle are located about $26°$ from the undertrimmed *ac*-plane, on both sides, and are arranged on minor circles (r $\sim 26°$). The centres of these minor circles lie in the *ac*-plane. The asymmetry is caused by the orientation of the cut. (c) *Diagram D3:* Laminated magnesite. Hochfilzen, $\perp c$ (*ab*-section) 636 *c*-axes: $> 2.5 - 2 - 1.5 - 1 - 0.5 - 0\%$, abs. 0%. Here, too, the broad, split *ac*-girdle with minima of axes in the *ac*-plane. Arrangement of the maxima in minor circles (r $\sim 26°$). Again slight accumulation around b can be seen

of comparable Triassic rock series near the deposit it would not be safe, however, to try to relate the tectonic formation to the Variscan or Alpidic orogeny.

6. Interpretation of Findings

Matter-concordant primary sedimentary apposition is no doubt indicated by the shape, position and spatial extent of the extraordinary — and so far unique — succession of laminated layers. The succession of beds together with the surrounding rock has been displaced to its present 80°-steep inclination. Internal sedimentation rather than external apposition fabric — syngenetic in the proper sense of the word — is, however, indicated by the sharp boundary contours all around and the difference in mineral composition, but even more importantly in fabric, from the immediately surrounding material. Within the massive magnesite body hematite and sericite are a frequent pigment, colouring the rock accordingly. Their distribution in the magnesite rock varies from layer to layer. In some massive magnesite beds laminated hematite and also sericite participation may be noticed. Nodules of hematite on the scale of cm also occur in places as intercalations, for example in the connecting deposit of Weißenstein. We are dealing here with a microngrained hematite — quartz — sericite felt containing coarse sparry concretions of magnesite.

The frequently found polarity and the partly outright colloform shape of the individual magnesite laminae in the presumed internal sediment are strong indications that the primary agent was a carbonate jelly, which was deposited in intercalation with ferrous clayey pelite. As the deposit lies in the influence region of "very low stage", maybe "low stage" metamorphism (*Winkler*, 1970), the present sericite felt could possibly have been formed by transformation of clayey (for example illitic-hydro-micaceous) material.

As primary jellies one could imagine aqueous carbonates such as trihydrocalcite $CaCO_3 \cdot 3H_2$), pentahydrocalcite $CaCO_3 \cdot 5H_2O$, huntite $CaMg_3 (CO_3)_4$, lansfordite $MgCO_3 \cdot 5H_2O$, nesquehonite $MgCO_3 \cdot 3H_2O$, hydromagnesite $Mg_5 (OH/(CO_3))_2 \cdot 4H_2O$, or artenite $Mg_2 ((OH)_2/CO_3) \cdot 3H_2O$, possibly also ferrous aqueous carbonates, a primary magnesia hydro-jelly being more likely considering the magnesite environment.

In this state of hydro-jelly, possibly still incompletely crystallized jelly, some of the laminated layers must have been subject to superficial external sliding, superficial with regard to the building zones of the laminated layer. Accretive crystallization (probably) increase in grain size and spathification could only have occurred after this sliding apart, the dating still being quite uncertain: between the early diagenesis and the orogeny, which finally caused the tectonic deformation along with the grain deformation and the orientation of the tectonite.

The succession of parallel surfaces with magnesite — mica — hematite was, therefore, transofrmed by two superimposed and symmetrologically at least partly distinguishable axial deformations: the systems of syndiagenetic, locally "external sedimentary" folds, probably in a void, and those of tectonically caused folds. The syndiagenetic mechanical stirring was accompanied by narrow monotropic foliation on the scale of cm due to the softness of still aqueous laminated layers, the tectonic stress, however, only be rather insignificant bend folds on the scale of dm to half a m.

The succession of laminated layers by itself reminds one of sedimentary magnesite fabrics as they occur, for example, in the tertiary deposits of Bela Stena (Southern Serbia) and Beli Kamen (near Strezovce, Kosovo) (*Ilić* and *Manojlowić*, 1968), spathification in the laminated layers, however, being generally absent there, I am indebted to Prof. Dr. Ing. *Karamata*, Belgrade, and Ing. *Zekic*, Kraljewo, for guiding me trhough these Yugoslavian deposits.

Considering the findings from the exposure at Bürglkopf near Hochfilzen with the laminated layers of magnesite, apparently of very local nature, enclosed in a massive to coarsely bedded magnesite seam rich in metasomatic replacement fabrics, the interpretation of the strange interbedded layer as an external sediment is possible, but not probable. Compared to the encasing rock, structure and fabric are quite different; there are no gradual transition, and lense-shaped extent may be observed here. All this suggests geopetal internal sediment. Regarding the shape and laminated structure there is some similarity to the bauxite internal sediment reproduced by *Schulz* (1976). The magnesite layers and the spatially varying rhythm have obviously been caused by continual deposition of highly dispersed Mg-hydrocarbonate, with layey matter and iron pigment being supplied intermittently.

Complete metasomatic exchange of matter by magnesite, replacing any carbonate, is a possible mechanism for selective reproduction of carbonate – clay laminated layers, but we do not think it is the actual mechanism. We regard the finding of colloform relic fabrics as very important and tend to associate it with a Mg hydro-jelly which participated already in the primary shaping of the laminated layers.

We have not been able to find features enabling us to date the formation of the laminated layers, other than the quite informative geopetal position. It is therefore possible to argue that the filling of the voids took place before the Variscan displacements. We consider the magnesite-hematite-sericite laminated layers to be a diagenetic internal sediment, which was formed near the ocean floor in a solutional void within a more or less completed magnesite bed, the latter being synsedimentary in a general sense, having originated from mostly metasomatic exchange of matter, that is still in the geosynclinal ocean sediment. Accretional crystallizations produced the weakly metamorphic carbonate-hematite-sericite rocks which we see today.

References

Clar, E.: Die Gefügeregelung eines Bändermagnesits. Jb. Geol. B.-A. LXXXI, 3–4, 387–402 (1931)

Clar, E.: Über Parallel-, Schräg- und Kreuzbänderung in Spatlagerstätten. Tscherm. Mineral.-Petrogr. Mitt. **4**, 1–4 (1954)

Höll, R.: Die Scheelitlagerstätte Felbertal und der Vergleich mit anderen Scheelitvorkommen in den Ostalpen. Bayer. Akad. Wiss., Math.-Naturw. Kl., Abh. 157, 1–114 (1975)

Höll, R., Maucher, A.: Genese und Alter der Scheelit-Magnesit-Lagerstätte Tux. Sitzungsber. Bayer. Akad. Wiss., Math.-Naturw. Kl., **1971/1**, 1–11, München 1967

Ilic, M., Manojlovic, D.: The Beli Kamen Magnesite deposit – a geological study of magnesite masses of the miocene basin at Strezovce near Kosovska Kamenica, and the relation with the surrounding and nearby rocks. Bull. Inst. Geol. Geophys. Res. Ser. A, 24–25, 29–122 (1968)

Ladurner, J.: Über ein geregeltes Magnesitgefüge. Tscherm. Mineral.-Petrogr. Mitt. 10, 1–4 (1965)

Leitmeier, H.: Die Genesis des kristallinen Magnesits. Zbl. Mineral. etc. 446–454 (1917)

Leitmeier, H.: Die Magnesitvorkommen Österreichs und ihre Entstehung. Montanztg. 67, 133–137 and 146–153 (1951)

Leitmeier, H.: Die Entstehung der Spatmagnesite in den Ostalpen. Tscherm. Mineral.-Petrogr. Mitt. 3, 305–331 (1953)

Leitmeier, H., Siegl., W.: Untersuchungen an Magnesiten am Nordrand der Grauwackenzone Salzburgs und ihre Bedeutung für die Entstehung der Statmagnesite der Ostalpen. Berg-Hüttenm. Monatsh. 99, 201–2-8 and 221–235 (1954)

Llarena, J.G.: Über die sedimentäre Entstehung des ostalpinen Magnesits „Typus Veitsch". Montanztg. 69, 55–62 (1953)

Rohn, Z.: Zur Frage der Entstehung des kristallinen Magnesits. Montanztg. 66, 1–5 (1950)

Rosza, M.: Über die primäre Entstehung des kristallinen Magnesits. Zbl. Mineral. etc., 195–217 (1925)

Sander, B.: Einführung in die Gefügekunde der geologischen Körper. Vols. I, II. Vienna: Springer 1948, 1950, Vol. I, 215 pp., Vol. II 409 pp.

Schulz, O.: Typical and Nontypical Sedimentary Ore Fabrics. Amsterdam: Elsevier 1976, Chap. 7, pp. 295–338

Siegl, W.: Zur Entstehung schichtiger und strahliger Spatmagnesite. Berg-Hüttenm. Monatsh. **100**, 79–84 (1955)

Siegl, W.: Entwurf zu einer salinär-sedimentären Entstehung der Magnesite vom Typ Entachen (Salzburg). Mineralium Deposita **4**, 225–233 (1969)

Siegl, W., Felser, K.: Der Kokardendolomit und seine Stellung im Magnesit von Hohentauern (Sunk bei Trieben). Berg-Hüttenm. Monatsh. **8**, 251–256 (1975)

Vavtar, F.: Gefügeanalytische Untersuchungen der Magnesitlagerstätte Bürglkopf-Weißenstein bei Hochfilzen, Tirol. Diss. Univ. Innsbruck (unpublished, 1974)

Vavtar, F.: Gefügeanalytische Untersuchungen der Magnesitlagerstätte Bürglkopf-Weißenstein bei Hochfilzen, Tirol. Verh. Geol. B.-A. **2**, 147–182 (1976)

Winkler, H.G.F.: Abolition of metamorphic facies, introduction of the four divisions of metamorphic stage, and of a classification based on isograds in common rocks. N. Jb. Min. Monatsh. 189–248 (1970)

Mesozoic Deposits

Comparative Reflections on Four Alpine Pb-Zn Deposits

L. BRIGO, Milano, L. KOSTELKA, Klagenfurt, P. OMENETTO, Padova,
H.-J. SCHNEIDER, Berlin (West), E. SCHROLL, Wien, O. SCHULZ, Innsbruck,
I. ŠTRUCL, Mežica

With 4 Figures

Contents

Summary

For three decades, modern investigations and interpretations on the genesis of the so-called "Alpine lead-zinc deposits" have stimulated international discussions as to the genesis of Pb-Zn deposits in carbonate sediments, which have been mostly summarized under the term "Mississippi-valley type". The four deposits of Bleiberg-Kreuth, Mežica, Raibll, and Salafossa, selected as significant representatives of the Mid-Triassic metallogenic province, are described and compared according to recent results.

The deposits are generally strata-bound and partly stratiform. Locally preserved sedimentary fabrics prove a primary stage of Alpine orogeny caused repeated remobilization of ore matter partly leading to unconformable ore bodies. The mineralogy and geochemistry of the ores reveal an extensive conformity.

A distinct correlation exists between the paleogeographic situation of the deposits, the facial and geochemical differentiation of their host rock sequences as well as the structure of the predominating ore bodies. The individual development of the four deposits is controlled obviously by their position in relation to the Periadriatic Line, the genetic influence of the geotectonic lineament, however, is still under discussion.

Preface

The four Pb-Zn deposits — Bleiberg-Kreuth, Mežica, Raibl and Salafossa — discussed in this paper belong to a Mid-Triassic metallogenic province and are concentrated within the south-eastern part of the Eastern Alps. Their ore production has amounted in recent years to more than 75% of the total lead and zinc production of the Alps. They are therefore generally of economic as well as of genetic interest.

Although there are a great number of deposits and occurrences of the same paragenesis and genesis in the Mid-Triassic sequences of the Eastern Alps, the ore reserves suitable for modern mining seem to be concentrated in the area under discussion. Because these important deposits have been subject to intensive scientific investigations for more than 100 years (the number of publications being practically impossible to survey), some comparative reflections on former and recent observations should be of common interest.

The basis for this discussion stems from the activity of the Working Group Eastern Alps, founded at Bled (1971) and stimulating the IGCP/IUGS-UNESCO Project No. 73/I/6, first titled: *Base metal occurrences in Permo-Triassic rocks in the Eastern Alps, adjacent platforms and Mediterranean.*

This paper is being presented as the first result of an international cooperation between Austrian, German, Italian and Jugoslavian scientific institutions. The financial support of our various national authorities is kindly acknowledged.

1. Introduction

The four deposits (Fig. 1) may be considered as the significant representatives of the paragenetic and genetic variations of the so-called "Alpine lead-zinc ores", mostly summarized under the term "Bleiberg type" and often compared with the Mississippi-valley type deposits (*Jicha,* 1951; *Schneider*, 1964; *Maucher* and *Schneider*, 1967). Although their predominating features of paragenesis, ore fabrics and host-rock facies are generally very similar, they exhibit significant differences in detail, indicating local variations of depositional formation. Even the relatively uniform mineralization is strat-bound, partly stratiform with typical sedimentary fabrics throughout the entire East Alpine geosyncline, there must be factors controlling the localization and concentration parameters of the ores, a still unsolved problem of general interest.

Overlooking the Pb-Zn distribution in the East Alpine Mid-Triassic province, it is possible to notice a significant economic ore concentration in its most south-eastern part: namely, in the area including the four deposits. In addition, this area is cut by an east-west-striking fault of regional importance: the "Periadriatic Line", which divides the Austroalpine nappes in the north from the South-Alpine (Dinaridic) block. According to regional interpretations, the geological units on both sides of this line have undergone deeply founded tectonic movements which suppose that the original locations of the deposits were formerly at greater distance. The present positions of Bleiberg-Kreuth and Mežica are north, and Raibl and Salafossa are south, of the Periadriatic line. The description of the deposits will follow according to their geotectonic appertainance.

In general, the lead-zinc mineralization is stratigraphically linked to a distinct time interval from Upper Ladinian to Lower Carnian, and paleogeographically connected

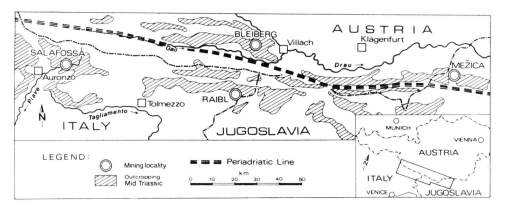

Fig. 1. Geotectonic position of the four Pb-Zn deposits within the SE Alpine realm in relation to the Periadriatic Line

with the transitional zone between the massive wall reef and the back-reef lagoon as sketched by *Schneider* (1964). Successively, this paleogeographic model was investigated in detail as regards the local features and peculiarities of the single deposits as listed below (Fig. 2).

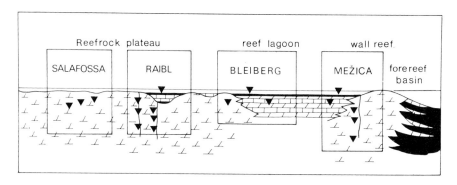

Fig. 2. Schematic reconstruction of the paleogeographic situation of the deposits without stratigraphic relation. (Predominant types of orebodies are indicated)

The Bleiberg-Kreuth lead-zinc deposit in the Gailtaler Alps is located in the high Bleiberg valley near Villach (SE Austria). Mining activity has been proved since 1333, but much older exploitation for lead appears to be probable. Since 1867 industrial exploitation has been managed by the Bleiberger Bergwerks Union. The present summary is synthetized from the papers of *Holler* (1936, 1953), *Schroll* (1953), *Schulz* (1960 a, b, 1966, 1968, 1975), *Kostelka* (1971), *Kanaki* (1972), *Schroll* and *Wedepohl* (1972), *Kranz* (1974), and *Bechstädt* (1975).

The Mežica lead-zinc deposit in Northern Karawankes lies 1 km south of Mežica (Slovenia, Jugoslavia). Exploited since 1665, the deposit has been extensively studied by several authors, e.g. *Duhovnik* (1954), *Zorc* (1955), *Grafenauer* (1958, 1959, 1966, 1969), *Strucl* (1966, 1970, 1971), *Drovenik M.* et al. (1970), and *Kranz* (1974).

The Raibl (Cave del Predil) lead-zinc deposit is located 7 km south of Tarvisio (province of Udine, North Italy), near the boundary with Austria and Jugoslavia. Exploited at present by the AMMI Company (EGAM group), the deposit has been known since the Middle Ages or earlier. The following description is summarized from the papers of *Colbertaldo* (1948, 1967), *Hegemann* (1960), *Schulz* (1964), *Romagnoli* (1966), *Dessau* (1967), *Assereto* et al. (1968), *Zeller* (1970), *Kranz* (1974), and *Brigo* and *Omenetto* (1976).

The Salafossa lead-zinc deposit is located near Sappada (province of Belluno, Italian Dolomites). Known for its outcrops since 1544, industrial exploitation by the PERTU-SOLA COMPANY began in the early 1960s with the discovery of the largest part of the orebody. The following description is summarized from the papers of *Colbertaldo* and *Franceschetti* (1960), *Dessau* (1967), *Lagny* (1969, 1974, 1975), *Lagny* et al. (1974), *Assereto* et al. (1976).

2. Stratigraphic and Paleogeographic Features

2.1 Deposits North of the Periadriatic Line (Bleiberg-Kreuth and Mežica)

In the Bleiberg-Kreuth Mid-Triassic series, the ore-bearing sediments belong to four stratigraphic levels, restricted to the upper 250 m (Cordevolian substage) of the 900 m thick Wetterstein dolomite/limestone sequence and to the overlying Raibl beds ((Julic) (Fig. 3). The depositional shallow lagoonal facies of the upper Wetterstein carbonate sequence is suggested principally by the intercalations of stromatolites, rhythmites, green marls (tuffites?), black resedimented breccias and calc-arenites. These features are particularly abundant in the uppermost 60 m of the sequence ("Bleiberg Facies") connected to cyclic and rhythmic sedimentation and possible emersion phases. In the overlying dolomitic and calcareous Raibl beds, 260 m thick and characterized by a basal shale-sandstone sequence, good stratigraphic markers are represented by three black slate niveaus (1., 2., 3. Raibler Schiefer = "Raibl markers"). More or less developed, syndiagenetic deformations and resedimented coarse breccias are observable both in the Wetterstein limestone and in the Raibl beds.

The Ladinian sequence of Northern Karawankes, in the Mežica region (Fig. 3), more than 1000 m thick, consists of three main units, characterized by differentiated facies development: (1) Wetterstein limestone (lagoonal facies); (2) Reef limestone; (3) "Partnachschichten" (basinal facies). The mainly lagoonal Wetterstein depositional facies shows analogies with that of Bleiberg. Within the massive, e.g., recrystallized and dolomitized micrites, intramicrites and intrasparites, abundant intercalations of stromatolites, resedimented breccias, oolitic limestones, green marls, etc, suggest very shallow marine sedimentation phases. Of particular interest is the presence of 5 to 15 cm thick black breccia layers, stratigraphically located at 10–15, 25 and 50–60 m below the base of Raibl beds, and thus representing valuable "markers" for stratiform ore exploration. In addition to these strata-bound types, in the Wetterstein limestone, larger, irregular, solution-collapse and, to a minor extent, resedimentation coarse breccia bodies can be

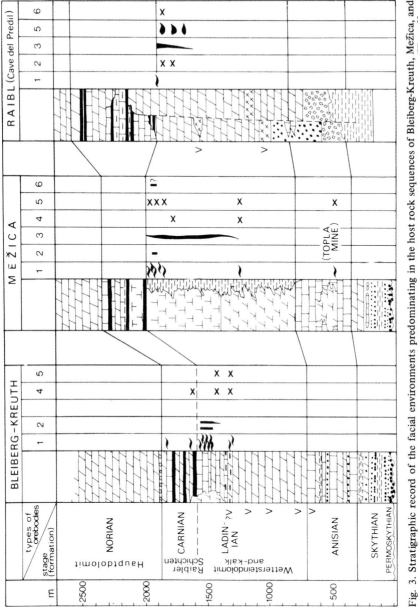

Fig. 3. Stratigraphic record of the facial environments predominating in the host rock sequences of Bleiberg-Kreuth, Mežica, and Raibl in relation to the main genetic types of orebodies. *Explanation: v:* volcanic activity, indicated by extrusions (X), tuffs or tufaceous marls. *Prevailing genetic types of orebodies: 1:* stratiform, mainly sedimentary ores; *2:* strata-bound veins and ore filled fissures; *3:* orebodies linked to disconformable faults of greater range; *4:* orebodies in brecciated parts, networks; *5:* irregular orebodies, oreshoots and stocks; *6:* karstic orebodies

observed, in particular at the transition between the Wetterstein limestone and the Reef limestone.

The Reef limestone (intrasparites and intramicrites) contains abundant, well-maintained coral reefs (with *Thecosmilia, Craspedophyllia, Oppelismilia*) and other sediment-binding organism. The presence of *Oppelismilia* and the stratigraphic relationships shows that reef facies developed during the Carnian up to the 2. Raibl marker niveau. To the south, the Reef limestone zone grades to basinal marly-clayey sediments (identical to the Partnachschichten).

The 300 to 350 m thick Raibl beds, conformably overlying the Wetterstein limestone, show laterally ubiquitous depositional facies development, often also in the dm-range.

As well as at Bleiberg, three Raibl markers (formed by argillites, marls, marly limestones, sandstones, and characterized by basal oolitic limestone niveaus) are recognizable in the sequence. The limestones between the 1. and 2. Raibl markers show analogies with the normal Wetterstein limestone facies (Pseudo-Wettersteinkalk).

In the same stratigraphic range, lateral transition to dolomitized zones or to reef-facies is observable. Dark limestones (Plattenkalk), covered by coarse bituminous dolostones, overlie the 2. Raibl marker. The Plattenkalk facies also forms the top of the Carnian sequence above the 3. Raibl marker and below the base of the Norian Hauptdolomit.

2.2 Deposits South of the Periadriatic Line (Raibl and Salafossa)

We summarize here the results of a recent review of the geology of the Raibl (Fig. 3) region, carried out by Prof. *R. Assereto* (†) in collaboration with *L. Brigo*. Above the Werfenian sandstones and limestones, the Anisian-Ladinian sequence in the area considered includes four lithostratigraphic units: (1) conglomerates and breccias; (2) shales, sandstones and tuffites; (3) tuffs and ignimbrites; (4) dolomites.

The conglomeratic and breccia unit (Ugovizza breccia) is referable to the base of the upper Anisian. For the next unit, of upper Anisian to lower Ladinian age, the old name of Kaltwasser Formation is preserved. This unit is overlain by the tuffs and ignimbrites of the Rio Freddo Volcanites, of lower Ladinian age.

The fourth unit is a carbonate buildup more than 1000 m thick (the so-called Dolomia Metallifera) covered by the Carnian units (Raibl Group). An upper and a lower Dolomia Metallifera can be individualized in the Raibl region by an intercalated carbonatic-tuffaceous formation (locally called "Buchenstein"), known at the same stratigraphic position both in the Raibl mine underground (*Romagnoli*, 1966; *Zeller*, 1970) and in the Valbruna area (*Brigo* and *Omenetto*, 1976). The Dolomia Metallifera is conformably overlain, at Raibl, by the sediments of the Raibl Group (Calcare del Predil, Rio del Lago, Rio Conzen and Tor formations).

The paleogeographic distribution and the mutual relationships of these units depend on the development, during the Carnian, of a W-E-trending basin between the Valbruna and the Rio del Lago (Raibl) valleys. The euxinic, marine sediments of the lower unit (Calcare del Predil Formation) up to 300 m thick in the centre of the basin, extend from the eastern (Raibl valley) to the western (Valbruna) borders of the basin. At the basin borders the unit appears to be reduced in thickness and is interrupted by faults. The lowermost part of the unit (well-stratified, dark bituminous marls and dolomites) is present only at the basin borders and appears to be heteropic with both the overlying black limestones of the Calcare del Predil upper portion and with the underlying Dolomia Metallifera.

The individualization of the Calcare del Predil's more subsident internal basin, with restricted circulation, was clearly controlled by a mainly N-S (and NE-SW, *Brigo* and *Omenetto*, 1976) striking syngenetic tectonics, of Lower Carnian age (*Assereto*, in *Assereto* et al., 1968). The syngenetic block-faulting is particularly well developed on the borders of the basin (Raibl and Valbruna areas), where it is emphasized by facies and thickness variations of the basal Carnian strata on both sides of the faults. Similar variations, along the same faults, of the "Buchenstein" intercalation in the Dolomia Metallifera show that the syngenetic tectonics was very probably already active during the Ladinian.

The entire Triassic sequence in the Raibl-Valbruna region, gently dipping towards the south and forming the northern flank of a regional syncline with an E-W-trending axial plane, underwent the effects of the alpine tectonics. In particular, the alpine movements in the Raibl lead-zinc district took place along the Triassic tectonic lines in the Dolomia Metallifera and in the Raibl beds, with mainly horizontal displacements along the N-S faults (*Colbertaldo*, 1948; *Zeller*, 1970). Recently *Brigo* and *Omenetto* (1976) discovered a NE-trending block faulting, with dominant vertical displacement, in the mineralized area of the Raibl mine limited to the north by the Bärenklamm fault (see below).

Following *Lagny* (1969, 1974, 1975) the Salafossa (Fig. 4) region was paleogeographically characterized, during the Mid-Triassic, by the continuous development of carbonate platform facies. The Serla Dolomite (Anisian) is overlain – above an assumed emersion surface (*Lagny*, 1975, Fig. 2) – by the Sciliar (Schlern) Dolomite of Ladino-Carnian age.

Following *Assereto* (*Assereto* et al., 1976, p. 175) in the larger Cadore and Carnian region three superimposed carbonatic platforms grew during the Mid-Triassic: (1) the Serla Dolomite, of lower Anisian age, unconformably overlain by late Anisian marly sediments; (2) the Schlern (Sciliar) Dolomite, of lower Ladinain age, showing transition to the Buchenstein basin sediments; (3) the "Infraraibliana" Dolomite of upper Ladin-

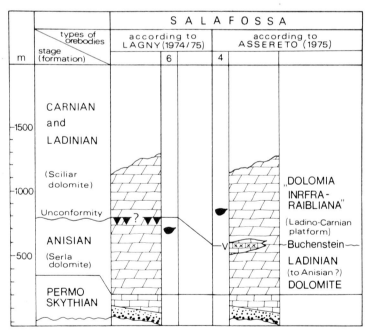

Fig. 4. Two different stratigraphic and genetic interpretations of the Salafossa deposit with respect to the facial development of the hostrock sequence. (Genetic types of orebodies: see explanation to Fig. 3)

ian to Carnian age, unconformably overlain to the south by continental and transitional sediments, and to the north grading to black bituminous limestones and shales (Raibl Group).

According to *Assereto*'s stratigraphy and in contrast with *Lagny*'s statement, the Salafossa region is characterized by the dominant development of the third platform ("Dolomia Infraraibliana").

3. Stratigraphic and Tectonic Relationships of the Orebodies

3.1 Bleiberg-Kreuth and Mežica (Fig. 3)

The economic lead-zinc mineralizations at Bleiberg-Kreuth are typically strata-bound, in the form of conformable layers and *s*-discordant veins and ore-cemented breccias. The lowermost ore-bearing horizon (about 250–200 m below the 1. Raibl marker) includes limited (100 m) stratiform bodies more than 1 m thick, comprised between marly-tuffaceous (?) layers and locally associated with ore-cemented deformation breccias. In the western Bleiberg-Kreuth section, replacement bodies formed within a probable stratigraphic range from 60 to 200 m below the 1. Raibl marker, under the selective (belteroporic) control of a deformation breccia fabric, as well as by diffuse metasomatic processes.

The second ore-bearing horizon is represented by the upper 120 m of the Wetterstein sequence, inside which are observable: (1) conformable ore "runs" (dimensions: length, several hundred m; width, up to 30 m; thickness, some m), and (2) veins (some hundred m long and up to 20 m thick, including lateral fissured zones) which developed within a 60–120 m-extending stratigraphic range (200 m at only one point). The veins show symmetrical and asymmetrical spatial connection with the "runs".

The third ore-bearing horizon is located in the Raibl beds, within the so-called 1. Zwischendolomit, some m below the 2. Raibl marker. The orebodies, conformable with subrounded outlines (diameter: some hundred m) show maximum thicknesses of more than 5 m. At the base, the stratiform ores grade to mineralized deformation breccias and reworked, coarsely clastic sediments. The uppermost horizon, characterized by stratiform mineralization and stratigraphically belonging to the dolomitic top of the Raibl beds, was recently discovered only over a limited part of the mine and proves to have been not completely defined up till now.

Taking into account the broad (10 km) E-W extent of the Bleiberg-Kreuth mining workings, the above-described four superimposed ore-bearing horizons appear to be only selectively mineralized in the different sections. Only in the western part of the Kreuth mine does the sequence show mineralization at every ore-favourable stratigraphic position, with the exception of the uppermost mineralization in the Raibl beds. From the exploitation viewpoint, the most important concentrations are at present the stratiform and breccia orebodies (respectively, in the Raibl beds, and in the Wetterstein dolomite).

Folding and block-faulting affected the whole sedimentary sequence (Wetterstein, Raibl beds and the overlying Norian Hauptdolomit), also involving the mineralization. In particular, two main Alpine tectonic phases are distinguishable: an older one, giving rise to deformation (with NW (WNW)-SE(ESE)-trending B-axes) and to the present NW-SE regional strike; and a more recent E-W-trending block-faulting, accompanied by over-thrusts, tensions and horizontal displacements along "diagonal lines". This tectonic evolution conditioned essentially the structural character of the suspended Bleiberg valley.

The lead-zinc ores of the Mežica district (stratiform bodies and s-discordant stocks and veins) are essentially strata-bound and spatially linked to well-defined horizons within the different stratigraphical units. Conformable ores appear to be in connection with horizons situated respectively at 10–15, 25, 50–60, 90, 140–150 and 650 m below the base of the Raible beds. The orebodies, with variable shape and size, form irregular networks of "runs" occupying a stratigraphic thickness of 2 to 4 m. The upper-most ore layers, and in particular those situated at 50–60 m below the Raibl beds, lie in the black breccia "reference" niveaus.

Unconformable ores can be distinguished, following their geometric features, such as:
a) strata-bound mineralized veins and fissures;
b) mineralizations linked to faults and fault-zones;
c) ore breccias;
d) mineralizations without any specific strata- or fault spatial connection.

Type (a) develops exclusively at the niveau of the black breccia layer situated 50–60 m below the 1. Raible marker. The mineralized fissures (cm- to 1 m thick) cut the stratification nearly at a right angle, with maximum vertical extension of 30 m. Type (b) occurs very frequently, cross-cutting numerous ore-bearing horizons in the Northern Karawankes area. Of great importance is the Union System ores, sinking across a strati-graphic sequence more than 600 m thick, below the 1. Raibl marker. Other and even deeper occurrences along similar fault systems are nevertheless devoid of ore-grade mineralization.

Type (c) is particularly common in the Graben mine, but is also present in Mučevo, Navršnik, Friedrich and Union mining sectors. The multiform (c) orebodies appear to raise a separate genetic problem for each occurrence. Finally, Type (d) also displays significant diffusion, in the form of extensive, highly irregular, massive sphalerite ore-bodies, with very low galena contents, developed both within the Wetterstein limestone and dolomite and in the dolomitized Reef limestone.

3.2 Raibl and Salafossa (Figs. 3 and 4)

The Raibl lead-zinc deposit is located within the Dolomia Metallifera, on the eastern border of the E-W-trending Valbruna-Raibl Carnian basin. Limited north-westwards and eastwards respectively by the Rinnengraben-Bärenklamm and Fallbach tectonic lines, the areal distribution of the ore-grade mineralization appears to coincide with the E-W-outcropping extension of the lowermost part of the Calcare del Predil Formation (dark, highly bituminous marls and dolomites).

The concentration of the ores took place along the main syngenetic faults: N-S-faults (Abendblatt-Morgenblatt, Struggl, Aloisi, Fallbach), NE-SW faults (Rinnengraben-Bären-klamm) and, to a lesser degree, NNW to NW-striking faults (Abendschlag, Vincenzo). The vertical distribution of economic ores appears to be limited to the upper Dolomia

Metallifera (thickness: some hundred m). This fact is particularly evident in the middle-southern part of the mine, where the characteristically developed "Buchenstein" sequence along the Aloisi fault clearly underlines the lower stratigraphic limit of the mineralization. In the northern part of the mine, the downwards-displaced block of the Aloisi Nord mineralized fault, south of the Bärenklamm zone, shows a lithology of the enclosing rock belonging to the upper Dolomia Metallifera facies, in spite of the reduced thickness of the underlying "Buchenstein" niveau crossed by the 17th Clara Level. In favour of this assumption, we also have the results of the geochemical research on the metalliferous series carried out by *Kranz* (1974) along the Aloisi fault at the 15th and 17th Clara Levels (Sect. 5.3).

In conclusion, the ore concentration at Raibl appears to be conditioned by the following paleogeographic and paleotectonic parameter: (1) the eastern border of the Valbruna-Raibl Carnian basin (ore-bearing area); (2) the upper Dolomia Metallifera (host rock); (3) the "Buchenstein" (lower stratigraphic limit of ore distribution); (4) the lowermost part of the Calcare del Predil Formation (upper stratigraphic limit of ore distribution); (5) the N-S (NNW) and NE-SW Triassic faults (ore concentration lines).

The positive influence of the above-mentioned parameters, taken as a whole, on the ore concentration process appears to be confirmed (with some minor differences) on the western border of the Carnian basin (Valbruna area).

Moreover, recent ore exploration in this area has permitted the discovery of geo-chemical anomalies and mineralized outcrops (with galena, sphalerite and dolomite) in the Dolomia Metallifera of the M. Nero (Jof di Miezegnot) and of the Torrente Carnizza, showing the same stratigraphic and paleotectonic situation as compared with the Raibl.

In the Raibl deposit, the following types of orebodies are distinguishable:

1. "stratiform" ores, of limited vertical extension, linked essentially to the Struggl fault zone at the contatto scisti;

2. stocks (columnar orebodies), roughly parallel to the general bedding-strike and located in the intermediate-upper portion of the upper Dolomia Metallifera. They represent the filling-up of the solution cavity systems and brecciated volumes. The typical example is the N-S-trending Colonna Principale (dimensions: length, about 1200 m; width, more than 50 m; thickness, some dozens of m). Peculiar vertical stocks are present in the northern sector of the mine (Bärenklamm zone) at the intersection between N-S and NE syngenetic faults (main example, the Cant. NE body, with melnicovite-pyrite + red Schalenblende-rich ores);

3. veins, along the main syngenetic faults and lateral fissured zones. The best example is the Aloisi vein, clearly underlined by the "Buchenstein" horizon. The veins, dm to some m thick, develop in length for several hundred m and vertically for some hundred m.

The Salafossa lead-zinc deposit (Fig. 4) appears to be enclosed within the massive Mid-Triassic dolomite, near the NNE-striking tectonic contact (defined as a triassic fault by *Lagny*) with the basement + Permian-Eotriassic units, these last very strongly reduced in thickness. The exploited orebody, columnar in shape and elongated parallel to the above-mentioned fault, shows the following dimensions: length, 700 m, width, 50–200 m, thickness, 50–100 m. The stratigraphic position of the deposit is under debate: according to *Lagny* the concentration took place within a karstic network of cavities at the top of the Anisian (Serla) Dolomite, below a supposed emersion surface directly overlain by the transgressive Ladino-Carnian (Sciliar) Dolomite. According to

Assereto, the Salafossa deposit belongs stratigraphically to the middle part of the "Dolomia Infraraibliana" buildup, overlying the original Buchenstein (Livinallongo) formation.

4. Ore Paragenesis and Fabrics

4.1 The Bleiberg-Kreuth and Mežica

The Bleiberg-Kreuth primary mineralization, with similar character in the four ore-bearing horizons, appears to be pretectonic as regards the alpine orogenesis, synsedimentary in origin and, thus, of Mid-Triassic age. Main components are: sphalerite, largely of colloform type (Schalenblende), galena, marcasite, pyrite, melnicovite, fluorite, quartz, barite, calcite, dolomite. To syndiagenetic mineral assemblage belong also anhydrite, celestite and strontianite (emmonite). Younger minerals (β-polygorskite, jordisite and molybdenite, probably also gypsum) appear to be crystallized within later syn- to post-tectonic fissures. The more or less deeply developed oxidized zone shows abundant, partly well-known for Bleiberg, crystalline varieties of secondary minerals: wulfenite, vanadinite, descloizite, ilsemannite, plumbocalcite, in association with the common supergenetic mineralization.

The average Zn/Pb ratio for the deposit is about 5/1. A primary, vertical zoning due to paragenetic variations is observable in the mineralized veins of the upper Wettersteinkalk. Moreover, the orebodies at different chronostratigraphic positions show consistent differences of paragenesis and fabric. Some particularities appear also on the regional scale.

At Bleiberg-Kreuth, the dominant ore fabrics (for example: bedded colloform fabric, collective crystallization aggregates, diagenetic replacements) belong to the types permitting alternative interpretation. Highly significant, therefore, are the less common but typical sedimentary fabrics of the stratiform bodies: i.e., mineralized lutites and arenites, showing rhythmic-, p.p. also transverse- and cross-lamination, as well as resedimentation and geopetal (top-bottom) features. Frequently, both mineralized and barren associated sediments display common behaviour during syndiagenetic deformation and fracturing.

Internal sedimentation is proved by the presence of laminated, mineralized lutites with depositional geopetal fabric inside the Triassic fissure system of the upper Wetterstein limestone at Bleiberg and in solution cavities within the ore-filled network of the deformation breccias. External mineralized sediments are, on the contrary, clearly recognizable in the few cases where, around the stratiform orebodies, gradual transition on all sides can be observed between the mineralized and respectively barren, carbonatic lutites of the "normal" sequence.

At Mežica, the very simple primary ore paragenesis shows differences in mineral abundances as regards the various mineralization types. Conformable ores consist chiefly of lead and zinc sulphides in dolomite "gangue", with minor amounts of pyrite, marcasite, fluorite, bitumen and some quartz. Among the unconformable ores, Type (a) appears to be particularly rich in galena, associated with abundant calcite and minor marcasite, pyrite and dolomite. Similar paragenesis with calcitic "gangue" also shows

Type (b) (Union System), whith minor fluorite linked to lead-zinc ores. Type (c) appears to be lead- and dolomite-rich with its galena-sphalerite-marcasite-pyrite ores, with some calcite and quartz. Type (d) is essentially zinciferous, with dolomite, calcite and minor barite and quartz. A peculiar feature of the Mežica deposit, when compared with Bleiberg-Kreuth, Raibl and Salafossa, is the abundance, even if not uniformly distributed, of lead ores; average Pb/Zn ratio is about 1/1. In the past, the Pb/Zn ratio changed from 3/1 to 2/1; the present 1/1 ratio will turn in favour of Zn.

In the important secondary paragenesis, of particular interest is the presence of wulfenite and descloizite, essentially linked to the Union System ores, in association with the common supergenetic mineralization.

4.2 Raibl and Salafossa

In the Raibl deposit, the fundamental paragenesis shows very similar features when compared with Salafossa. The Pb/Zn ratio is approximately 1/5. Sphalerite is present in crystal- and colloform varieties. The crystalline varieties are black, grey and rose in colour (*Colbertaldo*, 1948). The micro- and macrocolloform varieties (Schalenblende) are dominantly yellow and red. Detailed microscopic analysis showed evidence that the colour variations of the Schalenblende depend on the diagenetic evolution (crystallization) of the sulphide: the yellow varieties, although retaining their external, rhythmically grown fabric, are invariably crystalline; the red varieties are poorly crystalline to amorphous. Transition in crystallinity and colour between red and yellow Schalenblende is always observable.

Galena is normally coarse-grained, often in aggregates of large crystals. Iron sulphide (melnicovite-pyrite, pyrite, marcasite) is particularly abundant in the northern sector of the mine (Bärenklamm zone). Sulphosalts are rare: jordanite and probably gratonite (*Venerandi*, 1966) linked to red Schalenblende + fine-grained galena facies in the Bärenklamm and Fallbach zones. The presence of traces of pyrrhotite, chalcopyrite and cinnabar seems rather doubtful, never observed during modern microscopic studies. The well-developed secondary deposit at Raibl can be divided into an upper oxidized zone (rich in limonite) and a lower zone rich in carbonates (smithsonite). Among the "gangue" minerals, dolomite and barite prevail, associated with calcite in the secondary deposit. Fluorite seems to be completely absent.

The concentration of primary lead-zinc ores took place essentially per descensum, within a complex framework of syntectonic fissures and solution cavities, along the geometric pattern defined by the Triassic faults. Clearly connected also with the Triassic fault zones are the "stratiform" mineralizations at the contatto scisti (stratigraphic and/ or tectonic contact between the top of the upper Dolomia Metallifera and the Raibl beds). Depending on the intensity of the late displacements along the mineralized Triassic fault zones, the ores (already syntectonically deformed and transformed by diagenetic cry- and recrystallization, mobilization and, pro parte, metasomatic processes), underwent the effects of the alpine tectonics in variable amounts.

The following ore fabrics are recognizable:

a) Weak synsedimentary mineralizations (see also *Schulz*, 1964) in the basal Raibl strata, partly reworked in ore-grade dark syndiagenetic "stratiform" breccia (with a minimum thickness of some meters) at the contatto scisti, essentially along the Struggl fault zone.

b) Mineralized internal mechanical (sphalerite-, dolomite- and barite-bearing) and chemical (micro- and macrocolloform sphalerite + galena + iron sulphide + dolomite-bearing) sediments, representing the filling-up of the columnar and vertical stocks. The sulphide mineralization differs in diagenetic evolution degree and in corresponding sulphur isotopic composition (see Sect. 5.2) from older (Colonna Principale, Erzmittel) and younger (Bärenklamm zone) orebodies.

c) Dominant colloform lead-zinc ores in the veins (Aloisi, Fallbach, Struggl, p.p.). Fissures, filled with spathic galena and oxidized zinc ore, and karst cavities, filled with peculiar stalactitic ore [e.g., the "Röhrenerze", "tubular" (galena) ores], seem to be related to more recent phases of ore deposition, ill-defined up till now.

At Salafossa the dominant ore minerals are sphalerite (macro- and microcolloform), galena, pyrite (melnicovite) and marcasite. The Pb/Zn ratio in the crude ore is approximately 1/6. Geocronite is accessory as inclusions in galena (*Lagny* et al., 1974). Very rare are boulangerite and hutchinsonite. The nonmetallic minerals consist chiefly of dolomite and barite, with traces of detritic quartz (partly authigenic), acid feldspar, clay and tuffaceous fragments (Pietra Verde following *Lagny*).

Secondary minerals are poorly developed and only within the outcropping portion of the orebody. At Salafossa, a true gossan is lacking.

A detailed study on the fabrics of Salafossa lead-zinc ores has been carried out by *Lagny* (1974). The large pipe of ore-grade breccia grades to the massive dolomite through a fractured and slightly mineralized peripherical zone. The facies and paragenesis of the mineralization varies considerably between the lower part and the top of the orebody. At the base, the cavity network appears to be filled mainly with mineralized sediments (mechanical internal sediments sensu *B. Sander*): barite-bearing black dolomites, sphalerite-bearing dolomites and breccias with sedimentary mineralized matrix. At higher levels, within the collapse breccia and its periphery, the ore-fillings are essentially of chemical type (colloform sulphides and late spathic dolomites). In the deepest portion of the orebody, the paragenesis is characterized by abundant pyrite, marcasite and barite; in the collapse breccia, sphalerite is the main component. The different ore facies appear to be reworked and are collectively visible, in the form of broken fragments, within the cavity fillings at the base of the mineralized body.

5. Geochemical Data

Although some previous pilot works on different geochemical problems do exist (*Schroll*, 1953; *Hegemann*, 1960; etc), the data referred to below represent essentially the first results of still-operating extensive geochemical research programs.

5.1 Minor Elements in the Lead-Zinc Ores

In spite of the inhomogeneous distribution of minor elements in the different ore facies (linked also to the presence, for example, at Salafossa, of As-Sb-sulphosalt inclusions in galena) and to the lack of evidence of chemical equilibria in the co-existing lead and zinc sulphide microparagenesis, the four investigated deposits exhibit major common trends and some differences.

The principal minor elements in sphalerite are Cd, Tl and As (with maximum abundances of some thousand ppm) and Ge (some hundred ppm, up to 3000 ppm at Bleiberg-Kreuth). Ge-contents, together with Tl-, As- and Fe-contents (iron is generally very low) increase in the typical Schalenblende facies (for example: diagenetic Schalenblende in the Raibl beds at Bleiberg). Cu is present (max. 470 ppm) only in sphalerites from Mežica.

The galena of all deposits is significantly poor in Ag, with common minor contents of As and Tl at Bleiberg, Salafossa and Raibl. Sb is abundant at Salafossa (mainly in skeletal galena facies) and represents the sole minor element (together with traces of Cu) in the typically pure galena of Mežica.

The available data for "gangue" minerals, concerning only the Bleiberg-Kreuth paragenesis, do not permit an adequate comparative discussion in this paper.

5.2 Sulphur- and Lead-Isotope Compositions

The sulphur isotope compositions of the sulphides (sphalerite, galena and pyrite-marcasite) from Bleiberg, Mežica and Raibl are listed in Table 1. No available data exist for Salafossa.

Table 1. Range of sulphur-isotope compositions

Deposit	Range of $\delta^{34}S$ ‰ variation	Average $\delta^{34}S$ ‰
Bleiberg (1)	-4.4 to -30.4	-19.7
Mežica (2)	-1.70 to -20.93	-10
Raibl (3)	-6.43 to -25.60	-17.5

(1) *Schroll* and *Wedepohl* (1972); (2) *Drovenik, M.*, et al. (1976); (3) Analyst *J. Pežidic* (Jošef Stefan Institute, Ljubljana)

The ^{34}S ‰ values appear to be constantly negative (PbS > ZnS > FeS$_2$). A further correlation results between the Bleiberg and Raibl deposits: the minimum negative values belong, at Bleiberg, to galena and sphalerite of primary, stratiform mineralization and, at Raibl, to the strongly recrystallized ore facies. The maximum values show later marcasite and Schalenblende at Bleiberg, as well as the younger, early diagenetic red Schalenblende and melnicovite-pyrite assemblage at Raibl. These facts seem to argue both for a low-temperature deposition (large fractioning probably due to bacterial processes) and for a polyphasic evolution of the deposits.

Available comparative data for "gangue" minerals show sulphur isotope composition for the Bleiberg and Mezica syngenetic grey anhydrites ($\delta^{34}S$ ‰ respectively, + 16.1 and + 15.7) close to that of seawater during Cordevolian/Julic substages.

The composition of lead isotopes in the four deposits has recently been investigated by *Köppel* (see Table 2). According to a kind personal communication lead is of B-type with anomalous positive model age (300–350 my). As previously indicated by *Schroll* (1965), the model lead age seems to be 100 to 200 my older than the ore-bearing Triassic sediments.

Table 2. Ratios of lead-isotopes. (After *Köppel*, 1977)

Average values (number of analyzed samples)	$^{206}Pb/^{204}Pb$	$^{207}Pb/^{204}Pb$	$^{208}Pb/^{204}Pb$
Bleiberg-Kreuth (22)	$18.379 \pm .007$	$15.672 \pm .009$	38.556 ± 0.33
Mežica (4)	$18.362 \pm .012$	$15.669 \pm .010$	$38.532 \pm .025$
Raibl (4)	$18.386 \pm .006$	$15.685 \pm .004$	$38.621 \pm .017$
Salafossa (1)	$18.315 \pm .016$	$15.653 \pm .019$	$38.531 \pm .060$

5.3 Geochemistry of the Host Rock

Comparative data for Bleiberg-Kreuth, Mežica and Raibl Ladinian sequences are available according essentially to *Kranz* (1974), with the exception of fluorine distribution (see Sects. 4.1 and 5.4).

The dolomite content rises towards the upper part of the sequences: in the lagoonal areas dolomite occurs predominantly stratiform, indicating supratidal early diagenetic formation. In the massive wall reefs the dolomite represents essentially the product of late diagenetic alteration. Especially in this case, there is a positive correlation to Pb-Zn mineralization.

Sr contents seem to be indicative for the salinity degree of the depositional environment as well as for the range of diagenetic alteration. Extreme negative correlations between Sr- and dolomite contents are observable both in the lagoonal facies of Bleiberg-Kreuth and Mežica and in the Raibl reef complex.

Fe- and Mn contents increase towards the top of the Wettersteinkalk lagoonal facies, linked principally to clayey sediments and to the presence of disseminated iron sulphides.

Stratiform Zn anomalies in the lagoonal facies of Bleiberg and Mežica are restricted to a few stratigraphical niveaus corresponding to some ore-bearing horizons. Stratiform Pb anomalies are practically lacking. Zn anomalies also show some typical sediments of the so-called "Sonderfazies" (*Schneider*, 1964) as the green marls and black breccias.

In the Dolomia Metallifera at Raibl (*Kranz*, 1974) as well as at Salafossa (*Lagny*, 1975), primary stratiform Zn and Pb anomalies are lacking. Values above the background (*Zeller*, 1970) are indicated by the "Buchenstein" facies. At Raibl, stratiform anomalies are present, and likewise at Bleiberg and Mežika, only in the lowermost Raibl beds. Discordant anomalies in the Dolomia Metallifera close to orebodies show very rapid lateral downfall (in the range of 20–30 m) to background (15–30 ppm Zn, 2–5 ppm Pb: *Kranz*, 1974).

5.4 Recent Data on Fluorine Distribution

As previously mentioned, the occurrence of fluorspar in the ore assemblages shows clear differences: in contrast to the significant fluorspar mineralization at Bleiberg and the minor amounts quoted for Mežica, no recent findings are reported for Raibl and Salafossa. In accordance with this fact, the distribution pattern of fluorine contents in the host rock sequences appears to exhibit some distinct trends, indicating the congruent deposition of host rock and mineralization.

According to the first results of *Schneider* et al. (1977), three general fluorine distribution patterns are distinguishable:

1. Facies-Controlled Distribution. The intercalated marly beds reveal the highest geometric mean according to published data on similar rock types, due to the common behaviour of fluorine in clay-bearing sediments. The maximum fluorine anomalies are nevertheless linked to some of the limestone layers (see Table 3), where no megascopically or microscopically detectable fluorspar is observable. Because of the relative scarcity of dolomite layers within the analyzed sequences, the corresponding data on fluorine contents are not representative: in any case the geometric mean always seems to be higher than that of limestones, again in accordance with published data from other countries.

Table 3. Mean fluorine contents of the Ladinian sequences (in ppm) completed after *Zimmer* (1976)

Deposit	Rock type and number of samples	Range min. max.	Geometric mean
Lafatsch [a]	l (58)	52–13,500	106
	d (2)	75– 100	(87)
	m (7)	57– 1,050	268
Bleiberg	l (47)	84–29,300	227
	d (5)	183– 5,250	(738)
	m (11)	257– 1,900	785
Mežica	l (44)	72– 7,550	180
	d (12)	97– 733	316
	m (13)	433– 2,650	1,031
Raibl	l (48)	50– 712	102
	d (20)	57– 575	163
	m (3)	194– 1,188	(514)

[a] Representative for the northern part of the East-Alpine geosyncline (Northern Calcareous Alps). Rock type: (General mean fluorine content in the literature); *l:* limestone: (220 ppm); *d:* dolomite: (260 ppm); *m:* marl: (800 ppm).

2. Stratigraphic Correlation. Throughout the East-Alpine geosyncline, some distinct strata-bound anomalies are fairly well detectable, since they are independent on the absolute values of the local fluorine concentrations. In addition, a distinct correlation between the Zn- and F anomalies is observable. However, in opposition to the restricted extent of Zn anomalies (some tens of meters only in lateral distance), the fluorine contamination is extremely ubiquitous, even at places where neither zinc nor fluorine mineralizations are known. These anomalies are generally linked to the uppermost 100 m of the sequences, abundant − in the range of facies-controlled frequency − in every dominant rock type (limestones, dolomites and marls). Because of this fact, strata-bound F anomalies could become useful proximity indicators (like the marker beds) for stratigraphic orientation during local mining exploration as well as regional prospecting.

3. Paleogeographic Control. A comparison of the three analyzed host rock sequences of Bleiberg, Mežica and Raibl (data for Salafossa are not available) indicates a clear correlation of F distribution and paleogeography (Table 4). The mainly lagoonal, layered facies of Bleiberg and Mežica is characterized generally by higher geometric means and thresholds, and also by the magnitudes of the anomalies of limestones and dolomites. On the contrary, the massive reef facies of Raibl shows a smaller range of data.

Table 4. Range of local fluorine anomalies (facies group: limestone) completed after *Schneider* et al. (1977)

Lafatsch	1,300−13,500 (ppm)
Bleiberg	3,700−29,300
Mežica	5,450− 7,550
Raibl	600− 712

Furthermore, the frequency of fluorine contents in the three deposits reveals a significant variation, e.g., bimodal distribution at Bleiberg and Mežica (connected to lagoonal environment?) and a nearly unimodal distribution at Raibl, where the absolute contents are significantly lower in contrast to Bleiberg and Mežica.

The bulk fluorine content clearly decreases eastwards both in the Bleiberg *s.s.* district and from Bleiberg to Mežica, southwards, from Bleiberg to Raibl. This fact could be explained as the Bleiberg area remained close to the fluorine source, as indicated also by the REE contents in Bleiberg fluorites (*Schneider* et al., 1977).

6. Discussion

We decided to present this discussion as a comparison of the two deposits north of the Periadriatic Line (Bleiberg-Kreuth and Mežica) with the two deposits south of it (Raibl and Salafossa).

As the geotectonic function of the so-called "Periadriatic Line" is still under discussion, we prefer to summarize only some brief remarks on this problem, because it could become of paleogeographic importance for the genesis of the primary mineralization. According to the review of *Bögel* (1975) the majority of the geoscientists believe the Periadriatic Line to be a deep-rooted geotectonic lineament with paleogeographic influence since the end of Paleozoic: the outcrops indicate repeated tectonic movements in predominating vertical direction, on a regional scale of vast range. Caused by the intense tectonics, remarkable parts of at least the Paleozoic are eliminated.

On the other hand, the Peridadriatic Line is also interpreted as a transform fault: the block south of the lineament should be rotated towards west relatively to the block north of it. If we take into account the recent positions of the four great deposits on both sides of the lineament, an east-west distance is observable between Bleiberg (north) and Salafossa (south) of about 80 km, and between Mežica (north) and Raibl (south) of about 100 km (Fig. 1). The transport distance parallel to the lineament is suggested actually to be of this order.

The four deposits, surveyed under this scope, exhibit some significant individual records in mineralization and facies development in addition to the predominating common features, which may be interpreted as indications for a more or less N-S opposite position before the rotation took place. According to our knowledge of the deposits only we do not insist, however, on an interpretation for this coincidence to be a proof of the geotectonic function of the lineament.

Common features of the four deposits are:

1. Strata-bound mineralization linked to Upper Ladinian and Carnian sequences.

2. Local conservation of (external) sedimentary features of ore matter (rhythmic polar apposition, reworked breccia, slumping, cut-and-fill structures etc).

3. Correlation of paleogeographic patterns and ore structures/paragenesis.

4. Development of typical diagenetic patterns in accordance to the hostrock, well distinguishable into pre- and post-tectonic phases (e.g., vein fillings, breccias, ore shoots).

In order of their paleogeographic position north and south of the Periadriatic Line the deposits reveal some different features:

	North of Periadriatic Line Bleiberg-Mežica	South Raibl-Salafossa
Significant environmental conditions	Mainly lagoonal sediments (Cyclothemes) less recifal facies	Predominating reefrock facies
Stratigraphic position	Bleiberg: Upper Ladinian to Lower and upmost Carnian	Salafossa: *Assereto*: Ladinian *Lagny*: Anisian
	Mežica: Middle-Ladinian to Upper Ladinian	Raibl: Ladinian to Ladinian-Carnian
	Events leading to ore occurrence ascending in sequence from East to West	Events leading to ore occurrence synchronous respectively descending from East to West (*Lagny*)
Geometric shape of ore-bodies	Predominantly stratiform	Mainly discordant
Geochemistry of hostrock and environment	Bleiberg and Mežica: Rising Mg content and extreme negative correlation between Sr content and dolomite in lagoonal sediments	Raibl: Rising Mg content and extreme negative correlation between Sr and dolomite in lagoonal sediments
	The high Sr content indicate a weak evaporitic environment for the lagoonal sediments in both deposits	The low Sr content indicate an intense diagenetic alteration

	North of Periadriatic Line Bleiberg-Mežica	South Raibl-Salafossa
	The frequency and the absolute content of fluorine are connected with the lagoonal sediments. The F content seems to decrease from West to East and is in the northern deposits more abundant than in the southern ones.	The extremely low fluorine content at Raibl seem to be indicative for the distribution pattern of this
	Lagoonal sediments exhibit distinctive peaks of Zn in some beds	Distinctive peaks of Zn in "Dolomia Infraraibliana" limited to a small distance to the mineralized faults.
Mineralogy and geo-chemistry	Extensive conformity. No significant differences could be observed so far.	
Genesis	The authors agree that the events leading to the deposits started synchronous to the sedimentation phase in Mid-Triassic. Diagenetic processes played an important role, beginning immediately after sedimentation with a peak of intensity during Alpine orogenesis.	
	There is no accordance in respect to the sources among the authors. Two hypotheses are advocated:	
	1. The base metals have been carried by sea water within comparatively small distances. The source was connected with volcanic activities.	
	2. The base metals are derived from continental weathering processes over a greater distance.	

References

Assereto, R., Brigo, L., Brusca, C., Omenetto, P., Zuffardi, P.: Italian ore mineral deposits related to emersion surfaces – a summary. Mineralium Deposita 11, 170–179 (1976)

Assereto, R., Desio, A., Colbertaldo, D. di, Passeri, L.D.: Note illustrative della Carta Geologica d'Italia. Foglio 14 a Tarvisio, Napoli 1968, 70 p.

Bechstädt, Th.: Lead-zinc ores dependent on cyclic sedimentation (Wetterstein-Limestone of Bleiberg-Kreuth, Carinthia, Austria). Mineralium Deposita 10, 234–248 (1975)

Bögel, H.: zur Literatur über die „Periadriatische Naht". Verh. Geol. B.-A. Wien 163–199 (1975)

Brigo, L., Omenetto, P.: Le mineralizzazioni piombo-zincifere della zona di Raibl: Nuovi aspetti giacimentologici. L'Industria Mineraria 27, 49–56 (1976)

Cardich-Loarte, L.A., Schroll, E.: Die Verteilung und Korrelation einiger Elemente in einem Erzkalkprofil der Bleiberger Fazies (Bleiberg/Kärnten – Rudolfschacht). Tschermaks Mineral. Petrol. Mitt. 20, 59–70 (1973)

Colbertaldo, D. di: Il giacimento piombo-zincifero di Raibl in Friuli (Italia). Mem. presented 18th Sess. Intern. Geol. Congr., London, 1948, p. 149

Colbertaldo, D. di: Giacimenti Minerari-Giacimentologia generale e giacimenti di Pb-Zn (e Ag). Padua: CEDAM 1967, p. 383

Colbertaldo, D. di, Franceschetti, G.: Il giacimento piombo zincifero di Salafossa nelle Alpi Orientali Italiane. Intern. Geol. Congr., 21st Sess., Norden 16, 126–137 (1960)

Dessau, G.: Gli elementi minori nelle blende e nelle galene della miniera di Salafossa (S. Pietro e S. Stefano di Cadore, Alpi Orientali Italiane). Confronti con i giacimenti del Bergamasco e di Raibl. Atti Giorn. St. Geomin. Agordo, 123–134 (1967)

Drovenik, M., Leskovšek, H., Pezdič, J., Štrucl, I.: Sulfur-isotope composition on sulfides of some Yugoslav ore deposits. Min. Metall. Quart. 2–3, 153–173 (1970)

Duhovnik, J.: On the origin of molybdenum in the lead-zinc ore deposits of Mežica. Geologija-Razprave in poročila 2, 113–117 (1954)

Duhovnik, J.: Facts for and against a syngenetic origin of the stratiform ore deposits of lead and zinc. Econ. Geol. Mon. 3, 108–122 (1967)

Grafenauer, S.: Über Blei-, Zink- und Molybdänerzlagerstätten in Mežica, Jugoslawien. Min. Metall. Quart. 263–194 (1958)

Grafenauer, S.: Seltene natürliche Bleioxyde in Mežica (Jugoslawien). Geologijy-Razprave in poročila 5, 56–62 (1959)

Grafenauer, S.: The genetic classification of the lead and zinc deposits in Slovenia. Min. Metall. Quart. 2, 35–42 (1966)

Grafenauer, S.: On the Triassic minerogenetic epoch in Yugoslavia. Min. Metall. Quart. 3–4, 39–55 (1969)

Grafenauer, S., Ottemann, J., Strmole, D.: Über Descloizit und Wulfenit von Mežica (Mies) Jugoslawien. N. Jb. Mineral. 109, 25–32 (1968)

Hegemann, F.: Über extrusiv-sedimentäre Erzlagerstätten in den Ostalpen – II. Blei-Zink-Erzlagerstätten. Erzmetall 13, 122–127 (1960)

Holler, H.: Die Tektonik der Bleiberger Lagerstätte. Carinthia II 7, Sonderheft (1936)

Holler, H.: Die Stratigraphie der karnischen und norischen Stufe in den östlichen Gailtaler Alpen. Berg und Hüttenm. Mh. 69–75 (1951)

Holler, H.: Der Blei-Zink-Erzbergbau Bleiberg, seine Entwicklung, Geologie und Tektonik. Carinthia II 143, 35–46 (1953)

Jicha, H.L.: Alpine lead-zinc ores in Europe. Econ. Geol. 46, 707–730 (1951)

Kanaki, F.: Die Minerale Bleibergs (Kärnten). Carintha II 82, 7–84 (1977)

Köppel, V.: Bemerkungen zu den Bleiisotopen-Verhältnissen von PbS aus ostalpinen Pb-Zn-Lagerstätten (Personal written communication, 1977)

Kostelka, L.: Beiträge zur Geologie der Bleiberger Vererzungen und ihrer Umgebung. Carintha II, Sonderheft 28, 283–289 (1971)

Kranz, J.R.: Geochemische Charakteristik des erzhöffigen Oberen Wettersteinkalkes (alpine Mitteltrias) der Ostalpen. Diss. Freie Univ. Berlin (1974), p. 219

Kranz, J.R.: Strontium – ein Fazies-Diagenese-Indikator im Oberen Wettersteinkalk (Mittel-Trias) der Ostalpen. Geol. Rdsch. 65, 593–615 (1976a)

Kranz, J.R.: Stratiforme und diskordante Zink-Blei-Anomalien im erzhöffigen Oberen Wettersteinkalk (alpine Mitteltrias). Mineralium Deposita 11, 6–23 (1976b)

Lagny, Ph.: Minéralisation plombo-zincifère triasique dans un paléokarst (gisement de Salafossa, province de Belluno, Italie). C. R. Acad. Sci. (Paris) 268, 1178–1181 (1969)

Lagny, Ph.: Emersions médiotriasiques et minéralisations dans la région de Sappada (Alpes Orientales Italiennes): le gisement de Salafossa, un remplissage paléokarstique plombo-zincifère. Thèse Doct. Sci. Nat., Univ. Nancy I (1974), p. 366

Lagny, Ph.: Le gisement plombo-zincifère de Salafossa (Alpes Italiennes Orientales): Remplissage d'un paléokarst triasique par des sediments sulfurés. Mineralium Deposita 10, 345–361 (1975)

Lagny, Ph., Omenetto, P., Ottemann, J.: Géocronite dans le gisement plombo-zincifère de Salafossa (Alpes Orientales Italiennes). N. Jb. Miner. Mh. 12, 529–546 (1974)

Maucher, A., Schneider, H,-J.: The Alpine lead-zinc ores. Econ. Geol. Mon. 3, 71–89 (1967)

Romagnoli, P.L.: Contributo alla conoscenza del giacimento di Raibl. Atti Symp. Intern. Giac. Min. Alpi, Trento 1, 243–245 (1966)

Schneider, H.-J.: Facies differentiation and controlling factors for the depositional lead-zinc concentration in the Ladinian geosyncline of the eastern Alps. In: Sedimentology and Ore Genesis. Amstutz, G.C. (ed.). Amsterdam: Elsevier 1964, pp. 29–45

Schneider, H.-J., Möller, P., Parekh, P.P., Zimmer, E.: Fluorine contents in carbonate sequences and rare earth distribution in fluorites of Pb-Zn deposits in East-Alpine Mid-Triassic. Mineralium Deposita 12, 22–36 (1977)

Schroll, E.: Über die Anreicherung von Mo und V in der Hutzone der Pb-Zn-Lagerstätte Bleiberg-Kreuth in Kärnten. Verh. Geol. B.A. Wien 4–6, 1–20 (1949)

Schroll, E.: Über Minerale und Spurenelemente, Vererzung und Entstehung der Blei-Zink-Lagerstätte Bleiberg–Kreuth/Kärnten in Österreich. Mitt. Österr. Miner. Ges., Sonderheft 2, 1–60 (1953)

Schroll, E.: Anomalous composition of lead isotopes in the lead-zinc deposits of calcareous Alps sediments. Rud. Metal. Zbornik 2, 139–154 (1965)

Schroll, E.: Zur Korrelation geochemischer Charakteristika der Blei-Zink-Lagerstätte Bleiberg-Kreuth mit anderen schichtgebundenen Vererzungen in Karbonatgesteinen. Schriftenreihe der Erdwissensch. Komm. Band 3, Österr. Akad. Wiss. (in press, 1977)

Schroll, E., Wedepohl, K.H.: Schwefelisotopenuntersuchungen an einigen Sulfid- und Sulfatmineralen der Blei-Zinkerzlagerstätte Bleiberg-Kreuth, Kärnten. Tschermaks Mineral. Petrol. Mitt. 17, 286–290 (1972)

Schulz, O.: Die Pb-Zn-Vererzung der Raibler Schichten im Bergbau Bleiberg-Kreuth (Grube Max) als Beispiel submariner Lagerstättenbildung. Carinthia II, Sonderheft 22, 93 (1960a)

Schulz, O.: Beispiele für synsedimentäre Vererzungen und paradiagenetische Formungen im älteren Wettersteindolomit von Bleiberg-Kreuth. Berg- und Hüttenm. Mh. 105, 1–11 (1960b)

Schulz, O.: Mechanische Erzanlagerungsgefüge in den Pb-Zn-Lagerstätten Mežica-Mies (Jugoslawien) und Cave del Predil-Raibl (Italien). Berg- und Hüttenm. Mh. 109, 385–389 (1964)

Schulz, O.: Die diskordanten Erzgänge vom „Typus Bleiberg", syndiagenetische Bildungen. Symp. Intern. s. Giac. Miner. d. Alpi 1, 149–162 (1966)

Schulz, O.: Die synsedimentäre Mineralparagenese im Oberen Wettersteinkalk der Pb-Zn-Lagerstätte Bleiberg-Kreuth (Kärnten). Tschermaks Mineral. Petrol. Mitt. 12, 230–289 (1968)

Schulz, O.: Resedimentbreccien und ihre möglichen Zusammenhänge mit Zn-Pb-Konzentrationen in mitteltriadischen Sedimenten der Gailtaler Alpen (Kärnten). Tschermaks Mineral. Petrol. Mitt. 22, 130–157 (1975)

Štrucl, I.: Some ideas on the genesis of the Karavanke lead-zinc ore deposits with special regard to the Mežica ore deposit. Min. Metall. Quart. 2, 25–34 (1966)

Štrucl, I.: Die Zn-Pb-Vererzungen des Grabenreviers – ein besonderer Typ der Lagerstätte von Mežica. Geologija-Razprave in poročila 13, 21–34 (1970)

Štrucl, I.: On the Geology of the Eastern Part of the Northern Karawankes with Special Regard to the Triassic Lead-Zinc-Deposits. Sedimentology of Parts of Central Europe. Guidebook VIII, Intern. Sediment. Congr., 285–301 (1971)

Venerandi, I.: Sulla presenza di jordanite nel giacimento di Raibl. Atti Symp. Intern. Giac. Miner. Alpi, Trento 1, 243–245 (1966)

Zeller, M.: Tektonik, Gebirgsschläge und Vererzung im Blei-Zinkbergbau Raibl (Cave del Predil), Italien. Diss. Freie Univ. Berlin (1970), p. 114

Zimmer, E.: Zur Geochemie des Fluors im Oberen Wettersteinkalk (alpine Mitteltrias) der Ostalpen. Unpubl. Dipl.-Arbeit, Freie Univ. Berlin (1976), p. 78

Zorc, A.: Mining geological features of the Mežica ore deposit. Geologija-Razprave in poročila 3, 24–80 (1955)

Pb-Zn-Ores of the Westcarpathian Triassic and the Distribution of Their Sulphur Isotopes

J. KANTOR, Bratislava

With 5 Figures

Contents

Summary

Bimetallic lead-zinc deposits of the West Carpathians occur in dolomites and limestones of Anisian age. They are mostly confined to its lower part with the exception of Tiba, where the country rocks are probably of Upper Anisian age. At the Poniky deposit, the main mineralization is also located in Lower Anisian dolomites, whereas galena- and sphalerite-bearing ryhthmites are reported from Upper Anisian limestones. No lead-zinc deposits are known from Ladinian or Carnian, which are ore-bearing in the Alps.

Notwithstanding this position of the deposits, epigenetic and replacement textures dominate to such an extent that they are mostly interpreted as of hydrothermal-metasomatic origin and of Cretaceous or Neogene age.

External ore sediments deposited at the sea bottom may be represented by the mineralized dolomite-pellite rhythmites and certain ore types occurring at the Poniky deposit only.

In Anisian dolomites and limestones higher lead-zinc concentrations occur at places even outside the mineralized areas, whereas in Ladinian and Upper Triassic carbonates they are usually present in traces only (*Hanáček*, personal communication).

Evaporitic sedimentation is characteristic of the Scythian and Carnian.

Manifestations of volcanic activity are strongest near the Poniky deposit. In Upper Scythian sediments of the Drienok Nappe, intercalations of tuffaceous beds, as well as lava flows of paleorhyolitic (quartzporphyric) and paleoandezitic character, occur in the layer immediately under the deposit. Smaller extrusions of basalts and picrites were described from Upper Anisian of the Choč Nappe near Poniky. In the vicinity of the Ardovo deposit, tuffs and tuffites of quartz-porphyries and porphyrites occur at the base of the Ladinian.

In the West Carpathians the enclosing rocks of the lead-zinc deposits are prevailingly of Lower Anisian age. In the northern Limestone Alps the ore-bearing units comprise Upper Anisian to Upper Ladinian and in the Southern Limestone Alps mainly Upper Ladinian to Carnian.

The sulphur isotopic composition is relatively homogenous in some deposits. In others — Tiba and Poniky — the variation range is broad and negative $\delta^{34}S$ values prevail. The isotopic pattern indicates activity of sulphate-reducing bacteria. A certain analogy may be found in the deposit Gorno (*Fruth* and *Maucher*, 1966), and in the Binnen-Tal deposits of the Alps (*Graeser*, 1968) and others.

From geological observation and isotopic data, a Mid-Triassic age of emplacement of the ores is very probable.

1. Introduction

In the Westcarpathian system of Czechoslovakia the Mid-Triassic is represented by thick complexes of carbonate sediments.

In contrast to the Paleozoic and crystalline terrains of this mountain chain, which contain a large number of genetic types and many important mineral deposits, the metallogeny of these Mesozoic complexes is rather simple. It consists mainly of lead and zinc mineralizations scattered throughout the area in the form of small indications and a few larger accumulations which have been mined in the past.

Notwithstanding the apparent simplicity in mineralogical composition and geological setting, a great diversity of opinions regarding their genesis has been published. In this respect the genetic problems are the same as for similar lead-zinc deposits of the adjacent Eastern Alps, where different views have been vigorously discussed *(Maucher* and *Schneider,* 1957; *Schneider,* 1964; *Schulz,* 1967; *Holler,* 1971).

Whilst in the Alps the ore-bearing beds are confined to the Upper Anisian, Lower Ladinian, Upper Ladinian and Lower Carnian (*Schneider,* 1957, 1964), the Westcarpathian lead-zinc deposits, Poniky, Ardovo, Tiba, Malužiná and Píla (Fig. 1) are restricted to the Anisian stage only.

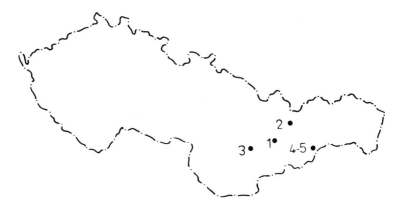

Fig. 1. Location map of deposits: *1:* Poniky; *2:* Malužiná; *3:* Píla; *4:* Ardovo; *5:* Tiba

Recently the author investigated samples of some of the Westcarpathian deposits. The aim of this report is to contribute some observations regarding the lead-zinc ores and the distribution of sulphur isotopes in their sulphides, to genetical problems of these mineralizations in Mid-Triassic carbonate rocks.

2. Poniky Deposit

2.1 General Data

This deposit lies in Central Slovakia, SE of the town Banská Bystrica, in a geologically complicated area.

Publications regarding recent observations on geology and ore deposits have been made mainly by *Losert* (1963, 1965), *Kantor* (1965), *Bystrický* (1966), *Slavkay* (1965, 1971) and others.

Triassic carbonate rocks occur in three tectonic units: the Križná-, the Choč- and the Drienok Nappe.

The lowermost Križná Nappe comprises Mid-Triassic (?) to Lower Cretaceous sediments.

In the overlaying Choč Nappe the Anisian is represented by dark "Guttenstein" limestones and dolomites with submarine effusions of paleobasalts and picrites (*Hovorka* and *Slavkay*, 1966); the uppermost part of the Anisian and the Ladinian by nodular, cherty limestones (Reifling type) and the Upper Triassic by Lunz beds and dolomites.

The Drienok Nappe was overthrust over various members of the lower tectonic units. Its base, the Lower Scythian, is formed by sandstones and variegated shales with intercalations of gypsum and anhydrite. The higher-lying marly schists, marls, limestones, as well as submarine effusive rocks, belong already to the Upper Scythian-Campillian.

Volcanic activity is manifested by a tuffaceous layer at the base, overlain by paleoandesitic to paleotrachytic lava flows and by younger paleorhyolites in the form of effusive bodies, as well as dykes cutting paleoandesites. According to *Slavkay* (1965, 1971) Campilian fauna was found in sediments alternating with the effusive and pyroclastic rocks, and the submarine volcanism is dated as Upper Campillian.

The Mid-Triassic of the Drienok Nappe is represented (*Bystrický*, 1964, 1966) from base to top by:

a) Gray dolomites.

b) Dark limestones of "Guttenstein type". Dolomitization is reportedly not very intensive and gray to dark gray dolomites abundant in "Guttenstein" limestones of other localities are restricted near Poniky to the polymetallic deposit of Drienok.

The dolomites are regarded as lateral, stratigraphic equivalents of the dark limestones and their epigenetic nature related to lead-zinc mineralization is possible (*Bystrický*, 1964, 1966).

The dark limestones are poor in fossils. Dasycladaceae, indicating Pelsonian age, were found in intercalations of light coloured limestones only in their upper part. The dolomites and the lower part of the dark limestones therefore belong to the Hydaspian and the upper part to the Pelsonian substage.

c) Light-coloured, massive limestones. Very rich in Dasycladaceae and other fossils. The facies is very similar to the "Wetterstein Kalk". As they are of Anisian (Pelsonian) age, they correspond to the "Steinalm" limestones of the Alps (*Bystrický*, 1966).

d) Gray ("sugar") dolomites. According to their position they are regarded as stratigraphical equivalents of "Steinalm" limestones.

e) "Reifling" limestones — typical dark, nodular limestones. In the upper parts intercalations of crinoidal limestones. Their appurtenance to the Ilyrian substage follows from their position.

f) "Wetterstein" limestones. Massive, organoclastic with Ladinian Diplopora.

g) The "Wetterstein " limestones are overlain by light dolomites of probably Upper Ladinian to Carnic age.

The immediate surroundings south of the Poniky area built up by thick complexes of Late Tertiary andesitic and subordinate rhyolitic rocks and their pyroclastics.

2.2 Lead-Zinc Mineralization

A dispersed lead-zinc mineralization occurs in Anisian limestones of the Choč Nappe. However, the largest accumulation of lead-zinc ores, the Poniky-Drienok deposit, is confined to the dark dolomites of Lower Anisian (Hydaspian) age of the Drienok Nappe, which occur in association with the dark limestones of the Guttenstein type (Fig. 2).

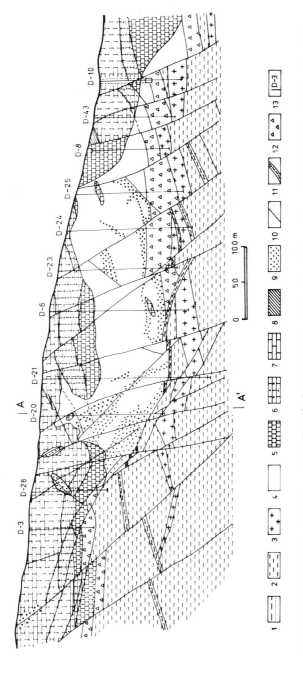

Fig. 2. Poniky-Drienok deposit. Cross-section after *Slavkay* (1971). *1*: schists and sandstones (Lower Scythian); *2*: marly limestones (Upper Scythian); *3*: paleoandesites and paleorhyolites (Upper Scythian); *4*: ore-bearing dolomites; *5*: "Guttenstein" limestones (Hydaspian-Pelsonian); *6*: light limestones with Dasycladaceae (Pelsonian); *7*: nodular and crinoidal limestones (Ilyrian); *8*: light "Wetterstein" limestones (Ladinian); *9*: Pb-Zn mineralization; *10*: tectonic lines; *11*: tectonic lines with mylonitized infilling; *12*: tectonic zone between Lower and Mid-Triassic; *13*: borings

The mineralization forms irregular, lense-shaped bodies located mostly in the lower parts of the dolomites. They consist of galena veins and veinlets with subordinate sphalerite; of dolomite breccias with pyrite, galena, sphalerite coatings and dolomite in the matrix; of small sulphide seams and of very fine impregnations in dolomites.

The ore breccias are cut by narrow veinlets of younger galena, sphalerite and pyrite in dolomite (*Slavkay*, 1971).

Losert (1963) distinguished the following ore types; metasomatic replacement bodies with gray dolomite, galena and minor amounts of other sulphides; gently dipping veins containing white Fe dolomite and some sulphides: typical open-space fillings. Dispersed impregnations in dark dolomites. Dolomite-rhythmites, slightly mineralized.

In the dolomite rhythmites very thin laminae of fine-grained dolomite alternate with laminae formed by clayey, bituminous substances. Graded bedding is accentuated by the distribution of grain size in space. Galena with small amounts of sphalerite and pyrite are also reported as constituents of the rhythmites.

The dolomite rhythmites attain thicknesses of up to 1 m. Their galena content is reported to be highest near the intersection of the rhythmites with the steep dolomite-galena replacement bodies gradually decreases from this boundary and within some 10 m or more a transition of the ore rhythmites into the barren dolomite-pelite rhythmites occurs.

The rhythmites are considered as primary sedimentary fabric with the exception of galena, which is interpreted as a later replacement phenomenon of the originally barren rhythmites (*Losert*, 1963).

Another interesting ore type was found on the dumps of the now-abandoned mines. It is rich in sulphides, especially in pyrite and galena, with lesser amounts of sphalerite, marcasite etc. The irregular sulphide crusts recall at places leaf-like algae (see Fig. 3). This mineralization probably corresponds with certain ore breccias mentioned by *Slavkay* (1971).

The most important minerals of the deposit are galena and sphalerite. Pyrite is widespread and abundant in the breccia ores; marcasite common in small amounts. Chalcopyrite occur sporadically, tennantite, bornite. Tennantite with quite subordinate chalcopyrite forms separate veins cutting the lead-zinc mineralization (*Slavkay*, 1971). The gangue minerals are dark dolomite, white dolomite, Fe dolomites and calcite.

Galena is found in the form of tiny, irregular grains dispersed in dark dolomite, as fillings of pores and vugs lined by dolomite crystals, in the breccia ores, in mineralized dolomite rhythmites, in typical epigenetic veinlets with Fe dolomites lining the walls.

The colloform textures of galena, sphalerite, pyrite and marcasite are well developed in the breccia ores (see Fig. 4a–d).

The following succession for the Drienok deposit is reported by *Losert* (1963):

Metasomatic (pre-ore) dolomites. Coarse-grained Fe dolomites of vein and stockwork type. Pyrite-galena-sphalerite mineralization/impregnation and stockwork-type; tennantite veins.

According to *Slavkay* (1971):

a) Gray, metasomatic (pre-ore) dolomites.

b) White dolomites and Fe dolomites, local calcites (stockworks) with deposition of sulphides in two periods:

1. Pyrite-galena period (succession: dolomite, subordinately quartz, pyrite, marcasite, sphalerite, galena, tetrahedrite, rhodochrosite.

Fig. 3. Pyrite-galena-sphalerite ore ("Algal textures"). δ^{34}S values at sites indicated by *arrows:*
A: − 12.2% pyrite, − 8,6% galena; *B:* − 15.4 py, − 10.8 gn; *C:* − 2.4 py; *D:* − 14.7 py, − 11.4 gn

2. Younger galena period (pyrite, sphalerite, galena, calcite, rhodochrosite, Au).

c) Tennantite period (pyrite, tennantite, chalcopyrite-calcite).

The deposit is considered as hydrothermal-epigenetic, partly metasomatic, genetically related to the Late Tertiary volcanism (*Losert*, 1965; *Slavkay*, 1971) or as possibly Mid-Triassic with later remobilization phenomena (*Kantor*, 1965).

Sulphur isotope investigations revealed a large spread in δ^{34}S values from − 18.1‰ to + 14.7‰ and prevailing enrichment in light sulphur (Fig. 5).

The distribution pattern in one sample of the rich pyrite-galena-sphalerite mineralization displaying algal-like textures is to be seen in Figure 3.

The mineralized dolomite rhythmites occurring in nodular limestones of Upper Anisian-Illyrian age (*Slavkay*, 1971) were not available for isotopic investigations, though their study could shed more light upon the genesis of the ores. They may represent external sedimentation of lead-zinc minerals.

The author and *M. Rybar* (*Kantor*, 1965; *Kantor* and *Rybar*, 1966. in: *Slavkay*, 1971) have studied the isotopic constitution of lead from the deposit. The following ratios were found for galenas:

$Pb^{206/204}$	$Pb^{207/204}$	$Pb^{208/204}$
18.40	15.67	38.57
18.40	15.66	38.51
18.37	15.65	38.51

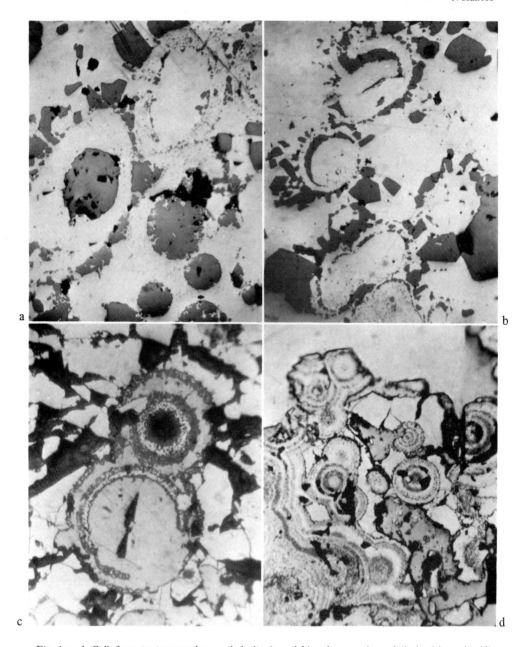

Fig. 4 a–d. Colloform textures: galena, sphalerite (a and b), galena, pyrite, sphalerite (c), pyrite (d)

The sulphur and lead isotopic patterns are contradictory to the ratios observed on hyrothermal deposits of undoubtable Neogene age.

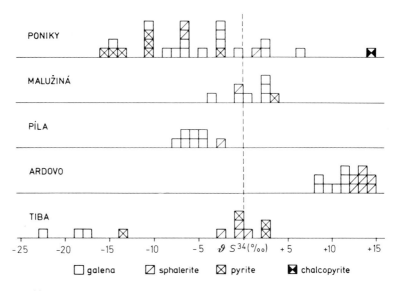

Fig. 5. Histogram of δ^{34}S values

3. Malužiná Deposit

Lead-zinc mineralization occurs near the village of Malužiná, in the Nízke Tatry Mts. Its surroundings are fomed by the Choč Nappe. The Meoszoic of this nappe is charac- terized by an intricate internal structure, according to which three tectonic subunits, slightly differing in lithological content, may be distinguished in a tectonic superposi- tion (*Biely*, 1963).

The lowermost one – the Boca Partial Nappe – comprises Carboniferous to Upper Triassic rocks. The Anisian is ore-bearing, represented by dark limestones ("Guttenstein type"), dark to gray dolomites and in the upper part by local intercalations of light- coloured limestones.

Sulphide mineralization is restricted to the dolomites. Sphalerite and galena with a variable amount of pyrite and subordinate marcasite are irregularly distributed in the form of thin, rapidly wedging out veinlets, and as feeble impregnations.

The distribution of sulphur isotopes is presented in Figure 5.

4. Píla Deposit

The Píla deposit is located on the eastern slopes of the Tribeč Mts in Central Slovakia near the boundary between older geological formations and the area built by mighty volcanic complexes of Neogene age.

The crystalline core is overlain by the mantle series; the higher tectonic units being represented by the Križná- and the Choč Nappes. Tectonic reduction of all the units is considerable at places (*Polák*, 1955; *Kantor* and *Biely*, 1965).

Lead ores occur in the Križná Nappe, which comprises sediments from Anisian to Albian times.

At the base of the Anisian, banked to massive, light-coloured limestones with local intercalations of dolomitic limestones and dolomites are developed. Their appurtenance to the Anisian follows from the position below Ladinain which is evidenced by fossils.

The Ladinian is represented by gray, massive to fine-grained dolomites with pelito-morph, organodetritic or pseudoolitic textures. Upper Triassic and Jurassic sediments are only rudimentarily preserved.

In the deposit, galena occurs near the base of the light Anisian limestones, overthrust here over migmatites along an NW-SE striking and gently NW-dipping overthrust.

Andesite dykes followed this overthrust and occupy at places a position similar to the ores between the crystalline schists and the Križna Nappe. These spatial relations lead some authors (e.g., *Polák*, 1955) to assume genetic links between the andesitic vol-canism and lead-zinc mineralization.

The mineralized zone is intensively dolomitized galena occurring as irregular nests and thin veinlets in the Fe dolomite. In the ancient galleries, rounded pieces of galena weighing up to several 100 kg were mined from a loamy, ochrous substance containing limestone and dolomite fragments.

Among primary minerals, galena is the only important one; sphalerite, pyrite and chalcopyrite occur in insignificant amounts.

Light sulphur is characteristic for all sulphides analyzed. In galena, $\delta^{34}S$ values range from -4.7 to $-7.3‰$ (see Fig. 5).

The isotopic constitution of common lead from the Píla deposit was found to have a considerably more ancient pattern than lead from the Pb-Zn mineralization of Neogene age (*Kantor* and *Biely*, 1965).

Neither sulphur-, nor lead isotopes lend support to the theory that the Píla deposit could be regarded as member of the Late Tertiary metallogeny.

5. Ardovo Deposit

The Ardovo, as well as the Tiba deposits, are localized in the Juhoslovenský Kras Mts (Southern Slovakian Karst) built up prevailingly by Mesozoic carbonate rocks. Their stratigraphical subdivision and geology has been treated by *Bystrický* (1964, 1973).

At the base of the Mid-Triassic, dark limestones and gray dolomites occur. In the upper part of the Guttenstein limestones, intercalations of white organodetritic lime-stones are found. In this ore-bearing complex of Lower Anisian age the dark and white limestones and dolomites intertongue due to different facial development, as well as to tectonic complications.

Higher in the sequence, light-coloured, massive organodetritic ("Steinalm") lime-stones with local lenses of light dolomites occur. They are very similar to Wetterstein limestones, but belong according to Dasycladaceae to the Upper Anisian: Pelsonian to Lower Illyrian (*Bystricky*, 1964).

The Upper Illyrian is developed in the facies of the "Schreyeralm" limestones of the Alps (dark and pink nodular limestones with cherts, organodetritic limestones etc).

In the Ladinian, two facies are distinguished; dark limestones and white "Wetterstein" limestones.

The dark limestones locally contain cherts, but in contrast to the similar "Reifling" limestones they are not nodular. At the base dark schists and green tuffs and tuffites of quartz porphyries to quartz-porphyrites occur at places, attaining thicknesses of up to 10–20 m (*Kuthan*, 1959).

In the "Wetterstein" limestone two facies were distinguished: with Sphinctozoa and with Dasycladaceae (*Bystrický*, 1973).

Data regarding the deposit are scarce. Rich lead-zinc ores are reported to have occurred at shallow depths in a loamy substance filling the open space parallel to the bedding planes of the enclosing dolomites. Impregnation ores are unknown from the Ardovo deposit.

The distribution of sulphur isotopes was studied on samples from the dumps. Results indicating strong enrichment in heavy sulphur are given in Figure 5. Sphalerite displaying always higher $\delta\ ^{34}$S values (by 2.3 to 3.4 ‰ with the average around 3.3‰) than galena.

6. Tiba Deposit

This deposit lies about 4 km SW of Ardovo and was in the past prospected from a shallow pit. The Mid-Triassic outcrops close to the pit consist of Ladinain ("Wetterstein") dolomites, and near Tiba also of dark limestones and dolomites of Hydaspian age, and light-cloloured limestones with intercalations of dolomites of Pelsonian to Lower Illyrian age (*Bystrický*, 1964).

Nodular and breccious limestones of probably Upper Anisian age with intense pyritization and local indications of galena and sphalerite are reported from a drilling near the deposit (*Ivanov*, 1971).

All available samples were dark, bituminous ore breccias with fragments of dolomite, limestone, galena, sphalerite and pyrite in a fine-grained mass of the same minerals.

In contrast to the Ardovo deposit, the galenas are highly enriched with light sulphur (see Fig. 5).

References

Biely, A.: Beitrag zur Kenntnis des inneren Baues der Choč-Einheit. Geol. Práce, Správy (Bratislava), 28, 69–78 (1963)

Bystrický, J.: Stratigraphie und Dasycladaceen des Gebirges Slovenký Kras. Bratislava: Ústredný ústav geologický 1964

Bystrický, J.: La stritigraphie et les Dasycladacées du Trias moyen de la serie du Drienok. Geol. Sbor. Slov. Akad. Vied (Bratislava), XVII, No. 2, 241–257 (1966)

Bystrický, J.: Triassic of the West Carpathian Mts. 10th Cong. Carpathian-Balkan Geol. Assoc. Guide to excursion D, Bratislava, 1973

Fruth, I., Maucher, A.: Spurenelemente und Schwefelisotope in Zinkblenden der Blei-Zink-Lagerstätte Gorno. Mineralium Deposita 1, 238–250 (1966)

Graeser, S.: Lead isotopes and minor elements in galenas and sulphosalts from Binnatal. Earth Planet. Sci. Lett. 4, 384–392 (1968)

Holler, H.: Gedanken zum Stand der Diskussion über das Alter der Pb-Zn-Vererzungen in der alpinen Trias und zur Zielsetzung noch erforderlicher Forschung. Carinthia II, Sonderheft 28, 273–282 (1971)

Hovorka, D., Slavkay, M.: Pikrite aus der Umgebung von Poniky. Geol Práce Správy 39, 41–51 (1966)

Ivanov, M.: Metasomatic ore occurrences in Mesozoic carbonates. Manuscript. D. Štúr Inst. Geol. Bratislava, 1971

Kantor, J.: Thermoluminiscence and lead isotopes at the Poniky lead-zinc deposit in the West Carpathians. Geol. Sbor. Slov. Akad. Vied XVI, No. 1, 211–223 (1965)

Kantor, J., Biely, A.: Ján Nepomucký deposit near Píla and its genesis according to lead isotopes. Geol. práce, Zprávy 37, 101–112 (1965)

Kuthan, M.: Spuren der vulkanischen Tätigkeit in der Mittleren Trias des Slovakischen Karstes.Geol. práce. Zošit 56, 67–74 (1959)

Losert, J.: Aufgabe der selektiven Metasomatose bei der Entstehung pseudosedimentärer Pb-Zn-Erze in der alpidischen Trias (am Beispiel der Lagerstätte Drienok bei Banská Bystrica, Tschechoslowakei), darglegt in: Symp. Prob. Postmagmatic Ore Deposition. Prague: Geol. Surv. Czechoslovakia, 1963, Vol. 1, pp. 572–577

Losert, J.: Ore deposits of the western part of the Ľubietová zone and of the adjacent Subtatricum (Slovakia). Sborník geol. vied ÚÚG, řada LG 6, 40–44 (1965)

Maucher, A., Schneider, H.-J. (eds.): Entstehung von Blei-Zink-Erlagerstätten in Karbonatgesteinen. Berg- u. Hüttenm. Mh. 9 (1957)

Polák, S.: Metasomatische Galenitvorkommen in mitteltriassischen Kalksteinen bei Píla, Nová Baňa. Geol. Práce Správy 2, 53–61 (1955)

Schneider, H.-J.: Diskussionen in Entstehung von Blei-Zink-Erzlagerstätten in Karbonatgesteinen. Berg- u. Hüttenm. Mh. 9, 238–240, 242–244 (1957)

Schneider, H.-J.: Facies differentiation and controlling factors for the depositional lead-zinc concentration in the Ladinian geosyncline of the Eastern Alps. In: Developments in Sedimentology. Amstutz, G.C. (ed.). Amsterdam: Elsevier 1964, Vol. 2, pp. 29–45

Schulz, O.: Die synsedimentäre Mineralparagenese im oberen Wettersteinkalk der Pb-Zn-Lagerstätte Bleiberg-Kreuth (Kärnten). Tschermaks Mineral. Petrogr. Mitt., 3rd Ser., 12, 230–289 (1967)

Slavkay, M.: Vulkanogénne horniny mezozoika na okolí Poník. Časopis pro mineral. a geol., roč. 10, c. 3 (1965)

Slavkay, M.: Polymetallic ore deposits near Ponika. Mineralia Slovaca III, 1, 209–213 (1971)

Strata-Bound Kies-Ore Deposits in Ophiolitic Rocks of the "Tauernfenster" (Eastern Alps, Austria/Italy)

K. DERKMANN, Clausthal-Zellerfeld, and D.D. KLEMM, München

Contents

Summary

The geology, petrography and geochemistry of the kies mineralization in the greenstones of the "Obere Tauern-Schieferhülle" (Austria and Italy) are described. The "Glockner facies" is regarded as a separate part of the Obere Schieferhülle; it consists of a sequence of serpentinite, greenstone, calcareous-mica-schist and black phyllite. The kies mineralizations are classified as a part of the ophiolite sequence. They are confined to the geological hanging wall of the meta-basitic sequence. Strata-bound, they can be traced horizontally with a sedimentary cover as leading level. The ore paragenesis consists mainly of pyrite, subordinate chalcopyrite and locally a little magnetite and pyrrhotite. Magnetite and pyrrhotite are secondary metamorphic to postkinematic mineralizations, which can be derived from displacement of pyrite. Deformation fabrics and, above all, prevailing recrystallization are common.

Results of petrographical and geochemical analyses support a division into two types of ore deposits:

1. North of the main Alpine ridge (Hauptkamm), the ore mineralization consists of basal, quartzitic, or massive ore seams and disseminated pyritic ores in thin-bedded chloritic-quartzitic composition (Feingebänderter Grünschiefer); because of geochemical criteria it might be regarded as a former volcano-sedimentary, tuffitic rock.

2. On the southern side of the Hohe Tauern, prasinitic rocks (as defined by *Niggli*, 1912) dominate. The geochemical data, i.e., extremely low K_2O content and small standard deviations of all the other elements, as well as highly corroded blast albites, indicate a porphyric basalt with tholeiite-basaltic composition, as educt. Kies mineralizations are found exclusively disseminated in prasinitic, highly chloritized gangue material

For rhythmic variations of Cr, Ni and Zn in foot wall, a connection with "eruptive phases" is held possible. The high Cr values in the disseminated ores (> 300 ppm in the ores, < 300 ppm in the wall rock) are considered to be a sign of ascendant, synsedimentary ore mineralization. The lack of Mn- and silica-enriched layers excludes a postvolcanic ore mineralization. The differentiation in K_2O-rich, thinly laminated greenschists in the northern "Tauernfenster" and in K_2O-poor prasinites in the south, permits a new classification of the eu-geosynclinal penninic trough in an axial (prasinites) and a marginal (Feingebänderte Grünschiefer) zone of formation. These results contrast with the classical classification, which regard an undifferentiated "Obere Schieferhülle", symmetrically arranged around the "Zentralgneis" cores as given.

1. Introduction

Until the late sixties, various theories existed on the genesis and age of the ore deposits in the Alpine region. Especially *Friedrich* (1936–1968) defended the theory of an epigenetic alpidic ore mineralization regarding the kies-ore deposits in meta-basites of the "Obere Tauern Schieferhülle"; this is a theory in contrast to *Hegemann* (1942) who postulates a syngenetic mineralization. Only *Schmid* (1973) could prove a synsedimentary ore mineralization for Prettau, Italy, with the aid of geochemical whole-rock analysis.

The research based on *Schmid* (1973) was carried out on ore mineralizations in comparable rock series (mesozoic tholeiit-basaltic meta-tuffites + meta-basalts) in the whole Tauern window. All deposits are no longer being worked and most have been laid still since the 19th century (Prettau since 1972).

2. Geological Environment

The Tauern window (tectonic window) consists of the "Zentralgneis" (allochthonous, *Tollmann*, 1975) as the lowermost series, and the schistose cover rocks which are subdivided into separate nappes, the lower (autochthonous to para-autochthonous in regard to the Zentralgneis), and the upper cover (allochthonous). The disseminated pyrite deposits studied belong to the Mesozoic upper cover and lie on the top in the Tauern window, being the youngest nappe.

The "Obere Schieferhülle" also called Glockner Nappe or "Glockner facies" (*Frisch*, 1974, 1976; *Tollmann*, 1975) is characterized by Trias "chips" with lime-free to lime-poor sediments (lower Jurassic, *Frisch*, 1976) or larger serpentinite bodies on top. A crystalline basement is lacking.

The sedimentation of black-shale facies (black phyllites) and the beginning of the "Bündnerschiefer development" (schistes lustrés) is due to the opening of the eugeosynclinal penninic trough.

During Jurassic times lime-rich sediments were deposited and tholeiitic-basaltic tuffs and basalts (greenschists, prasinites) extruded. The uppermost sediments of the "Obere Schieferhülle" consist of lime-free to lime-poor rock varieties.

3. Geology and Mineralogy of the Kies Deposits

The kies deposits within the "Obere Schieferhülle" are concentrated in the area north and south of the main Alpine ridge. According to this geographic classification, a division of the associated greenstone types (meta-tuffites, meta-basalts) is possible, as is also shown in the different structures of the ore mineralization.

3.1 Kies Deposits in the Northern Tauern Window

The most important deposits are in the upper Großarltal (near Hüttschlag); further deposits are in the Gastein-, Rauris- and Fuscher Valleys.

The sequence of the lime-rich, ophiolite-bearing cycle of the "Obere Schieferhülle" (maximum 1000 m) shows thin serpentinite lenses which are bedded within thinly banded greenschists (meta-tuffites) at the base. Thin calcareous mica schists are first intercalated towards the geological hanging wall in the thinly banded greenschists, later their thickness increases; the thickness of the greenschists, however, decreases.

The ore-bearing horizon, which is always within the finely banded greenschist, forms the boundary horizon of the reversing facies thicknesses. In the field it can be used as a perfect marker horizon, since the mineralization is strata-bound on the one hand and concordant on the other.

The ore horizon is always covered by calcareous mica schists with thin black phyllite intercalations at a distance of 2–5 m.

The finely banded greenschist is characterized by an interbedding of fine-grained (< 100 μm $\emptyset$ of single grains), partly monomineralic albite- and chlorite-muscovite-amphibole-epidote layers. Titanite is an important accessory mineral. This mineral paragenesis is typical for the greenschist facies.

This changing fine layering is regarded as primary volcano-sedimentary fine lamellation.

Kies-ore mineralization begins with a quartzitic-pyrite-(magnetite)-rich ore layer which is well-defined on both sides. Considerable variations of thickness (5–80 cm) are characteristic.

The few inclusions in pyrite and the mosaic texture indicate high-grade recrystallization (*Anger*, 1971). In the areas with "purified" pyrites (pyrite generation II) are pyrite blasts, especially in massive ore parts, which have silica-, sphalerite- or chalcopyrite inclusions and are regarded as pyrite generation I.

A new formation of magnetite which locally leads to complete "magnetitation" can be observed starting along grain margins and fissures.

Here relics of magnetite grains can be found, around which an idiomorphic zoning of the magnetite is obvious, and thus are due to a pseudomorphous replacement of pyrite after (100) or (210), due to oxidation differences in magnetite as a postkinematic event.

Pyrrhotite exists locally, without any indication of high-grade metamorphism. A context with a more intensive folding is rather seen as the only possible explanation (existence of "Böhm'sche Streifung" in quartz).

In the hanging wall a 0.5–2 m thick zone of kies impregnation in the quartzitic-chloritic matrix usually follows. Here, a distinct connection of the ore minerals (mainly pyrite next to chalcopyrite) with the chlorite phase is noticed.

The gangue material of the kies-impregnation type can be directly compared to the wall rock (finely banded greenschist) in texture.

With increasing pyrite content (variations of 2–15 vol% pyrite) an increasing quartz content is to be noticed; an increase in grain size of the gangue mineral chlorite is also observed, increased mobility and recrystallization being analogous to increased ore contents. Similar observations were made by *Bachinski* (1976), who maintains that the chlorites act as regulators for the Fe/Mg content during increasing metamorphism.

The sequence rich-ore layer/kies-impregnation zone can be repeated up to three times, barren greenstone being always intercalated.

3.2 Kies-Ore Deposits in the Southern Tauern Window

The most important deposits are in the upper Virgen-valley and Prettau/Ahrntal, South Tyrol The sequence of the rock series conforms to the north side of the main Alpine ridge, only the serpentinites at the basis of the nappe "Obere Schieferhülle" are thicker (e.g., "Heiligenblüter serpentinite complex").

The kies-ore mineralizations are also connected to the hanging rocks of the ophiolite series and lie underneath the sedimentary covering. With the aid of the ore-bearing horizon as marker it was here possible to prove an imbricate structure, already noticed by *Schmidegg* (1961).

The prasinitic wall rock, which is contrary to the finely banded greenschists because of poikiloblastic albites in a medium- to coarse-fibred, chloritic to amphibolitic matrix, is characteristic for the kies-ore deposits in the southern Tauern window.

The ore mineralization consists of a pyrite impregnation zone only. Chlorite is also the most important gangue mineral. This type of ore lacks finely banded gangue material; it is prasinitic and thus of irregular texture.

Muscovite content is lacking, in contrast to the finely banded greenschists. Epidote and titanite are important accessory constituents, however. A strata-bound and concordant ore mineralization can only be recognized macroscopically.

4. Geochemistry of the Ores

The analyses were made with XRF-analysis and AAS. Due to the immense data material (ca. 7500 single values), univariant and multivariant statistical methods were used. Comparisons of the impregnation ores in finely banded and prasinitic gangue material with factor analysis (main-component analysis) indicate equal characteristics. The distinct linkage of Co to Fe_2O_3 which is interpreted as ore mineralization elements is remarkable.

The gangue material-specific elements MgO, Al_2O_3, TiO_2 and Zr are concentrated with reversed sign. The association of Cr, Cu, and P_2O_5 with the ore indicators Co + Fe_2O_3 is of less importance. The dominant part of Co within the ore mineralization is especially stressed in the rich ores. Co-content can reach a maximum of 950 ppm, whereas it reaches an average of 73 ppm in the impregnation ores and only 59 ppm in the finely banded wall rock. The Co/Ni ratio of 0.78 in the finely banded impregnation ores clearly resembles the quotient of 0.62 in the finely banded wall rock. Co values of 98 ppm in the impregnation ore and 49 ppm in the prasinites should be mentioned for the southern deposits.

Characteristic maximum values in the mentioned impregnation ore types also exist for Cr. Cr and Co, but also Cu and Zn can be regarded as the most importante ore-indicating trace elements.

Discriminant analysis (R-mode), together with a new coordination of the given groups (impregnation ores in finely banded and prasinitic gangue material) shows a clear overlapping of the two types of ore. A large similarity in ore mineralization and its chemism is thus proven.

In both types of impregnation ore, a distinct impoverishment of CaO, Na_2O, Sr, Rb and Zr in respect to the wall rock is obvious. The degree of impoverishment is especially evident in the impregnation ores in finely banded gangue material. The CaO content decreases to 0.1 wt %, whereas the content in the finely banded greenschists reaches 10.2 wt %.

The CaO content in the ores linked to prasinitic gangue material decreases about 50% (wall rock 9.1%, ore 4.4% CaO). *Schmid* (1973) attributes this impoverishment to a secondary solution and transport of the easily soluble parts during ore feeding.

5. Geochemistry of the Greenstones

The already mentioned petrographically caused division into finely banded greenschists and prasinites is also shown geochemically. Whereas the finely banded greenschists are remarkable in the relative frequency distribution of main trace elements with broad scattering, the data of prasinites show a more restricted area. The values of standard deviation and variancy make this especially evident (Table 1). Very significant for the geochemical classification of both types of greenstones are the elements K_2O, Rb and Sr. Discriminant analyses shows differing element associations. Accordingly, the variants ignition lost, TiO_2, SiO_2, Al_2O_3 and MnO (as evaluated factors) contribute most to group division.

The extremely low K_2O and Rb contents, as well as the small range of variation in all elements in the prasinites is regarded as typical for magmatic rocks. The whole chemism corresponds to a tholeiitic basalt.

The broad variations, together with the petrographic and structural characteristics of the finely banded greenschists, indicate a primary volcano-sedimentary hybrid rock (tuffite). The bulk chemism also shows tholeiitic composition and is close to that of the prasinites.

6. Genetic Interpretation of the Greenstones

The first genetic interpretations of greenstones of the Glockner facies with the aid of geochemical data were made by *Bickle* and *Pearce* (1975).

Because of the rock sequence characteristic of ophiolites (definition after *Steinmann*, 1926; *Dickey*, 1975), *Frisch* (1974) and *Tollmann* (1975) speak of oceanic crust parts

Table 1.Average compositions for pyritic ores and their wall rocks

	quartzitic ore	disseminated pyrite ores		wall rocks	
		in thin–bedded gangue	in prasinitic material	thin–bedded greenschists	prasinites
SiO_2	44.3	47.3	49.7	48.6	49.6
Al_2O_3	6.3	10.9	11.8	15.5	13.9
TiO_2	.69	1.6	1.31	1.7	1.39
Fe_2O_3	42.1	25.2	20.1	10.8	10.8
MnO	.03	.12	.18	.17	.19
MgO	3.1	9.1	9.2	8.3	10.9
CaO	.5	1.5	4.4	10.2	9.1
Na_2O	.96	1.7	1.7	2.7	2.6
K_2O	.58	.92	.37	1.1	.32
P_2O_5	.17	.18	.31	.34	.24
Total	98.73	98.52	99.07	99.41	99.04
Cr	279	317	346	315	260
Co	390	73	98	59	49
Ni	57	94	107	96	83
Cu	12900	7725	2275	96	218
Pb	17	24	18	19	11
Zn	2325	1225	525	265	102
Sr	12	28	63	199	111
Zr	41	79	72	125	84

in the "Obere Schieferhülle". The interpretations of *Bickle* and *Pearce* (1975) made it possible to classify them for the first time as tholeiitic ocean-floor basalts.

Our own geochemical data were interpreted according to methods evolved by *Cann* (1970, 1971), later further developed by *Pearce* and *Cann* (1971, 1973), *Pearce* et al.

(1975), which are criteria for classification of basalts. In the variation diagram Ti-Zr-Y and Ti/100-Zr-Sr the single samples fall into the field of ocean-floor basalts. The high contents of sediments in the finely banded greenschists, and accordingly the postulation of a volcano-sedimentary hybrid rock, lead to the broad scattering in the variation diagram TiO_2-K_2O-P_2O_5. Equal results were gained by *Wagner* (1976) for the Pfitschtal.

7. Contributions to the Genesis of Kies-Ore Mineralization

According to the two-fold division of the wall rocks, a subdivision into kies-impregnation ores of finely banded gangue material, together with quartzitic rock-ore layers and kies ores in gangue material, could be made. The kies-ore mineralization which underwent the same metamorphism and tectonics as the wall rock and indicates, together with stratigraphic and petrographic results, a syngenetic ore mineralization (definition after *Anger*, 1971), as *Hegemann* (1942, 1956, 1958) and *Schmid* (1973) have already pointed out.

A possible answer to a classification — synsedimentary or syndiagenetic — of the feeding of material is seen in the behaviour of the trace elements Cr, Ni, Zn, Cu and Co.

Rhithmically repeating Cr maxima could be found in the footwall of the ore mineralization, as well as in the impregnation ores. Since comparative data from the literature are lacking, Cr can at the present only be regarded hypothetically as an indicator for volcanic "eruptive phases". Another proof for a direct connection between ore feeding and volcanic activity is the classification of ore mineralization by a quartzitic rich-ore layer and the lack of silica-rich, Mn-enriched horizons in the hanging wall of the kies mineralization. The latter characteristic, according to *Bonatti* et al. (1976) equals a syndiagenetic ore mineralization.

According to *Johnson* (1972), the Co content concentrated in the quartzitic rich ores can be used as indicator for the beginning of ore feeding. The solutions enriched in heavy metals are thus interpreted as ascendant and originate from hydrothermal, volcanic residual solutions of basaltic magmas. A synchronous deposition of the heavy metal solutions and gangue material can be supposed for ore mineralization in finely banded greenschists. The high impoverishments of Ca, Rb and Sr indicate near-surface precipitation and sedimentation.

According to the distribution of trace elements, as well as the structure of the ore mineralization, an abrupt and rhythmically repeating feeding is excluded for kies mineralization in prasinitic gangue material (southern side of the Tauern window).

Rather, a continuous deposition, in situ, but within the framework of a ceasing volcanism, is supposed.

8. Comparison with Other Kies-Ore Deposits

The connection between ore mineralization and pillow lavas and overlying silica-rich, Mn-enriched horizons is cited for the most important representatives of this group of ore deposits (Newfoundland, Skandinavia, USSR, Cyprus, Balkan Peninsula and Asia Minor; *Anger*, 1966; *Sillitoe*, 1972; *Karamata*, 1973).

The always strictly concordant ore mineralization in the "Obere Schieferhülle", as well as petrographical and partly geochemical results (finely banded greenschists), contradicts this.

Parallels are often drawn to the recent ore-deposit formation in the Red Sea (*Duke* and *Hutchinson*, 1973; *Evans*, 1975), but this comparison is not valid according to the latest research on Red Sea ore sludges by *Weber-Diefenbach* (1976), since these have to be interpreted as oxidic Fe ore deposits.

In summary, it can be said that a direct comparison with other kies-ore deposits linked to tholeiitic-basaltic rocks is not feasible. Geotectonic connections on a large scale have not yet been solved, although first models have been introduced with the help of plate tectonics (*Dietrich* and *Franz*, 1976; *Frisch*, 1976; *Lammerer* et al., 1976).

Acknowledgments. The authors are indebted to the Preussag AG for financial field support, and we also express our cordial thanks to Miss *Margret Viernstein*, who kindly translated the article into English.

References

Anger, G.: Die genetischen Zusammenhänge zwischen deutschen und norwegischen Schwefelkies-Lagerstätten unter besonderer Berücksichtigung der Ergebnisse von Schwefelisotopen-Untersuchungen. Clausth. H. Lagerst. Geoch. Mineral. Rohst. 3, 115 (1966)

Anger, G.: Microfabrics in geosynclinal sulfide deposits. Clausth. H. Lagerst. Geoch. Mineral. Rohst. 10, 42 (1971)

Bachinski, D.J.: Metamorphism of cupriferous iron sulfide deposits, Notre Dame Bay, Newfoundland. Econ. Geol. 71, 443–452 (1976)

Bickle, M.J., Pearce, J.A.: Oceanic mafic rocks in the Eastern Alps. Contr. Mineral. Petrol. 49, 177–189 (1975)

Bonatti, E., Zerbi, M., Kay, R., Rydell, H.: Metalliferous deposits from the Apennine ophiolites: Mesozoic equivalents of modern deposits from oceanic spreading centers. Geol. Soc. Am. Bull. 87, 83–94 (1976)

Cann, J.R.: Rb, Sr, Y, Zr and Nb in some ocean floor basaltic rocks. Earth Planet. Sci. Lett. 10, 7–11 (1970)

Cann, J.R.: Major element variations in ocean-floor basalts. Phil. Trans. Roy. Soc. London A268, 495–505 (1971)

Dickey, J.S.: A hypothesis of origin for podiform chromite deposits. Geochim. Cosmochim. Acta 39, 1061–1074 (1975)

Dietrich, V.J., Franz, U.: Alpidische Gebirgsbildung in den Ostalpen: ein plattentektonisches Modell (Kurzfassung). Geol. Rdsch. 65, 361–374 (1976)

Duke, N.A., Hutchinson, R.W.: Geological relationships between massive sulfide bodies and ophiolitic rocks near York Harbour, Newfoundland. Can. J. Earth Sci. 11, 53–69 (1974)

Evans, A.M.: Mineralization in geosynclines – the alpine enigma. Mineralium Deposita 10, 254–260 (1975)

Friedrich, O.: Zur Geologie der Kieslager des Großarltales. Sitzber. Akad. Wiss. Wien, math.-naturwiss. Kl. 145, Abt. 1, 121–152 (1936)

Friedrich, O.M.: Die Vererzung der Ostalpen, gesehen als Glied des Gebirgsbaues. Arch. Lagerst. Forschg. Ostalpen 8, 136, Leoben (1968)

Frisch, W.: Die stratigraphisch-tektonische Gliederung der Schieferhülle und die Entwicklung des penninischen Raumes im westlichen Tauernfenster (Gebiet Brenner–Gerlospass). Mitt. Geol. Ges. Wien 66/67, 9–20 (1974)

Frisch, W.: Ein Modell zur alpidischen Evolution und Orogenese des Tauernfensters. Geol. Rdsch. 65, 375–393 (1976)

Hegemann, F.: Die geochemische Bedeutung von Kobalt und Nickel in Pyrit. Z. angew. Mineral. 4, 121–239 (1942/43)

Hegemann, F.: Über extrusiv-sedimentäre Lagerstätten in den Ostalpen. Erzmetall 9, 305 (1956)

Hegemann, F.: Geochemische Untersuchungen über die Bildungsweise einiger ostalpiner Erzlagerstätten. Tschermaks Mineral. Petrogr. Mitt. 4, 432–438 (1958)

Johnson, A.E.: Origin of Cyprus pyrite deposits. 24th Intern. Geol. Congr., Sect. 4, 291–298, Montreal (1972)

Karamata, S.: Beziehungen zwischen den metallogenetischen, petrographischen und geochemischen Provinzen der Balkanhalbinsel und Kleinasien. Österr. Akad. Wiss., Erdwiss. Kommission 1, 106–119 (1973)

Lammerer, B., Fruth, I., Klemm, D.D., Prosser, E., Weber-Diefenbach, K.: Geologische und geochemische Untersuchungen im Zentralgneis und in der Greiner Schiefer-Serie (Zillertaler Alpen, Tirol). Geol. Rdsch. 65, 436–459 (1976)

Niggli, P.: Die Chloritoidschiefer und die sedimentäre Zone am Nordostrande des Gotthardmassivs. Beitr. Geol. Karte Schweiz, u.F. 36 (1912)

Pearce, J.A., Cann, J.R.: Ophiolite origin investigated by discriminant analysis using Ti, Zr and Y. Earth Planet. Sci. Lett. 12, 339–349 (1971)

Pearce, J.A., Cann, J.R.: Tectonic setting of basic volcanic rocks determined using trace element analyses. Earth Planet. Sci. Lett. 19, 290–300 (1973)

Pearce, J.A., Gormann, B.E., Birkett, T.C.: The TiO_2-K_2O-P_2O_5-diagram: A method of discriminating between oceanic and non-oceanic basalts. Earth Planet. Sci. Lett. 24, 419–426 (1975)

Schmid, H.: Geochemisch-lagerstättenkundliche Untersuchungen im Bereich der Kieslagerstätte Prettau, Ahrntal. Unveröff. Diss. Univ. München, 1973

Schmidegg, O.: Geologische Übersicht der Venedigergruppe nach dem derzeitigen Stand der Aufnahmen von F. Karl und O. Schmidegg. Verh. Geol. B.-A. Wien 1961, 34–54 (1961)

Sillitoe, R.H.: Formation of certain massive sulphide deposits at sites of sea-floor spreading. Trans. Inst. Min. Metall. Sci. 81, B141–B148 (1972)

Steinmann, G.: Die Ophiolithischen Zonen in den mediterranen Kettengebirgen. 14. Int. Geol. Congr. Fasc. 2, 637–668, Madrid (1927)

Tollmann, A.: Ozeanische Kruste im Pennin des Tauernfensters und die Neugliederung des Deckenbaues der Hohen Tauern. N. Jb. Geol. Paläont. 148, 286–319 (1975)

Weber-Diefenbach, K.: Geochemische Untersuchungen an Erzschlämmen des Atlantis-II-Tiefs (Rotes Meer). Unveröff. Habil.-Arb., Univ. München, 1976

Ore/Mineral Deposits Related to the Mesozoic Ophiolites in Italy

P. ZUFFARDI, Milano

With 1 Figure

Contents

Summary

A number of ore/mineral deposits/occurrences are contained in the Italian Mesozoic ophiolites; some of them show evidence of being syngenetic or in tight time- and space correlations with specific sections of the ophiolitic sequence.

Examples of this situation are: Cr and Ni occurrences in serpentinized ultra-maphites; Ti in cumulitic gabbroes; Fe-Cu-quartz veinlets in sheeted gabbroes; Fe-Cu-(Zn)-[(Au)] related to spilitic pillow-lavas; Fe-Cu-(Zn-Au)-[(Pb-Mo-C)] related to the "calc schists with greenstones series"; Mn in jaspers.

Other deposits/occurrences appear to be strata-bound but their time relation to the country rock is uncertain.

The Fe, asbestos, talc, magnesite deposits and the ophiolites held in serpentinites, mainly of the basal section. are examples.

Other deposits/occurrences show no evident bond to specific sections of the ophiolitic sequence: this is the case of the vein deposits of Au and of Ni-Co arsenides.

1. Introduction

Ore/mineral deposits related to ophiolitic belts may be considered as a particular type of time- and strata-bound accumulation.

There is, nowadays, a fairly general agreement among researchers on the composition of ophiolitic complexes and on the genesis and emplacement of their various sections.

One may say that a normal ophiolitic complex includes four sections, namely (from base to top):(1) ultra-maphites with tectonic structures; (2) gabbroes with cumulitic structures; (3) maphic dykes section; (4) maphic volcanics with pillow lavas structure (*Steinmann*, 1927).

Marine sediments, often including jaspers at the base and/or in the basal section, cover the ophiolitic complex.

The formation and emplacement of the four sections are thought to have been different, namely: crystallization in depth for both (1) and (2) and, vice versa, crystallization in subvolcanic/volcanic submarine environment for (3) and (4); emplacement at the solid (plastic) state for (1) and emplacement at the fluid state for the others.

The time of crystallization of (1) is consequently more ancient than the time of its emplacement; however the emplacement of the whole complex [and probably also the crystallization of (1)] is considered to have been controlled by a continuous series of geodynamic events, which lasted a relatively short time interval; about 100 my (from Lias to lower Cretaceous) in the case of Italian ophiolites. In this sense, the formation of an ophiolitic complex may be considered "time-bound".

On the other hand, each section of an ophiolitic complex is characterized by specific mineralizations which differ in the different sections, either in the location, or their parageneses and/or of their shapes and consistencies. In this sense, ore/mineral deposits related to an ophiolitic complex may be considered "strata-bound".

2. Geographical Distribution and Characterization of Italian Ophiolites

The Italian ophiolites are good examples of the above-mentioned scheme, as will be discussed here. The purpose of this paper is to present some factual observations supporting this statement. Genetic and geodynamic problems will not be discussed: they are dealt with in a number of published and to-be-published papers (see References).

The Italian ophiolites can be subdivided into two groups on the basis of their geographical locations, and — at the same time — of their lithological characteristics, and of the associated ore/mineral deposits; namely: (1) the Appenninic [1] type, (2) the Alpine [1] type.

The Appenninic-type ophiolites occur in Eastern Liguria, in Emilia, and in Tuscany; their main characteristics are: (a) absence of orogenic metamorphism; (b) severe plastic tectonics with, at places, formation of complicated reverse folds, and in other places, of olistostromes; (c) presence of economic cupriferous-pyritiferous deposits related to the volcanic section, and of manganesiferous deposits in the jaspers at the base of the overlying sediments.

Alpine-type ophiolites occur along the Alpine arc (particularly in the Western Alps), Western Liguria, Calabria; their main characteristics are: (a) intense orogenic metamorphism, (b) nappe structure, (c) presence of economic cupriferous-pyritiferous deposits related to meta-volcanics interbedded in the meta-sediments, overlying and/or heteropic to the ophiolitic complex; (d) the presence of economic ferriferous ores and amiantiferous concentrations related to serpentinized meta-ultra maphites of the basal section.

---

[1] "Appenninic" and "Alpine" are used to indicate localities, without any implication of Alpine-Appenninic orogenesis and magmatism.

3. The Appenninic Type

3.1 Geological Setting

The generalized lithological sequence in the Appenninic type (from base to top) as it can be established in Eastern Liguria [in the so-called "inner Ligurian zone", see *Decandia* and *Elter* (1972); *Galli* et al. (1972)], is the following:

1. Serpentinites with occasional remnants of lherzolites, (harzburgites, dunites) with tectonitic structure; their stratigraphic base is unknown.

2. Cumulitic and tectonitic gabbroes; they are absent in a number of sections.

3. (Occasional) ophicalcites, normally associated with serpentinites.

4. (Occasional) diabases, either massive and/or as dykes complex; volcanic breccias are at places interbedded and/or included in them.

5. Spilitic diabases with pillow lavas structure.

6. Sedimentary cover, normally beginning with various-shaded reddish jaspers, continuing with gray-greenish jaspers. Thin, discontinuous, pyroclastic beds with graded structure occur close to the diabasic section; thin, continuous beds of chloritic slates occur along the jasper section.

It should be pointed out that the exact age of zircons included in acidic differentiations of the gabbroes is Lias-Dogger, according to *Bigazzi* et al. (1971).

The jasper horizon is covered by the "Calpionella limestone" (also called "majolica" on the basis of its appearance), or, less frequently, by the "Palombini (= small doves) clays", which are a complex of gray-bluish slates including small lenses of grayish limestones.

The age of the transition zone between jaspers and the overlying sediments, on the basis of the fossils, is Berriasian-Valanginian (*Decandia* and *Elter*, 1972).

The arrangement of the ophiolitic complexes in Emilia and in Tuscany (that is in the so-called "outer Ligurian zone") is controlled by strong gravitative tectonics, so that they appear as olistolites embedded in clayey materials; each olistolite is normally made up of one lithotype: serpentinites and weathered pillow lavas are most frequent.

3.2 Associated Mineralization

The following mineralizations are related to the various sections of the Appenninic ophiolites:

1. Small, uneconomic occurrences of Cr are known in the basal section: Ziona and Canegreca are the only two localities where some mining activity was carried out in the past. Ni is present at the average grade of 0.18–0.20%, as in practically all ultramaphites of the Italian ophiolites.

2. Small, uneconomic Fe-Cu quartz veinlets are at places related to the gabbroidic section. Occasional gravitative accumulations of ilmenite and magnetite are also known in cumulitic gabbroes. The recently discovered rutile accumulation in eclogites of Val di Vara has to be considered as a metamorphic product of similar concentrations.

3. A number of generally small-tonnage Fe-Cu-(Zn) stratiform or strata-bound deposits, often with traces of Au, occur in the upper part of the spilitic pillow lavas

section, mostly along the boundary between them and the overlying sediments. The most common structure of these deposits is massive, up to 4–5 m thick strata of pyrite with 1–2% Cu and variable quantities of Zn (at places up to 3–4%). This structure is dominant when the location of the deposit is along the above-mentioned boundary. Closing in depth, cone-like, or wedge-like massive occurrences are not infrequent, along this same horizon.

Disseminated grains and/or stockwork structures prevail when the ore deposit is located inside the spilitic pillow lavas and/or around the massive accumulations.

A few displaced ore bodies originated by supergene remobilization under particular tectonic conditions, are at places present; they are composed either of oxidized or of sulphidic ores; of course they deviate from time- and strata-bound scheme.

Libiola in Eastern Liguria is the locality type as far as the "inner Ligurian zone" is concerned, whilst Montecatini in the Cecina Valley is the locality type of the "outer Ligurian zone", and is also the main example of supergenically enriched deposits.

4. Stratiform Mn ore deposits, with local deformations and enrichments by tectonic boudinage, are related to the lower part of the jasper section: Gambatesa in Eastern Liguria is the locality type.

5. Valuable ornamental stones are quarried in the ophicalcite section, and are commercially known under the name "rosso Levanto" (= Levanto Red) from their locality type (Eastern Liguria), and their dominant colour.

4. The Alpine Type

4.1 Geological Setting

The main character of the Alpine ophiolites is their orogenic metamorphism, so that it would be more appropriate to define them as "meta-ophiolites". The normal meta-ophiolitic sequence is fundamentally similar to the Appenninic type, including [from base to top: see *Brigo* et al. (1976), *Dal Piaz* (1971)] the following sections:

1. Largely or wholly serpentinized meta-ultra-maphites with tectonic structure; their stratigraphic base is unknown.

2. Meta-gabbroes with cumulitic structure.

3. Massive and vein-like meta-diabases.

4. Meta-diabases with pillow lavas structure. The thickness and horizontal development of this sequence are far less than in the Appenninic type.

5. Sedimentary cover, beginning with manganesiferous quartzites and continuing with marbles.

The age of the cover is estimated to be lower-middle Cretaceous on the basis of probable correlations to fossiliferous formations. No absolute age determination has been made on the ophiolites sequence.

The most important character of Alpine ophiolites, which makes them really different from the Appenninic ones, is the fact that the above-mentioned "normal sequence" and a part of its sedimentary cover are heteropically replaced, in large portions of the Alps, by a volcano-sedimentary sequence (called the "calc schists with greenstones series", which is equivalent to the "Schistes lustrés" and to the "Grünschiefer"), made of repeated beds/lenses of meta-volcanics (prasinites, eclogites), meta-jaloclastites, meta-tuffites (transformed in chloritoschists), calc schists, phyllites, mica schists, and occasional marbles.

Meta-gabbroidic blocks (olistolites?) are scattered along the whole series.

The stratigraphic base of this sequence, where recognizable, is made of marbles, calc schists, calc-mica schists, and intraformational breccias. The age of the upper part of this formation is Middle-Lower Lias.

4.2 Associated Mineralization

The following mineralizations are related to the various sections of the Alpine ophiolites:

1. The serpentinites of the basal section hold concentrations of magnetite (Cogne being the locality type), of asbestos (Balangero), and (with far less economic interest) of talc and magnesite.

Low-grade disseminations of Ni (0.17–0.18%) and veinlets of Ni-Co arsenides are also contained in this same section.

2. Stratiform accumulations of Fe-Cu-(Zn) with occasional traces of Au, are related to the spilitic pillow lavas: they are absolutely similar to those described at Sect. 3.2 (3) for the Appenninic ophiolites, both in location and paragenesis.

3. Kieslager type, Cu-Fe-(Zn), sometimes Au-bearing, deposits are fairly frequent in the "calc schists with greenstones series", in close connection to the meta-volcanics/meta-tuffites; Alagna is the locality type.

Other accessories (of no commercial interest, but with a probable bearing on genetic conditions) are Mo, Pb and graphite.

4. Au-bearing veins are known in ophiolites, either of the "normal series" (Ovada, western Liguria) and of the calc schists with greenstones series (Brusson, in part, Aosta Valley): remobilization from the underlying crystalline basement seems very probable.

5. Concluding Remarks

Tables 1 and 2 summarize the main characteristics of the deposits related to Italian ophiolites and Figure 1 shows their geographical distribution.

The univocal correspondence between some types of deposits and specific ophiolitic sections appears evident: this is particularly true if the whole paragenesis, instead of its main components, are considered.

A typical example is shown by the stratiform Cu-pyrite deposits: as pointed out by *Brigo* et al. (1976), those related to the upper part of the spilitic pillow-lavas section have paragenesis characterized by Fe-Cu-(Zn)-[(Au)], which is quite different from that of the "kieslager type" deposits held in the "calc schists with greenstones series", which is composed of Fe-Cu-(Zn-Au)-[(Pb-Mo-C)], and from that of the veinlets held in the gabbroidic section (Fe-Cu-quartz).

Moreover, the location of the Cu pyrite stratiform deposits of the Appenninic type (always along the upper part of the spilitic pillow lavas) enables to consider them as

Table 1. Ore/mineral substance accumulations in Italian Mesozoic ophiolites

Type 1.	Main metal: Cr Paragenesis: Cromite-(magnesite)-[(asbestos)] Type of accumulation: Subeconomic disseminations Country rocks: Serpentinized ultra-maphites close to the gabbroidic section Locality type: Ziona (29 [a]) in Eastern Liguria
Type 2.	Main metal: Ni Paragenesis: Native Fe-Ni (often: josephinite)-magnetite-pentlandite-(heazlewoodite-bravoite)-[(mackinavite-(calc)-pyrrotite-millerite-cobaltiferous pentlandite-valleriite)] Type of accumulation: Uneconomic disseminations Country rocks: Serpentinized ultra-maphites Locality type: Balangero (16) in Piedmont
Type 3.	Main metal: Fe Paragenesis: Magnetite-(hematite-martite)-[(pyrite-pyrrotite-calcopyrite-native Fe-Ni)]-(asbesto-brucite-clinohumite-olivine-calcite-dolomite) Type of accumulation: Economic massive concentrations and uneconomic disseminations Country rocks: Serpentinites derived from ultra-maphites Locality type: Cogne (10) in Aosta Valley
Type 4.	Main mineral: Asbestos Paragenesis: Crysotile-[(magnetite)] Type of accumulation: Economic lenticular stockworks Country rocks: Serpentinites derived from ultra-maphites Locality type: Balangero (16) in Piedmont
Type 5.	Main mineral: Magnesite Paragenesis: Magnesite-(opal-dolomite-aragonite)-[(quartz-sepiolite)] Type of accumulation: Economic tabular stockworks, crusts Country rocks: (Serpentinized) peridotite Locality type: Baldissero (14) in Piedmont
Type 6.	Main substance: Ophicalcite Paragenesis: Serpentinite-calcite-(aragonite)-[(magnesite-pyrite)] Type of accumulation: Economic tabular lenses with irregular footwall Country rocks: The upper boundary of serpentinites derived from ultra-maphites Locality type: Levanto (31) in Eastern Liguria
Type 7.	Main mineral: Talc Paragenesis: Talc-(chlorite)-[(asbestos)] Type of accumulation: Lenses, pipes, veins Country rocks: Serpentinites either derived (a) from ultra-maphites or (b) pertaining to the calc schists with greenstones series (c) olistolites of serpentinites Locality types: (a) Issogne (12) in the Aosta Valley; (b) Valmalenco (2) in Piedmont; (c) M. Albareto (25) in Emilia
Type 8.	Main metal: Ti Paragenesis: (a) Ilmenite-(magnetite), (b) rutile-(magnetite) Type of accumulation: Disseminations of uncertain economic value Country rocks: (a) Cumulitic gabbroes, (b) eclogites derived from the gabbroidic section Locality type: Val di Vara (22), in Western Liguria
Type 9.	Main metal/mineral: Cu-pyrite (Cu-Fe type) Paragenesis: Bornite-chalcopyrite-pyrite-quartz Type of accumulation: Uneconomic veinlets, disseminations Country rocks: Sheeted gabbroes Locality type: Bargone (28), Monte Rossola (30) in Eastern Liguria

[a] Numbers in parenthesis corresponds to Table 2 and Fig. 1.

Table 1 (continued)

Type 10. Main metal/mineral: Cu-pyrite (Cu-Fe-(Zn)-[(Au)] type)
Paragenesis: Pyrite-chalcopyrite-marcasite-(sfalerite-pyrrotite-mackinavite-hematite)-[(mag-netite-quartz-Au)]
Type of accumulation: Economic stratiform lenses (kieslager type)
Country rocks: (a) Upper part of spilitic pillow lavas, (b) olistolites of serpentinized diabases
Locality type: (a) Libiola (27) in Eastern Liguria, (b) Vigonzano (24) in Emilia

Type 11. Main metal: Cu
Paragenesis: Chalcocite-bornite-native Cu-(Chalcopyrite)-[(tetrahedrite)]-clayey materials
Type of accumulation: Economic pipes, veins, sub-horizontal lenses
Country rocks: Olistolites of weathered diabases
Locality type: Monecatini in the Cecina Valley (34) in Tuscany

Type 12. Main metal/mineral: Cu-pyrite (Fe-Cu-(Zn-Au)-[(Mo-Pb-C)] type)
Paragenesis: Pyrite-chalcopyrite-(pyrrotite-cubanite-mackinavite-sfalerite-Au)-[(magnetite-ilmenite-molibdenite-galena-graphite-linneite-millerite-rutile-goetite)]
Type of accumulation: Economic stratiform lenses (kieslager type)
Country rocks: Prasinites, anfibolites, calc schists of the "calc schists with greenstones series"
Locality type: Alagna (5) in Piedmont

Type 13. Main metal: Mn
Paragenesis: Braunite-[(rhodocrosite-rhodonite-pyrolusite-wad-pyrite-chalcopyrite)]
Type of accumulation: Economic stratiform lenses
Country rocks: Lower part of the jasper section
Locality type: Gambatesa (26), in Eastern Liguria

Type 14. Main metal: Au
Paragenesis: Pyrite-native Au-(pyrrotite-sfalerite-graphite)-[(marcasite-magnetite-"Ni-Co white ores"-ilmenite)]-dolomite-quartz-(calcedony)
Type of accumulation: Economic veinlets, subeconomic disseminations
Country rocks: Serpentinites derived from the ultra-maphites
Locality type: Ovada (23), in Western Liguria

Type 15. Main metals: Ni-Co (Ni-Co arsenide type)
Paragenesis: Smaltite-cobaltite-safflorite-rammelsbergite-arsenopyrite-(tetrahedrite)-calcite-dolomite-siderite-[(quartz)]
Type of accumulation: Subeconomic veins
Country rocks: Serpentinized ultra-maphites; prasinites, calc schists of the "calc schists with greenstones series"
Locality type: Cruvino (17) in Piedmont

---

a "key-bed", strictly time- and strata-bound. The same may be said for the manganesi-ferous jasper horizon.

Other examples of tight correlations between ores and ophiolitic sections are those of Ni, Cr, and Ti with the lower sections of the ophiolitic sequence: but one may speak only of "strata-bound occurrences" (not ore deposits) in such cases, for the Italian ophiolites.

On the other hand, the correlation between accumulations of Fe, asbestos, and talc with serpentinites may be considered suggestive, so far as strata bonds are considered, but not if time bonds are envisaged; as a matter of fact the formation of these materials is strongly controlled by the long and certainly complicated metamorphic events to which the complex was submitted certainly lasting well after the emplacement of the ophiolites, involving the Alpine orogenesis (*Dal Piaz*, 1971).

Table 2. Main deposits related to the Mesozoic ophiolites in Italy

Ref. No.	Name	Type according to Table 1
1	Predoi (Valle Aurina)	12
2	Val Malenco	7 b
3	Val Malenco	4
4	Preslong–Ollomont	12
5	Alagna	12
6	Praborna	13
7	Herin–Champdepraz–Mont Jovet	12
8	Brusson–Fenillaz–Camusera (in part)	14
9	Settarme–Chassant	4
10	Cogne	3
11	Mont Avic	3
12	Issogne–Mure–Mont Blanc	7 a
13	Fragne'–Chialamberto	12
14	Baldissero	5
15	Chicu–Brunetta	7 b
16	Balangero (S. Vittore)	4 and 2
17	Cruvino	15
18	Casellette	5
19	Beth–Ghinivert	12
20	Auriol–Sampeyre	4
21	Costa Aurello	7 b
22	Val di Vara	8
23	Ovada	14
24	Vigonzano	10 b
25	Monte Albareto	7 c
26	Gambatesa	13
27	Libiola	10 a
28	Bargone–Molin Cornaio	9
29	Ziona	1
30	Monte Rossola	9
31	Levanto	6
32	Cerchiara	13
33	Bisano	10 b
34	Montecatini in Cecina Valley	11

The presence of magnesite and/or ophicalcite accumulations has a paleogeographic/paleoclimatic significance rather than a time import. According to the most recent research (*Natale*, 1972; *Folk* and *McBride*, 1976), they should have been generated by supergene alteration of specific formations (serpentinites, especially of the basal section), in part during the same ophiolitic emplacement (especially: ophicalcites), in part afterwards (especially: magnesite): hence they could be considered "time- and strata-bound" only in the first case.

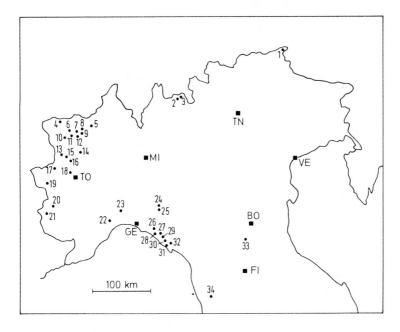

Fig. 1. Location map of the main deposits related to Italian ophiolites

The not neglibible Au vein deposits contained in ophiolites can be considered examples of remobilization from the Au-bearing crystalline basement, brought about by ophiolitic magmatism: however no particular bonds either to strata or to time can reasonably be established.

Finally, the Ni-Co arsenides vein type accumulations, occurring in Italian ophiolite, are too scanty and rare to allow any attempt at drawing general conclusions.

Acknowledgments. The author thanks Dr. *A. Ferrario* and Professor *L. Brigo* for assistance in preparing this report.

References

Bigazzi, G., Ferrara, G., Innocenti, F.: Fission track ages of gabbros from Northern Appennines ophiolites. Earth and Planet. Sc. letters **14**, 242–244 (1971)

Brigo, L., Dal Piaz, G.V., Ferrario, A.: Le mineralizzazioni cuprifere legate ai termini effusivi di alcuni complessi ofiolitici nell'area mediterranea. Boll. Ass. Mineral. Subalpina **13**, 352–371 (1976)

Dal Piaz, G.V.: Alcune osservazioni sulla genesi delle ofioliti piemontesi e dei giacimenti ad esse associati. Boll. Ass. Mineral. Subalpina **8**, 365–388 (1971)

Decandia, F.A., Elter, P.: La zona ofiolitifera del Bracco, nel settore compreso fra Levanto e la Val Graveglia (Appennino ligure), LVI Cong. Soc. Geol. Ital., Pisa 1972, p. 37–64

Folk, R.L., McBride, E.F.: Possible pedogenic origin of Ligurian ophicalcite: A Mesozoic calichified serpentinite. Geology **4** (6), 327–332 (1976)

Galli, M., Bezzi, A., Piccardo, G.B., Cortesogno, L., Pedemonte, G.M.: Le ofioliti dell'Appennino Ligure: un frammento di crosta-mantello oceanici dell'antica Tetide. LVI Cong. Soc, Geol. Ital., Pisa 1972, p. 1–36

Natale, P.: Recrystallization and remobilization in some pyrite deposits of the Western Alps. Symp. Remobilization of Ores and Minerals, Cagliari 1969, p. 129–148

Steinmann, G.: Die ophiolithischen Zonen in mediterranen Kettengebirgen. 14th Intern. Geol. Congr., Madrid, 2, 1927, p. 638–667

Additonal Bibliography

Brigo, L., Ferrario, A.: Le mineralizzazioni nelle ofioliti della Liguria orientale. Rend. Soc. Ital. Mineral. Petrol. 30, 305–316 (1974)

Castaldo, G., Stampanoni, G. (eds.): Memoria illustrativa della Carta Mineraria d'Italia scala 1:1000000. Serv. Geol. Ital., 1975, 213 pp.

Ferrario, A.: I giacimenti cupriferi a pillow-lavas della Liguria orientale. Rend. Soc. Ital. Mineral. Petrol. 29, 485–495 (1973)

Ferrario, A.: Ore/mineral deposits related to the Italian ophiolites. In: Seminar Metallogeny and Plate Tectonics, Belgrade 1976

Ferrario, A., Montrasio, A.: Manganese ore deposits of Monte del Forno. Its stratigraphic and structural implication (in press)

Mastrangelo, F., Natale, P., Zucchetti, S.: Mineralizzazioni metallifere e litoidi associate a rocce ultra-basiche delle Alpi Occidentali. Boll. Ass. Mineral. Subalpina 13, 341–351 (1976)

Natale, P.: Nuove osservazioni sull'origine dei giacimenti di magnesite delle Alpi Occidentali. Boll. Ass. Mineral. Subalpina 9, 107–124 (1972)

Zucchetti, S.: Ferro-Nichel nativo ed altri minerali nicheliferi in serpentiniti anche asbestifere delle Alpi Occidentali. Rend. Soc. Ital. Mineral. Petrol. 26, 377–397 (1970)

Zucchetti, S.: Prime osservazioni sui minerali nicheliferi (ferro-nichel nativo e solfuri) in serpentiniti della Liguria. Rend. Soc. Ital. Mineral. Petrol. 30, 3–20 (1974)

Zuffardi, P.: Les gisements metalliferes italiens en relation avec les roches maphiques et ultramaphiques. In: Deuxièmes Journées de l'Industrie Minérale, Bruxelles (in press)

Strata-Bound Intrusive Deposits

Genesis of Chromite in Yugoslavian Peridotite

S. GRAFENAUER, Ljubljana

With 21 Figures

Contents

Summary

Chromite-bearing ophiolitic masses in Yugoslavia are distributed in many zones as a result of static differentiation. The final position is due to the movements connected with the oceanic subduction under the western continental plate during the Jurassic age. The richest ores are situated in basal parts, and sometimes in upper parts of massifs, whereas the ores in the central parts of serpentinite massifs are usually poor. The origin of the ore deposits and their enrichment are directly connected with the origin of the parent rocks. The ore deposits began to accumulate in depth, but this may have been prolonged during the uprising of the ophiolites associated with continental drift. At this time there developed in the ore the same structures, textures, forms and deformation, which developed in rocks.

All chromite deposits, with rare exceptions, are associated with dunite. The beginning and sequence of crystallization was regulated by physicochemical conditions during the ore formation. Some deposits developed by crystallization in the liquid stage in the upper mantle; others were late magmatic and developed during the enrichment of a rich chromitite melt which intruded the already solid country rocks. The beginning of crystallization of single minerals depended on the oxygen fugacity and the content of volatile components in the rock, sometimes on the contamination mechanism with felsic inclusions. The influence of volatiles can be seen in many pegmatite veins in the neighbourhood of the ores.

1. Chromite-Peridotite Occurrences

Yugoslavian chromite deposits are mainly associated with Alpine-type gabbro-peridotite complexes (*Donath*, 1930), although sometimes they are present in secondary deposits in dark marine sands (*Grafenauer*, 1962). Primary chromite occurs almost exclusively in ultramafic igneous rocks in the form of great nidiform, lenticular or pipe-like concentrations, and in poorer impregnated streaks, bands, plates and lenses.

The locations of Yugoslavian ultramafic massifs which contain major deposits are shown in Figure 1. The largest chromite deposits are found in the southern part of Yugoslavia, where peridotite outcrops are most extensive. There are also large bodies of ultramafic rocks in Bosnia, where numerous, fairly small, chromite deposits occur; for instance Duboštica, Tribija, Rekovac, Grebečići and others (*Čermelj*, 1951; *Grafenauer*, 1957). Far more numerous are the occurrences in Serbia and Macedonia. The deposits stretch from Zapadna Morava to the southern border of Yugoslavia.

Fig. 1. Sketch showing Yugoslav ultramafic rocks and associated chromite deposits (*Janković*, 1967). Serpentinite massifs: *1:* Konjuh; *2:* Zlatibor; *3:* Suvobor; *4:* Jelica; *5:* Ibar; *6:* Sjenica; *7:* Deli Jovan; *8:* Gornji Lepenac (Brezovica); *9:* Ljuboten (Raduša); *10:* Djakovica; *11:* Rožden; *12:* Rabrovo

1.1 Chromite Deposits of the Ljuboten Peridotite Massif

The Ljuboten peridotite massif has the form of a sill with NW-SE strike and dip of less than 30° to NE. Magmatic layering can be seen more or less throughout the whole massif. The ultramafic rocks were probably formed by differentiation in the upper mantle of the earth. Subsequent emplacement in their present position was the result of tectonism accompanying the motion of Aegean Rhodopian continental plate (Fig. 2) in the south-western direction against the Pelagonian continental plate. During the movement, the rocks suffered kinetic differentiation and were partially melted (*Dickey*, 1970) forming many thin sheets and lenticular flow layers. The primary zoning (Fig. 3) was

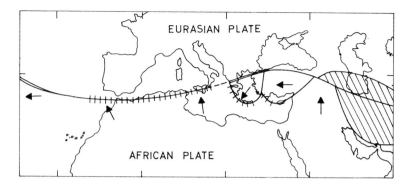

Fig. 2. Approximate positions of plate boundaries at present active, with *arrows* marking the directions of motion relative to the Eurasian plate. Boundaries creating lithosphere are shown with a *double line*, boundaries consuming plates with *short lines at right angles* to them. *Cross hatched* region in Eastern Turkey and Iran is seismically active throughout. Fault plane solutions are all overthrusts, and show that the crust is being thickened all over this region (*McKenzie*, 1970)

roughly preserved but overprinted by secondary flow layering, schistosity and brecciation. In the basal part dunite is most frequent and above this is the middle zone formed mainly of harzburgite with some dunite. Finally occurs the topmost barren zone with some pyroxenite and gabbro pegmatite, locally with rodingite (*Grčev* and *Milenković*, 1962).

The MgO:FeO relation in the Yugoslavian ultramafic rocks is usually greater than 8.1 and they beong to the ultramafic orogenic series of *Hess* (1938).

The orebodies in basal parts of ultramafic rocks are pipe-like with podiform enlargement at depth (Fig. 4). The contact between chromitite and dunite is sharp. The ramification of the orebodies from the central one, the sharp contacts, as well as cross-cutting relationships with the country rocks are evidence fo the kinetic differentiation (*Cloos*, 1936) of the orebodies in a semi-solid state. The characteristic fluidal structures, sometimes gneissic, or schistose, were developed during this period of emplacement. The ore was later stretched and fractured with silicate material migrating into transverse fractures. The pull-apart structures (*Thayer*, 1964) developed as a result of elongation of the chromite grains.

Sometimes we find on the border of chromitite a thin mantle of smaragdite, orthopyroxene, secondary actinolite and talc. In this mantle may sometimes be seen secondary octahedral crystals of chromite. This zone was developed probably during a high hydrothermal or late magmatic phase.

Chromite in the chromitite and in the chromite-bearing dunite is usually anhedral and, when surrounded by olivine, is almost rounded and magmatically corroded (Fig. 5). The chromite in the nodules is probably the result of collusion, the result of which is a structure, called "swimming" or "synneusis structure" (*Vogt*, 1921; *Bastin*, 1950) and is similar to the welding together of chromite grains as indicated by *G.E. Seil* mentioned in *Thayer* (1964, p. 145) (Fig. 6).

Ito and *Kennedy* (1967), using the natural peridotite assemblage, have demonstrated experimentally that Cr spinel can melt before olivine. As the peridotite mass rose, the chromite droplets were smeared out and then coalesced into a network of coarser-welded

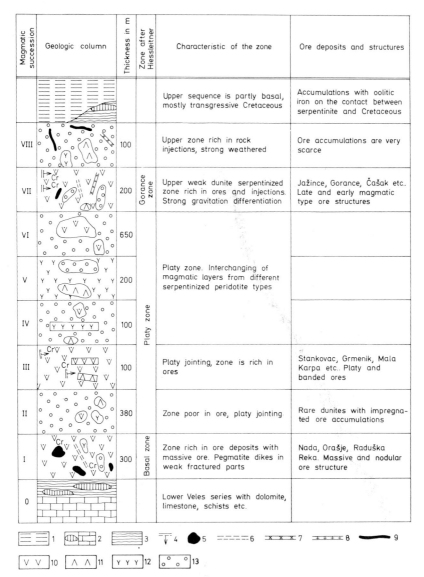

Magmatic succession	Geologic column	Thickness in m	Zone after Hiessleitner	Characteristic of the zone	Ore deposits and structures
				Upper sequence is partly basal, mostly transgressive Cretaceous	Accumulations with oolitic iron on the contact between serpentinite and Cretaceous
VIII		100		Upper zone rich in rock injections, strong weathered	Ore accumulations are very scarce
VII		200	Gorance zone	Upper weak dunite serpentinized zone rich in ores and injections. Strong gravitation differentiation	Jažince, Gorance, Čašak etc.. Late and early magmatic type ore structures
VI		650			
V		200		Platy zone. Interchanging of magmatic layers from different serpentinized peridotite types	
IV		100	Platy zone		
III		100		Platy jointing, zone is rich in ores	Stankovac, Grmenik, Mala Karpa etc.. Platy and banded ores
II		380		Zone poor in ore, platy jointing	Rare dunites with impregnated ore accumulations
I		300	Basal zone	Zone rich in ore deposits with massive ore. Pegmatite dikes in weak fractured parts	Nada, Orašje, Raduška Reka. Massive and nodular ore structure
0				Lower Veles series with dolomite, limestone, schists etc.	

Fig. 3. Schematic geologic column through the Ljuboten peridotite massif: *1:* upper Cretaceous; *2:* paleozoic limestones; *3:* phyllite; *4:* chromite streaks and disseminated ores; *5:* chromite podiform orebodies and chromitite veins; *6:* gabbropegmatite; *7:* porphyroid peridotite; *8:* websterite, diallagite; *9:* amphibolite, rodingite; *10:* serpentinized dunite; *11:* serpentinized pyroxenite; *12:* serpentinized harzburgite; *13:* serpentinized enstatite dunite

grains and finally into spherical-shaped nodules that have the lowest energy interfaces. The relative sequence of crystallization of spinels or olivine from the melt was still dependent on oxygen fugacity as shown by *Ulmer* (1969). At lower oxygen fugacities, olivine would initially crystallize ant then chromite. The presence of volatile components in the peridotite melt was thus very important in determining the crystallization

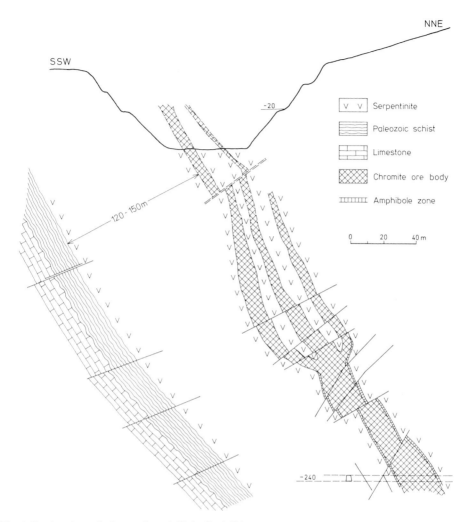

Fig. 4. Section through the ore deposit Nada (Raduša)

and concentration of different mineral components during the upward migration of the mass. The presence of volatile components in the mineralized chromitites is indicated by the occurrence of smaragdite, kaemmererite, and partly uvarovite (Fig. 7), which appear to have developed often in the sequence following the crystallization of chromite. The temperature necessary for the formation of uvarovite is 900° to 1400° C at 10 to 30 kb (*MacGregor*, 1970). The temperature during the emplacement of the crystalline dunite mush (after *Bowen* and *Tuttle*, 1949) was between 1100° and 1200° C, which is quite sufficient for partial fusion of the chromite. It is possible that small amounts of volatiles seriously lowered the solidus of the different mineral components. The rounded anhedral form of chromite grains in compact ore is evidence against the concentration by gravitational crystallization, but is more likely as proof for melting and growing of the individual

Fig. 5. Chromite grain, corroded and fractured with two fracture generations, slightly altered into ferit-chromite, containing one magnetite veinlet. Sample Albania, Belčiza 2. Refl. pol. light (x 200) oil

Fig. 6. Welding texture of chromite grains. Drenovac. Refl. pol. light (x 50)

Fig. 7. Chromite (*white*), Cr diopside (*light gray*), gangue minerals (*gray*) as magnesite, kaemmererite, serpentine etc. Stari Čaf Pruša-Djakovica massif. Refl. pol. light (x 200)

grains during the kinetic differentiation. Many times small, round or euhedral olivine inclusions were still tectonically deformed, and the resulting fractures were filled with serpentine.

The investigations of *Karamata* et al. (1964) on the contact of ore and country rock show the expected differences between chromite and country rocks, but the evidence is insufficient to give definite proof for or against late magmatic origin of the ores.

Very often in the ore nodular structures can be found, rarely negative nodular (Fig. 8) and orbicular chromite (Fig. 9).

Fig. 8. Negative nodular ore from the Kafe Odjak mine, Raduša

The disseminated chromite ore is most common in all middle magmatic zones. Rhythmic banded seams are most likely due to gravitational settling of chromite under the influence of magmatic currents (*Jackson*, 1961).

"Great difference in thicknesses of podiform deposits, grain size, composition of chromite and associated rocks and tremendous volumes of pure dunite and olivine-rich peridotite in the Alpine complexes, require a different environment", than the stratiform chromite deposits (*Thayer*, 1964). *Thayer* supposed the concentration occurred in the upper mantle of the earth and believed that the nodular and orbicular ores of podiform deposits must also reflect magmatic conditions that are not attainable in the upper crust of the earth, for none are known in stratiform complexes. The textures and particularly the grain size demand an environment that cooled more slowly than the stratiform chromite.

It is agreed that the first concentration occurred in the upper mantle, but we believe that all the described structures and textures could develop with partial melting during

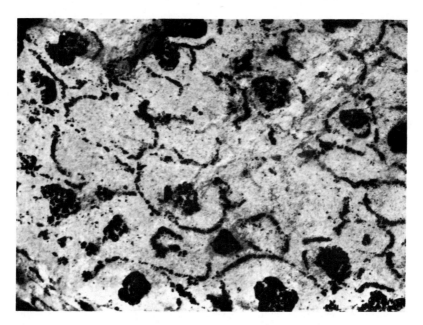

Fig. 9. Orbicular ore from the Kafe Odjak mine, Raduša

the migration of the ophiolites to the present position. The ores show lineation and flowage structures, the ore layers are lenticular and wedge out in short distances, but alternate rhythmically with barren serpentine. The grains of chromite are uniformly or irregularly disseminated or form thinner streaks or thicker platy ore.

A considerable number of the largest occurrences of chromite in the Ljuboten massif occur in the upper parts of the Gorance zone. The ore may appear in streaks and rich plates and lenses or sometimes in dykes, as in Čašak. In the margins of the upper parts of ophiolitic massif there are numerous dykes of pyroxenite, gabbro-pegmatite and rodingite, which probably indicate relative enrichment of volatiles in this section. The ore is always arranged in fluidal, lineated and foliated forms and the characteristic structures of podiform deposits are rarely observed. Platy concentrations of chromite with sharp contact against serpentinized country rock are frequently seen. In the enriched ores synneusis, chain and network arrangements of the chromite grains are commonly seen. The ore body at the Gračane mine, in the vicinity of Čašak, has the characteristic form of a dyke (Fig. 10). The ore dyke fractured the solid country rock and blocks and boulders of the country rock occasionally can be found in the chromitite dyke. The ore structure is generally nodular and massif. Chromite nodules are lineated in the direction of the long axis of the orebody. The nodules were later fragmented by processes which have resulted in pull-apart structures and finally filled with serpentine. The chromite grains are hypidiomorphic to xenomorphic, and are magmatically corroded and fractured. Concentration of the nodules has resulted in the development of the compact ores. The high-temperature mineral uvarovite can often be found replacing chromite. The form of the orebodies and the lineation of the nodules indicate that the orebody was enriched during kinetic differentiation. It is not possible to understand the mechanism of the lineation of the nodules by only considering static differentiation in upper mantle of the earth as proposed by *Thayer* (1964). The important factor here, as in the basal zone, is the oxygen fugacity, which regulated the sequence of crystallization of both chromite and olivine. The injection of ore dykes into the solid rocks was completed, however, before the final solidification of the whole peridotite magma, since later differentiates of the magma have intruded the orebodies. For example, pyroxenite and gabbro pegmatite dykes broke through the upper weakened zones in already solidified dunitic and harzburgitic country rocks.

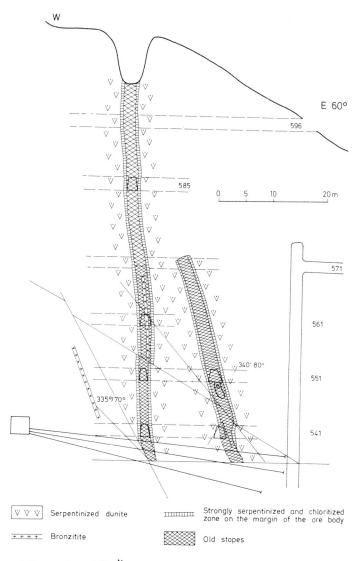

Fig. 10. Cross-section through the ore deposit Gračane

1.2 Other Chrome-Bearing Districts in Yugoslavia

The ultramafic, chromite-bearing massif of Brezovica, with many chromite occurrences on the north slope of Ljuboten, has the form of a sill with a dip about 15 to 20° eastwards (*Majer*, 1956; *Karamata*, 1968, 1969). The Brezovica intrusive mass suffered partial melting during intrusion, with the temperature being approximately 1200° C (*Karamata*, 1969). The chromite grains show all the characteristics observed in the Raduša occurrences. The ore displays very good structures of magmatic and late magmatic flowage, for example pull-apart texture with extension fractures, bending, folding and deformations of chromite bands, and late magmatically formed dunite and pyroxenite veins in chromitite are all visible.

Sometimes the dunite material was denser, but less competent than the chromite, although there are examples which show that the dunite was more competent than the chromite-rich dunite or chromitite (*Karamata*, 1968). In this case the fragments of dunite are included in chromite-rich dunite rock.

In the ores, the common association of chromite with some secondary minerals such as kaemmererite, smaragdite, fuchsite and uvarovite, is evidence of relative enrichment on volatiles. Chromite ore is rich in accessory sulphide minerals such as pyrrhotite, chalcopyrite, sphalerite, valeriite, mackinavite and millerite.

The ore is found in banded and layered bodies or in podiform deposits that are enriched in nodular ore. Fractures in the chromite are usually older than serpentinization of the associated rocks. All the textures and structures prove that the ultramafic mass of Brezovica was initially differentiated at depth, and was subsequently deformed by tectonic emplacement. During this movement most of the mass flowed in a laminar fashion, although some obvious turbulence occurred (*Karamata*, 1968). During the emplacement the ultramafic rocks and chromite differentiates underwent kinetic differentiation with all appearances of partial fusion, concentration, formation of nodules, etc. The chrome ore is very rich in chrome and the chromite is only weakly transformed into feritchromite and magnetite. The country rocks are strongly serpentinized.

The chromite-bearing ultramafic massif of the Djakovica district is a part of larger ultramafic mass which in Albania covers over 1200 km². The rocks are mostly harzburgites and contain much more iron than the corresponding rocks in the Ljuboten serpentinite massif (*Donath*, 1930). In general the chromite in these rocks is also more rich in Fe, and the average ratio Cr:Fe is only about 2.3 (*Janković*, 1967), compared with the chromite from Raduša or Brezovica, which has a ratio Cr:Fe greater than 3.0. The ores are mostly platy or stock-forming streaks and bands, with minor poor disseminated ores. Dunite is invariably the matrix of the chromite-bearing rock, and there can often be found a few-cm thick mantle of dunite at the margin of the orebody. The most important deposit is Deva. Chromite, in rich plates, is hypidiomorphic to xenomorphic, magmatically corroded, and often shows welding structures. Occasionally a sideronitic structure of chromite in olivine occurs (Fig. 11). This is believed to be the first report of the occurrence of sideronitic structure in any chromite deposit.

In all types of ore from Djakovica features of flow layering, foliation and lineation can be seen, proof of ore formation during the movement of ultramafic rocks from depth.

The chrome ores in peridotites from the southern Yugoslavian-Albanian border are important with respect to comparison of ore genesis in most Yugoslavian regions.

Fig. 11. Sideronitic chromite in serpentine, earlier olivine. Sample from Djakovica massif, location Steva 7. Refl. pol. light (x 50)

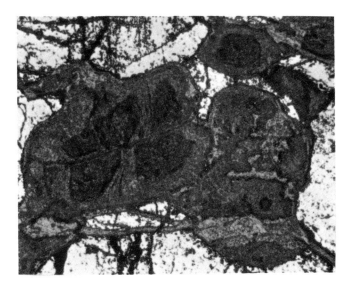

The ores have the same structures and textures as in Yugoslavia, but they are rich in sulphide minerals, containing small inclusions of pentlandite, chalcopyrite, pyrrhotite, rarely heazlewoodite, mackinawite and valeriite. In fractures awaruite and platinum can be found. In the ore of Vlahna, chrome aggregates appear very unusually. Extremely porous chromite grains are full of olivine inclusions with the porosity being greater at the periphery of the grains (Fig. 12). The ore was probably brecciated due to fast cooling during obduction into water-bearing sediments. Chromite was in eutectic relation with olivine and developed myrmekitic texture (Fig. 13). It appears likely that chromite crystallized so fast that it was inhomogeneous and later transformed into feritchromite and magnetite. Some grains are separated by contraction fractures.

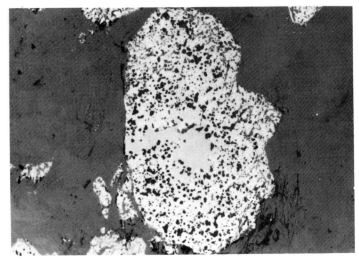

Fig. 12. Porous chromite, full of olivine inclusions with porosity being greater at the periphery of the grains. Sample Albania 10. Refl. pol. light (x 40)

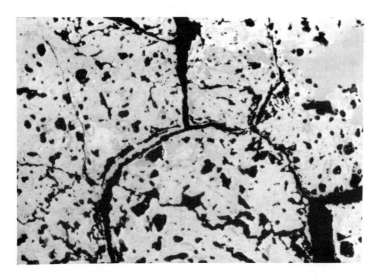

Fig. 13. Contraction fractures in chromite ore altering to feritchromite. Eutectic type intergrowing from chromite and olivine, now serpentine. Sample Albania 10. Refl. pol. light (x 200), oil

The conditions during the formation of these ores were very unusual, but they indicate that chromite and olivine at the time of movement of ophiolite rocks (or magmas) were still plastic and in eutectic relation.

The chromite-bearing rocks between the Kosovo field on the north-east and Djakovica on the south-west form the north-western extension of the Ljuboten and Brezovica massifs. Chromite appears in two regions: in the Orahovac mass and in the Drenovac mass. The ore grades from network textures into disseminated ore. Chromite grains of Drenovac are hypidiomorphic to xenomorphic and the nodules are plastically deformed and elongated in the direction of flowage. Chromite grains are magmatically corroded and sometimes show typical welding textures (Fig. 6). The ore of the Orahovac ultramafic mass is characterized by the same textures as in the Drenovac district. Sometimes excellent idiomorphic inclusions of olivine in chromite are visible (Fig. 14). Apparently olivine and chromite crystallized simultaneously from the melt. At the grain borders between chromite and olivine, small zig-zag lines are common, probably developed during late magmatic corrosion of olivine by chromite.

The mineral occurrences in Bosnia are relatively small and they are mostly present in harzburgite, rarely in dunite. The rocks are rich in iron and silica and so also are the ores. The ratio of Cr:Fe in chromite is usually about 2 to 2.3. Accessory spinel is mostly picotite, chromite is present in the deposits but often chromespinel is more common. The largest deposits occur at Rakovac and Krivaja, near Duboštica.

In the district of Deli Jovan, there are small lens-like concentrations of chromite in ultramafic rocks, which form the margin of the Deli Jovan gabbro. Primary chromite in Deli Jovan is very rich in alumina and is mostly chromespinel. The ratio of Cr:Fe is only 1.6. The chromite grains are hypidiomorphic to idiomorphic (Fig. 15) and they contain many inclusions of idiomorphic, orbicular and corroded olivine (Fig. 16), sometimes in micro-orbicular texture. Synneusis textures observed between chromite grains, inclusions

Fig. 14. Idiomorphic inclusion from olivine in chromite showing on interfaces small zig-zag lines. Southern part of Šip-Orahovac massif. Refl. pol. light (x 170)

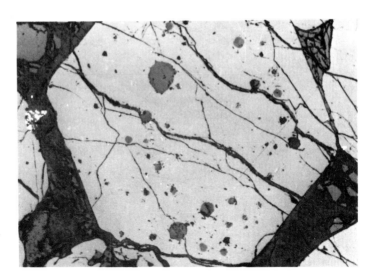

Fig. 15. Idiomorphic chromite with idiomorphic, corroded olivine inclusions, rare magnetite grains in chromite. Sample from Deli Jovan. Refl. pol. light (x 35)

of olivine in chromite and partly idiomorphic grains of olivine indicate superposition in the crystallization sequence, and fast changes in physical and chemical conditions in localized areas. This may be considered as proof for a small temperature range and for high viscosity during the crystallization. The equilibrium between olivine and chromite probably altered due to changing pressure during obduction, or as a result of changes in fO_2 in different parts of the ultramafic body. Chromite and olivine probably had a

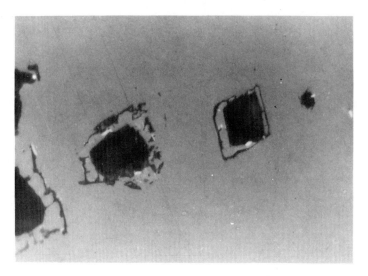

Fig. 16. Micro-orbicular corroded olivine inclusions in chromite, rare magnetite grains. Sample from Deli Jovan. Refl. pol. light (x 400), oil

cotectic or peritectic relationship (*Irvine*, 1967). Chromite enriched in alumina usually crystallized later and thus is present as a reaction rim associated with olivine. It may be significant that experimentally spinel proper ($MgAl_2O_4$) coprecipitates with forsterite in much larger amounts than does picrochromite. A similar relation in natural magmas could be a reason for relatively extensive interstitial overgrowth on cumulus grains of aluminous spinel as compared to those of Cr-rich chromite. In the chromite of Deli Jovan small inclusions of ilmenite and rutile can sometimes be seen.

The chromite-bearing ultramafic massif at Lojane, north-east of Skopje, is composed basically of lherzolite and dunite, often intruded by dacitic andesites and slightly older granites and syenites. The chromite concentrations appear in schlieren or plates. The central Lojane deposit is a typical podiform concentration. Disseminated ores form the aureole of the enriched ore stock. The ore is characterized by the association of uvarovite and Cr grossularite, which both replace chromite. In the parts intruded by syenite, chromite is strongly altered to an aggregate of feritchromite, magnetite, goethite and many Cr silicates. Analysis with a microprobe showed that there was an enrichment of Cr and Fe and a decrease in Al and Mg in the feritchromite. Similar changes are known in the öccurrences of chromite from Australia (*Golding* and *Bayliss*, 1968). The oxidation of ferric to ferrous iron caused the expulsion of Al and Mg cations from the spinel structure (*Wells*, 1962; *Evans*, 1964; *Golding* and *Bayliss*, 1968) which then resulted in the formation of kaemmererite, smaragdite, fuchsite etc. Cr grossularite is known only from the Lojane deposit and probably originated by the influence of granite intrusion on the chromite ore.

2. Principal Characteristic of Associated Minerals

2.1 Chromite

The composition of chromite in Yugoslavian peridotites is given in Figures 17 and 18. The majority of Yugoslavian chromites belong to alumochromite. Alpine-type chromites differ from stratiform chromites in having smaller amounts of iron and by higher quantities of chromium, magnesium, and partly aluminium. The content of iron in chromites

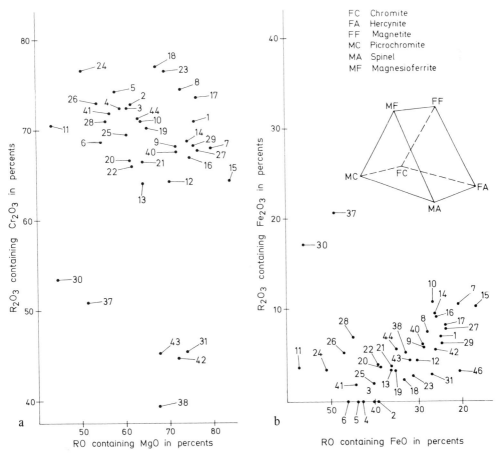

Fig. 17. Plot of analyses of Yugoslavian chromite with the projection on the plane (a) MC-FC-MA-FA and (b) MC-FC-MF-FF of the triangular compositional prism. Mol %. *FC:* chromite; *FA:* hercynite; *FF:* magnetite; *MC:* picrochromite; *MA:* spinel; *MF:* magnesioferrite. Samples are as follows: *1—6:* rich ore from Nada mine (Raduša); *7—9:* rich ore from Orašje mine (Raduša), *10:* compact ore from Gorna Raduša; *11:* ore from Kafe Odjak mine (Raduša); *12:* Stankovac mine (Raduša); *13 and 14:* ore from Gorance mine (Raduša); *15:* ore from Staro Selo (Raduša); samples *1—16* are from Ljuboten massif. *17:* banded schlieren ore from Jezerina; *18:* concentrate from Jezerina; *19:* disseminated ore from Mušitište (Ostrovica); *20—22:* ore from Lojane mine; *23:* ore from the mineralization at Novičani near Veles; *24:* ore from the mineralization Krečana at Alšar; *25:* ore from the mineralization Arničko near Alšar; *26—28:* rich ore from Orahovac; *29 and 30:* ore from Petković; *37:* accessory chromite from Ohrid; *38:* chromite from Deli Jovan; *40 and 41:* chromite from Albania; *42 and 43:* chromite from Kozarevo; *46:* picotite from Bešlinac (Bosnia)

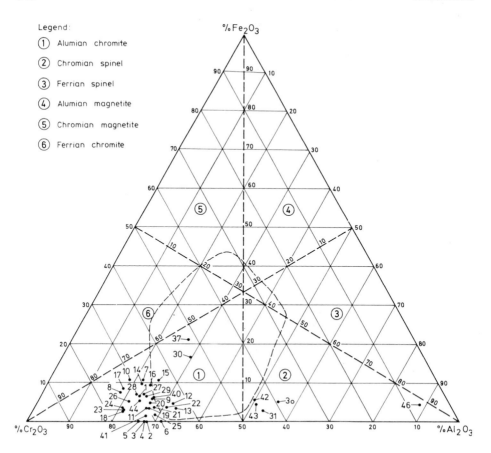

Fig. 18. Variation of Cr_2O_3, Al_2O_3 and Fe_2O_3 (in mol % of total R_2O_3) in Yugoslavian chromites. *Stippled area* shows the composition of 99 chromites from Stillwater Complex, Montana, USA. (After *Thayer*, 1964)

changes markedly as a consequence of varying conditions during formation, but there is a tendency for an increase in iron from the basal to higher parts of the serpentinite massifs.

The composition of accessory chromite also changes drastically. The chief reason for this change is a consequence of the limited mass of chromite. The cumulus chromite and the mafic silicate minerals were initially uniform in composition regardless of their relative concentrations. The minerals tended to mutual equilibrium as the cumulates cooled from their temperature of formation.

As discussed in *Irvine* (1965), the distribution coefficient $K = Mg/Fe_{sil} \cdot Fe/Mg_{spinel}$ increases with decreasing temperature. Thus, as shown in Figure 19, where chromite is a minor constituent in the silicate rock, it should become appreciably richer in ferrous iron as the rock cools, while the silicate mineral remains constant in composition. Where the silicate is subordinate, it should become more magnesian, while chromite changes only slightly in composition. To obtain the best estimates of the initial temperature of

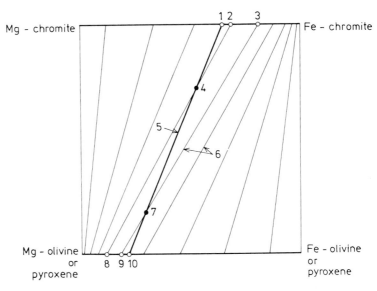

Fig. 19. Possible interpretation of the MgO-FeO relations observed between the minerals in adjacent units of chromitite and peridotite or pyroxenite (*Irvine*, 1965, 1967). The vertical axis represents the relative proportions of spinel and silicate; the horizontal axis, their Fe^{2+}/Mg ratios. It is assumed that the absolute value of the distribution coefficient, K = (Mg/Fe^{2+}) silicate $(Fe^{2+}/Mg)_{spinel}$ increases with falling temperature (the high-temperature tie line corresponds to lnK = 2; the lower-temperature ones, to lnK = 3), and that the minerals undergo local re-equilibrium as they cool. Thus the final composition of each mineral is a function of its initial concentration. *1:* initial chromite; *2:* final concentrated chromite; *3:* final disseminated chromite; *4:* silicate-bearing chromitite; *5:* high-temperature tie line; *6:* lower temperature tie lines; *7:* chromite-bearing silicate unit; *8:* final interstitial silicate; *9:* final concentrated silicate; *10:* initial silicate

crystallization of chromite and olivine assemblages, one should use coefficients that compare the composition of the minerals as they occur at maximum concentration in adjacent units (that is layers, bands, pods etc) and that also belong to the same stage of paragenesis, rather than considering adjacent grains in a sample of one rock type.

Yugoslavian chromites are rich in Cr and the ratio Fe^{3+}/Cr + Al + Fe rarely exceeds 12%. Sometimes, in peridotites chromite rich in Al occurs, which is usually younger than the olivine. *Irvine* (1965) believes the Alpine-type chromites crystallized at low fO_2 and it is likely that variations rich in Cr crystallized at lower fO_2 than those with high Al. The crystallization of Alpine-type chromite began at greater depth than that occurring in stratiform deposits. If a mass of mantle rock is displaced upwards, it will follow a gradient between the adiabat and the geotherm (Fig. 20). Eventually the rising mantle rock may pierce its solidus and begin to melt (*Dickey*, 1970). As long as all of the original solid phases are present and the pressure remains constant, the composition of the melt will not change. During gradual decompression at nearly constant temperature, the Mg:Fe and Ca:Na ratios of the melt will increase and the concentrations of Ti (*MacGregor*, 1966) and Al in the melt will decrease. These changes will be gradual, but as soon as one of the solid phases is consumed, the composition of the melt will begin to change more rapidly. If liquid and solid remain in equilibrium, the change in melt composition will be continuous. If the liquid is removed, however, the next melt from

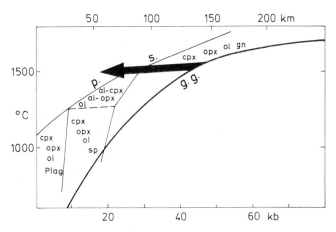

Fig. 20. Mantle rock (pyrolite = 1/4 tholeite basalt + 3/4 dunite), rising adiabatically from some point on the geothermal gradient (*g.g.*) would follow a P-T trajectory parallel to the *arrow*. The stability fields are for pyrolite composition. Eventually the rising mantle rock might pierce its solidus (*p.s.*) and begin to melt. The mineral assemblages which crystallized from such a melt would depend upon its composition and the relative rates of cooling and decompression. *px:* pyroxene; *ol:* olivine; *sp:* spinel; *plag:* plagioclase; *cpx:* clinopyrocene; *opx:* orthopyroxene; *gn:* garnet. (After *Dickey*, 1970)

the solid residue will be quite different in composition. The latter process has been called fractional fusion (*Presnall*, 1969). As the peridotite mass rises, it deforms, and the droplets of melt are smeared out. As deformations continues (*Dickey*, 1970), droplets coalesce into thin sheets, parallel to the flow lines of the deforming peridotite. With prolonged deformation the thin sheets coalesce, forming thicker, more widely spaced layers. As the layers grow, equilibrium between the melt and the residual solid can be approached along a decreasing number of interfaces. Along these interfaces olivine and pyroxene in the peridotite become richer in Fe as cooling proceeds.

It is very interesting that during partial fusion chrome spinel melts before olivine (*Ito* and *Kennedy*, 1967). The melting temperature at 1 atm is about $1150°$ C, and small quantities of water lower the temperature of melting. It is not surprising that by these processes the chromite masses were also partially fused.

Texture and Structure of Chromite. All the chromite orebodies in Yugoslavian peridotites show more or less clear signs of fluidal texture. Flowage textures, which demonstrate the moving of the chromite-bearing mass, have been described by *Grafenauer* (1956, 1975), *Karamata* (1969, 1970) and others. Pull-apart textures, and the development of transverse silicate-filled fractures by elongation or extension of solid chromite, characterize most medium- to high-grade ores. The chromite grains from all chromitites are usually xenomorphic and are corroded. They are usually disrupted and granulated, sometimes locally elongated in the direction of late magmatic flowage. Sometimes the dunitic material was impressed and crystallized at the time of the intrusion, and occasionally chromite has been injected into dunite.

Most of the rich ores contain nodular-type structures and the nodules are elongated in the direction of flowage. Some plastic deformation also occurred in these nodules. The ores in alpino-type deposits are essentially different from those in stratiform deposits.

Chromite grains are xenomorphic and in the network-type of ore are often welded together. In the chromite there are often idiomorphic inclusions of olivine, sometimes myrmekitic, micrographic and emulsion textures are visible. In the negative nodular ore sideronitic textures can sometimes be seen. In orbicular ore idiomorphic olivine crystals are sometimes in contact with xenomorphic chromite grains.

The range of structures in alpino-type ore deposits is given in Table 1. The disseminated ores with indistinct flowage features are relatively rare, this characteristic is more often seen in rich chromite ore. In the massive ores there are chiefly nodular and platy structures, the nodules having diameters of 5 to 44 mm. In the massive ores gneissic and schistose structures sometimes occur, which are a consequence of metamorphic influence during flowage of the deposit. Transitional eutectic structures appear in association with network, orbicular and negative nodular textures. In such transitional structures there are often myrmekitic and micrographic structures present, which have formed as a result of fast cooling.

Table 1. Structures of alpinotype chromite ores

I. Homogeneous structure with indistinct flowage features	
A. Massive ore	B. Disseminated ore
1. uniformly distributed ore 2. nodular ore	1. uniformly distributed ore 2. nodular ore
II. Taxitic structure, chromite grains are oriented under the influence of flowing	
A. Platy ore	B. Banded ore
1. uniformly disseminated ore 2. irregular disseminated ore 3. nodular ore	1. uniformly disseminated ore 2. irregular disseminated ore 3. nodular ore 4. orbicular ore 5. negative nodular ore 6. network ore

On the basis of the analysis of different structures in specific deposits the following can be stated: chromite in Yugoslavian ophiolites accumulated with gravitational settling in the upper mantle; during the uprising of the rocks accompanying obduction in the Mediterranean region they generally lost, to some extent, their primary structures. During the movement they were partially fused, enriched with partial melts, corroded and arranged in flow layers. Rich orebodies during the kinetic differentiation were sometimes enlarged in the lower parts of the mass, where the podiform ores developed. In the last stages of the movement, the chromite ore and rock were solidified and the characteristic pull-apart structures were formed.

Besides large corrosion shapes small tooth-formed shapes also appear under the influence of hydrothermal effects or as a consequence of fast cooling.

2.2 Minerals Associated with Chromite

Olivine is the most common mineral associated with chromite. It appears as inclusions in chromite or as a matrix in the interstices between the chromite grains. It is idiomorphic, corroded or skeletal. The development of olivine in emulsion intergrowth with chromite is unusual. Olivine in the rich ore has slightly larger dimensions than in the disseminated chromite ores, where it is about ten times larger than chromite.

Serpentine appears in network-like texture as chrysotile, serpophite and bastite. It contains remnants of olivine and small particles of magnetite "powder".

Other silicates include enstatite, diopside, sometimes chromdiopside, smaragdite, kaemmererite, fuchsite and uvarovite.

Magnetite appears as small rims at the margins of chromite grains or as filling in fractures. Magnetite can grade sharply into chromite or gradually into feritchromite, where chromite is slowly enriched in iron especially along cracks or in the porous parts of the body.

Exsolutions are relatively rare. Hematite in Balkanic chromite is very rare, ilmenite and rutile occur sometimes as exsolved phases. Rutile is found in chromite from Deli Jovan as oriented inclusions such that $(100)_{ru} /(111)_{Cr}$.

Pyrrhotite is the chief sulphide found in chromite in association with other sulphides.

Pentlandite appears occasionally in pyrrhotite in strings, bands and wedge fillings between hypidiomorphic, rounded pyrrhotite grains. It sometimes forms flame-shaped exsolutions in pyrrhotite, which are partly transformed into bravoite. Pentlandite flames in pyrrhotite may at first consist of chalcopentlandite which then further decomposes (*Ramdohr*, 1975). Together with pyrrhotite, sphalerite, chalcopyrite, mackinawite, valeriite, heazlewoodite and millerite may rarely occur, very rarely nickolite and maucherite. In nearly all chromite ores there are small quantities of native metals, especially native platinoids. Awaruite is present in nearly all the chromite and occurs mostly together with serpentine as very small grains in cracks in chromite.

Besides chromite, carbonate-calcite, magnesite and exceptionally dolomite can occasionally be found. The latest minerals are talc, opal and chalcedony. In Figure 21 the mineral association and paragenesis of single minerals is presented. The diagram agrees

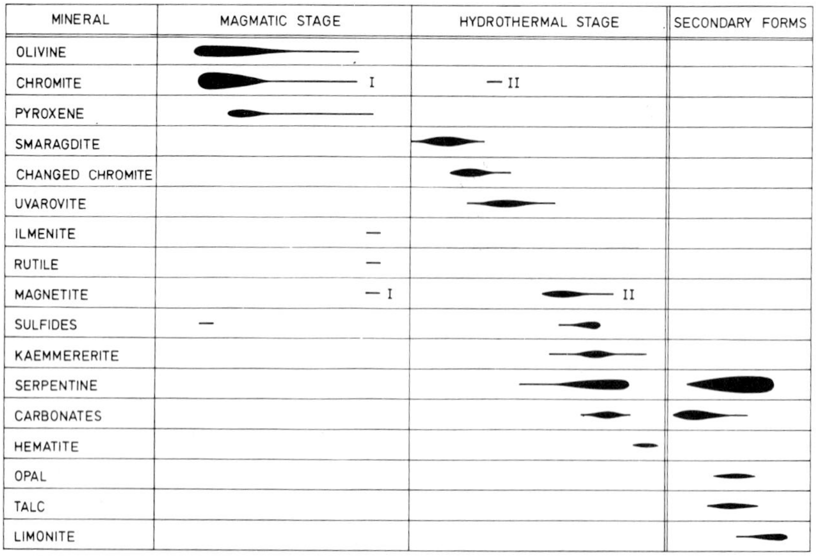

MINERAL	MAGMATIC STAGE	HYDROTHERMAL STAGE	SECONDARY FORMS
OLIVINE	━━━━		
CHROMITE	━━━ I	─ II	
PYROXENE	━━		
SMARAGDITE		━━	
CHANGED CHROMITE		━━	
UVAROVITE		━━	
ILMENITE	─		
RUTILE	─		
MAGNETITE	─ I	━━ II	
SULFIDES	─	━	
KAEMMERERITE		━	
SERPENTINE		━━━	━━
CARBONATES		━	━━
HEMATITE		─	
OPAL			━
TALC			━
LIMONITE			━

Fig. 21. The mineral association and succession of single minerals in Yugoslavian peridotite

with the succession of *Krause* (1958) and *Kern* (1968). It is probable that this paragenetic succession is characteristic for all types of alpino-type chromite deposits.

3. Genesis of Chromite in Yugoslavian Peridotite

Chromite deposits started to develop during chromite enrichment in static gravitational
differentiation in pyrolitic rocks within the upper mantle of the earth. Here also was
developed the larger layer of the chromitite zones. Structures, textures and forms corresponding with these features are expected to occur in anorogenic chromite ores. The
difference, however, is that the anorogenic chromite ores were formed in a less deep
environment. Orogenic ophiolites are much richer in ultramafic components.

During the upward movement kinetic differentiation took place. As the result of
decreasing pressure partial fusion occurred (*Dickey*, 1970; *Ito* and *Kennedy*, 1967). The
peridotite rock was deformed and sheets, drops and layers of partial chromite (dunite
rich in chromite) melts were formed. Droplets of chromite melt merged together and by
the process of liquation the nodules were formed. The arrangement of the ores corresponds with the flow lines of the peridotite. As deformation continued, the smaller streaks
coalesced to form thicker plates which were often rhythmically repeated. The essential
zonal arrangement formed at depth could be saved much better in the bigger ultramafic
massives than in the small ones. Especially important in the alteration was the control of
oxygen fugacity during the obduction.

The oxygen fugacity probably changed more quickly on the marginal parts of the
massif, where it came in contact with water-bearing sediments, and where also gases
could excape through cracks in the country rock. This produced the enrichment of
chromite in Mg, and the silicate minerals in Fe. Pulsation of partial oxygen pressure
probably caused very different rhythmic successions of chromite composition and thus
affected the shape of the ore deposits. The ore has a flow-layering structure, and the
rich orebodies enlarge their surface areas at depth. This resulted from the spreading of
large chromitite masses in a partly liquid mush with a crystalline matrix of silicates. The
orebodies wedge out very sharply at depth.

In the poor ore concentrations, bands or disseminated chromite grains welded together as a result of settling and fusion, with repeated crystallization of the energetically
most ideal form. The mutual relationship between chromite and silicate minerals was
dependent upon changing circumstances during kinetic differentiation. Sometimes the
olivine corroded chromite, and sometimes the reverse.

In the marginal parts of the ophiolite mass there were relatively more concentrated
volatile components, and as a result, high-temperature silicates (uvarovite, smaragdite,
etc) developed by the replacement of chromite. The mineralization was completed by
a hydrothermal sulphide phase of formation. Together with chromite a small quantity
of native Pt was precipitated and subsequently awaruite. Chromitite was occasionally
more liquid than the neighbouring silicate minerals. In these circumstances ore pipes or
ore veins and tree-like forms are developed. Similar forms are also found in veins in
dunite. After the emplacement of the orebodies was accomplished, there sometimes
formed, at their margins with country rocks, thin aureoles containing smaragdite,

chrome-chlorite and talc. These aureoles suggest the influence of chromite on the neighbouring rocks, as does also the ophiolite with glaucophane facies rocks.

All chromite deposits can thus be arranged within a large group of alpino-type deposits:

Singular genetic types are given in Table 2.

Table 2. Morphologic and genetic classification of alpinotype chromite and other deposits in Yugoslavian peridotites

---

I. Chromite deposits formed with static gravitational differentiation, deformed and shaped during obduction

    a) chromite disseminated in the rock as an accessory mineral
    b) chromite streaks and chromite-enriched peridotite in flow-layering and other shapes

II. Chromite deposits formed from static differentiated masses by their fusion and enrichment under the influence of volatile components during kinetic differentiation associated with obduction

    a) chromite podiform bodies with more or less clear contacts in the neighbourhood of primary static differentiated types; typified by nodular, orbicular and eutectic structures
    b) hysteromagmatic ore bodies with sharp contacts, enriched during the injection between more solid silicate masses, sometimes with exometamorphic features

        1. hysteromagmatic pipes and ramifications
        2. hysteromagmatic veins

III. Metamorphosed Fe, Ni residual deposits developed by weathering of peridotite during serpentinization

IV. Secondary ore deposits developed by mechanical transfer

    a) oolitic Ni and Fe ores
    b) placers and sands with chromite

Two principal groups are distinguished and differentiated by the degree of crystallization, structure, texture and form of the orebodies:

1. Chromite deposits which were mechanically altered in form and which were very slightly enriched during kinetic differentiation.

2. Podiform deposits, which were kinetically enriched during obduction.

All the deposits were transported and fractured after the initial emplacement. The serpentinization of ophiolitic rocks was most intensive along the weak marginal parts of the masses as a result of autohydratation and assimilation of water from the adjoining water-bearing sediments.

The ophiolitic masses were relative rich in volatile components and serpentinization was prolonged in all parts as an autometamorphic process. The gravitative differentiation can be shown by the regular zonation of the serpentinite massifs, whereas the kinetic differentiation is indicated principally by flow layering, foliation and lineation, gneissic and schistosic structures. The ores often show textures due to welding of grains. Chromite in poor ores may occur as accessory chromite in dunite or may form banded, sometimes platy ores.

The relation Fe/Mg in chromite and neighbouring minerals is variable, usually the chromite is little more rich in Mg than olivine. The ratio depends on the oxygen fugacity and on the pressure during the kinetic differentiation. Usually in these ores the chromite is older than the olivine, but it continued to crystallize even after the beginning of precipitation of olivine. Sometimes chromite grains were repeatedly fused and welded together, and as a result network-like structures were formed.

Chromite in the rich ores was kinetically differentiated during obduction. During partial fusion and movements, parts of the associated rocks and the chromitites were separated into single bands, lenses and pipe-like forms. Enrichment in chromite may have occurred during the kinetic differentiation by processes of segregation, liquation, intrusion and splashing of relatively submergent chromitite into zones of crystalline dunitic material.

The most important Yugoslavian ore deposits are hysteromagmatic, situated usually in the basal or the upper parts of serpentinite massifs. The shape of the ore deposits depends upon the character of the weakened zones into which the semi-molten chromitite was injected.

The contact between the rich orebodies and the country rock is usually sharp, often with secondary smaragdite, kaemmererite and uvarovite, that indicates the original presence of volatile components in the hysteromagmatic masses.

The contact between the rich ores and the country rock would not be so sharp if the orebody had not separated at depth because of the kinetic differentiation in the parent rock. Textures of different ore deposits are very variable, and can grade from one type to another. Sometimes chromite segregated before olivine, and sometimes simultaneously, and as a result emulsion, myrmekitic, orbicular and negative nodular structures are developed.

The volatile component in the magma was chiefly water. *Kushiro* (1967/68) has shown experimentally the possible effects of water when present in the system forsterite-diopside-silica. The opinions about the role of volatiles during the crystallization of chromite are still argumentative, but observations made both in the field and experimentally approach the same conclusion, that this kind of formation was most important for the development of alpino-type chromite deposits.

4. Conclusion

Chromite-bearing ophiolitic masses in Yugoslavia are distributed in many zones as a result of static differentiation. The final position is due to obduction. The richest ores are situated in basal parts, and sometimes in upper parts of massifs, whereas the ores in the central parts of serpentinite massifs are usually poor. The origin of the ore deposits and their enrichment are directly connected with the origin of the parent rocks. The ore deposits began concentration in depth, but this may have been prolonged during the uprising of the ophiolites associated with continental drift. At this time there developed in the ore the same structures, textures, forms and deformation, which develop in the rocks.

All chromite deposits, with rare exceptions, are associated with dunite. The beginning and sequence of crystallization was regulated by physicochemical conditions during the ore formation. Some deposits developed by crystallization in the liquid magmatic

stage in the upper mantle; others were late magmatic and developed during the enrichment of a rich chromitite melt which intruded the already solid country rocks. The
beginning of crystallization of single minerals depended on the oxygen fugacity and the
content of volatile components in the rock. The influence of volatiles can be seen in
many pegmatite veins in the neighbourhood of the ores.

Acknowledgments. This work was carried out and completed while the author was
employed as a Visiting Professor at Purdue University. I should like to thank the Head
of the Department of Geosciences, Professor *G. Kullerud* for the provision of facilities
for research in the laboratories of the department.

I am grateful to Professor *H.O.A. Meyer* of the same department at Purdue University
for the electron microprobe analyses and for reading the entire text. He eliminated
many of the language errors that inevitably creep in. However, the scientific views
expressed are those of the author alone.

Gratefully acknowledged is the help of Mr. *E. Geist* in preparing the polished sections.

Grants and aid of research and field work from the Yogoslavian Federal Fund for
Research in Sciences are gratefully acknowledged.

References

Bastin, E.S.: Interpretation of ore textures. Ithaca, New York: Geol. Soc. Am. Memoir **45**, 79 (1950)
Bowen, N.L., Tuttle, O.F.: The system $MgO-SiO_2-H_2O$. Bull. Geol. Soc. Am. **60**, 439–460 (1949)
Čermelj, S.: Some practical experience in the chromite deposits of Yugoslavia (Serbian). Rudarstvo
    i metalurgija **1**, 12–23 (1951)
Cloos, H.: Einführung in die Geologie. Ein Lehrbuch der inneren Dynamik. Berlin: Bornträger 1936,
    p. 903
Dickey, J.S.: Partial fusion products in Alpine type peridotites: Serrania de la Ronda and other
    examples. Mineral. Soc. Am. Pap. **3**, 33–49 (1970)
Donath, M.: Geologisch-mineralogische Studien an Serbischen Chromitlagerstätten. Freiburg 1930,
    p. 102
Evans, R.C.: An Introduction to Crystal Chemistry. 2nd ed. Cambridge: Univ. Press 1964, p. 687
Golding, H.G., Bayliss, P.: Altered chrome ores from Coolas serpentinite belt, New South Wales,
    Australia. Am. Mineral. **53**, 162–183 (1968)
Grafenauer, S.: Geology and origin of chromite deposits of Raduša. Ljubljana, Rudarsko geološki
    zbornik, 81–105 (1956)
Grafenauer, S.: Geology and origin of chromite ore deposits of Yugoslavia (Serbian). 2nd Cong.
    Yugoslav Geologists, Sarajevo, 246–250 (1957)
Grafenauer, S.: Mineralogical and chemical alterations of chromite in magmatic and hydrothermal
    phases (Serbian). 5th Cong. Yugoslav Geologists, Beograd, **2**, 285–293 (1962)
Grafenauer, S.: Ore petrology of ultramafic associations in Yugoslavia . (ed.) Ljubljana: Academia
    Scientiarum et artium Slovenica Classis IV, Opera 22, 1975, p. 152
Grčev, K., Milenković, P.: Development of the roof of the Ljuboten part of the Ljuboten serpenti-
    nite massif (Serbian). Proc. 5th Cong. Yugoslav Geologists, Belgrade **2**, 275–284 (1962)
Hess, H.H.: A primary peridotite magma. Am. J. Sci. **209**, 321–344 (1938)
Hiessleitner, G.: Serpentin- und Chromerzgeologie der Balkanhalbinsel und eines Teiles von Klein-
    asien. Wien: Jahrb. geol. Bundesanstalt, Sonderband 1, I. und II. Teil, 1951/52, p. 683
Irvine, T.N.: Chromian spinel as a petrogenetic indicator, Part I, Theory. Can. J. Earth Sci. **2**, 648–
    672 (1965)
Irvine, T.N.: Chromian spinel as a petrogenetic indicator, Part II, Petrologic application. Can. J.
    Earth Sci. **4**, 1–103 (1967)
Ito, K., Kennedy, G.C.: Melting and phase relations in natural peridotite to 40 kb. Am. J. Sci. **265**,
    510–538 (1967)

Jackson, E.D.: Primary textures and mineral associations in the ultramafic zones of the Stillwater Complex, Montana. U.S. Geol. Surv. Prof. Paper 358, 1–106 (1961)

Janković, S.: Metallogenic epochs and ore bearing districts in Yugoslavia (Serbian). Beograd: Prosveta 1967, p. 197

Karamata, S.: Zonality in contact metamorphic rocks around the ultramaphic mass of Brezovica (Serbia, Yugoslavia). 23th Intern. Geol. Cong. 1, 197–207 (1968)

Karamata, S.: Characteristics of Brezovica chromitites and their genesis (Serbian). Geol. anali Balk. Poluostrva 34, 461–473 (1969)

Karamata, S.: Chromitite Jugoslawiens. Gefüge, Chemismus und genetische Beobachtungen. Fortschr. Mineral. 48, Beiheft I, 11–13 (1970)

Karamata, S., Milenković, P., Čuturić, N.: A contribution to the geochemistry of chromite ores in the Raduša Mine. Inst. Rech. geol. geoph., Bull., Ser. A., 22/23, Beograd, 67–75 (1964)

Kern, H.: Zur Geochemie und Lagerstättenkunde des Chroms und zur Mikroskopie und Genese der Chromerze. Clausthaler Hefte zur Lagerstättenkunde und Geochemie der mineralischen Rohstoffe. H. Borchert (ed.). Berlin, 6, 1968, p. 236

Krause, H.: Erzmikroskopische Untersuchungen an türkischen Chromiten. N. Jb. Mineral. Abh. 90, 305–366 (1958)

Kushiro, I.: Liquidus relations in the system forsterite-diopside-silica-H_2O. Annual report of the director of Geoph. Lab. (Washington), 158–161 (1967/68)

MacGregor, I.D.: The effect of pressure on the minimum melting composition in the system MgO-SiO_2-TiO_2. Trans. Am. Geoph. Union 47, 207–208 (1966)

MacGregor, I.D.: The effect of CaO, Cr_2O_3, Fe_2O_3 and Al_2O_3 on the stability of spinel and garnet peridotites. Phys. Earth Planet. Interior., Spec. Vol. Phase transformations and the earth interior. Amsterdam: North Holland Publ. Co. 1970, pp. 372–377

Majer, V.: Petrography and petrogenesis of Brezovica ultrabasic rocks on the northern slope of Šar Mountains (Croatian). Doctor thesis. Yugoslav Acad. Sci., Class of Earth Sci., Acta geol. 1, 89–142 (1956)

McKenzie, D.P.: Plate tectonics of the Mediterranean region. Nature (London) 226, 239–243 (1970)

Presnall, D.C.: The geometrical analysis of partial fusion. Am. J. Sci. 267, 1178–1194 (1969)

Ramdohr, P.: Die Erzmineralien und ihre Verwachsungen. 4th ed. Berlin: Akademie-Verlag 1975, p. 1277

Thayer, T.P.: Application of geology in chromite exploration and mining. Symp. Chrome Ore, Ankara, 197–223 (1961)

Thayer, T.P.: Principal features and origin of podiform chromite deposits and some observations on the Guleman Soridag district, Turkey. Econ. Geol. 59, 1497–1524 (1964)

Thayer, T.P.: Gravity differentiation and magmatic reemplacement of podiform chromite deposits. In: Magmatic Ore Deposits. Symp. Econ. Geol. Monogr., 4th ed. Washington D.C.: H.D.B. Wilson 1969, pp. 132–146

Ulmer, G.C.: Experimental investigation of chromite spinels. In: Magmatic Ore Deposits. Symp. Econ. Geol. Monograph 4. (ed.) Washington D.C.: H.D.B. Wilson 1969, pp. 114–131

Vogt, J.H.: The physical chemistry of the crystallization and magmatic differentiation of igneous rocks. J. Geol. 29, 318–350 (1921)

Wells, A.F.: Structural inorganic chemistry. (ed.) London: Oxford Univ. Press 1962, p. 732

Oxygen Fugacity and Its Bearing
on the Origin Chromitite Layers in the Bushveld Complex

R. SNETHLAGE, München, and G. VON GRUENEWALDT, Pretoria

With 7 Figures

Contents

Summary

Essential to the understanding of layered intrusions is an explanation of the mechanism that accounts for the repetitive occurrence of chromitite layers. For that purpose *Ulmer*'s (1969) hypothesis, according to which the repetitive occurrence of chromite layers could be caused by pulsations in the oxygen fugacity, is discussed. The results obtained on synthetic systems are compared with f_{O_2} measurements of natural chromites from the Zwartkop Chrome Mine western Bushveld Complex.

With the aid of *Hill* and *Roeder*'s (1974) experimental results on natural basalt, a range of temperature (1175–1200°C) and f_{O_2} ($10^{-6.35}-10^{-7.20}$ atm) conditions can be derived for the formation of the chromitite layer LG4. For chromitite layer LG6, the range of temperature and f_{O_2} are 1162–1207°C and $10^{-6.20}-10^{-7.50}$ atm. respectively.

An isothermic profile at 1200°C for all f_{O_2}-T curves of all chromites supports *Ulmer*'s (1969) thesis because the resulting f_{O_2} profiles are reflected in the mode. The f_{O_2} values of chromites from massive chromitite are higher than those of disseminated chromites, so that isothermic f_{O_2} changes could have been responsible for the formation of massive chromitite layers.

The feasibility of fluctuations of f_{O_2} to account for the origin of chromitite layers is subsequently reviewed in the light of some of the more recent information that has become available on the Bushveld Complex.

1. Introduction

In view of the low solubility of chromium in liquids of basaltic composition, chrome spinel usually crystallizes during the early stages of fractional crystallization of basaltic magmas and occurs as a disseminated mineral or concentrated as layers in the lower parts of many of the layered intrusions throughout the world.

The effect of this low solubility and its tendency to crystallize early can be illustrated more quantitatively with the aid of investigations on systems containing Cr_2O_3, such as MgO-Cr_2O_3-SiO_2 (*Keith*, 1954). Recently this system was enlarged by adding the iron oxide component (*Arculus* et al., 1974) in order to establish a better correlation between the experimental results and the natural phase relations. This system (Fig. 1) is characterized, firstly, by the large volume of primary spinel crystallization, which extends from the system MgO-Cr_2O_3-SiO_2 on the left face, through the tetrahedron to the primary phase field of magnesioferrite on the base and, secondly, by the very narrow primary phase fields of olivine and pyroxene, amounting to a maximum of only about 1.4 wt % Cr_2O_3 in the liquid. The maximum Cr_2O_3 content of olivine and pyroxene in equilibrium with liquid in this system was determined by *T.N. Irvine* to be 0.5 per cent and 0.3 per cent respectively (*Arculus* et al., 1974, p. 317). A liquid containing about 2 wt % Cr_2O_3 always starts to crystallize chromite, which is followed by olivine or orthopyroxene, and a massive chromitite layer can be the result, irrespective of whether equilibrium or fractional crystallization is assumed. Alternatively, a magma with a considerably lower Cr_2O_3 content and a composition in the primary phase fields of either olivine or orthopyroxene may, on fractional crystallization and fractional segregation of the phases, also yield chromitite layers. Significant in this respect is the observation by *Arculus* et al. (1974) that small changes in temperature and composition during crystallization of the liquid may result in a much wider range of spinel compositions than coexisting olivine or pyroxene.

Although the above-mentioned relationships in the MgO-iron-oxide-Cr_2O_3-SiO_2 system and the experimental work on natural basalts by *Hill* and *Roeder* (1974) have demonstrated that chrome spinel crystallizes from liquids with extremely low Cr_2O_3 content, mechanisms to explain the large number of chromitite layers in many of the large layered intrusions has been the subject matter of discussion for a considerable time. In the Critical Zone of the western Bushveld Complex *Cousins* and *Feringa* (1964) have traced 15 individual chromitite layers for many km along strike. *Cameron* and *Desborough* (1969) have recorded as many as 29 chromitite layers from the Critical Zone in the northern sector of the eastern Bushveld Complex, several of which have been followed along strike for up to 90 km. Therefore, apart from briefly reviewing data in synthetic systems pertinent to spinel stability and presenting new information on oxygen fugacity determinations on Bushveld chromite, the meachanisms by which the recurrence of massive chromitite layers in the sequence can be explained are discussed in the light of the more recent field and laboratory investigations.

2. Experimental Results

2.1 Synthetic Systems

From many experimental investigations in the system MgO-iron oxide-Cr_2O_3-Al_2O_3 at 1300°C, it has become evident that the stability of spinel depends on the oxygen fugacity conditions. The results of these investigations were summarized diagrammatically on the spinel prism by *Ulmer* (1969). In these diagrams, reproduced here in Figure 2, the volumes of the spinels stable at the indicated oxygen fugacities are outlined by

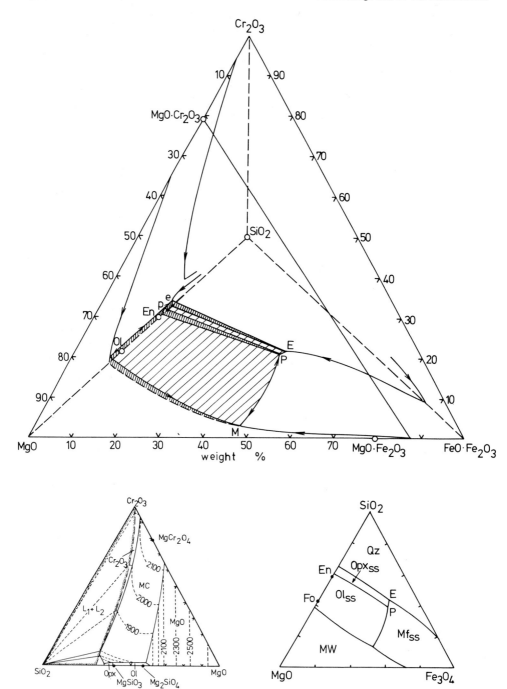

Fig. 1. Liquid-phase relationships in the system MgO-iron oxide-Cr_2O_3-SiO_2 in air (after *Arculus* et al., 1974). For clarity the left face of the tetrahedron (MgO-SiO_2-Cr_2O_3, after *Keith*, 1954) and the basal plane (MgO-FeO · $Fe_2O_3SiO_2$, after *Muan* and *Osborn*, 1956) are added in separate figures. *Ol:* olivine; *Opx:* orthopyroxene; *Mf:* magnesioferrite; *Qz:* quartz; *En:* enstatite; *MW:* magnesio-wustite; *L:* liquid

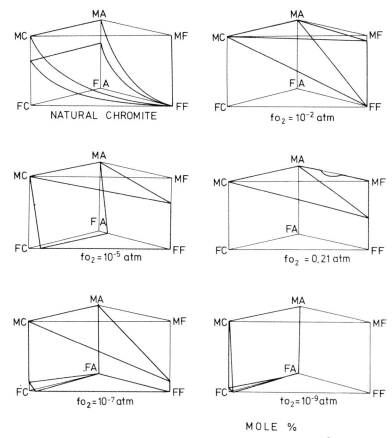

MOLE %

Fig. 2. The spinel compositional prism showing the spinel solid solutions stable at 1300°C and at the indicated oxygen fugacity. For comparison the figure contains analyses of natural chromites after *Stevens* (1944). *MC:* $MgCr_2O_4$; *MA:* $MgAl_2O_4$; *FC:* $FeCr_2O_4$; *FA:* $FeAl_2O_4$; *MF:* $MgFe_2O_4$; *FF:* $FeFe_2O_4$. (After *Ulmer*, 1969)

thick lines. These diagrams illustrate that the volume of the spinel stability has a maximum in the range 10^{-5} to 10^{-7} atm and comparison of these volumes with the compositional distribution of natural spinels given by *Stevens* (1944) shows reasonably good agreement. Therefore, at a realistic geological temperature of 1300°C, an oxygen fugacity range of 10^{-5} to 10^{-7} atm seems to be indicated during formation of chromitite layers.

Similar results are obtained from liquidus information containing spinel silicate assemblages. Figure 3 a–c shows the liquid-phase relations of suitable sections of the complex system MgO-FeO-Fe_2O_3-SiO_2 at three different oxygen fugacity conditions (from *Muan* and *Osborn*, 1956). By comparing these three diagrams, it can be seen that the spinel field decreases with decreasing oxygen fugacity. The experiments conducted at constant CO_2/H_2 ratio of 40 show that an oxygen fugacity of 10^{-7} atm is attained at the invariant point at 1225°C in Figure 3b. For the extreme reducing conditions of coexisting metallic iron (Fig. 3c) the spinel field disappears completely and the oxygen fugacity reached at the invariant point at 1305°C is $10^{-10.5}$ atm.

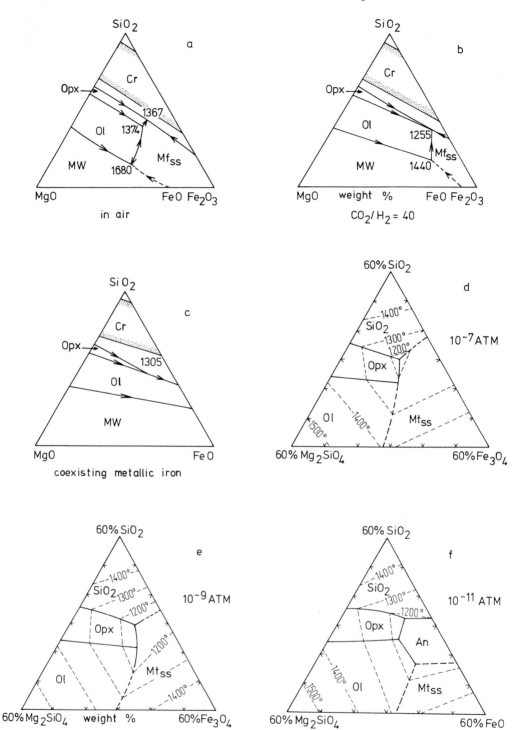

Fig. 3 a—f

Comparable results were obtained by *Roeder* and *Osborn* (1966) in the system MgO-FeO-Fe$_2$O$_3$-CaAl$_2$Si$_2$O$_8$-SiO$_2$. Figure 3 d–f shows the liquid-phase relations of this system at constant An content of 40 wt % at different oxygen fugacities. Again, the size of the spinel field is reduced by a decrease in the oxygen fugacity, but an additional important feature of geological significance is that pyroxene and spinel become incompatible at an f_{O_2} value between 10^{-9} and 10^{-11} atm. From these diagrams the phase assemblage pyroxene + anorthotite + spinel seems to be stable only at oxygen fugacities above 10^{-11} atm.

The remarkable influence of the oxygen fugacity on the liquid-phase relations in the two above-mentioned systems is quite evident. It must however be borne in mind that these three diagrams are sections through the complex quinary system MgO-FeO-Fe$_2$O$_3$-CaAl$_2$Si$_2$O$_8$-SiO$_2$ and it is therefore not possible to relate the complete crystallization sequence from them. They do, however, suffice to be qualitatively related to the early stages of fractional crystallization of a basaltic magma. It is therefore quite feasible that as a result of an increase in the oxygen fugacity, a liquid during fractional crystallization is shifted away from a cotectic boundary line to become isolated in the spinal phase volume. The resulting crystallization of spinel will cause a decrease in the oxygen fugacity and the return of the liquid to the cotectic line. Fluctuation of oxygen fugacity during fractional crystallization is therefore a very attractive mechanism by which chromitite layer formation has been explained (*Roeder* and *Osborn*, 1966; *Ulmer*, 1969), a mechanism which also seems to explain satisfactorily many of the intricate features of the chromitic intervals in the Critical Zone of the eastern Bushveld Complex (*Cameron* and *Desborough*, 1969).

In the system MgO-FeO-Fe$_2$O$_3$-SiO$_2$ (*Muan* and *Osborn*, 1956) and MgO-iron oxide-Cr$_2$O$_3$-SiO$_2$ (*Arculus* et al., 1974) referred to in the preceding sections, the spinel forms complete solid solution series between MgFe$_2$O$_4$ and Fe$_3$O$_4$ and MgCr$_2$O$_4$ and MgFe$_2$O$_4$ respectively at high temperatures and indicated oxygen fugacity conditions. Therefore all the liquid-crystal equilibria are either univariant or bivariant, so that the composition of the solid phases has some degree of freedom regarding temperature and/or oxygen fugacity. Bearing in mind that Mg spinels are stable at higher oxygen fugacities than lower Mg spinels (Fig. 2; *Arculus* and *Osborn*, 1975) it follows for both systems that, independent of any other coexisting phases, the Mg/Mg + Fe^{2+} ratio in the spinel phase decreases with decreasing oxygen fugacity (isothermic conditions) and also decreases with decreasing temperature and oxygen fugacity. The latter conditions would correspond to those prevailing during crystallization at constant total composition (*Osborn*, 1959). On the other hand *Arculus* and *Osborn* (1975) suggest that the Fe^{3+}/Fe^{2+} ratio of the spinel increases with decreasing temperature, so that two parameters are available for relating the experimental and the natural conditions of formation. However, quantitative correlation between the experimental results and natural conditions are not possible, even though the Fe^{3+}/Fe^{2+} ratio for a particular magma composi-

---

◄ Fig. 3 a–f. Liquid-phase relationships under varying oxygen fugacity conditions in the system MgO-FeO-Fe$_2$O$_3$-SiO$_2$ (after *Muan* and *Osborn*, 1956) and in the system MgO-FeO-Fe$_2$O$_3$-CaAl$_2$Si$_2$O$_8$-SiO$_2$ (after *Roeder* and *Osborn*, 1966). *Ol:* olivine; *Opx:* orthopyroxene; *Mf:* magnesioferrite; *Mt:* magnetite; *MW:* magnesiowustite; *Cr:* cristoballite; *An:* anorthite

tion is fixed by temperature and oxygen fugacity (*Fudali*, 1965). This was elaborated upon in the discussion by *Osborn* on the paper by *Ulmer* (1969), where it was pointed out that different magmas, sufficiently comparable in composition to be classified as the same rock type, e.g., basalt, may have considerably different Fe^{3+}/Fe^{2+} ratios for any given temperature and f_{O_2}. It follows that the influence of the availability of elements on the spinel composition must be taken into consideration as an important factor during the formation of natural spinels; a statement which is also valid for other components, such as Cr_2O_3 and Al_2O_3. Because of these unknown factors, valid quantitative conclusions on the conditions of formation of spinel with the aid of cation or oxide ratios cannot be made. This also applies to the $Mg/Mg + Fe^{2+}$ ratio, even though it can probably be considered as the most reliable indicator for the oxygen fugacity and temperature conditions.

2.2 f_{O_2} Measurements of Bushveld Chromitite

2.2.1 Laboratory Investigations and Petrography

Information about the influence of the oxygen fugacity on the formation of chromitite layers can be obtained by means of direct P_{O_2} measurements on natural chromitites. These measurements can be carried out in the temperature range between 600 and 1300°C with the aid of galvanic cells using ZrO_2 (CaO) as solid electrolyte. Since the f_{O_2} of the reference buffer as a function of temperature is known, it is possible to determine the unknown oxygen fugacity f_{O_2} of the sample by means of EMF measurements using the Nernst Equation

$$EMF = \frac{RT}{4F} \ln \frac{f_{O_2^I}}{f_{O_2^{II}}} \quad (V, K, atm),$$

where R is the Gas constant and F the Faraday constant.

The possibility of measuring oxygen fugacities of oxide samples is due to the fact that the chemical potential of oxygen μ_{O_2} in any oxide is a function of pressure, temperature and composition, which here means the metal/oxygen ratio. If, under experimental conditions, the quantity of the sample exceeds the gas phase in the sample chamber, the chemical potential μ_{O_2} of oxygen can be measured, because the chemical potentials of oxygen in the gas phase and in the solid phase under prevailing equilibrium must be equal.

From the above it is evident that every postmagmatic process may influence the chemical potential of oxygen in the oxide and consequently this will restrict the conclusions which can be drawn from the experimental results. It is therefore necessary to bear the following in mind before interpreting the results.

a) Directly after settling, equilibration of coprecipitated chromite and pyroxene or olivine will result in an increase of the Fe in chromite and a decrease of Mg, because the equilibrium constant of the reaction

$$Fe^{2+}_{px} + Mg^{2+}_{chr} = Fe^{2+}_{chr} + Mg^{2+}_{px}$$

increases with decreasing temperature. If this reaction takes place to any great extent, oxygen fugacity measurements reflect those f_{O_2} and T conditions at which the equilibration stopped.

b) Electron microprobe measurements on natural chromite grains in different environments from the eastern Bushveld Complex (*Cameron*, 1975) do not show any decrease of Mg or increase of Fe in the rims. This led *Cameron* to conclude that subsolidus equilibration probably did not occur, and that the compositional variations associated with the variations in the modal proportions of the cumulus phases stem from the cumulus stage. Consequently the chemical potential of oxygen could also not have been affected. Furthermore if the coprecipitated phases were covered soon after settling and the system was closed, then these phases would act as buffers, so that the chemical potential could no longer have been influenced by the magma.

If the chromite has not undergone any significant alteration after deposition as deduced above, then it should be possible to obtain from the various f_{O_2} curves of chromites from a particular layer a picture of the relative isothermal f_{O_2} variation during the formation of the particular layer.

For the f_{O_2} measurements, continuous chromite samples across two layers from the Zwartkop Chrome Mine, western Bushveld Complex were available. Both layers, LG4 and LG6, belong to the lower pyroxenitic part of the Critical Zone in which the thicker chromitite layers are numbered LG1 to LG7 from the base upwards (notation by *Cousins* and *Feringa*, 1964). The limited amount of chemical and modal data presented here is extracted from a large quantity of data which is presently being evaluated for a more comprehensive study of the chromitite layers at the Zwartkop Chrome Mine (*von Gruenewaldt* and *Snethlage*, in preparation). All the chemical analyses were done by electron microprobe and the Fe^{3+} was calculated on the basis of 24 cations to 32 oxygen.

The hanging and footwall rocks of LG4 consist of olivine cumulates in which the chromite content is usually less than 5%. Olivine in these rocks has been replaced to varying degrees by reaction replacement orthopyroxene. 67 cm below the top of LG4 is an abrupt increase in the amount of disseminated chromite from less than 3% to about 40%. In this olivine chromite cumulate, which is about 40 cm thick, the chromite grains clearly outline the original grain margin of the olivine prior to replacement by orthopyrocene. The LG4 chromitite layer at Zwartkop consists of an upper massive chromitite some 14 cm thick and a lower, 12 cm thick, laminated chromitite. Within this laminated portion are five thin chromitite layers 1 to 2 cm thick and separated from one another by thin silicate-rich layers. Chromite is the only cumulus mineral in the chromitite layer of LG4, and orthopyroxene, plagioclase and clinopyroxene occupy the spaces interstitial to the chromite grains. The Cr/Fe ratio of 2.3 is the highest of all chromitite layers and the associated olivine cumulates are the most basic rocks in the layered sequence of the western Bushveld Complex.

Chromite is a much more abundant constituent of the bronzite associated with LG6 than in the rocks associated with LG4. The basal contact of the 80-cm thick LG6 is sharp, whereas the top contact is gradational. One disseminated "leader" occurs in the immediate hanging, and two higher up in the sequence, although the LG6A, often referred to as the Leader Seam, is about 7 m above LG6 in the Zwartkop area. No olivine was observed in any of the thin sections from LG6 and the associated rocks, and plagioclase is by far the most abundant interstitial mineral.

CHROMITE SEAM LG 4

ZWARTKOP CHROME MINE (Western Bushveld Complex)

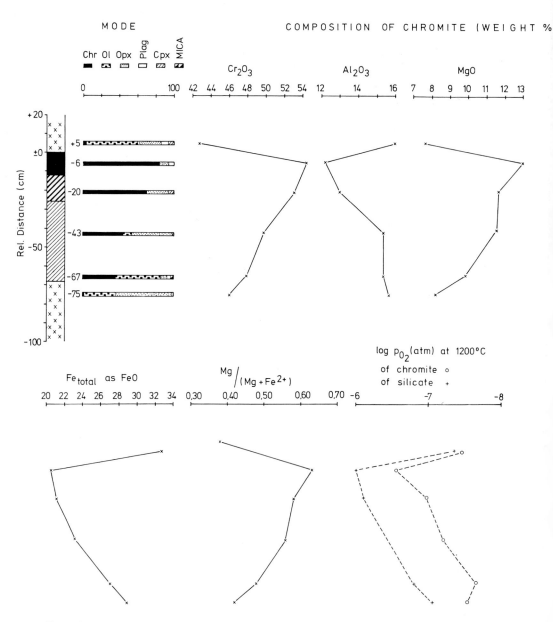

Fig. 4. Petrographic data and chemical analyses of the chromites of the chromite seam LG4 (Zwart-kop Chrome Mine, western Bushveld Complex). The Mg/(Mg + Fe^{2+}) ratio of chromite is calculated on the basis of the assumption that RO:R$_2$O$_3$ = 1:1. Attention must be paid for the interdependence between the modal analysis and the oxygen fugacities of chromites and silicates measured at 1200°C. Further explanations please see text. *Ol:* olivine; *Chr:* chromite; *Opx:* orthopyroxene; *Cpx:* clino-pyroxene; *Plag:* plagioclase

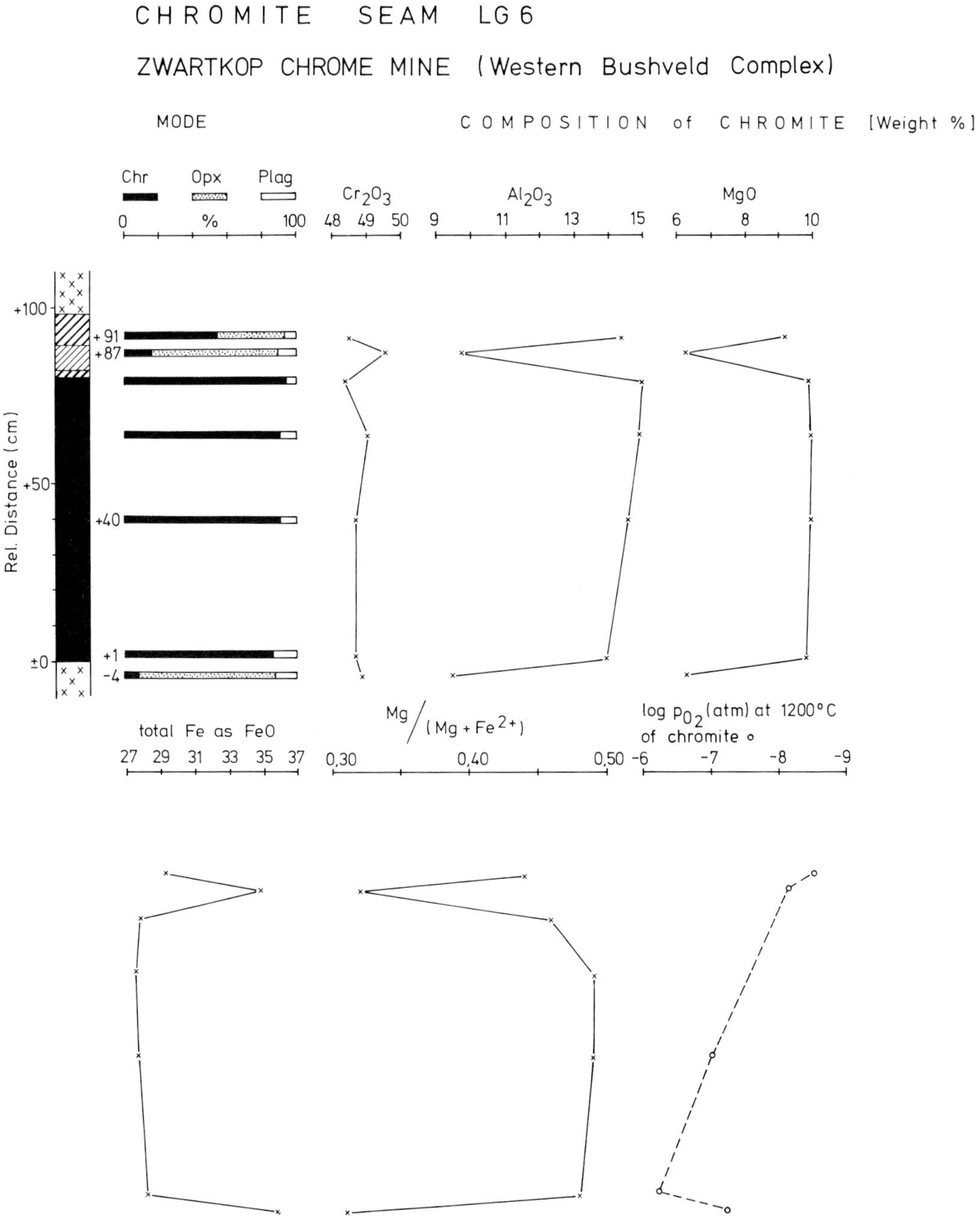

Fig. 5. Petrographic data and chemical analyses of the chromites of the chromite seam LG6 (Zwartkop Chrome Mine, western Bushveld Complex). The Mg/(Mg + Fe^{2+}) ratio of chromite is calculated on the basis of the assumption that RO:R$_2$O$_3$ = 1:1. Attention must be paid for the interdependence between the modal analysis and the at 1200°C measured oxygen fugacities of chromites. Further explanations see text. *Chr:* chromite; *Opx:* orthopyroxene; *Plag:* intercumulus plagioclase and clinopyroxene

In the chromitite of both layers, the MgO content increases and the Fe content decreases with an increase in the modal percentage of chromite, which is consistent with many other chromitite layers of the Bushveld Complex as well as other layered intrusions (*Cameron*, 1964; *McDonald*, 1967; *Jackson*, 1969; *Henderson* and *Suddaby*, 1971). The behaviour of the other elements, however, differs considerably in these two layers. For LG4 an increase in the amount of chromite is accompanied by an increase in the Cr_2O_3 content and a decrease in the Al_2O_3 content of the chromite. In LG6, the Cr_2O_3 content shows no noticeable variations, whereas the Al_2O_3 content, in contrast to LG4, increases abruptly at the sharp footwall contact.

The positions of the samples selected for the f_{O_2} determinations are indicated next to the columnar sections of Figures 4 and 5. The preparation of the samples consisted of crushing, magnetic separation and pulverizing under acetone. Readings were carried out at various temperatures in the range 1000 to 1200°C and were controlled by approaching the measuring points from lower and from higher temperatures during heating and cooling respectively (*Sato*, 1972; *Klemm* and *Kostakis*, 1976). The accuracy of the f_{O_2} measurements is estimated as ± 0.2 log units. The technique of measurements will be described more fully in *von Gruenewaldt* and *Snethlage* (in preparation).

2.2.2 Interpretation of the Results

Interesting deductions can be made when the results of this investigation are combined with those obtained on basaltic liquids by *Hill* and *Roeder* (1974). In their experiments they found that spinel is a liquid phase only from a definite oxygen fugacity value onwards (see Fig. 6 and 7), and that the spinel liquidus is shifted to higher temperatures with increasing Cr content. In their experimental work *Hill* and *Roeder* also noted a significant increase in the amount of spinel crystallization at temperatures and oxygen fugacities corresponding to the area in the vicinity of A–B in Figures 6 and 7. Although this observation is of qualitative nature only, it is very significant in that there seems to exist a T-f_{O_2} range at which spinel crystallization is enhanced, irrespective of the initial Cr concentration of the magma, and it may very well be that a temperature and an oxygen fugacity corresponding to values along A–B are a prerequisite for chrimitite layer formation.

Diagrams of this nature can only be applied to one particular basalt composition as the liquid curves and the various assemblage fields change with a change in the composition of the basalt. Although the crystallization conditions and the bulk composition of the Bushveld magma are unknown and certainly different from the experimental conditions, especially with respect to pressure, from the basalt types investigated by *Hill* and *Roeder*, their diagram for an olivine tholeiite from Hawaii will be used in the ensuing discussion. A corresponding diagram for the Bushveld magma will look somewhat different, although the important features, particularly the area of enhanced chromite crystallization, will be similar.

The experimentally determined f_{O_2}-T curves of the chromites from the massive parts of the layers (samples LG4, − 6 and 20; LG6, + 1 and + 40) and of the chromites from the host rocks (LG4, − 75 and + 5; LG6, − 4) are plotted on to Figures 6 and 7. In the case of the chromites of the massive parts of layers the intersection of the f_{O_2}-T curves

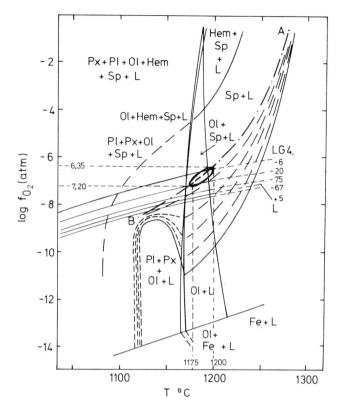

Fig. 6. Liquid relationships of a basaltic liquid in dependence of oxygen fugacity and temperature, showing conditions of enhanced chromite crystallization corresponding to the line $A-B$ (simplified after *Roeder* and *Hill*, 1974). The experimentally determined f_{O_2} curves of the chromites from the massive part of the layer LG4 (-6 and -20) intersect line $A-B$ in the outlined area. If this area can be regarded as representative for the conditions of the chromite formation, the massive parts of the layer LG4 should have formed at the indicated f_{O_2}- and T-conditions. The curves of the chromites from the host rocks of LG4 (-75 and $+5$) plot below the above mentioned area, intersecting, at constante temperature, the trend lines of less chromite crystallization, which is in agreement with the observations in the nature and supports *Ulmer's* thesis (1969). *Ol:* olivine; *Px:* pyroxene; *Pl:* plagioclase; *Sp:* spinel; *Hem:* hematite; *L:* liquid

with *Hill* and *Roeder's* line of enhanced chromite crystallization A–B, is outlined. If this area on the diagram is regarded as reprensentative for the chromite formation, then the massive parts of layer LG4 should have formed in the temperature interval of 1175–1200°C and an f_{O_2} interval of $10^{-6.35} - 10^{-7.20}$ atm (Fig. 6). For the LG6 similar values result (1162–1207°C, $10^{-6.20} - 10^{-7.50}$ atm f_{O_2}; Fig. 7) so that it may be concluded that both layers, in spite of their different appearance and composition, could have formed under very similar conditions. It has to be kept in mind that, although these derived values are very realistic for magmatic conditions, they must be regarded with caution because of the uncertainty at this stage whether the experimentally determined range of enhanced chromite crystallization is also a characteristic feature of magmatic environments.

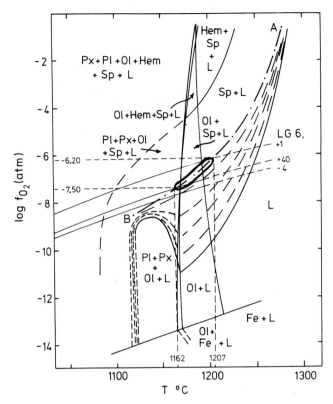

Fig. 7. Liquid relationships of a basaltic liquid in dependence of oxygen fugacity and temperature, showing conditions of enhanced chromite crystallization corresponding to the line $A-B$ (simplified after *Roeder* and *Hill*, 1974). The experimentally determined f_{O_2} curves of the chromites from the massive part of the layer LG6 (+ 1 and + 40) intersect the above explained line $A-B$ in the outlined area. If this area can be regarded as representative for the conditions of the chromite formation, the massive parts of the layer LG6 should have formed at the indicated f_{O_2}- and T-conditions. The f_{O_2}-T-curve of the chromite from the host rock (LG6, − 4) plots below the above-mentioned area, intersecting, at constant temperature a trend line of less chromite crystallization. This result is in agreement with the observations in the nature and supports *Ulmer*'s thesis (1969). *Ol*: olivine; *Px*: pyroxene; *Pl*: plagioclase; *Sp*: spinel; *Hem*: hematite; *L*: liquid

    The curves of the host rock chromites plot below the above-mentioned, bordered area of the chromites from the massive parts and consequently, at constant temperature (see above), these curves intersect the trend lines of less chromite crystallization at lower oxygen fugacities. Both observations agree with the findings in nature and support *Ulmer*'s hypthesis (1969).

    It might be appropriate at this stage to discuss isothermic f_{O_2} profiles across the chromitite layers. This idea is influenced and supported by the observation that the curves for all the chromites from a particular layer (except one which is not included here) have a nearly identical inclination and therefore do not intersect (cf. Figs. 6 and 7). If any isothermic change of the oxygen fugacity should have triggered the chromite crystallization, the f_{O_2} variation across any particular layer should reflect a characteristic course. According to the temperatures derived from Figures 6 and 7, a temperature of 1200°C seems to be appropriate for this purpose for LG4 as well as for LG6.

The oxygen fugacities obtained at $1200°C$ for the various chromite specimens are plotted in Figures 4 and 5, as well as the oxygen fugacities of the coecisting silicates in the case of LG4.

The oxygen fugacities of the LG4 specimens increase continually from the base up to the massive part which is also reflected in the modal analyses by an increase in the amount of chromite and in the chemical composition of the chromite by an increase in the $Mg/(Mg + Fe^{2+})$ ratio. At the top, between specimens LG4 – 6 and LG4 + 5, a sharp drop in the f_{O_2} is accompanied by an abrupt change in modal and chemical composition. A similar trend is observed for the silicates. Because of the lack of chemical analyses of the silicates at present, this may tentatively be interpreted by the fact that both chromite and silicates are cumulus minerals, except in LG4 – 6 where the silicates are intercumulus. For LG6 a similar interdependence can be observed between the f_{O_2}-profile and the modal composition as well as the chemical composition of the chromite. The $1200°C$ isothermic section (Fig. 5) indicates a sudden increase in the f_{O_2} at the sharp footwall contact.

From the hitherto obtained results it is astonishing to observe no increase in oxygen fugacity between the specimen LG6 + 87 (disseminated chromite) and LG6, + 91 (chromite of the satellite layer), although there are sharp discontinuities in the chemical composition. The drop in f_{O_2} from + 87 tò + 91 implies that the increase in chromite in this satellite leader was probably brought about by mechanical means of concentration.

Unfortunately no oxygen fugacity measurements on other specimens from the western Bushveld Complex exist for comparison. f_{O_2}-measurements on chromite specimens from the Leader Seam (eastern Bushveld Complex) at $1300°C$ (*Ulmer*, 1969, p. 131) show that the oxygen fugacity of the massive chromite is $10^{-6.3} \pm 10^{-0.5}$ atm and $10^{-7.7} \pm 10^{-0.5}$ atm for the disseminated chromite from the overlying pyroxenite. It is difficult to compare these measurements with those reported here, because, firstly, the distance between the two localities in the Bushveld Complex is very great and, secondly, the exact positions of the samples in the layer are unknown. Nevertheless, the massive chromites from LG4 and LG6 show nearly identical f_{O_2} values when their curves are extrapolated to $1300°C$. For the massive part of LG4 a f_{O_2} interval of $10^{-5.80}$ to $10^{-6.15}$ atm is obtained and for LG6 a f_{O_2} interval of $10^{-5.0}$ to $10^{-5.80}$ atm. Disseminated chromite from LG4 and LG6, however, shows higher f_{O_2}-values at $1300°C$ than those reported by *Ulmer* (1969, p. 131), namely $10^{-6.10}$ atm and $10^{-6.60}$ respectively.

These differences should not be overvalued, firstly, because the f_{O_2} values obtained by extrapolation depend on the inclinations of the curves, which are different for LG4 and LG6 chromites, and, secondly, because the rock successions in the Critical Zone of the western and the eastern Bushveld Complex are not identical, so that differences of this magnitude are to be expected.

3. Discussion

Available data from synthetic systems containing spinel as a primary phase, as well as direct measurements of f_{O_2} on natural chromite, present a strong case for oxygen fugacity as a major, if not a controlling factor for chromitite layer formation. Processes by which changes in the f_{O_2} could be achieved in a crystallizing magma were summarized by *Ulmer* (1969). Of these, periodic assimilation of CO_2 and H_2O from wall rocks of the intrusion, and cyclic build-up and release of total pressure through a vent are probably the most important mechanism to account for a periodic increase in the f_{O_2}. Mechanisms of this nature, rather than the contamination model proposed by *Irvine* (1975) are considered more feasible to explain the tremendous lateral extent of the chromitite layers in the Bushveld Complex. *Cameron* and *Desborough* (1969) place much emphasis on wall rock assimilation to account for periodic f_{O_2} increase and present evidence that such processes were operative. It can be envisaged that conditions for assimilation of wall rock material would be favoured during periods of tectonic readjustment of the magma chamber. *Cameron* (1971) has related several features such as disconformities within the Critical Zone to such movements of the magma. However, important additional evidence for tectonic disturbances simultaneously with the accumulation of the layered sequence has recently come to light along the western flank of the eastern Bushveld Complex (*Hunter*, 1975; *Marlow* and *Van der Merwe*, 1977). In the area investigated by *Marlow* and *Van der Merwe* in the northwestern part of the eastern Bushveld, rocks of the Critical Zone, together with the underlying floor-rocks of the Complex, were updomed to such an extent that they are now directly overlain by magnetite-bearing diorites of the top of the Upper Zone. The rocks correlated with those of the Cirtical Zone and the lower part of the Main Zone display ample features of deformation at high temperature. Contamination by wall rock material is evidenced by anorthosite containing up to 10 modal % quartz and by the development of noritic rock with up to 20 modal % primary, intercumulus hornblende. Similar rocks are not encountered in the normal rock sequences elsewhere in the Complex.

However, not all the compositional features of the chromite in the Bushveld can at present be explained solely by variations in oxygen fugacity. *Cameron* and *Desborough* (1969) have, for instance, pointed out that the Al_2O_3 variations in Bushveld chromite do not concur with experimental results in the system FeO-Fe_2O_3-Al_2O_3-SiO_2 investigated gated by *Muan* (1957). It is also of interest that *De Waal* (1975) has shown that components in solid solution in the spinel display various trends superimposed on the broad compositional trend which has resulted from fractional crystallization of the magma. He has pointed out that the chrome spinels are not of the simple $M^{2+}M_2^{3+}O_4$ type, but that they they contain varying amounts of $M_2^{2+}M^{4+}O_4$ and $M_2^{3+}O_3$ components in solid solution. According to his calculations, the chromite in pyroxene-rich cumulate environments contain a fairly constant amount of the ulvospinel type of solid solution ($M_2^{2+}M^{4+}O_4$). A pronounced change occurs in the chromite of the first layer in a plagioclase-rich cumulate environment, in that the chromite contains, apart from the ulvospinel type solid solution, also 5% of $M_2^{3+}O_3$ in solid solution, which decreases rapidly upwards in the succession to 0% in the uppermost chromitite layer. Although the experimental data summarized by *Ulmer* (1969) points towards an increase of the $M_2^{3+}O_3$ component in solid solution in spinel at increasing oxygen fugacity, the variation in the amount of this component in Bushveld chromite with changes in the type of coexisting cumulus silicates is probably also a reflection of the changing composition of the fractionating magma, as a result of which the coprecipitating phases changed, and also the availability of certain elements.

A mechanism of variable depth convection was proposed by *Jackson* (1961) to account for the origin of cyclic units and the chromitite concentration within them, in the ultramafic zone of the Stillwater Complex. Although *Jackson* (1970) applied this mechanism to explain certain cyclic features in the lower part of the Critical Zone in the western Bushveld Complex, *Cameron* and *Desborough* (1969) pointed out that cyclic units of the Stillwater type are only to be found in parts of the Lower Zone where a repetitive stratigraphy involving dunite and bronzitite without chromitite is developed. They discard this hypothesis on the grounds of lack of any cyclical pattern or repetition of chromitic rocks and silicate rocks in the eastern Bushveld.

A cyclicity in the rock units is however encountered near the top of the Critical Zone where the sequence chromitite, bronzitite, norite, anorthosite is repeated several times, and which includes layers such as the Merensky Reef, the Bastard Reef as well as the UG1 and UG2 chromitite layers (*Feringa*, 1959; *Wager* and *Brown*, 1968). In this respect it must be noted that repeated addition of fresh, undifferentiated magma into the magma chamber during crystallization could result in rock sequences very similar to those at the top of the Critical Zone and the cyclic units described by *Jackson* from the Stillwater Complex. Such repeated additions of magma were in fact proposed by *Cooper* as early as 1936 to account for the repetitive stratigraphy in the Bay of Islands Complex, and subsequently by *Brown* (1956) for the rhythmic units in the ultrabasic complex of Rhum and by *Irvine* and *Smith* (1967) for the cyclic units of the Muskox Intrusion. *Irvine* (1977), in reconsidering his contamination hypothesis, has demonstrated that a mechanism of repeated introduction of primitive magma and blending with the differentiated liquid in the magma chamber is a more feasible explanation for the origin of the chromitite layers of large layered intrusions.

Intermittent heaves of undifferentiated magma have also been suggested by *Wager* and *Brown* (1968) to account for the alternating cumulates of the Critical Zone and Lower Zone of the Bushveld Complex, and even *Hess* (1960), who considered the Bushveld as essentially one surge of magma, conceded that a quantitatively small influx of fresh magma could have taken place at the level of the Merensky Reef. *Cameron* (1971) concluded that a progressive increase in the En content of orthopyroxene upwards in the M unit of the Critical Zone was what would be expected of a slow influx of undifferentiated magma and slow mixing with the partially differentiated magma in the chamber. From a similar but more extreme reversal of the composistion of coexisting cumulus minerals in the Main Zone *von Gruenewaldt* (1973) calculated that at least 10% of the total volume of Bushveld magma was added to the magma chamber immediately prior to crystallization of the Pyroxenite Marker of the Main Zone.

The problem of successive influxes of magma has been complicated by more evidence recently presented for the existence of two parental magmas of different composition. Although already suggested by *Willemse* (1969) the two magma hypotheses was elaborated upon by *Gijbels* et al. (1974) who concluded on the basis of a partitioning model for platinum metals between magma and precipitated phases that the Lower Zone represents 2/3 crytallization of the first batch of magma. According to this argument 2/3 of the bulk composition of the Lower Zone (defined here as including the Lower and Transition Zone of *Cameron*, 1971) should yield the minimum content for each oxide in the first magma; a minimum value of 21% for MgO as calculated from modal analyses and mineral compositions by *Hamilton* (1977).

Surprising, in view of the low solubility of Cr in basaltic liquids, is the absence of chromitite layers in a sequence of close to 1500 m of ultrabasic rocks which constitute the Lower Zone. For this reason *Gijbels* et al. (1974) interpreted the first, ultrabasic magma to be Cr-poor in contrast to the second, basic magma which must have contained considerably more chromium in order for chromite to appear in large quantities at the base of the Critical Zone. Apart from this evidence, the presence of ultrabasic bodies at Burgersfort (*Willemse*, 1969) and several offshoots of the Complex in the Potgietersrus area (*Van der Merwe*, 1970), including the Grasvally occurrence (*De Villiers*, 1970), strongly favour the existence of such an early, ultrabasic magma.

Several chromitite layers are, however, developed in the ultrabasic rocks of the Potgietersrus area (*De Villiers*, 1970; *Van der Merwe*, 1976) and a composition of En_{94} enstatite, Fo_{86-87} for cumulus olivine and a Cr/Fe ratio of between 2.7 and 3.0 makes the rocks at Grasvally the most basic of the Bushveld Complex. Conceivably, these rocks can be regarded as being representative of a stratigraphic position well below the Lower Zone as it is exposed in the eastern and the western Bushveld Complex, and could possibly indicate the presence of additional chromitite layers in the deeper, hidden parts of the intrusion, i.e., closer to the feeders of the Bushveld magmas.

It is hoped that the above discussions have illustrated to some extent the variety of the processes which could have been operative during the fractional crystallization of the Bushveld magma, each one of which undoubtedly had, to a larger or lesser extent, an effect on the formation and the composition of the chromite concentrations. Only by means of detailed study of the nature and composition of the coexisting minerals in complicated rock sequences such as those containing the chromitite layers in the Bushveld Complex, and relating this to features of experimentally investigated systems can it be hoped to gain more insight into the course of crystallization which produced the intricate features of the layering in the Complex. In this respect direct measurements of the oxygen fugacity for mineral assemblages promises to become an increasingly important tool in the evaluation of the processes responsible for chromitite layer information.

Acknowledgments. We wish to express appreciation to Mr. *A. Leeb-du Toit* of Union Corporation Ltd. and the Management and staff of the Zwartkop Chrome Mines Ltd. for assisting with the underground sampling of the chromitite layers. Both a research followship of the Alexander von Humboldt Foundation which enabled one of us (G.v.G) to spend 1974 at the Institute für Allgemeine und Angewandte Geologie der Universität München, and a research fellowship of the Deutsche Forschungsgemeinschaft (file number Sn IV/1) are gratefully acknowledged.

References

Arculus, R.J., Gillberg, M.E., Osborn, E.F.: The System MgO-Iron Oxide-Cr_2O_3-SiO_2: Phase Relations among Olivine, Pyroxene, Silica, and Spinel in Air at 1 Atm. Ann. Rep. Geophys. Lab. Carnegie Inst., Washington, 1974, p. 317–322

Arculus, R.J., Osborn, E.F.: Phase Relations in the System MgO-Iron Oxide-Cr_2O_3-SiO_2. Ann. Rep. Geophys. Lab. Carnegie Inst., Washington, 1975, p. 507–512

Brown, G.M.: The layered ultrabasic rocks of Rhum, Inner Hebrides. Phil. Trans. Roy. Soc. Lond. B 240, 1–53 (1956)

Cameron, E.N.: Chromite Deposits of the Eastern Part of the Bushveld Complex. In: The Geology of Some Ore Deposits in Southern Africa. Haughton, S.H. (ed.). Geol. Soc. S. Africa, Johannesburg 1964, Vol. II, pp. 131–168

Cameron, E.N.: Problems of the eastern Bushveld Complex. Fortschr. Mineral. **48**, 86–108 (1971)

Cameron, E.N.: Postcumulus and subsolidus equilibration of chromite and coexisting silicates. Geochim. Cosmochim. Acta **29**, 1063–1075 (1975)

Cameron, E.N., Desborough, G.A.: Occurrence and characteristics of chromite deposits – eastern Bushveld Complex. In: Magmatic Ore Deposits, a Symposium. Wilson, H.D.B. (ed.). Econ. Geol. Monogr. **4**, 23–40 (1969)

Cooper, J.R.: Geology of the southern half of the Bay of Islands Igneous Complex. Newfld. Dep. Nat. Res. Bull. **4**, 62 p. (1936)

Cousins, C.A., Feringa, G.: The chromite deposits of the western belt of the Bushveld Complex. In: The Geology of Some Ore Deposits in Southern Africa. Haughton, S.H. (ed.). Geol. Soc. S. Afr. Johannesburg 1964, Vol. II, pp. 183–202

Feringa, G.: The geological succession in a portion of the northwestern Bushveld (Union Section) and its interpretation. Trans. Geol. Soc. S. Afr. **62**, 219–232 (1959)

Fudaldi, R.F.: Oxygen fugacities of basaltic and andesitic magmas. Geochim. Cosmochim. Acta **29**, 1063–1075 (1965)

Gijbels, R.H., Millard, H.T., Desborough, G.A., Bartel, A.J.: Osmium, ruthenium, iridium and uranium in silicates and chromite from the eastern Bushveld Complex, South Africa. Geochim. Cosmochim. Acta **38**, 319–337 (1974)

Gruenewaldt, G. von: The main and upper zones of Bushveld Complex in the Roossenekal area, Eastern Transvaal. Trans. Geol. Soc. S. Afr. **76**, 207–227 (1973)

Hamilton, J.: Sr isotope and trace element studies of the Great Dyke and Bushveld mafic phase and their relation to early Proterozoic magma genesis in Southern Africa. J. Petrol. **18**, 24–52 (1977)

Henderson, P., Suddaby, P.: The nature and origin of the chromespinel of the Rhum Layered Intrusion. Contr. Mineral. Petrol. **33**, 21–31 (1971)

Hess, H.H.: Stillwater Igneous Complex, Montana. A quantitative mineralogical study. Geol. Soc. Am., Mem. **80**, 230 p. (1960)

Hill, R., Roeder, P.: The crystallization of spinel from basaltic liquid as a function of oxygen fugacity. J. Geol. **82**, 709–729 (1974)

Hunter, D.R.: The regional geological settling of the Bushveld Complex. (An adjunct to the provisional tectonic map of the Bushveld Complex). Econ. Geol. Res. Unit, University of the Witwatersrand, Johannesburg, 1975, 18 pp.

Irvine, T.N.: Crystallization sequences in the Muskox intrusion and other layered intrusion – II. Origin of chromitite layers and similar deposits of other magmatic ores. Geochim. Cosmochim. Acta **39**, 991–1020 (1975)

Irvine, T.N.: Origin of chromitite layers in the Muskox intrusion and other stratiform intrusions: a new interpretation. Geology **5**, 25, 273–277 (1977)

Irvine, T.N., Smith, C.H.: The ultramafic rocks of the Muskox Intrusion, North-west Territories, Canada. In: Ultramafic and Related Rocks. Wyllie, P.H. (ed.). New York: Wiley & Sons 1967, pp. 38–49

Jackson, E.D.: Primary textures and mineral associations in the Ultramafic Zone of the Stillwater Complex, Montana. U.S. Geol. Surv. Prof. Pap. **358**, 106 p. (1961)

Jackson, E.D.: Chemical variation in coexisting chromite and olivine in chromitite zones of the Stillwater Complex. In: Magmatic Ore Deposits, a Symposium. Wilson, H.D.B. (ed.). Econ. Geol., Monogr. **4**, 41–71 (1969)

Jackson, E.D.: The cyclic unit in layered intrusions – a comparison of repetitive stratigraphy in the ultramafic parts of the Stillwater, Muskox, Great Dyke and Bushveld Complexes. Geol. Soc. S. Afr. Spec. Publ. **1**, 391–424 (1970)

Keith, L.M.: Phase equilibria in the system $MgO-Cr_2O_3-SiO_2$. J. Am. Cer. Soc. **37**, 490–496 (1954)

Klemm, D.D., Kostakis, G.: Messung des Sauerstoffpartialdrucks über einzelnen Oxidphasen mittels galvanischer Hochtemperaturzlellen. N. Jb. Miner. Abh. **126** (2), 146–157 (1976)

Marlow, A.G., Van der Merwe, M.J.: The geology and the potential economic significance of the Malope area, northeastern Bushveld Complex. Trans. Geol. Soc. S. Afr. **80** (in press, 1977)

McDonald, J.A.: Evolution of part of the Lower Critical Zone, Farm Ruighoek, Western Bushveld. J. Petrol. **8**, 165–209 (1967)

Merwe, M.J. Van der: The layered sequence of the Potgietersrus limb of the Bushveld Complex. Econ. Geol. **71**, 1337–1351 (1976)

Muan, A.: Phase equilibrium relationships at liquidus temperatures in the system $FeO-Fe_2O_3-SiO_2$. J. Am. Cer. Soc. **40**, 420–431 (1957)

Muan, A., Osborn, E.F.: Phase equilibria at liquidus temperatures in the system $MgO-FeO-Fe_2O_3-SiO_2$. J. Am. Cer. Soc. **39**, 121–140 (1956)

Osborn, E.F.: Role of oxygen pressure in the crystallization and differentiation of basaltic magma. Am. J. Sci. **257**, 609–647 (1959)

Roeder, P.L., Osborn, E.F.: Experimental data for the system MgO-FeO-Fe$_2$O$_3$-CaAl$_2$SiO$_2$O$_8$-SiO$_2$ and their petrologic implications. Am. J. Sci. **264**, 428–480 (1966)

Sato, M.: Intrinsic oxygen fugacities of iron-bearing oxide and silicate minerals under low total pressure. Geol. Soc. Am. Mem. **135**, 290–307 (1972)

Stevens, R.E.: Composition of some chromites of the Western Hemisphere. Am. Mineral. **29**, 1–34 (1944)

Ulmer, G.C.: Experimental investigations of chromite spinels. In: Magmatic Ore Deposits, a Symposium. Wilson, H.D.B. (ed.). Econ. Geol., Monogr. **4**, 114–131 (1969)

Villiers, J.S. de: The structure and the petrology of the mafic rocks of the Bushveld Complex south of Potgietersrus. Geol. Soc. S. Afr. Spec. Publ. **1**, 23–35 (1970)

Waal, S.A. de: The mineralogy, chemistry and certain aspects of reactivity of chromitite from the Bushveld Igneous Complex. Nat. Inst. Metall. Rep. 1709, Johannesburg 1975, 80 pp.

Wager, L.R., Brown, G.M.: Layered Igneous Rocks. Edinburg-London: 1968, 588 p.

Willemse, J.: The geology of the Bushveld Igneous Complex, the largest repository of magmatic ore deposits in the world. In: Magmatic Ore Deposits, a Symposium. Wilson, H.D.B. (ed.). Econ. Geol., Monogr. **4**, 1–22 (1969)

Geochemical and Minerogenetical Problems

Einige überraschende Beobachtungen zum Problem der Entmischung der „Titanomagnetite"

P. RAMDOHR, Heidelberg

Mit 5 Abbildungen

Inhaltsverzeichnis

Summary

The unmixing of titanomagnetite contains many problems. Experimental studies always show first an unmixing of ulvospinel, in spite of the fact that natural occurrences carry mostly only unmixing of ilmenite. Rocks of the well-known *"larvikite"* show in the immediate neighbourhood four or more types of unmixing: old unmixing of ilmenite followed by ulvospinel, unmixing of ulvospinel alone, breakdown to ilmenite alone, unmixing of contemporaneous ulvospinel and ilmenite, and further variability – that is in a rather acid rock.

In Smålands Taberg "titanomagnetite" may contain unmixing of ulvospinel, partly preserved, partly, often spot-like, altered into ilmenite. Often, perhaps always, it contains some ilmenite, dissecting the network of ulvospinel and surely older than it. Sometimes the whole network of ulvospinel itself is altered without remnants into ilmenite. In a special case, from a locality "Russian Karelia", an olivine-spinellite contains titanomagnetite with old coarse lamellae of ilmenite and a filling consisting of a second generation of primary ilmenite which in its turn is (nearly) contemporaneous with hercynite unmixing. Experiments, till now, by no means give an explanation of all these complicated natural observations.

Vorbemerkung

Mein lieber Herr Maucher!

Nun kommen Sie ebenfalls ins Triarier-Alter und auch ich, Ihr Lehrer im schönen Fach Mineralogie und Lagerstättenkunde von etwa 1926–1928, soll etwas zur Festschrift beitragen, möglichst über Ihr Arbeitsgebiet, die schichtgebundenen Lagerstätten. Nun, in diesem Gebiet hab' ich, wie ich glaube, durch meine Arbeiten über den Rammelsberg, Witwatersrand, Broken Hill, mein Soll erfüllt, und so wird diese Arbeit einem ganz anderen mineralogisch-lagerstättenkundlichen Problem gewidmet sein, dem meine Arbeit schon 1924 galt, dem Problem der Entmischung titanreicher Magnetite.

1. Einleitung

Das Problem besteht, seitdem — wohl zum ersten Mal — *Singewald* (1913) die eigentümlichen orientierten Verwachsungen von Magnetit und Ilmenit erkannte, die dann auf Jahrzehnte unwidersprochen in Dutzenden, wenn nicht Hunderten von Arbeiten verschiedenster Autoren als Entmischungen angesehen wurden. Bedenken kamen erstmalig, als *Mogensen* (1946) den Ulvöspinell[1] beschrieb, den ich zwar selbst schon mehrfach erkannt und, z.T. mit falscher Deutung, auch abgebildet hatte (*Ramdohr*, 1940), den ich aber wegen seiner außerordentlich feinen Verwachsung weder isolieren konnte noch auch röntgenographisch identifizieren. Es wurden bald Zweifel laut, ob der Ilmenit nicht erst durch nachträgliche Umlagerung aus einem originalen Ulvitnetz bei erheblich sinkender Temperatur oder durch Oxidation ($3Fe_2TiO_4 + O = Fe_3O_4 + 3FeTiO_3$) entstanden sein könnte. Das umsomehr, als ich selbst (*Ramdohr*, 1953) das Vorkommen von reichlichem und für die Verarbeitung höchst störendem Ulvöspinelle in großen „Titanomagnetitvorkommen" hoher Bildungstemperatur beschrieben und auch den Übergang bzw. die Pseudomorphosierung des überaus feinen Ulvitnetzwerkes in noch feinere Ilmenitpartien erwähnt hatte.

Eine mit größter Sorgfalt durchgeführte synthetische Arbeit von *Buddington* und *Lindsley* (1964) schien diese Annahme zu bestätigen. Nach ihren Versuchen sollte, ganz im Gegensatz zu anderen Spinellen, Magnetit für das „Sesquioxyd" Ilmenit überhaupt kein Lösungsvermögen besitzen.

Mir blieben stets Zweifel, weil viele Dutzende von Vorkommen, die zweifellos relativ niedrig temperiert waren, ausschließlich Ilmenit // (111) des Magnetits, oft in recht grober Form, zeigten, ohne daß selbst bei größter Beobachtungssorgfalt auch das geringste Anzeichen von Reliktstrukturen des Ulvits vorhanden gewesen wäre. Und das wohlverstanden auch in Paragenesen, bei denen jede „Metamorphose", abgesehen von normalem Abkühlungsverlauf, höchst unwahrscheinlich war. Das wäre vielleicht verständlich, wenn es sich dabei nur um Gesteine gehandelt hätte, die, bei hohem Na_2O-Gehalt, z.B. sowohl Ägirin als Magnetit geführt und die damit schon hohe Fe_2O_3 ausgewiesen hätten — es gibt sehr wohl solche! Es trifft aber auch zu auf Gesteine bzw. Erze, die durch Graphit- oder sogar reichlichen Sulfidgehalt gekennzeichnet sind. Bei den sehr häufigen Teilverdrängungen von Titanomagnetiten durch spätmagmatische Sulfide (z.B. Sudbury, Sohland usw.) haben die Sulfide stets nur bereits Titanomagnetite angetroffen, die Ilmenit-Magnetit-Netzwerke waren, und das wohlverstanden auch in Fällen, wo sie sehr hochtemperierte Sulfidverdrängung — fast stets ist das Erstverdrängte ja Magnetit — durch Relikte von z.B. Mooihoekit ausgewiesen ist. Aber auch sicher ganz hochtemperierte Erzgesteine wie „Titanomagnetitspinell-Olivinite" (Smålands Taberg) oder auch manche sicher hochtemperierte Gabbros oder Norite zeigen oft Ilmenitentmischung, manchmal allein, manchmal mit wenig oder viel Ulvöspinell zusammen.

Im letzten Fall zeigt sich, daß die Ilmenitentmischungen oft älter sind als das Ulvitnetzwerk. Die daneben vorhandenen „Hercynitkörnchen" oder Lamellen sind in der Regel jünger als sowohl Ilmenit, mindestens als der eben genannte frühe, als auch Ulvöspinell. Es gibt aber reichlich Ausnahmen, wo Spinell älter ist als das Ulvitnetz, ja Fälle,

---

[1] Ich benutze oft die Bezeichnung Ulvit, die mir *Mogensen* (schon 1948) selbst als erwünschten Ersatz für „Ulvöspinell" vorgeschlagen hat.

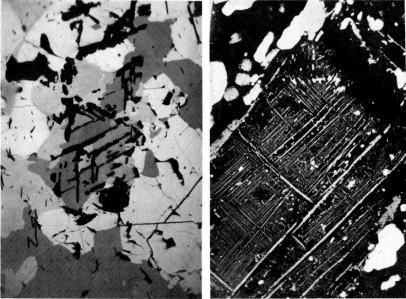

Abb. 1   Abb. 2

Abb. 1. Ilmenitnetzwerk, sehr dunkelgrau, eines entmischten Titanomagnetits. Der eigentliche Magnetit ist völlig weggelöst und durch Pyrrhotin, mittelgrau, Pentlandit und Kupferkies ersetzt (reinweiß bzw. mattweiß) (x 250, 1 Nic.)

Abb. 2. „Titanomagnetit" aus Metadiabas, Kalgoorlie, Australien. Der Magnetit ist völlig weggelöst, der entmischte Ilmenit ist in Anatas übergeführt, der alle Struktureigentümlichkeiten der Entmischung abbildet (x 250, 1 Nic.)

wo zweifellos zwei gut getrennte Entmischungsgenerationen des Hercynits vorliegen, älter als das Ulvitnetz und jünger als dieses, oft sogar jünger als der von *Buddington* und *Lindsley* als Umlagerungsprodukt des Ulvitnetzes angesehene junge Ilmenit.

Es erhebt sich also eine Fülle von Fragen über das Warum, Wie, bei welchen Temperaturen, durch welche Lösungsgenossen beeinflußt, in welchen Paragenesen, usw.

Trotz sehr reichen Vergleichsmaterials — aus magmatischen Gesteinen verschiedenster Art, Klein- und Großlagerstätten von Titanomagnetit, auch Proben aus der Mittelatlantischen Schwelle, Gesteinen von vermutlicher Herkunft aus dem Upper-Mantle, aus Meteoriten — hat der Verfasser noch kein einigermaßen befriedigendes Bild oder "Modell" gewinnen können.

2. Verschiedenes Verhalten im selben Gestein

Eigentlich zufällig und routinemäßig untersuchte Proben aus einer großen Steinschleiferei brachten nun neue Gesichtspunkte, die, auf eine kurze Formel gebracht, mehr oder minder sagen: Die Natur weiß es selbst nicht, wie und warum sie es so oder so machte! Dieses Resultat ist natürlich nicht befriedigend. Ich will also zuerst nur meine Beobachtungen darstellen.

Aus dem bekannten und z.B. für Geschäftshausfassaden in Europa und den USA viel verschliffe-
nen, in großen Steinbrüchen aufgeschlossenem „hellen Larvikit" von Larvik, Südnorwegen, werden
riesige Blöcke in Quaderform geliefert und in Deutschland verarbeitet. Unter diesen Blöcken, von
den Verarbeitern mit dem Handelsnamen „Toiwa" bezeichnet, war ein ganz gleichartig zusammen-
gesetztes Gestein, das reichlich „Titanomagnetit" führt. Ob die Bezeichnung „Toiwa" ein bestimm-
tes Gebiet in den Steinbrüchen bezeichnet oder vielleicht nur ein bedeutungsloses Codewort ist, ist
mir unbekannt.

Jedenfalls ist es ein sehr einheitliches grobes Tiefengestein, das aus Na-reichem, fein entmischtem
und dadurch labradorisierendem Kalifeldspat (Mikroperthit), sehr wenig Plagioklas, schon recht
wenig Biotit und Diopsid, selten stark grau-blau gefärbter Hornblende und vielleicht, schlierig auf
geringste Entfernung etwas wechselnd, 5–10% „Titanomagnetit" besteht. Wie andere Alkalisyenite
führt das Gestein in kleinen Mengen Accessorien wie Apatit, sehr wenig Titanit u.ä. Auffallend
darunter ist eine kleine Menge Graphit in winzigen Kügelchen. Insgesamt sei auf *Rosenbusch-Osann*
(1923, S. 152–155) verwiesen. Ausdrücklich sei aber vermerkt, daß die Rhomben-Textur in unserem
Gestein völlig zurücktritt, was aber vielleicht nur dadurch bedingt ist, daß die Mode wenig texturierte
Gesteine bevorzugt. Irgendwelche Tektonisierung fehlt durchaus.

Der *Titanomagnetit* verblüfft durch seine Vielgestaltigkeit. Fast alle von ihm im fol-
genden in diesem Abschnitt gezeigten Bilder stammen aus *einem* Anschliff und *einem*
Areal von vielleicht 6 x 6 mm, wo man wirklich an grundlegende Verschiedenheit der
Bildungsbedingungen wie Temperaturgefälle, Redoxpotential, Pufferung, verschiedenem
Zeitablauf etc., sicher nicht denken kann. Übrigens zeigen andere Proben desselben
Blockes genau dieselben Verhältnisse. Andere, z.T. Dutzende von Jahren zuvor und vom
Autor selbst gesammelte Stücke von Larvik zeigen ähnliche, wenn auch von Fall zu Fall
weniger variierende Verhältnisse.

Ein Zonenbau fehlt be größerer Übersicht den abgebildeten Körnern durchaus. Er
kann also nicht Anlaß zu den so wechselnden Strukturen gewesen sein, ebenso fehlen
in näherer Umgebung der Körner des behandelten „Titanomagnetitnestes" etwa lokal
verteilte reduzierende oder oxidierende Komponenten!

Die nun gezeigten Photos sind mit sehr starker Vergrößerung (meist 1400fach) und Ölimmersion
aufgenommen und stammen von drei sehr eng benachbarten Mineralkörnern. Bei der Scharfeinstel-
lung, die mir wegen einer Augenkrankheit jetzt Schwierigkeiten macht, unterstützte mich in freundlicher Weise Dr. *O. Medenbach*.

Fast alle Magnetitkörner sind durch grobe durchlaufende Ilmenitlamellen nach (111)
in kastenförmige Räume von Oktaeder- oder Tetraeder-Form aufgeteilt, wobei jeder
„Kasten" sich ganz verschieden verhalten kann.

Wahrscheinlich sind diese „groben" Ilmenitlamellen ursprünglich sehr viel feiner und
ganz früh angelegt, sind aber im Verlauf der Gesteinsabkühlung zu dickeren Lamellen
angewachsen. Bei sehr sorgfältiger Beobachtung findet man jedenfalls in den durchlaufenden, die „Kasten" veranlassenden Lamellen dünne „Seelen", die eine Spur dunkler
und etwas mehr blau-, nicht braungrau gefärbt sind. Da diese „Seelen" von weiterem
Ilmenit in gleicher Orientierung überwachsen sind, ist eine Täuschung durch verschie-
dene Orientierung ausgeschlossen, der sehr feine Farbunterschied also sicher reell.
Photographisch ist aber leider wenig zu sehen.

In den einzelnen „Kästen" sind nun die Verhältnisse ganz verschieden:

1. Es geht wohl der ganze Titangehalt in grobe, dem kastenbildenden Ilmenit ganz
ähnliche, aber nicht durchlaufende Lamellen hinein, die in den einzelnen Lamellen
schon merklich wechselnde Reflexionsfarben zeigen (Abb. 3A in den Randpartien,
Abb. 3B allverbreitet, aber eng mit Ulvit durchwachsen, Abb. 3C dicke Lamellen,
Abb. 3D dünne Lamellen – die drei breiteren sind es nur durch die Schnittlage).

2. Es bildet sich ein Netzwerk von Ulvit // (100), das meist, keineswegs immer, durch eine dünne Zone fast reinen Magnetits von den groben Lamellen des Ilmenits getrennt ist (Abb. 3A Zentralteil, Abb. 3B allverbreitet, Abb. 3C in sehr kompakter Form, Abb. 3D sehr locker, Abb. 3E, Hauptteil des Gesichtsfeldes einnehmend). – Bemerkenswert ist im letzten Fall noch, daß trotz der sehr großen primären Ilmenitkörner keine „kasten"bildenden Lamellen entstanden.

3. Es bildet sich ein – zunächst außerordentlich leicht zu übersehendes Netzwerk von wellig diffusen Lamellen, die keinerlei Ähnlichkeit sowohl mit dem Lamellenwerk von Ilmenit wie auch dem von Ulvit haben. Es sieht bei nicht ganz so starker Vergrößerung fast so aus, als ob homogener Magnetit vorläge, vielleicht eine Spur dunkler als der ± reine Magnetit dieser Paragenese. Bei fast gekreuzten Nicols erkennt man aber, daß Flecken vorliegen, die arealweise verschieden anisotrop sind und solche, die völlig isotrop und einheitlich bleiben. Mit äußerster Sorgfalt erkennt man, daß die ersteren aus einem losen Aggregat ehemaliger Ulvöspinelle bestehen, der jetzt in Ilmenit übergeführt ist. Daß diese Ilmenit- (früher Ulvit-) Aggregate weniger anisotrop erscheinen als geschlossener Ilmenit und auch weniger braun und so fast im Magnetit verschwinden, beruht auf ihrem lockeren, stets mit Magnetit durchwachsenen Aufbau (Abb. 3F), ebenso auf der sehr feinen (111) Orientierung des Netzwerkes.

4. Wieder liegen in den „Kasten" Verwachsungen von Ulvit und Magnetit vor, die aber viel feiner sind als im Fall 2, und, ganz im Gegensatz zu dort, unregelmäßig vielleicht eine Vorstufe von 3 darstellen. Dabei ist aber der Ulvitanteil (bzw. frühere Ulvitanteil) (Abb. 3C, Mitte) ebenfalls größer als in Abbildung 3A (Mitte), B und D.

Überraschend ist, daß unter Verhältnissen wie in Fall 3 oder in den seltenen Fällen, wo völlig ungegliederter noch intakter Magnetit vorzuliegen scheint oder schließlich besonders grobes Ulvitnetz vorliegt, die „Kästchen"struktur wenig auffallend ist, ja fehlen kann. Auffallend ist dann, daß die großen primären xenomorphen Ilmenitkörner keinerlei Anreiz zur Bildung der Ilmenitkästen lieferten (Abb. 3F).

Stücke aus anderen Teilen der sehr ausgedehnten Larviksteinbrüche führen oft keinerlei Ulvit, sondern reine und ziemlich grobe Magnetit-Ilmenitverwachsungen, etwa wie oben in Abbildung 3A. Ich hatte solche Fälle ja schon früh (1925) abgebildet. Andere Proben können aber auch genau das gleiche, was hier dargestellt ist, wenn auch in geringerer Vielgestaltigkeit, zeigen. Ich sagte schon oben, daß ich mir kein befriedigendes Bild über diese merkwürdigen Befunde geben kann.

Abbildung 3E wäre vielleicht unter der Annahme verständlich, daß die großen selbständigen Ilmenitkörner durch Wanderung auf kleinste Entfernung die Gehalte an Mn, evtl. auch an Mg, an sich gezogen haben, so daß gewissermaßen die „Initialzündung" für die Misch-Ilmenite, die Anlaß zur Kastenbildung geben, sich hier nicht auswirken konnten, so daß unmittelbar Ulvitbildung einsetzte, oder aber der Anreiz soweit fehlte, daß überhaupt eine Entmischung unterblieb. Daß im ersteren Fall der Ulvit so relativ grob ist, braucht gegenüber dem letzteren keinen Widerspruch zu bedeuten, da ja geringe Keimzahl (Kernzahl) immer relativ erhebliche Korngröße bedeutet. Nur wenig herabgesetzte Keimzahl hätte es dann überhaupt nicht zur Entmischung kommen lassen. Auch der Fall in Abbildung 3F wäre einigermaßen erklärbar, wenn auch eigentümlich bliebe, daß das Netzwerk gegenüber Abbildung 3E doch wesentlich feiner ist – wohlverstanden wieder bei unmittelbarer Nachbarschaft.

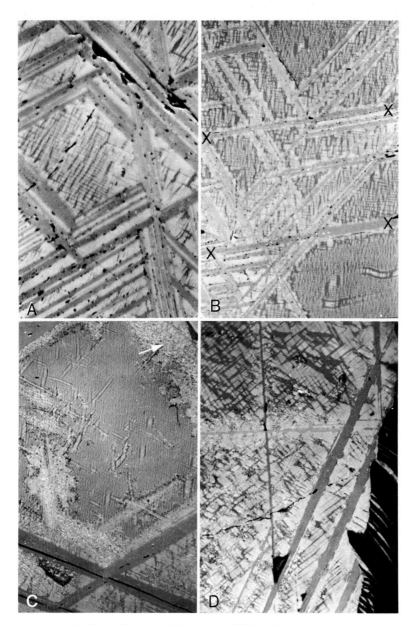

Abb. 3 A–D. (A) Larvik-Toiwa. „Titanomagnetit", hier besonders häufiger Typ. Dicke Lamellen von Ilmenit, von denen einige weit durchlaufen und die Begrenzung der sogenannten Kästchen bilden. Diese enthalten oft etwas hellere Seelen (*unten, schräg links-rechts*). Die Lamellen *im oberen Drittel* sind recht einheitlich und scheinen ohne eine Zwischenstufe von Ulvit gebildet. Im *Zentrum* ein Kästchen mit nur Magnetit + Ulvit. Diese Lamellen enthalten viel mehr Spinellkörper als die frühgebildeten (x 1200, 1 Nic.). (B) Das Bild ist unruhiger, da durchlaufende Lamellen der „Kästchen" von den anderen nicht scharf getrennt sind. Man hat den Eindruck, als ob Lamellen und Ulvitnetzwerk ± gleichalt sind. Immerhin sind einige Lamellen (*xx*) sicher alt. (C) Grobe Kästchenlamellen mit einem Füllwerk von Ulvit, das randlich in hellere Lamellen, vielleicht „Titanomaghemit" (→) im Sinne von *Basta* übergeht (nur x 600). (D) Ilmenitlamellen, lang durchlaufend und dünn, bilden Kästchen (die dicken Lamellen erscheinen durch flache Schnittlage verbreitert). Das Ulvitnetzwerk ist hier lockerer als z.B. in (B) *links unten* oder (C) *Mitte*. Die in die Gangart eindringenden, whisker-ähnlichen Lamellen sind anscheinend reiner Magnetit, der übrigens roh // einer Schar der Ulvitlamellen orientiert ist

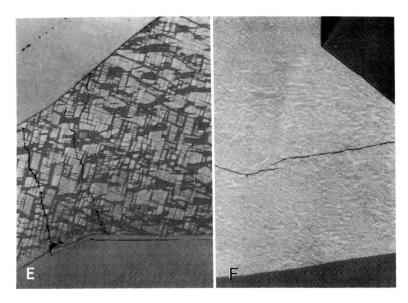

Abb. 3 E und F. (E) Oben links und unten rechts große ungegliederte Körner von Ilmenit, ohne Entmischungen. Der „Titanomagnetit" besteht aus Magnetit und Ulvitnetzwerk etwa im Verhältnis 3:2. Trotz der unmittelbaren Nachbarschaft zum Ilmenit haben sich hier Ilmenitlamellen in keiner Form gebildet. Vielleicht hat der grobe Ilmenit die Nebenelemente (Mn + Mg) an sich gezogen, so daß sie keine Impfwirkung im Magnetit ausübten. (F) Fast homogen erscheinender Titanomagnetit, bei dem nur eine ganz leichte „Welligkeit" auf Inhomogenität hinweist. Die wahre Komplexität ist im Text geschildert

Daß die Auscheidung der kastenbildenden Altilmenite mit der Anwesenheit von „Sesquioxyd"-Komponenten in anomaler Mischkristallbildung zu tun hat, ist nicht zu bezweifeln; das ist im Falle Chromite-Ilmenit, Spinell-Korund, aus zerfallendem Chromit entstandenen Restspinell + Eskolait in Meteoriten oder mißratenen Glasschmelzen, Hämatit im Magnesioferrit u.v.a. nicht überraschend — warum sollte es hier ander sein!?

Daß der experimentelle Befund dem scheinbar widerspricht, ist wohl darauf zurückzuführen, daß dem Experiment die Variabilität und die „Unsauberkeit" der Naturbedingungen fehlte.

2.1 Messungen mit der Mikrosonde

Durch eine zufällige und unter ganz anderen Gesichtspunkten durchgeführte Messung eines „Titanomagnetits" wurde ich angeregt zu versuchen, ob sich vielleicht in den ersichtlich früh ausgeschiedenen „Ilmenitlamellen" Wanderungen von Spurenelementen würden nachweisen lassen, die die frühzeitige Bildung von Ilmenit begünstigen könnten. Gedacht war besonders an Mg und Mn, die in Gestalt von Geikielith- bzw. Pyrophanit-Molekülen ja gerne ins Ilmenitgitter eingehen, während Al_2O_3, V_2O_3, Cr_2O_3 u.a. eher im Magnetitrest zu erwarten wären.

Die Zahl der Messungen ist vorläufig noch klein; einmal zeitbedingt, dann aber vor allem bedingt durch die Tatsache, daß die als frühgebildet erkennbaren Lamellen bzw.

„Lamellenseelen" sehr dünn sind, so daß „Verschmierungen" durch enge Verwachsungen immer möglich erscheinen. Man wird sie kaum vermeiden können. Sie wurden also bewußt in Kauf genommen.

Das Ergebnis bestätigte unsere Überlegungen ganz. Drei Messungen an dünnen durchlaufenden Lamellen ergaben: 4.527, 5.82, 5.69% Mn, während die ersichtlich späteren erst durch Impfwirkung enstandenen, aber deutlich dickeren und viel voluminöseren Lamellen kaum meßbare Mn-Gehalte (< 0.1) zeigen.

Auch eine Wanderung der Mg-Gehalte in die Frühlamellen scheint vorzuliegen. 0.31 bzw. 0.34, was bei dem sehr niedrigen Mg-Gehalt des Gesteins schon hoch erscheint! — Demgegenüber liegen die Gehalte an V mit 0.03 kaum über der Fehlergrenze, an Cr, das allerdings im Gestein sehr niedrig ist, darunter.

Dies Ergebnis erklärt zwar die Bildung der Kastenlamellen und ihrer Impfwirkung, die oft zur völligen Unterdrückung der Ulvitbildung führt, recht gut, es erklärt auch das Fehlen der Früh-Ilmenite in Abbildung 3E, weil ja auf die kurzen Wege Mn und evtl. Mg sehr früh zu den großen Primärilmeniten ausgewandert sind, erklärt aber nicht, warum in unmittelbarer Nähe diese Effekte völlig ausbleiben, d.h. die frühen Lamellen so wenig Impfwirkung haben oder die Bildung der frühen Lamellen ganz ausbleibt (Abb. 3F) und es dann zu feinst netzförmiger Bildung von Ulvit kommt, der sich dann aber sofort quantitativ in fast submirkroskopische Magnetit-Ilmenitnetzwerke umsetzt.

Besonders die weitere Untersuchung mit der Mikrosonde an sicher sehr hochtemperierten „Titanomagnetiten" verspricht Erfolge, wenn man vielleicht ausnahmsweise grobkörnig zerfallene Proben auffindet, etwa von dem in Abb. 4 gezeigten Typ.

Die Messungen mit der Mikrosonde wurden in freundlicher Weise durch Prof. *A. El Goresy* durchgeführt, dem der Verfasser auch sonst manche Hilfe verdankt.

3. Einige weitere Bemerkungen zu unserem Problem

Einen interessanten Beitrag zu den vorliegenden Problemen liefert eine gerade erschienene Arbeit von *B. Halfén*. In dieser wird versucht zu erklären, ob bei entmischten Mischkristallen durch Homogenisierung ein Reflexionsvermögen und eine Farbe entstehen, die gewissermaßen als Summation aus den Zerfallskomponenten zu deuten ist. Als Beispiel wird der bekannte Titanomagnetit von Norra Ulvö vom Locus typicus für Ulvöspinell gewählt. Das Ergebnis der Arbeit, die mit sehr großer Sorgfalt, viel Fleiß und mit allen — manchmal etwas übertriebenen — experimentellen Vorsichtsmaßregeln durchgeführt ist, ist im wesentlichen negativ. Exakte Schlüsse auf die zu erwartenden Reflexionswerte aus dem Gemisch für das Produkt der Homogenisierung und damit wohl auch umgekehrt auf die Zerfallsprodukte einea analytisch bestimmten Mischkristalls sind nicht zu ziehen. Im fraglichen Beispiel weicht das Homogenisierungsprodukt übrigens besonders im roten Teil des Spektrums von dem der Summierung der Entmischungskomponenten ab.

Für uns ist etwas anderes von größerem Interesse. Dieser „Titanomagnetit" von Ulvö scheint gar kein Mn, aber ca. 2.2% MgO zu enthalten. Soweit ich Proben von Ulvö ansah — sie sind stets in Ulvit + Magnetit entmischt! Es scheint also tatsächlich ein Zusammenhang der Frühentmischung von Ilmenit mit Gehalt an Mn vorzuliegen!

In Smålands Taberg liegen die Verhältnisse wieder anders, wie es bei der Größe des Erzkörpers an den verschiedenen Begleitgesteinen vielleicht auch zu erwarten ist — Temperaturgefälle, lokale Einflüsse des Nebengesteins, auch der Chemismus (hoher oder niedriger Anteil von Spinell oder Plagioklas) sind sicher verschieden. Die relativen Anteile von erhaltenem oder bereits zu Ilmenit zerfallenem Ulvitnetzwerk, von frühen Ilmenitlamellen und von dunkelgrünem Hercynit weichen ganz voneinander ab — von Areal zu Areal, von Handstück zu Handstück. Die Abbildung 4A, B gibt charakteristischen Aufschluß. Eine Mikrosondenanalyse über die in Abbildung 4A gezeigte Gesamtprobe wie von deren Einzelkomponenten war noch nicht durchführbar. Die hier untersuchte Probe fiel im Gegensatz zu Dutzenden anderer Stücke des Vorkommens dadurch auf, daß sie keinerlei Ulvitreste mehr enthielt, sondern bereits nur mehr Magnetit und Ilmenit zeigte, wie ja aus Abbildung 4B — mit + Nicols — sofort erkennbar ist. Auffallend ist wieder, daß sie als ältestes Entmischungsprodukt relativ grobe Ilmenitlamellen führt (besonders diagonal, Ecke unten rechts in Abbildung 4A. Es folgen dann, im Aussehen einander völlig gleichend und offenbar auch gleichalt, Lamellen // (100), im Aussehen völlig den Ulvitlamellen anderer Vorkommen ähnelnd, und solche // (111), vielleicht jeweils etwas länger verfolgbar als die erstgenannten. Beide Scharen bestehen aber jetzt ausschließlich aus Ilmenit. Dabei ist aber zu beachten, daß die (100)-Lamellen nicht, wie es sonst oft bei ungewandelten Ulvitlamellen der Fall ist, treppen- oder zickzackartig in Ilmenit verwandelt sind, sondern jeweils in ein Individuum übergehen, das dann gewissermaßen 45° schief auslöscht, also ⧄, nicht ⧅. Es ist demnach auch in den (100)-Lamellen eine geheimnisvolle Art ⧄ ⧅ der Vororientierung vorhanden, die die ja „selbstverständlich" zu erwartende Wahrscheinlichkeit der Orientierung // (111) und // (11$\bar{1}$) unterdrückt.

Abb. 4 A und B. (IN bzw. +N) Smålands Taberg, Schweden. „Titanomagnetit", fast genau // (100) geschnitten. Unten Entmischung einiger grober und ganz früher Ilmenitlamellen, dann mit grobem Netzwerk von Spinell-Lamellen // (100), schwarz. Die Füllmasse ist sehr eigenartig; sie besteht aus Lamellen // (100), die wie Ulvöspinell aussehen, aber ebenso wie die feinen (111)-Lamellen ausschließlich aus Ilmenit bestehen. Entmischung von kleinkörnigem Spinell tritt hier sehr zurück. Die dunklen Stellen in (B) sind gerade in Dunkelstellung

Von besonderem Interesse ist schließlich eine Probe, die einer uralten Sammlung entstammt und, reichlich unklar, nur mit "Russisch Karelien" bezeichnet war. Es ist ein Olivin-Ilmenit-Spinellit mit nur recht untergeordnet (< 1/5) Titanomagnetit. Auch hier wurde für die Photos (Abb. 5A und B) wieder zum Vergleich ein Schnitt // (100) ausgewählt. Das hat natürlich andererseits den Nachteil, daß der Pleochroismus des Ilmenits sich nicht bemerkbar macht. Hier ist, höchst bemerkenswert bei einem sicher sehr hochtemperierten Gestein, auch nicht die geringste Andeutung vorhanden, daß der Titangehalt ganz oder teilweise einmal das Stadium des Ulvits durchlaufen hätte. Ilmenit // (111) haben sich anscheinend etwa gleichzeitig entmischt, Spinell vielleicht etwas später, da mehrfach Spinell die Ilmenitlamellen „diskordant durchsetzt" und häufig zudem der Spinell in die schon irgendwie vorgezeichnete Richtung der Ilmenitlamellen abbiegt, was ja in dem herausvergrößerten Bild (Abb. 5B) an vielen Stellen sichtbar ist.

Schlußwort. Dem Verfasser war daran gelegen, möglichst ohne vorgefaßte Meinungen und möglichst ohne Kritik einige Naturbefunde auch an sicher z.T. sehr hochtemperiert gebildeten Gesteinen darzustellen. Sie repräsentieren, wenngleich sehr verschieden in den Einzelheiten, die außerordentliche Kompliziertheit beim Entmischungszerfall der „Titanomagnetite". Eines scheint sicher: Die Experimente zu dieser Frage reichen durchaus noch nicht, die Probleme zu lösen!

Literatur

Buddington, A.F., Lindsley, P.H.: Iron titanium oxide minerals and synthetic equivalents. J. Petrol. 310–356 (1964)

Halfén, B.: Difference in spectral reflectivity between grains of homogeneous and exsolved titanomagnetite. Mineral. Mag. 40, 843–851 (1976)

Mogensen, E.: A ferrotitanite ore from Södra. Ulvön. Geol. Foren. Förh. 68, 578–588 (1946)

Ramdohr, P.: Die Erzmineralien in gewöhnlichen magmatischen Gesteinen. Abhandl. Math-nat. Kl. No. 2, Abh. 4a und 4b, Berliner Akademie (1940)

Ramdohr, P.: Ulvöspinel and its significance in titaniferous iron ores. Econ. Geol. 48, 677–688 (1953)

Ramdohr, P.: Die Beziehungen von Fe-Ti-Erzen aus magmatischen Gesteinen. Bull. Comm. Geol. Finland 73, 1–18 (1956)

Rosenbusch, H., Osann, A.: Elemente der Gesteinslehre. Stuttgart: E. Schweizerbarthsche Verlagsbuchhandl. 1923

Singewald, I.T.: Titaniferous iron ores of the U.S. U.S. Dept. Mines Bull. 64 (1913)

Abb. 5. (A) „Russisch Karelien". Aus einem Olivin-Ilmenit-Spinellit. „Titanomagnetit" geschnitten sehr nahe // (100), so daß die Orientierung // (100) des Spinells (schwarz) erkennbar ist. Das Füllwerk besteht ausschließlich aus Magnetit und Ilmenit (etwas dunkler) ohne die geringste Spur von Ulvöspinell oder etwaigen Reliktstrukturen. Die „Flecken" *rechts unten* und *links oben* sind gerade in (100) geschnittene Spinelle. Die Entmischung des Ilmenits muß etwa gleichzeitig mit der des Spinells erfolgt sein. (Etwa x 1400, 1 Nic.). (B) „Russisch Karelien". Ähnliches Objekt wie in (A), aber nachvergrößert auf x 2000. Man erkennt, daß manche Spinell-Lamellen in die Richtung (111) abgelenkt sind, zweifellos entlang einem bereits vorhandenen Ilmenitnetz und daß oft Spinell-Lamellen ältere Ilmenite „querschlägig" durchsetzen (z.B. *nahe dem Südrand, Mitte*)

Isotopic Composition of Sulphur
and the Genesis of Hydrothermal Sulphates

F.V. CHUKHROV, L.P. ERMILOVA, and L.P. NOSIK, Moscow

Summary

The mantle and the lower crust of the earth in connection with vadose waters are believed to be the sources of sulphur of hypogene sulphates. This paper discusses the nature of the sulphate sulphur of various hypogene minerals, such as barite, gypsum, and anhydrite, as well as the sulphur in some volcanic areas, both on the basis of their isotopic compositions. Temperature appears to be the most decisive factor for the δ^{34}S balance in the system $SO_4^{2-} - H_2S$.

The influence of bacterial reduction of sulphate and of sulphur isotope exchange reactions for δ^{34}S are described.

A decisive part in forming the isotopic sulphur composition of hypogene sulphates is played by the mixing of sulphate sulphur of different isotopic compositions and by the isotope exchange between different valent forms of sulphur.

Mixing of sulphate sulphur of different δ^{34}S values is frequently observed in the circulation zone of vadose waters. The sulphate sulphur in these waters is derived from two main sources — marine (evaporites, scattered sulphur aggregations in sedimentary rocks, and buried waters), and atmospheric. The isotopic composition of marine sulphate sulphur and, consequently, of the sulphur of evaporites has been changing throughout the geological history of the earth.

A participation of marine sulphate sulphur (including evaporite sulphur) in hypogene mineral formation is highly probable. In all likelihood, it formed part of those hypogene sulphates whose δ^{34}S values were near to that of the corresponding evaporites and could be up to + 30‰. It should be borne in mind, however, that, because of bacterial sulphate reduction, sulphate sulphur in buried waters may be heavier than marine sulphate sulphur of a given period.

Water of present atmospheric precipitation on continents mainly has δ^{34}S values below + 10‰, the usual values being + 5‰ or less. Surface run-off waters have similar δ^{34}S. When the sulphate of continental waters gets mixed with marine sulphate, these values increase. Hypogene sulphates which might derive from continental waters would be characterized by low δ^{34}S.

In those waters where hydrogen sulphide is a product of incomplete bacterial reduction of sulphate, sulphide sulphur has lower δ^{34}S than sulphate sulphur. Yet when such waters curculate in rocks including lighter sulphur, the δ^{34}S of sulphate may become less than in hydrogen sulphide. A case in point is the thermal brines of the Neogene red series at Cheleken which come from different water-bearing horizons. At the Kyr-Kyzyl-tepe Site, water from well V-1 flowing into a collecting tank contains sulphide sulphur and sulphate sulphur with $\delta^{34}S$ equal to + 13.8 and + 19.4‰, respectively; water from well P-164 flowing into the same tank is free of hydrogen sulphide and contains sulphate with $\delta^{34}S = 6.6‰$. When sulphide-free and sulphide-containing water are mixed in the tank, the $\delta^{34}S$ of sulphate sulphur drops to + 6.8‰. At the Toyunly Site, water of well P-166 from the depth of 705 m has $\delta^{34}S$ of SO_4^{2-} and H_2S equal to + 18.5 and + 7.3‰, respectively.

In many volcanic areas, there are favourable conditions for acid sulphate solutions to mix with atmospheric waters, as well as with waters containing marine sulphate sulphur; in the former case this may lead to a slight, and in the latter to a great increase of $\delta^{34}S$.

Sulphur isotope exchange reactions responsible for changes of $\delta^{34}S$ evolve under various conditions.

The sulphide sulphur of rocks developed from plutonic magmas and free of sulphur from sedimentary series has an isotopic composition similar to mantle sulphur.

When sulphate sulphur or sulphide sulphur from sedimentary series was absorbed, the sulphide sulphur of magma would have a $\delta^{34}S$ increased or decreased respectively. Parts of palingene magmas and rocks developed from them may also be enriched in the heavy or light isotope, as exemplified by Rapakivi granites ($\delta^{34}S$ up to + 14.3‰) and granites near Outokumpu in Finland ($\delta^{34}S$ up to + 28.7‰). The reduction of sulphate sulphur trapped by magma may be incomplete. In an intrusive rock from El-Salvator, $\delta^{34}S$ of anhydrite and sulphide are + 7.3 and − 3.9‰, respectively; in this case, anhydrite prevails over sulphide (*Jensen*, 1972). At the Norilsk-1 deposit, margins of some large sulphide phenocrysts in picritic gabbro-dolerite and separate aggregates contain anhydrite with $\delta^{34}S$ from + 14.8 to + 17.9‰; sulphides associating with the anhydrite are also enriched in the ^{34}S isotope ($\delta^{34}S$ from + 7.7 to + 9.8‰). Devonian gypsum-anhydrite beds are found in the area and they are believed to be the source of the sulphate sulphur in anhydrite. It is assumed that after a large proportion of sulphate sulphur had converted to sulphide sulphur, there was an isotopic exchange between the two, which resulted in a decreased $\delta^{34}S$ of sulphur in the residual anhydrite (*Grinenko* and *Grinenko*, 1967).

In hot volcanic gases, there is isotopic exchange between H_2S and SO_2, so that the former is enriched in the ^{32}S isotope, and the latter, in the ^{34}S; accordingly, sulphuric acid of the volcano is also enriched in the ^{34}S isotope. In volcanic gas, SO_2 prevails over H_2S at temperatures above 450°C, while at lower temperatures H_2S greatly prevails over SO_2. Isotopic exchange is responsible for a lower ^{34}S content in H_2S compared with SO_2.

Reduction of sulphate sulphur is widespread in the hypergenesis zone, but at the zone's typical temperatures there can be no isotopic exchange between sulphide (lighter) sulphur and sulphate (heavier) sulphur of solutions. The exchange is enabled when the solution are heated by the plutonic heat to high temperatures.

Temperature is the decisive factor for $\delta^{34}S$ balance in the SO_4^{2-}-H_2S system. For example, the theoretical isotopic fractionation in that system at 200°C is 32‰; at higher temperatures, it declines (down to 10‰ at 600°C). The absolute equilibrium values of $\delta^{34}S$ depend on the isotopic composition of sulphate sulphur and hydrogen sulphide sulphur and on the proportion of SO_4^{2-} and H_2S contained in hydrothermal solutions (*Rye* and *Ohmoto*, 1974).

With a predominance of SO_4^{2-}, exchange leads to enrichment in the ^{32}S isotope, while a predominance of H_2S results in a heavier SO_4^{2-}. At a temperature under $200°C$, it is, however, very difficult for the SO_4^{2-}-H_2S system to reach isotopic equilibrium, as the isotopic exchange is extremely slow.

At sufficiently high temperatures, similar processes take place after hydrogen sulphide coming from plutonic sources mixes with marine sulphate sulphur. As a result, sulphide sulphur may have a $\delta^{34}S$ content slightly or greatly higher than that of mantle sulphur. Presumably, this happens where hot hydrogen sulphide emanates through the sea floor. A case in point is the $\delta^{34}S$ content of pyrite from ocean floor bedrock in the Hess Deep (Pacific), being as high as $+7.9‰$.

Sulphate-sulphide pairs are likely to arise from oxidation of a part of sulphide sulphur in solutions which were originally sulphate-free and contained no SO_2. At a sufficiently high temperature, isotopic exchange reactions would evolve in such pairs and eventually enrich sulphide sulphur in ^{32}S and sulphate sulphur in ^{34}S.

Data collected at the Mendeleyev Volcano in the Kunashir Island (the Kuriles) are helpful in judging about the variations of the isotopic composition of sulphate sulphur of the thermal fields of volcanoes and their genetic role.

Near the Mendeleyev Volcano top, vapour, gas jets and hot springs well up from the slope near massive pyrite ore bodies and volcanoclastic rocks impregnated with pyrite and marcasite, and around the outlets there are aggregations of authigene sulphur. It is beyond any doubt that pyrite originated from an action of acid H_2S-containing solutions on iron-containing volcanic rocks. The values of sulphur in iron sulphides (from -1.0 to $-3.6‰$) indicate its juvenile (mantle) origin.

For authigene sulphur observed at outlets of vapour and gas jets and in hot springs. the following values of $\delta^{34}S$ $(‰)$ have been determined:

Upper solfataric field, sublimate from vapour and gas jet	-2.9
Northeast solfataric field, aggregates in pyrite ore	-3.5
Middle solfataric field, right wall of Kisly Creek, from sulphurated rock	-1.5
Middle solfataric field, right wall of Kisly Creek, segregations in boiling mud pot	-3.6
Spring 1–2 of the Lower Mendleyev Springs, deposit near the spring	-6.4
Stolby Headland, at vapour and gas jet outlets (now run dry)	-1.4

Table 1 gives data on the isotopic composition of the sulphate sulphur in waters of the thermal field of the Mendeleyev Volcano; pH and water temperature were measured by *L.M. Lebedev* and *I.B. Nikitina*.

Within the solfataric field of the Mendeleyev Volcano, where numerous vapour and gas jets spurt out, sulphate sulphur of waters with $\delta^{34}S$ varying from -8.0 to $-3.4‰$ is likely to have been formed by oxidation of sulphur of plutonic hydrogen sulphide and partly of sulphurous anhydride. Any influence of marine sulphate upon the isotopic composition of sulphur of the solfataric field is ruled out, for the field is situated at the top part of the volcanic structure. Far beneath it, on a rather gentle slope, are the Lower Mendeleyev Springs, to which an underflow of sea water by convection is quite probable. However, marine sulphate is there strongly diluted by very acid sulphate water with greatly lightened sulphate flowing down from the solfataric field; $\delta^{34}S$ of sulphate sulphur are up to $+13.4‰$. The proportion of marine sulphate is even higher in the Lower Doktorskie Springs, which are situated near the Goryachiy Beach in the bottom-most part of the Mendelyev Volcano at the Pacific shore.

Table 1. Isotopic composition of sulphate sulphur from the Mendeleyev Volcano

Site of sampling	$\delta^{34}S, \%o$
North-east solfataric field, boiling spring discharging into Kisly Creek; pH 2.60, elevation datum over 300 m	− 3.0
Same, mud volcano; pH 2.25	− 3.2
Middle solfataric field; elevation about 300 m, boiling mud pot	− 3.4
Lower Mendeleyev Springs; elevation 70 m	+ 10.2
1−2; pH 1.70, t = 93°C	+ 10.5
1−3; pH 2.05, t = 73.5°C	+ 10.9
1−4; pH 1.85, t = 85°C	+ 13.4
Kisly Creek, upstream from mouthing into Lesnaya River	+ 7.2
Lesnaya River, upstream from Kisly Creek mouth	+ 5.4
Goryachiy Beach area, Lower Doktorskie Springs, elevation 19−21 m	
No 1 pH 2.40, t = 58°C	+ 14.6
No 3 pH 7.0, t = 54°C	+ 15.7
No 4 pH 2.55, t = 59°C	+ 10.8
No 5 pH 2.60, t = 51°C	+ 10.6
Goryachiy Beach, Pacific shore, Well 5, hot water with little hydrogen sulphide, pH 6.25	+ 18.9
Lechebnaya River, upstream from Well 5; pH 6.75, cold water	+ 1.7
Pacific Ocean water; pH 7.60	+ 19.4
Upper Stolbovskie Springs:	
No 1 pH 7.25, t = 85°C	+ 22.6
No 2 pH 7, t = 83°C;	+ 22.1
Stolby Creek; pH 6.80, cold water	+ 22.1
Sea of Okhotsk at Stolby Headland	+ 19.8

Over and above the sulphate sulphur composition, the greater importance of marine sulphate is indicated by chlorine content and total sulphate sulphur content in solutions. The difference between chlorine and sulphate sulphur content in solfataric field waters and in the underneath springs is illustrated by the following data of *L.M. Lebedev* and *I.B. Nikitina* (mg/l):

	Cl^-	SO_4^{2-}	HSO_4^-
Middle solfataric field, Kisley Creek	221.2	1677.2	535.4
Lower Mendeleyev Springs	977.3−1467.6	489.6−807.4	266.0−669.3
Lower Doktorskie Springs	1221.2−1400.5	620.2−622.8	73.7− 83.0

Judging by pH and the isotopic composition of sulphur, Spring 3 of the Lower Doktorskie group receives the smallest amount of sulphate sulphur from the solfataric field. Waters of the Lower Mendeleyev Springs and atmospheric precipitation run down into the Kisly Creek, and waters of atmospheric precipitation into the Lesnaya River. Well 5 apparently yields mainly bottom sea water, whose sulphate becomes involved in bacterial reduction. Water of the Lechebnaya River (falling into the Pacific at the Goryachiy Beach), with pH 6.75 and $\delta^{34}S = + 1.7\%o$, seems to contain atmospheric sulphate with an admixture of solfataric field sulphate.

Near the Stolby Headland at the shore of the Sea of Okhotsk, there are the Upper Stolbovskie hot springs of hydrogen sulphide-containing water. There is no doubt as to the possibility of a convective underflow of sea water to these springs. A slightly increased weight of marine sulphate should be attributed to bacterial sulphate reduction in the coastal zone. Water of the cold spring at Stolby is apparently sea water strongly diluted with atmospheric precipitation.

Table 2 gives data on the isotopic composition of sulphur of the waters of the Golovin Volcano. The general characteristics are given according to *A.V. Zotov*. Samples of hydrogen sulphide and sulphate sulphur were taken by *V.A. Volchenkova*. In the pots, sulphate sulphur shows no sharp difference in isotopic ocmposition from the sulphide sulphur from which it had mainly originated.

Table 2. Isotopic composition of sulphur from the Golovin Volcano

Test objects	Water characteristics	$\delta^{34}S$, ‰	
		H_2S	SO_4^{2-}
Sulphate water pots, north bank of crater lake Kipyashchee			
No 1	pH2.40, t = 97°, H_2S = 5 mg/l HSO_4^-= 1085 mg/l, SO_4^{-2} = 1085 mg/l Cl = 127 mg/l	+ 3.8	+ 4.0
No 2	pH 1.60, t = 80°, HSO_4^-= 3020 mg/l SO_4^{2-}= 1570 mg/l, Cl^-= 181 mg/l	+ 3.5	+ 4.2
No 3	pH 2.15, t = 59°, H_2S = 13 mg/l HSO_4^-= 1090 mg/l, SO_4^{2-}= 540 mg/l	+ 2.8	+ 4.7
Crater lake Kipyashchee (sulphate-chloride water)	pH 2.7, t = 30–35°, H_2S = 3.2 mg/l, HSO_4^-= 10 mg/l, SO_4^{2-}= 220 mg/l Cl^-= 692 mg/l	+ 2.7 + 4.5	+ 18.4 + 18.4
Spring at north-east bank of Kipyashchee Lake	pH 6.4, t = 82°, H_2S found, SO_4^{2-}= 260 mg/l, Cl^-= 105 mg/l	+ 4.3	+ 11.2

The high $\delta^{34}S$ of sulphate sulphur of the Kipyashchee Lake are a sign of sea water inflow into it, with $\delta^{34}S$ of hydrogen sulphide and pH of waters being determined by abyssal income of sulphur. Sulphur of alunite found in the lake's bottom mud has nearly the same isotopic composition as that of sulphate dissolved in lake water. In the spring with pH 6.4, the share of sea sulphate is larger, but $\delta^{34}S$ of sulphate sulphur (+ 11.2‰) suggests a big admixture of rainfall sulphur.

In the samples studied by the authors, the $\delta^{34}S$ values of hydrogen sulphide vary from + 2.7 to + 4.5‰. Earlier (*Vinogradov*, 1965), for authigene sulphur and pyrite samples from the Kipyashchee Lake, $\delta^{34}S$ had been determined as + 8.5 and + 1.2‰, respectively. In contrast to the Golovin Volcano, the values of $\delta^{34}S$ of hydrogen sulphide and authigene sulphur from the Mendeleyev Volcano are negative. This may be due to different conditions of isotopic exchange between SO_2 and H_2S in volcanic gases.

In the Uzon caldera on the Kamchatka, which is a young depression 10 km across, waters of various compositions discharge, including chloride-sodium waters. Deep thermal waters mix with those from higher levels. A part of sulphate sulphur forms from H_2S under the influence of atmospheric oxygen (*Pilipenko*, 1974). In one spring, $\delta^{34}S$ of sulphate increases to + 22.4‰ (*Ozerova* et al., 1971), which is an indication of a marine origin of a large part of it. The same is true of the solfataric field of the Yellowstone National Park in the United States (*Schoen* and *Rye*, 1970). Different origins of sulphide sulphur and sulphate sulphur are recorded for quartz veins in andesite deposits at Tui (New Zealand), where during one mineralization stage the deposition took place of cinnobar with sulphur ($\delta^{34}S$ = + 1.3‰) derived from a plutonic source and of barite ($\delta^{34}S$ from + 16 to + 19.5‰) containing marine sulphate sulphur brought from underlying greywackes (*Robinson*, 1974).

The $\delta^{34}S$ content of acid volcanic waters suggests that the variations of the isotopic composition of their hypogene sulphates' sulphur may stem from the mixing in different proportions of heavy marine sulphate sulphur and light sulphate sulphur formed from plutonic hydrogen sulphide.

Alunite is a typical hypogene sulphate of the areas of recent and ancient volcanic activity. Samples from the Kamchatka and Kurile volcanoes have $\delta^{34}S$ varying from + 12.1 to + 22.3‰ (*Vinogradov*, 1967), which corresponds with the values for sulphate of acid hot springs whose water contains relatively little plutonic sulphur.

In hydrothermally altered rocks of the Taupo volcanic region (New Zealand), where thermal fluids are significantly chloride-sodium-containing and have temperatures of up to 265°C at the depth of 12oo m, one sometimes observes hypogene anhydrite ($\delta^{34}S$ = + 16.1‰). In contrast to the sulphate sulphur of the H_2S-containing fluid ($\delta^{34}S$ = + 22.0‰), the sulphur of this anhydrite is diluted by that of light sulphate formed by oxidation of H_2S sulphur ($\delta^{34}S$ = + 3‰). The same volcanic units contain some alunite with $\delta^{34}S$ from – 1.4 to + 6.9‰ (*Steiner* and *Rafter*, 1966). The different $\delta^{34}S$ values of anhydrite sulphur and alunite sulphur should be accounted for by different proportions of plutonic and biogene sulphur in them. This is also true of Zaglik (Azerbaijan), where alunite confined to tuffogene-sedimentary rocks has $\delta^{34}S$ ranging from – 2.9 to + 25.9‰ (*Shipulin* and *Vinogradov*, 1967).

In the Goldfield deposit of a Tertiary volcanic region in the United States, finds were made of hypogene alunite with $\delta^{34}S$ from + 11.6 to + 23.3‰ and of hypergene alunite with $\delta^{34}S$ from – 2.5 to + 1.7‰; sulphides of that deposit have $\delta^{34}S$ varying from – 2.5 to + 1.7‰ (*Jensen* et al., 1971). In contrast to hypogene alunite, the hypergene (supergene) alunite was formed without any contribution of marine sulphate sulphur.

Variations of alunite $\delta^{34}S$ within one secondary quartzite block show the alunite sulphur to be either wholly plutonic or mixed with biogene sulphur, which sometimes is predominant. For example. in the Karabas deposit (North Balkhash area), $\delta^{34}S$ of hypergene alunite varies from – 1.6 to + 16.6 and that of the deposit's sulphides studied thus far, from – 0.6 to – 1.6‰ (*Gavrikova* and *Vinogradov*, 1967). In Soviet Central Asia, $\delta^{34}S$ of alunite (+ 6.9; + 7.0‰) and barite (+ 7.1‰) from the Gushsai deposit, by contrast, proves a limited participation of biogene sulphur in hypogene sulphate formation; this is even more so with the Trans-Carpathian deposits, where alunite $\delta^{34}S$ varies between – 2.1 and + 5.7‰ (*Vinogradov*, 1967). A marine sulphate origin is also indicated by the δ^{34}S values of barites from kuroko-type Miocene ores in Japan and from thermal fields of New Zealand (*Sakai* and *Matsubaya*, 1971).

The preceding examples suggest that the δ^{34}S values of hypogene sulphates in volcanic series may be accounted for by their being formed from solutions containing sulphate sulphur of a different origin — biogene (heavy) carried into active volcanic zones by convective sea water or ground-water flows, and volcanic (light) brought into oxidizing medium in the form of hydrogen sulphide and, partly sulphurous anhydride. Hypergene sulphates formed due to volcanic activity have δ^{34}S values determined by the proportions of biogene to abiogene sulphate sulphur in solutions.

Barite is the most typical hypogene sulphate of deposits developed without any connection to the more or less simultaneous volcanic events; less widespread are celestine, anhydrite, and thaumasite.

Barite is highly typical of a great number of low- and medium-temperature sulphide and oxide ore deposits; it often occurs together with fluorite, or as almost monomineral bodies. In many deposits, a later deposition has been established for barite as compared with sulphides and oxide minerals; barite bodies are often isolated from sulphide mineralization areas.

Barite precipitation from thermal waters may be due to a drop in $BaSO_4$ solubility after a drop in temperature. A case in point is the deposition of barite from hot brine of Well G-39 at Cheleken; the barite has δ^{34}S equal to $+ 16.2\%o$. On the other hand, barite can definitely be deposited from mixing chloride and sulphate solutions. There can be no doubt that barium-bearing waters could have taken part in forming deposits whose sulphate sulphur derives either from plutonic sources or from sea water. In either case, sulphate sulphur may have a lower δ^{34}S than sea water sulphates or evaporites of the corresponding geological period. The prime reason seems to be dilution of heavier marine sulphate with lighter surface water sulphate. At higher temperatures, a drop in sulphate sulphur δ^{34}S may be caused also by isotopic exchange between SO_4^{2-} and H_2S.

Let us take a look at several examples illustrating δ^{34}S of hypogene barite and hypogene celestine. Table 3 gives data on the isotopic composition of sulphur of barites and sulphides from skarns of the Karagaily lead-zinc deposit (Central Kazakhstan). The data indicate that, over and above plutonic sulphide sulphur, biogene sulphate solutions had been supplied to the ore zone. The same applies to barite of Lower Kairakty deposit (Central Kazakhstan), which makes up bodies of a very low sulphide mineral content; δ^{34}S in four barite samples varies from $+ 20.4$ to $+ 15.2\%o$. Similar data have been obtained for a great number of deposits in other regions.

The variations of the isotopic composition of the sulphur of barite show no certain dependence on whether or not sulphide minerals associate with the barite and on the minerals' sulphur isotopic composition. The assumption of a single plutonic source of the sulphur of barite and that of sulphides is also contravened by sharply different values of their δ^{34}S in some hydrothermal deposits, which rules out their having derived from the same solutions. In particular, at the Vyshkovo mercury deposit (Trans-Carpathian area), the difference between δ^{34}S of sulphide sulphur (from $- 5.9$ to $0.0\%o$) and barite (up to $+ 38.1\%o$) attains $44\%o$ (*Ozerova* et al., 1967). In the Bleiberg-Kreuth deposit (Austria), sulphides have δ^{34}S ranging from $- 6.9$ to $- 25.9\%o$ and barite, from $+ 14.8$ to $18.9\%o$, i.e., the difference is as great as $44.8\%o$ (*Schroll* and *Wedepohl*, 1972). In ores of the lead-zinc deposit Kurgashinkan at Karamazar (Soviet Central Asia),

Table 3. Isotopic composition of sulphur of sulphides and barite: the Karagaily deposit

Sample no.	Mineral	$\delta^{34}S$, %o	Sample no.	Mineral	$\delta^{34}S$, %o
1	Galenite	+ 1.8	3	Barite	+ 12.7
	Chalcopyrite	+ 3.5	4	Barite	+ 16.1
	Barite	+ 15.5	5	Barite	+ 16.8
2	Galenite in barite	+ 1.8	6	Barite	+ 16.8
			7	Barite	+ 16.7
	Barite	+ 14.5	8	Barite	+ 15.2
	Galenite in barite	+ 0.4	9	Barite	+ 16.6
	Sphalerite in barite	+ 1.1			

Table 4. Isotopic composition of sulphur of barite and of sulphide minerals: the Bakal deposit

Section	Sample no.	Mineral	$\delta^{34}S$, %o
Irkuskan	1	Pyrite	+ 8.6
	2	Pyrite	+ 41.0
	1	Chalcopyrite	+ 14.7
	2	Barite	+ 27.8
		Barite	+ 29.0
New Bakal		Pyrite	+ 24.2
Old Bakal		Pyrite	+ 4.2
Ditto	1	Barite	+ 34.4
		Barite	+ 28.3
Imeni 15-letia VLKSM		Pyrite	+ 18.4

Table 5. Isotopic composition of sulphate sulphur: Dzhezdy and Naizatas

Deposit	Sample no.	Mineral	$\delta^{34}S$, %o
Dzhezdy	1	Celestine	+ 10.7
	2	Celestine	+ 8.4
Dzhezdy	1	Barite	+ 12.3
	2	Barite	+ 16.5
	3	Barite	+ 16.7
Naizatas	1	Barite	+ 13.6
	2	Barite	+ 14.8
	3	Barite	+ 15.3

galenite has δ^{34}S (from $-$ 17.2 to $-$ 23.0‰) lower than that of barite (+ 20.3‰) by 43.3‰ (*Badalov* and *Vinogradov*, 1967). According to our data, δ^{34}S of the rather scarce sulphide minerals and barite from the Bakal deposit of magnesial siderites (Table 4) differ by up to 37‰.

The δ^{34}S values of sulphate sulphur of oxide sulphide-free ores of the Dzehezdy and Naizatas deposits in Central Kazakhstan (Table 5) likewise can be accounted for by an influx of marine sulphate sulphur; barite and celestine in those ores segregated later than iron oxides and manganese oxides.

Participation of marine sulphate is particularly evident in the formation of the thermal-sedimentary deposits. A typical example is the Meggen barite-lead-zinc deposit in West Germany, where δ^{34}S of sulphides and barite varies between + 11.9 and + 24.1‰, respectively, and between + 20.8 and + 26.8‰ (*Buschendorf* et al., 1963). In another thermal-sedimentary deposit in West Germany, Rammelsberg, δ^{34}S varies for sulphides from + 7 to + 20‰ and for barite from + 19.0 to 28.8‰ (*Anger* et al., 1966).

Available data indicate that biogene sulphate sulphur which contributes to the forming of ore bodies may often either be heavier or lighter than marine sulphate sulphur. Apparently, it becomes heavier because of bacterial sulphate reduction in solutions circulating in sedimentary series. It is this process that is responsible for the forming of sulphates with a largely heavier sulphur-celestine (δ^{34}S from + 17.5 to + 43.8‰) and barite (δ^{34}S = + 52.5‰) – in kimberlite pipes in Yakutia (*Vinogradov* and *Ilupin*, 1972). The lowest positive values of δ^{34}S in hypogene sulphates are due to dilution of heavier marine sulphate with lighter sulphate of subsiding surface waters. The dilution may occur, primarily, through freshening of sea water in coastal zones and estuaries. For example, δ^{34}S of six studied barite samples from the ore-bearing series at Dzhezkazgan which developed in freshened waters varies from + 4.7 to + 8.2‰ (average + 5.9‰); in sulphur of celestine making up a veinlet in sandstone of that series, the content of ^{34}S isotope is somewhat higher (δ^{34}S = + 10.0‰).

There can hardly be any doubt that lowered and low δ^{34}S of many hypogene barite from subsiding continental waters or from their mixture with marine sulphate sulphur solutions. Typically, δ^{34}S varies between + 3 and + 10‰. The idea that the sulphate sulphur of hypogene barite could have been formed from plutonic sulphide sulphur is at variance with the following facts.

1. Concentrated barite aggregates are frequently observed in separation from sulphide ores and are practically free of minerals characteristic of them. Besides, barite mineralization often exceeds in scope sulphide mineralization. In Azerbaijan, a barite-bearing band 10 to 16 km wide stretches for 100 km, while sulphide minerals (galenite, sphalerite, chalcopyrite) have been found in minor quantities only in some of the deposits (*Kerimov* and *Efendiev*, 1957).

2. The temperature at which barite is formed is not high enough for intensive isotopic exchange between SO_4^{2-} and H_2S or between SO_2 and H_2S take place. In the Caucasian and Central Asian deposits, it was below 150°C (*Uchaimeshvili* et al., 1971), and in Durham County (U.K.), from 110 to 70° and less (*Hirst* and *Smith*, 1974).

3. Variations of δ^{34}S of sulphate sulphur show no regular dependence on δ^{34}S of sulphide sulphur in the same deposits.

4. Formation temperatures of sulphides and barite as determined on the δ^{34}S equilibrium values in sulphide pairs and in sulphide-barite pairs suggest different deposition

times for sulphides and barite. This is confirmed by data on barite segregation after tectonic influences on sulphide ores.

Ahydrite occurs in several hypogene deposits. Sulphur of all anhydrites studied thus far is of an increased weight.

In the Koktenkul molybdene-tungsten deposit (Central Kazakhstan), anhydrite (δ^{34}S = + 22 and + 23‰) occurs as lenses and veinlets in silicified tuffs at a depth of 650–690 m separately from orebodies whose sulphides are characterized by δ^{34}S from − 1.0 to + 0.9‰. In the Alekseyev copper ore deposit (Urals), anhydrite with δ^{34}S from + 12.3 to + 16.6‰ developed in the post-ore stage and is observed in wall rock, skarns, and orebodies. Sulphides of the ore stage in that deposit have δ^{34}S varying from − 3.3 to + 0.8‰. Maximum δ^{34}S, from + 14.7 to + 17.7‰, have been recorded at a depth of 375.5–470 m in the Levikhino copper pyrite deposit (Urals), where δ^{34}S of sulphide minerals varies from − 3.8 to + 1.6‰ (*Vinogradov* et al., 1969). Magnetite skarn deposits of the Turgai trough contain some sulphide minerals with δ^{34}S from + 0.9 to − 6.1‰ and anhydrite with δ^{34}S from + 12.1 to + 13.8‰ (*Sokolov* and *Pavlov*, 1974).

Marine sulphate seems a very probable source of the sulphur of hypogene anhydrites; it was partly mixed up with lighter sulphate of surface waters. Particularly obvious is the formation of anhydrite (δ^{34}S from + 10.5 to + 21.5‰) from evaporite sulphate sulphur in the Almalyk ore field (Karamazar), where the anhydrite (δ^{34}S from + 10.5 to + 21.5‰) also occurs in sedimentary Devonian-Carboniferous series (*Badalov* and *Vinogradov*, 1967).

Thaumasite in ores is usually formed during a later, sulphide-free stage. Sources of its sulphate sulphur are vadose waters heated in the depth. When in them atmospheric sulphur prevails, thaumasite has low δ^{34}S. Examples are: the Kusinskoe titanium-magnetite deposit in the Urals (+ 1.3‰), the Akdzhal lead-zinc deposit (from + 1.3 to + 4.4‰), and the Akmaia molybdene-tungsten deposit (from + 1.1 to + 6.1‰) in Central Kazakhstan. Dissolution of evaporites by vadose waters favours the formation of ^{34}S enriched thaumasite; a case in point is thaumasite from Willerois in Canada (our data).

Findings of the study of the isotopic composition of sulphate sulphur of solutions circulating in volcanic matter show it to derive from various sources. The lightest sulphate develops from oxidation of sulphur coming from the depths (mainly as hydrogen sulphide) without any great change of its isotopic composition. In volcanic areas, authigene aggregates of sulphur develop from hydrogen sulphide or the sulphur forms part of the newly formed iron disulphide. In top parts of volcanic structures, plutonic sulphur converted to sulphate form mixes up with a subordinate amount of atmospheric sulphur characterized by a low or lowered ^{34}S content. Therefore, sulphate sulphur of hot acid solutions of solfataric fields has typically low δ^{34}S. Where underground vadose waters, including sea water, flow by convection into the thermal fields of a volcano, the values of sulphate sulphur of solutions grow, while the solutions retain acid reaction and capacity for alunitization of volcanic rock. Alunite δ^{34}S mostly depends on proportion in acid solution of sulphate sulphur formed from juvenile (plutonic) sulphur, on the one hand, and vadose water sulphur, on the other. In hypogene sulphates formed outside volcanic areas at various depths, δ^{34}S also varies but, in contrast to alunites, it is normally never negative. The isotopic composition of sulphur of barite, anhydrite, celestine and thaumasite is predetermined by the proportion in them of heavy marine sulphate and lighter sulphate derived from the atmosphere or from weathered sulphides mainly from sedimentary series. Correspondence of the δ^{34}S values of hypogene sulphates to those of evaporites for a given period should not be presumed.

In major deposition zone of hypogene sulphates in hydrothermal deposits, there are not conditions for lighter sulphate sulphur to be formed from plutonic hydrogen sulphide, as oxygen normally has a minor part to play in that zone.

References

Anger, G., Nielsen, H., Puchelt, H., Ricke, W.: Sulfur isotopes in the Rammelsberg ore deposit (Germany). Econ. Geol. **61**, 3 (1966)

Badalov, S.T., Vinogradov, V.I.: On sources of sulphur in endogene deposits of Northwestern Karamazar. In: Izotopy sery i voprosy rudoobrazovaniya. Moscow: Nauka 1967 (in Russian)

Buschendorf, F., Nielsen, H., Puchelt, H., Riecke, W.: Schwefel-Isotopen-Untersuchungen an Pyrit-Sphalerit-Baryt-Lager Meggen (Lenne) und an verschiedenen Devon-Evaporiten. Geochim. Cosmochim. Acta **27**, 5 (1963)

Gavrikova, S.N., Vinogradov, V.I.: On isotopic composition of sulphur in minerals of the copper deposit Karabas. In: Izotopy sery i voprosy rudoobrazovaniya. Moscow: Nauka 1967 (in Russian)

Grinenko, L.N., Grinenko, V.A.: On genesis of purple anhydrite in Norilsk-1 intrusion (on the basis of isotopic composition of sulphur). Geokhimiya 1 (1967) (in Russian)

Hirst, D.M., Smith, F.W.: Controls of barite mineralization in the Lower magnesian Limestone of the Ferrihill area, County Durham. Trans. Inst. Min. Metall., Sect. B.: Appl. Earth Sci., May 1974

Jensen, M.L.: Application of isotopic data to studies of the origins of intrusives. IV. Vsesoyuznyi simpozium po geokhimii stabilnykh izotopov. Tezisy dokladov, Moscow, 1972 (in Russian)

Jensen, M.L., Ashley, R.P., Albers, J.P.: Primary and secondary sulfates at Goldfield, Nevada. Econ. Geol. **66**, 4 (1971)

Kerimov, G.I., Efendiev, G.Kh.: Barite. Azerbaijan geology. Non-ore mineral resources. Baku: Azerbaijan Academy of Sciences Press 1957 (in Russian)

Ozerova, N.A., Aidinyan, N.Kh., Vinogradov, V.I.: Isotopic composition of sulphur of mercury deposits. In: Izotopy sery i voprosy rudoobrazovaniya. Moscow: Nauka 1967 (in Russian)

Ozerova, N.A., Naboko, S.I., Vinogradov, V.I.: Sulphides of mercury, antimony, and arsenic, forming from the active thermal springs of Kamchatka and Kuril Islands. Proc. IMA-IAGOD Meet. 1970, Joint Symp, Tokyo (1971)

Pilipenko, G.F.: Hydrotherma of the Uzon caldera. In: Gidrotermalnye rastvory oblastei aktivnogo vulkanizma. Novosibirsk: Nauka 1974 (in Russian)

Robinson, B.W.: The origin of mineralization at the Tui mine, Te Aroha, New Zealand, in the light of stable isotopes studies. Econ. Geol. **69**, 6 (1974)

Rye, R.O., Ohmoto, H.: Sulfur and carbon isotopes and ore genesis: a review. Econ. Geol. **69**, 6 (1974)

Sakai, H., Matsubaya, O.: Sulfur and oxygen isotopic ratios of gypsum and barite in the black ore deposits of Japan. Proc. IMA-IAGOD Meet. 1970, Joint Symp., Tokyo (1971)

Schoen, R., Rye, R.O.: Sulfur isotope distribution in solfataras, Yellowstone National Park. Science **169**, 3950 (1970)

Schroll, E., Wedepohl, K.H.: Schwefelisotopenuntersuchungen an einigen Sulfid- und Sulfatmineralien der Blei-Zink-Erzlagerstätte Bleiberg-Kreuth, Kärnten. Tschermaks Mineral. Petrogr. Mitt. B. 17 (1972)

Shipulin, F.K., Vinogradov, V.I.: On sulphur isotopes in Dashkesan ores. In: Izotopya sery i voprosy rudoobrazovaniya. Moscow: Nauka 1967 (in Russian)

Sokolov, G.A., Pavlov, D.I.: On sources of ore-forming substances of postmagmatic iron ore deposits. Sources of ore matter of endogene deposits. Abstract of paper, Moscow (1974) (in Russian)

Steiner, A., Rafter, T.A.: Sulfur isotopes in pyrite, pyrrhotite, alunite and anhydrite from steam wells in the Taupo volcanic zone, New Zealand. Econ. Geol. **61**, 6 (1966)

Uchaimeshvili, N.E., Malinin, S.D., Khitarov, N.I.: Physico-chemical principles of the formation of hydrothermal barite in some typical barite-containing associations. 1st Intern. Geochem. Cong. Abstr., Moscow (1971) (in Russian)

Vinogradov, V.I.: New data on the origin of lake sulphur on the Golovin Volcano. Geokhimiya 1 (1965) (in Russian)

Vinogradov, V.I.: Role of sulphates in ore formation. In: Izotopy sery i voprosy rudoobrazovaniya. Moscow: Nauka 1967 (in Russian)

Vinogradov, A.P., Grinenko, L.N., Grinenko, V.A., Stolyarov, Yu.M.: Isotopic composition of sulphur of sulphides and sulphates of Alekseyev copper ore deposit (Middle Urals) and some problems of its genesis. Geokhimiya 8 (1969) (in Russian)

Vinogradov, V.I., Ilupin, I.P.: Isotopic composition of sulphur in kimberlites. Doklady AN SSSR **204**, 6 (1972) (in Russian)

Carbon and Oxygen Isotope Studies in Rocks of the Vicinity of the Almadén Mercury Deposit (Province of Ciudad Real, Spain)

R. EICHMANN, Mainz, F. SAUPÉ, Vandoeuvre, and M. SCHIDLOWSKI, Mainz

With 1 Figure

Contents

Summary

Carbonate constituents from 36 samples of spilitized basalts, dolerites, possible lamprophyres, pyroclastic tuffs and hydrothermal vein fillings which occur in the sedimentary fram of the Almadén mercury deposit show consistently low $\delta^{13}C$ values of -6.6 ± 2.2 ‰ vs. PDB, and a corresponding $\delta^{18}O$ average of $+17.9 \pm 2.5$‰ vs. SMOW. With the $\delta^{13}C$ values of the rare intercalations of sedimentary limestones and dolomites lying in the normal range (-0.2 ± 0.3‰ vs. PDB), the igneous carbonate constituents are unlikely to be of sedimentary pedigree, but rather represent genuinely "hypogenic" material derived from primary magmatic sources.

Zusammenfassung

Karbonatbestandteile von insgesamt 36 magmatischen Paragenesen (spilitisierte Basalte, Dolerite, fragliche Lamprophyre, pyroklastische Tuffe, hydrothermale Gänge) aus dem sedimentären Rahmen der Quecksilberlagerstätte Almadén besitzen durchgehend niedrige $\delta^{13}C$-Werte von -6.6 ± 2.2‰ [PDB] und ein zugehöriges $\delta^{18}O$-Mittel von $+17.9 \pm 2.5$‰ [SMOW]. Da die relativ seltenen Einschaltungen sedimentärer Kalke und Dolomite innerhalb der Schichtfolge die normalen Werte mariner Karbonate zeigen ($\delta^{13}C = -0.2 \pm 0.3$‰ [PDB]), kann der Karbonatanteil der magmatischen Bildungen nicht aus den umgebenden Sedimenten hergeleitet werden, sondern repräsentiert mit größter Wahrscheinlichkeit „hypogenes" Karbonatmaterial magmatischer Provenienz.

1. Introduction

During a recent reinvestigation of the well-known mercury deposit of Almadén (*Saupé*, 1973) some questions have been raised which, as yet, lack a satisfactory answer. Among these problems some can be solved with the help of C and O isotope studies.

2. Geological Background

Almadén ist situated in the southern part of the Iberian Meseta, bordered by the Sierra Morena to the South and the Montes de Toledo to the North. The sedimentary sequence consists of two different series separated by the Sardic ("Iberic") unconformity which are, in turn, overlain by sediments of Tertiary to Recent age. The older series is of pre-Ordovician age (Upper Precambrian), consisting of shales and graywackes altogether between 6000 and 7000 m thick, with rare intercalations of limestone and conglomerate horizons. The younger series represents the Paleozoic between Arenig and Visé (with a break during Middle Devonian) in epicontinental facies, having a thickness of some 4000 m. Starting off with the Armorican Quartzite, this Paleozoic sequence continues with shales, alternating shales and arenites, and three major quartzite horizons, one of these latter (of Llandovery age) being the principal host rock of the mercury deposit ("cuarcita del criadero"). Minor intercalations of sedimentary carbonates are confined to the Ordovician (Ashgill) and Lower Devonian.

During the Paleozoic, three major phases of mafic volcanism (of the tholeitic type, provided that the criteria used for the distinction of normal basalts can be applied to rocks that underwent such strong transformations) have resulted in the formation of numerous sills and flows interspersed in the sedimentary sequence. This magmatism was, as a whole, characterized by a weak differentiation, having yielded basalts during the earlier stages, and dolerites during the later stages, with a subsequent differentiation of the dolerite members towards quartz-andesites. Both basalts and dolerites were consequently subjected to large-scale deuteric processes involving volatiles (H_2O and CO_2) and producing rocks rich in albite, chlorite, carbonates and iron-hydroxydes (epidote and sericite remain subordinate). This low-temperature mineral assemblage may have resulted either from direct crystallization of a fluid-rich magma or from replacement of primary high-temperature mineral phases. Further, submarine explosions have led to the formation of pyroclastic tuffs, interstratified in the sedimentary sequence. Basic dykes, having some chemical similarities with lamprophyres, intruded after the folding. The Paleozoic formations have experienced a weak (anchizonal at the most) metamorphism that in no instance even reaches the greenschist facies.

Thus, carbonates are fairly wide-spread in the different types of igneous rock. It is of interest for understanding the genesis of the mercury deposit to trace the origin of the CO_2 in the different groups of these rocks, as well as that of the hydrothermal dolomites and of the carbonate cement of the pyroclastic rocks.

The intercalations of sedimentary carbonates at the base of the Ashgill are lenticular but apparently coeval. In the vicinity of the orebody, although separated from the latter by about 100 m of shales, this intercalation happens to be a dolomite, whereas elsewhere it is usually a limestone. A microscopic examination (*Saupé*, 1973) led to the conclusion of syngenetic dolomitization followed by a diagenetic recrystallization. Because of the vicinity of the ore deposit, it seemed appropriate to obtain also isotopic evidence. For comparison, carbonates from the cement of a sandstone and from a diagenetic nodule in the host rock were also analyzed.

3. Experimental

The samples have been processed by the conventional techniques in use for isotopic work on carbonates (*McCrea*, 1950) and reduced carbon (*Craig*, 1953). The generated carbon dioxide was analyzed on a Varian-MAT CH5 spectrometer equipped with a double inlet system and two Faraday collectors. Results are reported as $\delta^{13}C$ values relative to the PDB standard. All values are corrected for ^{17}O (*Craig*, 1957). The mass spectrometric correction factors applied were determined after *Deines* (1970). The standard deviation found (covering the whole processing and analytical procedure) was ± 0.3 ‰ for carbonate carbon and less than ± 1.0‰ for organic carbon. The $\delta^{18}O$ are reported relative to the SMOW standard.

4. Analytical Results

The results of our isotope measurements are listed in Tables 1 and 2, differentiating between carbonates from apparently igneous and those from sedimentary environments. In Table 2, three measurements carried out on reduced (organic) carbon have been included. A graphic synopsis of all carbonate values is given in Figure 1, while Table 3 lists the averages (with standard deviations) of the principal carbonate groups. Since no obvious correlation exists between the isotopic composition and the dolomite content of sedimentary carbonates (*Schidlowski* et al., 1976), we did not determine the calcite dolomite ratio of each sample investigated (chemical data on some of the carbonates are reported in Table 4).

As is obvious from Table 1 and the graphic representation of the respective isotope data (Fig. 1), the $\delta^{13}C$ values of the „igneous" carbonates fall, as a rule, within the range between − 3.5 and − 9.0‰ [PDB], with an overall mean of − 6.6 ± 2.2‰. The averages for basalts (− 5.8 ± 1.3‰), dolerites (− 7.1 ± 1.4‰), possible „lamprophyres" (− 7.0 ± 0.3‰) and pyroclastic rocks (− 5.4 ± 1.9‰) lie fairly close to each other, with only the mean for hydrothermal vein carbonates (− 10.0 ± 3.4‰) distinctly shifted in negative direction (see also Table 3). Two samples of this latter group fall outside the general field, approaching closely the − 14 permil line (Fig. 1). They do, however, not belong to the volcanic association of Almadén and were formed by an independent and later event. The $\delta^{18}O$ values of all "igneous" carbonates lie − with two minor exceptions − between + 13.0 and + 20.0‰ vs. SMOW, with an overall average of + 15.8 ± 1.9‰.

With $\delta^{13}C$ values close to 0‰, the Ashgill carbonates (Table 2A) intercalated in the sequence display the normal isotopic composition of sedimentary limestones and dolomites (cf. *Keith* and *Weber*, 1964; *Eichmann* and *Schidlowski*, 1975; *Schidlowki* et al., 1975). On the other hand, the $\delta^{18}O$ values yielded by these samples (+ 17.9 ± 2.5‰ vs SMOW) are smaller by some 6‰, as compared with an average close to + 24‰ [SMOW] (recalculated from PDB) reported by *Keith* and *Weber* (1964, p. 1796) for 35 marine limestones of Ordovician age, this shift towards the negative side resulting most probably from a post-depositional oxygen exchange with isotopically light meteoric waters (cf. *Clayton* and *Degens*, 1959; *Keith* and *Weber*, 1964). The exceptional $\delta^{13}C$ value of

Table 1. Carbon and oxygen isotope composition of carbonate constituents of igneous rocks or of carbonates of apparently igneous affinities

No.	Description of sample	$\delta^{13}C$ [‰, PDB]	$\delta^{18}O$ [‰, SMOW]
A. Basalts			
1.	Finely dispersed carbonate from groundmass and altered pheno-crysts (mostly pyroxene) of a primary pyroxene basalt now changed into a spilite	− 5.6	+ 15.3
2.	Carbonate specks from groundmass, altered phenocrysts and vesicles of a primary basalt now completely spilitized	− 3.6	+ 17.3
3.	As No. 2, with carbonate material stemming mainly from vesicles	− 4.1	+ 16.7
4.	As No. 2	− 4.8	+ 13.1
5.	As No. 2 (carbonate mainly from groundmass and vesicles)	− 6.0	+ 15.8
6.	Carbonate from vesicles of a completely spilitized basalt	− 8.9	+ 15.1
7.	Carbonate from completely spilitized basalt exposed in under-ground workings of Almadén	− 5.0	+ 12.8
8.	As No. 7	− 6.1	+ 16.0
9.	As No. 7	− 7.1	+ 14.2
10.	As No. 7	− 7.1	+ 16.4
11.	Carbonate from largely altered basalt exposed in underground workings of Nueva Concepción (Almadenejos)	− 5.5	+ 13.5
12.	As No. 11	− 5.1	+ 13.8
13.	As No. 11	− 5.6	+ 13.5
14.	As No. 11	− 6.5	+ 14.0
15.	As No. 11	− 6.0	+ 15.4
B. Dolerite sills			
16.	Finely dispersed carbonate in strongly altered dolerite	− 5.5	+ 16.0
17.	Carbonate (stemming from alteration of pyroxene) occupying net of tiny rock fissures	− 7.1	+ 19.9
18.	Carbonates from fissures and altered phenocrysts	− 6.0	+ 14.8
19.	Carbonates from pseudomorphs, minute fissures and finely disseminated inclusions within altered dolerite	− 9.0	+ 14.8
20.	Carbonate from fissure fillings in altered dolerite	− 7.7	+ 17.7
C. Altered mafic dykes ("lamprophyres")			
21.	Carbonate (dolomite, ankerite) in fine-grained mafic rocks occupying fault planes of Hercynian age	− 6.7	+ 15.3
22.	As No. 21	− 6.9	+ 16.1
23.	As No. 21	− 6.7	+ 16.2
24.	As No. 21, but from other dyke	− 7.1	+ 15.1
25.	As No. 21, but from different dyke	− 7.4	+ 16.8

Table 1 (continued)

No.	Description of sample	$\delta^{13}C$ [‰, PDB]	$\delta^{18}O$ [‰, SMOW]
D. Pyroclastic rocks			
26.	Lapilli tuffs within an argillite-carbonate matrix. The carbonates analyzed represent a mixture of matrix material and that contained in vesicles within the lava	− 3.9	+ 18.2
27.	As No. 26	− 4.0	+ 16.1
28.	As No. 26 (mainly carbonate matrix)	− 5.3	+ 14.0
29.	Mixture of matrix carbonates and vesicle fillings in rocks as No. 26	− 8.6	+ 16.2
30.	As No. 26	− 3.9	+ 15.3
31.	Carbonate constituents in the pyroclastic "Frailesca" cross-cutting sedimentary quartzite [the clastic material is composed of angular lava fragments and sediment debris (shale, quartzite)]	− 6.4	+ 17.9
E. Hydrothermal carbonates (dolomite)			
32.	Crystals of dolomite (with inclusions of cinnabar on growth-planes) from vein transecting Hg-mineralization at Almadén	− 8.3	+ 17.6
33.	Same as No. 32	− 7.1	+ 19.3
34.	Same as No. 32	− 7.2	+ 18.9
35.	Dolomite crystals from Carmen (Sta. Ana) exploration diggings on a galena vein (right bank of Rio Esteras near Garlitos, Province of Badajoz)	− 13.9	+ 12.7
36.	Dolomite crystals from exploration diggings on a sphalerite-bearing quartz-dolomite vein (near Chillón, Province of Ciudad Real)	− 13.3	+ 18.1

Table 2. Isotopic composition of carbonates and reduced carbon from sedimentary environments

No.	Description of sample	$\delta^{13}C$ [‰, PDB]	$\delta^{18}O$ [‰, SMOW]
A. Sedimentary carbonates			
1.	Ashgill limestone from La Calera quarry (2 km SW Fontanosas, Province of Ciudad Real)	− 0.3	+ 15.1
2.	Ashgill limestone from obsolete quarry E of railway station of Chillón (Province of Ciudad Real)	− 0.4	+ 18.8
3.	Ashgill dolomite with pyrite inclusions from Chorillo (south of Almadén)	+ 0.1	+ 19.8
4.	As No. 3, but from exposure 15 km to the east. Slightly mineralized (galena, sphalerite) and transected by numerous veinlets of secondary carbonate	− 5.0	+ 16.7

Table 2 (continued)

No.	Description of sample	$\delta^{13}C$ [‰, PDB]	$\delta^{18}O$ [‰, SMOW]
	B. Carbonates from interstices, concretions, etc.		
5.	Carbonate constituents of black bituminous Llandovery shales overlying mineralized quartzite in underground workings of Almadén	− 7.9	+ 15.7
6.	Ellipsoidal dolomite concretion (with internal lamination, enclosed by seam of pyrite) contained in black shales of No. 5	− 8.1	+ 11.5
7.	Carbonate contained in black bituminous sandstone overlying S. Pedro ore horizon (Almadén)	− 4.2	+ 13.9
	C. Reduced (organic) carbon		
8.	Reduced carbon from carbonate concretion of No. 6	− 26.5	− −
9.	As No. 8	− 28.2	− −
10.	Reduced carbon from black shales of No. 5	− 34.4	− −

Table 3. Mean $\delta^{13}C$ and $\delta^{18}O$ values and standard deviations of major carbonate groups listed in Tables 1 and 2

Group No.	No. of samples	Group description	$\delta^{13}C$ [‰, PDB]	$\delta^{18}O$ [‰, SMOW]
1	15	Carbonates in basalts (Table 1A)	− 5.8 ± 1.3	+ 14.9 ± 1.4
2	5	Carbonates in dolerites (Table 1B)	− 7.1 ± 1.4	+ 16.6 ± 2.2
3	5	Carbonates in "lamprophyrs" (Table 1C)	− 7.0 ± 0.3	+ 15.9 ± 0.7
4	6	Carbonates in pyroclastics (Table 1D)	− 5.4 ± 1.9	+ 16.3 ± 1.6
5	5	Hydrothermal carbonates (Table 1E)	− 10.0 ± 3.4	+ 17.3 ± 2.7
Σ 1–5	36	Average "igneous" carbonates	− 6.6 ± 2.2	+ 15.8 ± 1.9
6	3 [a]	Sedimentary carbonates (Table 2A)	− 0.2 ± 0.3	+ 17.9 ± 2.5
7	3	Concretions and interstitial carbonates in sediments (Table 2B)	− 6.7 ± 2.2	+ 13.7 ± 2.1
8	3	Sedimentary organic carbon (Table 2C)	− 29.7 ± 4.2	− −

[a] In calculating the average for this group, sample No. 4 has been omitted on account of being obviously affected by post-depositional alteration (cf. sample description in Table 2).

Table 4. Calculated mineral composition of the carbonate phases in some samples

No.	CaCO₃ %	MgCO₃ %	FeCO₃ %
1	90.52	1.87	1.22
2	89.54	0.77	0.55
3	52.19	34.29	3.85
4	48.35	24.32	16.97
5	0.93	1.01	2.59
7	0.48	0.04	1.85
32	51.12	30.64	13.11
33	48.06	28.77	9.95
34	42.92	23.89	8.71
36	44.41	26.04	16.41

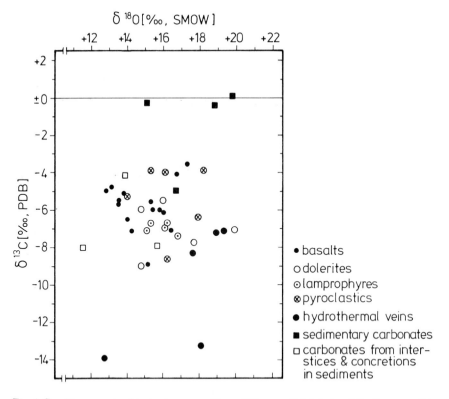

Fig. 1 Graphic synopsis of isotopic composition of "igneous" (*circles*) and "sedimentary" (*squares*) carbonates listed in Tables 1 and 2. The two extremely light hydrothermal carbonates in the lower part of the graph come from lead-zinc veins in the district (Table 1, Nos. 35 and 36)

− 5.0 ‰ [PDB] exhibited by No. 4 is certainly due to secondary processes as testified by abundant veinlets of reconstituted carbonates and the presence of ore minerals (galena, sphalerite) in this sample.

Interstitial carbonate and a dolomite concretion from non-carbonate sediments (Table 2B) have yielded a $\delta^{13}C$ average of − 6.7 ± 2.2‰ [PDB], with a corresponding $\delta^{18}O$ of + 13.7 ± 2.1‰ [SMOW]. The isotopic composition of sedimentary organic carbon (Table 2C) falls within the common range ($\delta^{13}C$ = − 29.7 ± 4.2‰ vs. PDB).

5. Discussion of Results

With the isotope data listed in Tables 1 and 2 at our disposal, a genetic interpretation of the carbonates originating from the mafic igneous suite of the Almadén mercury district is considerably facilitated.

The marked difference between the $\delta^{13}C$ averages yielded by the "igneous" carbonate constituents (− 6.6 ± 2.2‰ vs. PDB) and the scarce occurrences within the sequence of sedimentary limestones and dolomites (− 0.2 ± 0.3‰ vs. PDB) makes it unlikely that the former were derived by assimilation of sedimentary material. This applies to all groups of "igneous" carbonates investigated and most decidedly to the hydrothermal samples (Table 1E) characterized by a mean of − 10.0 ± 3.4‰ (cf. Table 3). Without exception, the $\delta^{13}C$ averages of the individual groups fall within the range previously reported for carbonates of unquestionable hypogenic affiliation, notably carbonatites (see, inter alia, *Friedrichsen*, 1968; *Deines* and *Gold*, 1973) and hydrothermal vein calcites (*Harzer* and *Pilot*, 1969; *Schoell* and *Stahl*, 1970, 1971). Also, the $\delta^{18}O$ mean of the "igneous" group as a whole (+ 15.8 ± 1.9‰ vs. SMOW, Table 3) coincides with the respective isotope spreads displayed by magmatogenic carbonates. The absence of intermediate values between the field of "igneous" $\delta^{13}C$ values and that of the ambient sedimentary limestones and dolomites (Fig. 1) would well-nigh exclude the possibility that the igneous members have evolved from the sedimentary ones as a result of metamorphism (which may account for limited isotope shifts in sedimentary carbonates towards the negative side, see *Deines* and *Gold*, 1969; *Shieh* and *Taylor*, 1969). These findings are also consistent with the low-grade metamorphism of the region in general (*Saupé* et al., in press 1977).

The above conclusion is decidedly supported by the results of previous investigations which have shown that a distinction between genuinely magmatic and sedimentogenic carbonates in igneous environments is indeed possible by means of carbon isotope work. For instance, in a large suite of rocks from several European spilite and ophicalcite provinces *Schidlowski* et al. (1970) and *Schidlowski* and *Stahl* (1971) were able to discriminate between carbonate material of sedimentary and magmatic pedigree. As a result of this investigation, the sedimentary derivation of the carbonate fractions of the "weilburgites" (cf. *Lehmann*, 1949) of the Rhenish Mass and the Arosa spilites (Swiss and Austrian Alps) could be firmly established, while on the other hand, the respective constitutents of the Verrucano spilites (Helvetian nappes of Switzerland) proved to be of magmatic origin. Incidentally, the similarity of the "igneous" Almaden values with those of the Helvetian Verrucano province ($\delta^{13}C$ = − 5.3 ± 0.5‰ [PDB], $\delta^{18}O$ = 17.0

± 0.8 ‰ [SMOW] is indeed striking. In a similar way, *Fritz* (1969) and *Stahl* (1971) have utilized the bulk isotopic composition of vein carbonates and calcareous joint fillings respectively for determining the origin of these materials. Among the hydrothermal vein carbonates (Table 1E), three show values falling in the field of the volcanic assemblage of Almadén from which they were derived by local remobilization during a later tectonic event. The two others (Nos. 35 and 36) represent hydrothermal carbonates from different occurrences and are listed here for comparative purposes.

The "normal" $\delta^{13}C$ values close to 0‰ of the Ashgill carbonates (Table 2A) provide clear-cut proof that the dolomite ("dolomite du Chorillo") of this discontinuous layer (usually a few m thick) is a genuine sediment, thus substantiating previous conclusions based on field evidence alone (*Saupé*, 1973). As mentioned before, the exceptional value yielded by sample Nr. 4 clearly reflects secondary processes, most probably hydrothermal alteration.

As for the isotopic composition of diagnetic carbonate from interstices and concretions (Table 2B), it is obvious that the respective values coincide with the spread displayed by the igneous carbonate constituents. This coincidence is, however, a fortuitous one. As has been shown by several investigators (*Berner*, 1968; *Hoefs*, 1970; *Sass* and *Kolodny*, 1972) and others) carbon dioxide stemming from the decomposition of organic matter usually contributes significantly towards the formation of diagenetic carbonate, notably carbonate concretions. With the participation of major amounts of isotopically light biogenic CO_2 in the process of diagenetic carbonate formation, the $\delta^{13}C$ values of the resulting carbonates should lie somewhere half-way between the values of organic (− 25‰) and inorganic carbon (± 0‰), depending on the relative contributions of both sources. Thus, the similarity in isotopic composition between diagenetic and magmatogenic carbonates is the result of some "isotopic convergence", reflecting no genetic ties whatsoever between the two carbonate groups.

Acknowledgments. This work was initiated while one of the authors (F.S.) stayed at the University of Heidelberg thanks to a grant of the Alexander von Humboldt Stiftung. The two others would like to acknowledge financial support rendered by the Deutsche Forschungsgemeinschaft (SFB 73)

References

Berner, R.A.: Calcium carbonate concretions formed by the decomposition of organic matter. Science 159, 195–197 (1968)

Clayton, R.N., Degens, E.T.: Use of carbon isotope analyses of carbonates for differentiation of fresh-water and marine sediments. Bull. Am. Ass. Petrol. Geol. 43, 890–897 (1959)

Craig, H.: The geochemistry of the stable carbon isotopes. Geochim. Cosmochim. Acta 3, 53–92 (1953)

Craig, H.: Isotopic standards for carbon and oxygen and correction factors for mass-spectrometric analysis of carbon dioxide. Geochim. Cosmochim. Acta 12, 133–149 (1957)

Deines, P.: Mass-spectrometer correction factors for the determination of small isotopic composition variations of carbon and oxygen. Intern. J. Mass Spectrom. Ion. Phys. 4, 283–295 (1970)

Deines, P., Gold, D.P.: The change in carbon and oxygen isotopic composition during contact metamorphism of Trenton limestone by the Mt. Royal pluton. Geochim. Cosmochim. Acta 33, 421–424 (1969)

Deines, P., Gold, D.P.: The isotopic composition of carbonatites and kimberlite carbonates. Geochim. Cosmochim. Acta 37, 1709–1733 (1973)

Eichmann, R., Schidlowski, M.: Isotopic fractionation between coexisting organic carbon-carbonate pairs in Precambrian sediments. Geochim. Cosmochim. Acta 39, 585–595 (1975)

Friedrichsen, H.: Sauerstoffisotopen einiger Minerale der Karbonatite des Fengebietes, Süd-Norwegen. Lithos 1, 70–75 (1968)

Fritz, P.: The oxygen and carbon isotopic composition of carbonates from the Pine Point lead-zinc ore deposits. Econ. Geol. 64, 733–742 (1969)

Harzer, D., Pilot, J.: Isotopenchemische Untersuchungen an Ganglagerstätten des Harzes. Ber. deut. Ges. Geol. Wissensch. (B) 14, 129–138 (1969)

Hoefs, J.: Kohlenstoff- und Sauerstoff-Isotopenuntersuchungen an Karbonatkonkretionen und umgebendem Gestein. Contrib. Mineral. Petrol. 27, 66–79 (1970)

Keith, M.L., Weber, J.N.: Isotopic composition and environmental classification of selected limestones and fossils. Geochim. Cosmochim. Acta 28, 1787–1816 (1964)

Lehmann, F.: Das Keratophyr-Weilburgit-Problem. Heidelberger Beiträge zur Mineral. Petrogr. 2, 1–116 (1949)

McCrea, J.M.: The isotopic chemistry of carbonates and a paleotemperature scale. J. Chem. Phys. 18, 849–857 (1950)

Sass, E., Kolodny, Y.: Stable isotopes, chemistry and petrology of carbonate concretions (Mishash Formation, Israel). Chem. Geol. 10, 261–286 (1972)

Saupé, F.: La géologie du gisement de mercure d'Almadén (Province de Ciudad Real, Espagne). Nancy, Sciences de la Terre, Mémoirs 29, 342 p. (1973)

Saupé, F., Dunoyer de Segonzac, G., Teichmüller, R.: Étude du métamorphisme régional par la cristallinité de l'illite et la réflectance de la matière organique dans la zone du gisement de mercure d'Almadén (Province de Ciudad Real, Espagne). (Submitted for publication to Sciences de la Terre)

Schidlowski, M., Eichmann, R., Junge, C.E.: Precambrian sedimentary carbonates: carbon and oxygen isotope geochemistry and implications for the terrestrial oxygen budget. Precambrian Res. 2, 1–69 (1975)

Schidlowski, M., Eichmann, R., Junge, C.E.: Carbon isotope geochemistry of the Precambrian Lomagundi carbonate Province, Rhodesia. Geochim. Cosmochim. Acta 40, 449–455 (1976)

Schidlowski, M., Stahl, W.: Kohlenstoff- und Sauerstoff-Isotopenuntersuchungen an der Karbonatfraktion alpiner Spilite und Serpentinite sowie von Weilburgiten des Lahn-Dill-Gebietes. N.J. Mineral. Abh. 115 (3), 252–278 (1971)

Schidlowski, M., Stahl, W., Amstutz, G.C.: Oxygen and carbon isotope abundances in carbonates of spilitic rocks from Glarus, Switzerland. Naturwissenschaften 57, 542–543 (1970)

Schoell, M., Stahl, W.: Kohlenstoff- und Sauerstoff-Isotopenanalysen an hydrothermalen Kalziten aus dem Harz. Fortschr. Mineralog. 48, Beiheft 1, 87–90 (1970)

Schoell, M., Stahl, W.: Kohlenstoff- und Sauerstoff-Isotopenanalysen an hydrothermalen Kalkspäten aus St. Andreasberg, Harz. In: Isotope Titles 1. Leipzig: Sonderheft ASTI 1969, 1971, pp. 242–253

Shieh, Y.N., Taylor, H.P.: Oxygen and carbon isotope studies of contact metamorphism of carbonate rocks. J. Petrol. 10, 307–331 (1969)

Stahl, W.: Isotopen-Analysen an Carbonaten und Kohlendioxid-Proben aus dem Einflußbereich und der weiteren Umgebung des Bramscher Intrusivs und an hydrothermalen Carbonaten aus dem Siegerland. Fortschr. Geol. Rheinld. u. Westf. 18, 429–438 (1971)

Origin of Amphibolite and Associated Magnetite Ore from the Pocheon Iron Mine (Korea)

CHIL-SUP SO, Seoul

With 7 Figures

Contents

Summary

Ten previously supposed para-amphibolites from the Pocheon magnetite deposit, Korea, were analyzed for 21 elements by wet-chemical and emission-spectroscopic methods. Trends of all the chemical variations were compared with those of ortho-amphibolites from the Gyeonggi metamorphic complex, and consistently indicate that all the amphibolites analyzed are of igneous parentage and probably represent metamorphosed basic extrusive rocks. Banding in the amphibolites has resulted from chemical differentiation during metamorphism. The field and trace-element data for Pocheon magnetite ore, which is always intimately associated with the amphibolites, also suggest a volcanic origin.

1. Introduction

Among the magnetite deposits reported in Korea, only a few are being mined, due to a gradual exhaustion of the existing high-grade ore reserves and an increase in production costs caused by a more difficult mining situation. The Pocheon mine, which is situated in the northern part of the Gyeonggi province, is one of the largest iron producers in Korea with a total of about 10 m.t. of reserves, which is increasing because of present active prospecting. The mining area is underlain mainly by pre-Cambrian gneisses, schists and lime-silicate rocks of the Gyeonggi metamorphic complex and, to the northeast, by Mesozoic biotite granite. Previously the origin of the deposit was considered to be contact-metasomatic by granite intrusion (*Kanda*, 1969), based mainly on qualitative observations.

The magnetite ore in the deposit is strata-bound with a lens-like form, the biggest body attaining dimensions of 180 x 40 x 200 m, and is nearly always closely associated with amphibolites, which show partly gradational relations with the surrounding meta-sediments. The amphibolites were supposed previously to be metasedimentary rocks intercalated in the surrounding metamorphic sequence. The magnetite ore shows commonly a distinct layered structure alternating with lime-silicate bands and rarely, in the lower part of the orebody, with amphibolite. The orebody has suffered minor folding, although this is not clear in the high-grade part. The amphibolites also commonly display similar tectonic features. The objective of the present study is to determine the origin of amphibolites at the Pocheon iron deposit by examination of its chemical features, in order to contribute to an understanding of the genesis of the deposit. The methods of *Evans* and *Leake* (1960) and *Leake* (1964) are used, in which major elements are treated as Niggli numbers. Trace-element data are also useful in determining the origin of the rocks. Furthermore, a comparison of the trace-element abundances of the magnetites from the Pocheon mine with those of magnetites from elsewhere in the world (*Hegemann* and *Albrecht*, 1954; *Frutos* and *Oyarzún*, 1975) is presented.

2. Sampling and Treatment

The distribution of the rock specimens collected from the − 3 and − 5 levels of the Pocheon mine is shown in Figure 1. Sampling was designed to obtain materials representative of the amphibolite. Specimens, each weighing 2 to 3 kg, were collected. Magnetite ore was also collected for trace-element study from different parts of the mine.

All specimens were examined petrographically, and modal analyses of approximately 1000 counts each were carried out with a Swift automatic point-counter; results are presented in Table 1. It should be noted that although the compositions of some samples is such that they can hardly be called amphibolite (samples 4 and 5 in Table 1), they are included in the discussion because of their close association with more typical amphibolite. Plagioclase compositions were measured, using extinction angles and refractive indices. Alteration to sericite is commonly strong and hampers measurement of composition in several thin sections. Measurement of refractive indices of hornblende was carried out by the use of standard liquids on cleavage fragments. The index of refraction of the liquid was checked on an Abbe refractometer immediately after each match was obtained.

Samples were prepared for chemical and spectroscopic analysis by slicing the rocks across the banding and breaking them into 8-cm chips. The chips were broken to sand-size grains in a jaw-crusher prior to being powdered in a shatter-box with a tungsten carbide grinding container. The powders were ground until they passed through a 150-mesh sieve.

The separation of magnetite (− 100/+ 120 mesh) from the other silicate and opaque minerals of the powdered ore samples ready for trace-element analysis was carried out by successive heavy-liquid treatments and then by using a Franz isodynamic magnetic separator with variable magnetic flux. The final purification was accomplished by hand-picking under a high-power binocular microscope; a very sharp needle with a slight

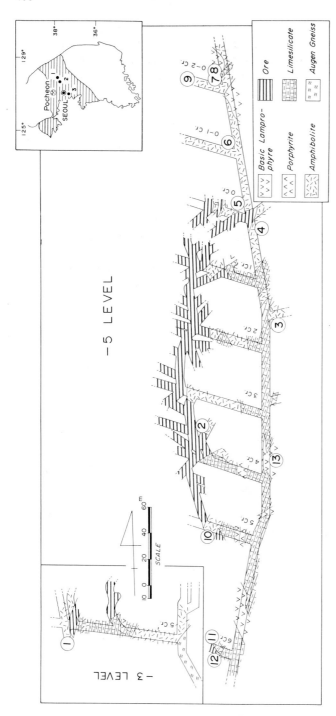

Fig. 1. Underground map of the Pocheon mine showing sampling localities, with index map (*inset*) of the Gyeonggi metamorphic complex (*lined area*: *1*: Chuncheon area; *2*: Cheongpyeong area; *3*: Anyang area)

bend near the tip served the purpose. The sharpened tip of the needle was immersed in foam soaked with vegetable oil. The oil-coated tip readily pinpointed and picked up any unwanted grains. Since for comparative studies of trace elements, only completely separated mineral grains can be successfully analyzed, portions of the prepared samples were mounted in araldite, and the polished and thin sections studied for any impurities; none were found.

3. Analytical Methods

3.1 Major Elements

The oxides of silicon, iron (III), calcium, magnesium and aluminum were determined using a gravimetric method, following alkaline fusion. Ferrous oxide was extracted into a saturated aqueous solution of boric acid and was titrated with permanganate solution. Determination of phosphorus and manganese was conducted by colorimetry using vanadomolybdate and permanganate, respectively. Potassium and sodium were analyzed using a Perkin-Elmer Model 303 atomic absorption spectrophotometer.

3.2 Trace Elements

Quantitative trace analysis of all the treated samples of amphibolite was made for 11 elements by emission spectrography. Each prepared sample was mixed with a buffer (KCl + carbon powder SP-3) free from spectroscopic impurities. Internal standards (specpure BeO and GeO_2) were added to the buffer. A crater in the lower electrode (National SPK L-3703) of highest purity graphite was filled with the sample mixture. Arc spectra of the prepared samples were then recorded on Eastman Kodak SA-1 plates using a 3.4 m grating spectrograph, Jarrell-Ash Ebert mounting, with dispersion in the first order of 5.1 Å per mm. Specpure silicon dioxide, aluminum oxide, magnesium oxide, calcium carbonate and ferric oxide were used to prepare a pure base substance, which was used to make standards ranging from 1000 to 1 ppm for all the interesting elements. The spectral plates were evaluated photometrically with the Jarrell-Ash console microphotometer. The content of each element was determined from the working curves (with background correction) using emulsion calibration curves obtained by a seven-step filter method. For the separated magnetite grains, nine trace elements were determined with the spectroscopic method described above. For this, specpure lanthanum (as La_2O_3) and yttrium (as Y_2O_3) were added to the buffer as internal standards, and specpure ferric oxide was used to prepare a base substance. Each analysis is the mean of three replicates per sample.

4. Petrography of the Amphibolites

The amphibolites at the Pocheon magnetite deposit occur as conformable layers, well foliated with the surrounding metasedimentary lime-silicate rocks and schists. They vary from a few tens of cm to 10 m in thickness and up to about 200 m in length, and are nearly always intimately associated with the magnetite ore (Fig. 1).

Table 1. Modal analyses of amphibolites from the Pocheon mine

Minerals (vol. %)

	Horn-blende	Plagio-clase	Pyro-xene	Biotite	Sphene	Epidote	Apatite	Quartz	Chlorite	Calcite	Musco-vite
1	65.0	22.2	–	2.0	5.1	1.0	2.0	<0.5	–	–	–
2	51.0	31.0	–	<0.5	7.1	2.2	2.0	<0.5	<0.5	–	–
3	60.1	15.0	1.1	3.0	3.2	10.0	3.0	1.0	–	–	–
4	21.0	37.0	26.0	1.0	2.1	8.1	2.0	<0.5	–	<0.5	<0.5
5	1.0	18.0	–	62.1	0.5	8.0	4.1	<0.5	–	–	–
6	61.1	28.0	–	2.0	2.0	<0.5	–	–	1.0	–	<0.5
7	46.2	11.0	2.0	–	4.0	31.0	3.0	–	–	<0.5	–
9	49.3	17.0	<0.5	14.0	<0.5	8.2	3.0	–	–	–	<0.5
10	61.0	19.0	1.0	<0.5	3.0	5.1	4.2	–	–	–	–
11	44.3	22.0	<0.5	2.0	7.3	16.0	3.0	<0.5	<0.5	–	–

Specimens are numbered as in Figure 1

The predominant minerals are hornblende and plagioclase, accompanied, in order of abundance, by biotite, pyroxene, sphene, apatite, ilmenite and quartz, with muscovite and opaques and accessory minerals. Secondary epidote, chlorite and calcite are found along the hornblende cleavages (Table 1). Common subhedral hornblende is light-greenish or greenish (z-direction) and has an average grain-size ranging from 0.1 x 0.3 mm to 0.2 x 0.4 mm. Refractive indices of hornblende are 1.644–1.668 (N_α) and 1.662–1.692 (N_γ). The composition of plagioclase could be measured in eight of the ten specimens, and was found to vary from An_{35} to An_{42}. Some amphibolites contain appreciable amounts of pale-greenish pyroxene. Biotite, reddish-brown in color, has a random distribution. The amounts of pyroxene and biotite increase as hornblende decreases. Opaque minerals comprise mainly ilmenite, small amounts of magnetite and pyrite, and uncommon chalcopyrite. Ilmenite scales occur throughout the rock. The grains are always extensively decomposed to titanite and rutile. Large euhedral magnetite grains are commonly surrounded by thin bands of fine-grained sphene, and elongated magnetite is arranged along hornblende cleavages.

5. Geochemistry

In areas of increased metamorphism and metasomatism, the identification of banded amphibolites and their association with carbonates in the field is not strong evidence for determining their origin, because of the chemical equivalence of para- and ortho-amphibolites for nearly all the major elements, together with other characteristics such as mineralogy, texture and rock magnetism (*Evans* and *Leake*, 1960). On account of this, much of the previous chemical work has been aimed at discovering differences in abun-

Table 1 (continued)

Minerals (vol. %)				Counts pointed	Traverse sp. mm	An % of plagioclase	Ref. index of hornblende	
Opaques								
Ilmenite	Magnetite	Pyrite	Chalco-pyrite				$N\alpha$	$N\gamma$
1.5	–	0.5	–	1016	1/6 x 2	35	1.646	1.682
2.0	2.0	0.5	0.5	1021	1/6 x 2	38	1.666	1.684
3.0	–	0.5	0.5	1030	1/6 x 2	40	1.663	1.688
0.5	0.4	0.5	–	1092	1/6 x 2	41	1.664	1.680
4.0	1.0	1.0	–	1105	1/6 x 2		1.644	1.662
4.0	0.5	0.5	–	1173	1/6 x 2	37	1.668	1.690
2.5	–	0.5	–	1030	1/6 x 2		1.660	1.681
5.0	0.5	0.5	1.0	1042	1/6 x 2	40	1.668	1.692
2.0	2.0	1.0	1.0	1046	1/6 x 2	42	1.664	1.690
3.0	0.5	–	0.5	1142	1/6 x 2	39	1.665	1.685

dance levels of certain elements between ortho- and para-amphibolites, and special attention has been focused on trace elements. Recently, *Leake* (1964) emphasized that although an individual analysis cannot determine the origin of an amphibolite, the nature of the variation trends given by plotting groups of analyses and their relationship to known igneous and sedimentary trends are the most valuable distinctions between ortho- and para-amphibolites.

All the chemical data for the amphibolites studied were compared with the results outlined by *Evans* and *Leake* (1960) and *Leake* (1964), and compared with basic dyke rocks and a lime-silicate rock in the Pocheon mining area. Major- and trace-element analyses for a suite of eight known ortho-amphibolites from Chuncheon, Cheongpyeong and Anyang areas in the Gyeonggi metamorphic complex, Korea (Fig. 1) given by *So* (1977) are also presented for comparison. All elements in the rocks studied are presented in Table 2, and major elements are recalculated into Niggli values. All the analytical data have been used to study variation trends in the compositions of the amphibolites as a means of determining their origin.

Plots of the Niggli numbers clearly indicate variation trends in the Pocheon amphibolites similar to those for basic igneous rocks and for the ortho-amphibolites from the Gyeonggi metamorphic complex, Korea. In plots of Niggli *c* against *mg*, and Niggli 100 *mg*, *c* and *al–alk* (Figs. 2 and 3), the amphibolites from the Pocheon magnetite deposit follow fairly closely the basic igneous trend of the Karroo dolerites (*Walker* and *Poldervaart*, 1949), plotting together as a well-defined group within the field given by the Connemara striped amphibolites and other ortho-amphibolites (*Leake*, 1964). The scatter of data points about the Karroo trend may be due to chemical rearrangements during metamorphism. The basic dyke rocks in the mining district plot in the same field, and the lime-silicate rock analyzed has much higher *mg* and *c* values and is comparable to a normal dolomite composition. Niggli *al–alk* versus *c* (Fig. 4) also shows the similar

Table 2. Chemical and spectrographic analyses, Niggli values and oxidation ratios of the rocks studied from the Pocheon mine

	1	2	3	4	5	6	7	8	9	10	11	12	13
SiO_2 (wt %)	49.00	54.60	43.15	48.40	36.10	46.00	46.10	17.20	49.10	44.50	48.40	55.60	57.00
Al_2O_3	15.70	17.70	22.90	20.70	21.40	21.70	21.60	4.37	20.60	22.00	20.35	19.10	18.00
Fe_2O_3	1.75	2.10	4.03	2.47	3.36	2.02	4.79	1.23	3.51	2.81	6.02	1.21	1.68
FeO	6.17	7.49	7.50	7.50	11.50	12.50	7.93	tr.	8.81	7.64	6.84	4.41	2.94
MnO	0.23	0.19	0.27	0.20	0.37	0.28	0.34	0.06	0.20	0.19	0.37	0.11	0.08
MgO	11.70	4.77	6.22	5.30	9.56	4.60	5.38	19.10	4.32	6.38	5.60	5.34	3.17
CaO	9.60	6.83	7.03	10.10	8.47	8.92	6.89	30.20	7.40	10.10	6.54	4.63	6.19
Na_2O	1.64	4.00	3.84	2.40	1.78	2.20	3.12	0.18	3.20	2.35	2.89	2.50	1.68
K_2O	2.56	0.84	0.65	1.90	3.80	1.34	0.54	0.96	1.74	1.92	1.45	4.40	3.50
P_2O_5	tr.	0.03	0.02	0.03	0.04	tr.	tr.	tr.	tr.	tr.	0.03	tr.	tr.
Ig. loss	1.44	1.20	4.22	0.74	3.65	0.65	3.34	26.60	1.13	1.80	1.24	2.37	5.46
Total	99.79	99.75	99.83	99.74	100.03	100.21	100.03	99.90	100.01	99.69	99.73	99.67	99.70
Ba (ppm)	720	289	403	1649	335	439	201	211	515	466	250	1225	340
La	4	4	4	4	5	4	5	7	4	7	5	13	5
Sc	26	21	63	18	17	20	51	8	26	20	49	15	10
V	23	16	15	55	7	155	13	4	61	46	11	13	8
Rb	250	60	80	120	650	100	90	20	20	100	70	200	200
Zr	36	178	115	179	41	90	123	2	87	87	103	55	40
Ni	240	84	65	90	134	104	44	–	84	145	56	78	68
Cr	510	76	50	70	35	40	28	31	104	90	42	250	158
Co	88	94	72	58	104	84	68	–	100	130	64	88	50
Cu	23	68	198	106	17	120	214	10	50	80	236	80	56
Sr	220	58	140	260	30	170	120	490	104	190	130	100	70

Table 2 (continued)

	1	2	3	4	5	6	7	8	9	10	11	12	13
al	20	28	31	29	24	29	30	4	30	28	29	33	36
fm	52	41	43	38	51	43	44	45	41	40	45	37	29
c	22	20	17	25	17	21	18	49	19	24	17	15	22
alk	7	12	9	8	8	7	8	1	10	8	9	15	13
si	104	146	98	113	69	104	110	26	120	98	116	163	192
p	0	0	0	0	0	0	0	0	0	0	0	0	0
k	0.51	0.12	0.10	0.34	0.58	0.29	0.11	0.77	0.27	0.34	0.24	0.54	0.58
mg	0.73	0.47	0.50	0.49	0.54	0.36	0.44	0.97	0.39	0.52	0.44	0.63	0.55
Oxid. ratio	20.4	20.0	32.5	22.4	20.8	13.0	35.3	100.0	26.5	25.4	44.4	20.8	34.9
Field occurrence	▲	▲	▲	▲	▲	▲	▲	▣	▲	▲	▲	△	△

Specimens are numbered as in Figure 1. ▲ Amphibolites; ▣ Lime-silicate rock; △ Dyke rocks

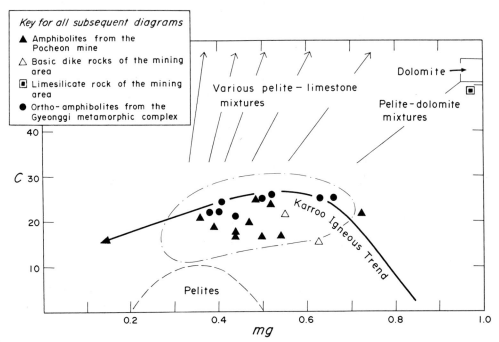

Fig. 2. Diagram of Niggli *c* values plotted against *mg*. —·—·— field of Connemara striped amphibolites and other ortho-amphibolites (*Leake*, 1964); — — — field of pelitic rocks (*Shaw*, 1956) (key for all subsequent diagrams)

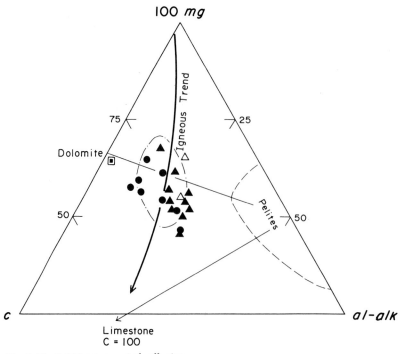

Fig. 3. Niggli 100 *mg, c,* and *al—alk* plot

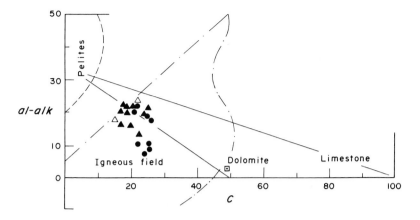

Fig. 4. Niggli *al–alk* plotted against *c* showing the igneous-rock field

features. In *alk* and *k* versus Niggli *mg* diagrams (Fig. 5), again nearly all the data points of the Pocheon amphibolites fall within the ortho-amphibolite field, together with basic dyke rocks from the mining area and the Gyeonggi ortho-amphibolites, but possess no apparent variation trends. The lime-silicate rock has lower *alk* and higher *k* values. The *ti/mg* and *p/mg* plots are not useful for distinguishing the origin of the amphibolites in the present study.

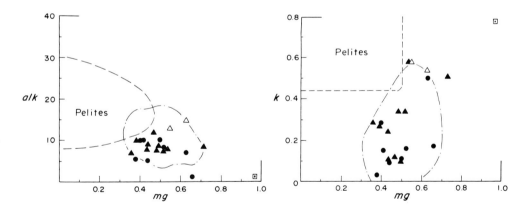

Fig. 5. Niggli *alk* and *k* plotted against *mg*

The trace-element data give a further indication of the meta-igneous origin of the studied amphibolites. In the differentiation of basic igneous rocks, a decrease in *mg* is accompanied by a decrease in Cr and Ni, whereas in mixtures of pelite with limestone or dolomite an increase in Cr and Ni with a decrease in the *mg* value can be observed. On this basis, *Evans* and *Leake* (1960) and *Leake* (1964) have shown that Cr and Ni are the most useful trace elements for distinguishing igneous from sedimentary trends. The Cr and Ni concentrations plotted against *mg* for the rocks studied (Fig. 6) show pronounced positive correlations similar to those for basic igneous rocks, including the Gyeonggi ortho-amphibolites.

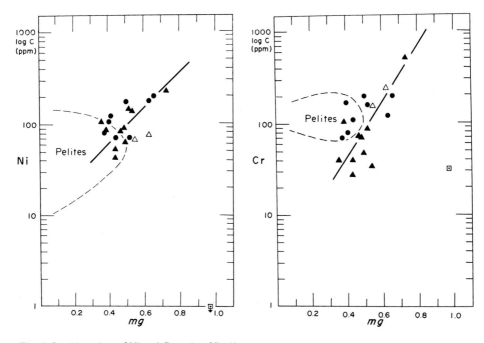

Fig. 6. Semi-log plots of Ni and Cr against Niggli *mg*

Table 3. Oxidation ratio of the Pocheon amphibolites, compared with those of intrusive and extrusive basic rocks from various igneous provinces (*Elliott* and *Cowan*, 1966; *Preto*, 1970)

	Fe_2O_3	FeO	Oxidation ratio
Average Pocheon amphibolite	3.29	8.39	26.1
Average Pocheon basic dykes	1.45	3.68	27.85
Average Palisades Dolerite	1.6	8.7	14.2
Average Watchung Basalt	3.4	8.6	26.1
Average Spitzbergen intrusions	3.4	10.3	22.7
Average Spitzbergen extrusions	4.8	10.1	30.0
Average Karroo basic intrusives	1.2	9.3	11.0
Average Karroo basic extrusives	2.7	7.9	23.6
Average of above intrusives	2.1	9.4	16.5
Average of above extrusives	3.6	8.9	26.8
Average Grand Forks amphibolite	1.7	8.72	14.8

6. Origin of the Amphibolite and the Magnetite Ore

The magnetite ore of the Pocheon iron deposit is always closely associated with layered amphibolites, which have been supposed to have originated as metasedimentary rocks within the surrounding metamorphic sequence. The chemical similarity of the amphibolites to basic igneous rocks is clearly indicated by major-element abundances and variations, and trace-element trends. The banding of the amphibolites resulted from chemical differentiation during metamorphism. The amphibolites studied also show definite chemical similarities to basic dyke rocks from the Pocheon mining district and to ortho-amphibolites from the Gyeonggi metamorphic complex, Korea (So, 1977) in all major- and trace-element diagrams presented. A lime-silicate rock analyzed is similar to normal dolomite in composition.

The foregoing presentation is based on considerations of the bulk compositions of the amphibolites and on the presupposition that during metamorphism, individual rocks acted essentially as closed systems and that metasomatism did not take place to any significant degree. *Orville* (1969) presented a model for the metasomatic origin of thinly layered amphibolites and listed a number of features for determining their origin. Of these, the presence of muscovite or of aluminosilicate-bearing assemblages within, or near to, amphibolite layers are two readily determined characteristics which would strongly indicate that metasomatic processes did not play a significant part in their formation. In this study, muscovite is always present as a minor constituent in the surrounding metasedimentary rocks, together with rare sillimanite, and in small amounts even in some amphibolite specimens (Table 1). *Preto* (1970) also applied these features for determining the origin of amphibolites from the Grand Forks Quadrangle, British Columbia, Canada. The other features outlined by *Orville* (1969) as suggestive of a metasomatic origin from interbedded carbonates and pelites were not observed in this study.

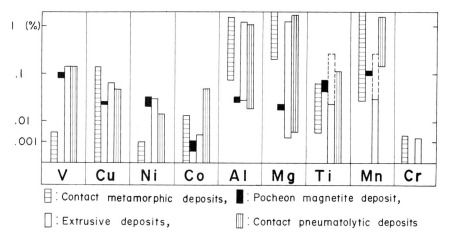

Fig. 7. Ranges of trace-element contents in magnetites of contact- and extrusive origin (after *Hegemann* and *Albrecht*, 1954) with addition of trace-element contents (range obtained from five samples) of magnetites from the Pocheon mine

The oxidation ratios, that is, mol $2Fe_2O_3 \times 100/2Fe_2O_3 + FeO$, for the individual amphibolite samples, are shown in Table 2, and that for the average composition is shown in Table 3. If the average oxidation ratios obtained for intrusive and extrusive basic rocks from various igneous provinces (*Elliott* and *Cowan*, 1966) are compared, as in Table 3, it can be seen that the oxidation ratios of the amphibolites from the Pocheon iron deposit are similar to those of basic extrusives.

A comparison of the trace-element concentrations in the magnetites from the Pocheon deposit with those of magnetites from contact and extrusive origins collected from other parts of the world, and analyzed by *Hegemann* and *Albrecht* (1954), indicates characteristic features (Fig. 7). The range of the trace-element values in the Pocheon magnetites is more comparable to that from those of known extrusive deposits than that from those of contact-metamorphic and contact-pneumatolytic origin. Their composition also corresponds to that of magnetite from Quaternary andesitic volcanic rocks from the El Laco area, Chile [V = 1000–1300, Cr = 400–500 and Ti = 20,000–36,000 in ppm, *Frutos* and *Oyarzún* (1975)], although the Ti content of the Pocheon magnetites is somewhat lower.

Bearing in mind the extrusive origin proposed for the amphibolites closely associated with the deformed strata-bound magnetite ore, and the trace-element chemistry of the magnetite itself, it is suggested that the Pocheon deposit originated from volcanism in a sedimentary environment. There is no evidence to support the general opinion that the Pocheon magnetite is related to granite intrusion in the area.

Acknowledgments. The writer gratefully acknowledges Dr. *R.H. Sillitoe*'s critical review of the manuscript. Thanks are also due to Mr. *D.S. Son* for assistance with sample preparation, and to Mr. *D.G. Choi* for assistance with major-element analysis.

References

Elliott, R.B., Cowan, D.R.: The petrochemistry of the amphibolites of Holleindalen Greenstone Group, Jotunheimen, Norway. Norsk. Geol. Tiddskr. **46**, 309–326 (1966)
Evans, B.W., Leake, B.E.: The composition and origin of the striped amphibolites of Connemara, Ireland. J. Petrology **1**, 337–368 (1960)
Frutos, J.J., Oyarzún, J.M.: Tectonic and geochemical evidence concerning the genesis of El Laco magnetite lava flow deposits, Chile. Econ. Geol. **70**, 988–990 (1975)
Hegemann, F., Albrecht, F.: Zur Geochemie oxydischer Eisenerze. Chemie der Erde **17** (1954)
Kanda, Y.: Geology and ore deposit of Pocheon Iron Mine, Korea. J. Korean Inst. Mining Geol. **2**, 53–67 (1969)
Leake, B.E.: Chemical distinction of ortho- and para-amphibolites. J. Petrology **5**, 238–254 (1964)
Orville, P.M.: A model for metamorphic differentiation origin of thin layered amphibolites. Am. J. Sci. **267**, 64–86 (1969)
Preto, V.A.G.: Amphibolites from the Grand Forks Quadrangle of British Columbia, Canada. Geol. Soc. Am. Bull. **81**, 763–782 (1970)
Shaw, D.M.: Geochemistry of pelitic rocks, Part III. Major elements and general geochemistry. Geol. Soc. Am. Bull. **67**, 919–934 (1956)
So, C.-S.: Geochemistry and origin of amphibolite and magnetite from the Yanyang Iron Deposit in the Gyenonggi Metamorphic Complex, Republic of Korea (submitted for publication to Mineralium Deposita, 1977)
Walker, F., Poldervaart, A.: Karroo dolerites of the Union of South Africa. Geol. Soc. Am. Bull. **60**, 591–706 (1949)

Geochemistry and Diagenesis of Recent Heavy Metal Ore Deposits at the Atlantis-II-Deep (Red Sea)

K. WEBER-DIEFENBACH, München

With 8 Figures

Contents

Summary

The ore sludges of the Atlantis-II-Deep (Red Sea) were examined geochemically and it was attempted to determine the influence of early diagenesis.

The evaluation of the analysis data yielded the following results: chemical variation is large; even sediment areas of the same facies often differ considerably. Main element is Fe; the Zn- and Cu-content is rather small, it only increases considerably in the sulphidic area. Partly higher Cd, Pb and Ag content can be noted. From the geochemical point of view it can be concluded that the hydrothermal solutions are predominantly volcanic. Diagenesis already starts near the surface. The development of the layering and the crystallinity of the minerals increase with depth, at the same time the percentage of adsorptive Zn decreases.

1. Geology

The continental plates Africa and Arabia are drifting apart. Geological, petrographical and mainly geophysical research indicates that the Red Sea is an originating ocean.

This theory is supported by the results of refraction seismic (*Girdler*, 1970; *Haenel*, 1972) and the knowledge of gravity anomalies and magnetic anomalies (upper mantle material, *Phillips* et al., 1969; *Allan*, 1970) in the region of the central trench of the Red Sea.

A narrow strip, about 10 km wide, in the area of the central trench consists predominantly of lava and pyroclastics of theoleiitic-basaltic chemism.

According to *Bäcker* and *Richter* (1973) a large part of the sediments consists of Miocene evaporites, which partly reach a thickness of more than 2000 m.

On the African side, *Carella* and *Scarpa* (1962) describe a series (Cretaceous-Tertiary) of marl, sandstones, evaporites, arenites and clays which have a thickness of 3000 m, and are interlain by basaltic lavas and tuffs.

Since Miocene Times the Red Sea sediments have been deposited into an active trench system and presumably have been enriched with metal-bearing precipitates (*Bäcker* and *Richter*, 1973).

The Atlantis-II-Deep, a basin filled with layered hot brines, lies in the central trench of the Red Sea at a depth of 2100 m underneath the water surface. Ore sludge forms here out of hydrothermal springs which supply heavy minerals. These muds of a thickness of 10—25 m overlie young basalts, which had intruded because of the drifting apart of the African-Arabian plate.

The hydrothermal springs are highly saliniferous; they leached their salt content out of thick evaporites which lie laterally of the central trench. Due to the geothermal anomaly in the rift zone the salt-rich waters are heated up and enriched with heavy metals. The origin of the heavy metals is regarded as magmatic according to isotope

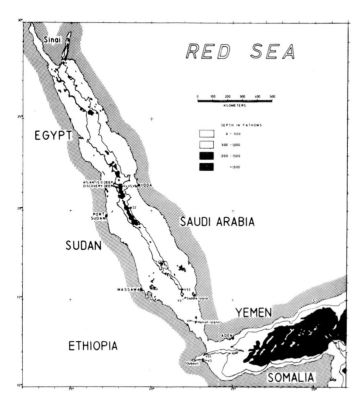

Fig. 1. Position of Atlantis-II-Deep in the Red Sea. Bathymetric conditions and positions of volcanite samples are shown. (After *Laughton*, 1970)

measurements; certain parts, however, could have been leached from host rocks (schists and basalts) by the aggressive saliniferous solutions. The hot solutions ascending in the area of the central trench emerge as springs in the deepest parts of the Red Sea. Due to the morphological trap structure, a dilution of the hot brines through ordinary sea water is to a large extent prevented. The brines have different densities due to their different salt contents and temperatures: they form two layered brine bodies, lying on top of each other and containing different solution content. The connection to the ordinary sea water of the Red Sea is formed by an overlying transition zone.

The sediments examined here com from a 12-m box plumb line, taken by the German research ship Valdivia 1973 in the northern passage of the Atlantis-II-Deep. The sediments of the Atlantis-II-Deep are highly water-carrying (up to 95%) extremely fine-grained ($> 60\% < 2\mu$ Ø) muds.

The layering is characterized by thin lamellae and bands of strong colouring. Different facies can be identified: detritic biogenetic facies, oxide facies, sulphide facies, silicia facies, sulphate facies and carbonate facies. These, mostly intermixed facies, appear in four sedimentological zones (*Bäcker* and *Richter*, 1973): amorphous-silica zone, AM (geochemic research has shown that the AM zone chemically corresponds here to a large extent to the SU_2 zone, and differs mainly in its amorphous state of the precipitates; the (1) upper sulphide zone SU_2; (2) the middle oxide zone CO; (3) lower sulphide zone SU_1; (4) detritic-oxide pyritic zone DOP (not cut by this core).

Since conventional sediment-petrographical research methods are mostly not practicable due to the small grain size of the precipitates/minerals and due to the partly X-ray-amorphous and/or badly crystallized state, the classification into sedimentological zones and ranges of facies had to be done mainly by means of geochemical methods and special ways of mineral determinations (SEM plus energy-dispersive analysis).

Because of the very narrow grid of sampling and therefore the high amount of chemical analyses data, the sedimentological classification can be drawn in a very exact way. Especially those ranges of facies which have a thickness of only a few cm in some zones can here be well determined.

2. Mineralogy

By using modern analyzing methods (Scanning Electron Microscope, SEM; Energy-Dispersive Analysis) it was for the first time possible to determine the mineral content of X-ray-amorphous parts of the ore sludges of the Atlantis-II-Deep. The more exact determination of the chemical structure of the sulphide minerals also brought new results: Pure mineral phases are rarely found, usually they are Fe sulphides (pyrrhotine) with Zn and Cu content; Zn is partly bound adsorptively. The most important minerals are Fe oxides/hydroxides and Fe-bearing clay minerals (Fe montmorillonite). The sulphides only appear to a larger extent in thin layers of the sulphide zone. Framboids. spheroids and mineral agglomerates have here for the first time observed in the Atlantis-II-Deep. An evolutionary series spheroid-framboid agglomerate can be observed with increasing depth.

The theory of a formation of the framboids by sulphide-bacteria can be excluded, their genesis is a "chemical" one out of hydrothermal solutions. The observation of oxide/hydroxide framboids is new — only sulphide framboids have been reported in the literature.

The evolutionary series FeS-FeS_2 observed in laboratory experiments (*Sweeney* and *Kaplan*, 1973) can be confirmed for the Atlantis-II-Deep.

3. Geochemistry

116 samples of the analyzed box plumb lines were chemically analyzed. These make up $> 40\%$ of the total core. Considering only the lower 10 m, which were closer sampled, the percentage even rises to $> 75\%$. Thus the requirements for a detailed chemical characterization and for the following of trends are met.

Contrary to former publications which were aiming at geochemical analysis of the sediments (e.g., *Bischoff*, 1969), for the first time a complete sediment package of these ore sludges of the Atlantis-II-Deep (from the basalt in the bottom wall to the sea floor has been carefully examined geochemically, with closely set sample grids.

To avoid changes in mineral constituents and solution of adsorptively bound content of heavy ore minerals, desalted samples were analyzed, as well as slowly dried salt-carrying samples. Thus the adsorptive contents of Zn and Cu could be determined; they reach higher concentrations in some layers. Also an observation of the otherwise water-soluble Fe chlorides is made possible by this method of sample preparation.

H_2O is a main constituent of the ore sludges. In the literature, a content of $> 90\%$ is reported (e.g., *Bischoff*, 1969). Here a maximum of 70% H_2O content could be noticed. This discrepancy is probably due to the long time elapsing between sample-taking and research. In spite of careful packing and storage predominantly in refrigerating rooms, H_2O evaporated.

The brines of the Atlantis-II-Deep are an almost NaCl-saturated, sometimes over-saturated solution. Apart from the extremely H_2O-rich muds in the upper parts of the core, the water content in the other parts is rather constant (Fig. 2), excluding some thin bands. Only under $- 800$ cm, and more so under $- 1100$ cm is a decrease noted (see diagenesis).

The salt content, mainly NaCl, is shown in Figure 2, which shows the salinity of the pore solutions and the brines. The values are constant as in the H_2O profile, only under $- 1100$ cm can a decrease be noted (Fig. 2).

3.1 Main and Accessory Components

In Table 1, important data for statistics of the chemical analysis data are combined: Average value, modus, median, minimum, maximum, standard deviations and variance.

SiO_2 appears in the samples with very differing contents; it is enriched in the sulphide zones SU_1 and SU_2, but has lower values in the oxide zone CO. The SiO_2-histogram shows various maxima: at the highest at 9% SiO_2, accessory maxima are at 5%,

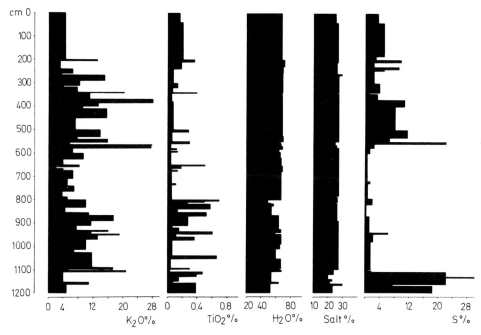

Fig. 2. Chemical profiles of the ore sludges: K_2O, TiO_2, H_2O, salt and S. H_2O and salt-free substance, excluded profiles: H_2O and salt. Due to the generally higher content of pore solutions in the upper parts of the core

12% and 18%. The positive inclination of the histogram, also documented by the highly differing data of average value and modus, is due to a partly strong contribution of diatoms. Si was also found in framboids.

TiO_2 (Fig. 2) has the highest values between − 800 cm and − 1000 cm; especially in the upper parts of the CO zone it is only represented with low content.

Al_2O_3 is enriched in the CO zone and has high values in the very muddy samples in the top of the core. Al is predominantly built into the clay minerals, a small part is probably due to detritic feldspars.

Fe_2O_3 calculated as total Fe to Fe(III) oxide, is the best represented element in the ore sludges with an average value of 52.1 vol% (dried substance). It is especially well-represented in the CO zone which consists predominantly of the geothite and montmorillonite facies. The Fe_2O_3 histogram has its only maximum at 66%. The inclination is negative, average value, modus and median differ relatively little (Table 1) − a sign for a generally differing, but in some parts ± regular feeding during sedimentation history. The lowest Fe-contents are directly connected with the highest of the elements of value and sulphur. This can partly be observed in the upper sulphide zone (SU_2) already, but it becomes much more evident in the SU_1 zone, where lower Fe concentrations are more abundant. Therefore a certain falsification of the real conditions has to be considered here.

The increase of the elements of value and of sulphur at the cost of iron is due to a large extent to different solution contents of the hydrotherms, to a lesser extent to changed conditions of precipitation.

Table 1. Some statistical data to the geochemical analysis. (Concentration in ppm. Calculation by Computer-Programm SPSS)

Variable	Average value	Modus	Median	Minimum	Maximum	Standard deviation	Variance
Thickness (cm)	10.03	10.0	9.55	1	99	11.59	134.4
SiO_2	136,892	180,000	127,916	29,000	307,000	69,211	n.b. [a]
TiO_2	1,733	800	878	300	69,000	2,287	n.b. [a]
Al_2O_3	66,106	157,000	45,800	9,900	244,000	53,653	n.b. [a]
Fe_2O_3	521,456	598,000	539,250	89,000	786,000	159,115	n.b. [a]
MnO	23,038	3,700	7,950	500	510,000	63,266	n.b. [a]
CaO	56,500	39,600	42,250	12,000	285,000	60,790	n.b. [a]
MgO	14,994	19,200	12,950	3,800	36,900	9,023	n.b. [a]
Na_2O	16,836	7,000	11,250	2,700	75,000	15,144	n.b. [a]
K_2O	92,662	40,000	77,750	19,000	280,000	61,431	n.b. [a]
P_2O_5	1,535	1,500	1,487	100	5,500	793	n.b. [a]
CO_2	5,104	0	950	0	200,000	10,363	n.b. [a]
S	45,052	13,400	16,950	1,200	300,000	62,236	n.b. [a]
Zn	13,168	1,000	2,866	400	132,000	22,976	n.b. [a]
Cu	3,703	2,000	1,950	20	27,000	5,046	n.b. [a]
Ag	19.1	6	9.8	6	200	29.7	884
Cd	77.9	2	10.1	2	990	160.2	25683
Co	153	63	73	20	760	156.2	24422
Cr	74.4	68	71.1	42	215	22.5	508
Ni	53	27	32.7	14	330	51.8	2685
Pb	524	60	190	25	2,400	625.5	391312
Rb	34.1	1	36.6	1	98	19.4	378
Sr	100.7	90	89.2	40	380	51.1	2615

[a] Not calculated.

If the solution contents had always stayed constant, similar chemical conditions within these sediment areas would prevail at similar conditions of precipitation (e.g., as in the two sulphide zones). This does not apply. Different conditions of precipitation at ± constant supply of solution would lead to different mineral phases, the total chemism, however, would only change little.

MgO increases with depth. Comparing the two sulphide zones it becomes evident that Mn is especially enriched in the SU_1 zone — not considering the thin band of the manganite facies in the CO zone. This indicates different conditions of precipitation in the two sulphide zones, probably due to a displacement of the springs (see genesis of ore deposits).

CaO content is mostly due to biogenetic material. A calcite-rich layer can be observed at about − 800 m in the CO zone.

MgO is represented to a low degree in the sludges. The values are especially low in the upper parts of the core. Since not enough dolomite is present, chlorite can be the cause for the Mg contents.

Na_2O is analyzed in the de-salted and dried samples with differing contents. A dependence on the sediment zones cannot be seen. In the shock-dried samples large, idiomorphous halite crystals can sometimes be observed.

K_2O shows distinct maxima in the range − 200 cm to − 650 cm of the box plumb line (Fig. 2). Another increase is in the SU_1 zone at − 1000 cm. The high content is mostly due to clay minerals (e.g., illite).

P_2O_5 shows changing, low values of concentration. A dependence on facies areas cannot be seen.

CO_2 is represented with higher contents in only one thin layer. It is especially bound to the manganosiderites and calcites.

S as sulphide belongs to the most important ore-forming elements. The S histogram (Fig. 3) shows clearly, however, that the large majority of the samples (35%) has S contents of < 2%. Still, small accessory maxima are at 8% and 19%. Sulphur is in the first place built into pyrite, sphalerite and also chalcopyrite. The correlation of the metals of value with sulphur also indicates a sulphidic bond of these elements. In the columnar section of the sedimentary facies areas are outlined. The SU_1 zone (− 1200 cm to − 1070 cm) has high S values; these decrease in the CO zone and increase strongly at the beginning of the SU_2 zone (− 570 cm).

Zn is, due to the high content, one of the economically most interesting elements of the sludges. The Zn concentrations of the died substances are shown in Figure 4. Zn correlates well with S and appears like it especially in the sulphide facies of the box plumb line. The highest Zn contents are observed in the older SU_1 zone, but higher values appear also in the SU_2 zone and even in the amorphous zone (AM). It seems to be characteristic that the Zn peak values always appear in only very thin layers, and that after a maximum, an abrupt change to very low Zn contents always follows. The average value of the Zn contents is at 1.31% ("adsorptively" bound Zn is not considered). Very high Zn values only appear in thin layers. This fact causes the strong positive inclination in the development of the Zn histogram (Fig. 3). The maximum is at < 1% for Zn contents.

X-ray scanning images of Zn by means of SEM and analyses of the water of the Zn-rich samples showed partly considerable contents of "adsorptively" bound Zn. Those Zn contents which can be dissolved in water are here considered together as "adsorptive" Zn.

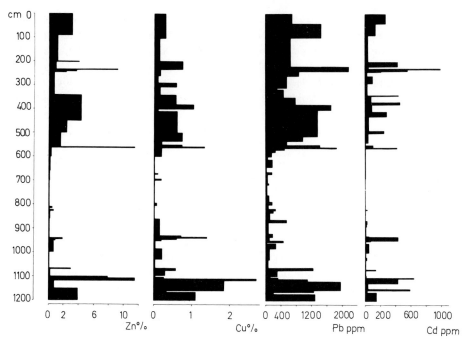

Fig. 3. Chemical profiles of the ore-sludges: Zn, Cu, Pb, and Cd. H_2O and salt-free substance. Adsorptive contents are not considered

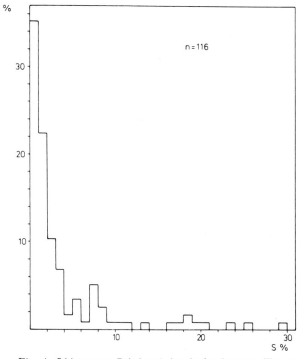

Fig. 4. S-histogram. Dried and de-salted substance. The maximum at very low S values and the strongly positive inclination are an indication that sulphides only contribute little to the structure of the ore sludges

Naturally the content of Zn in the partly higher solubility showing Zn gel modifications can thus not be determined. The Zn "adsorptive" values consist here of the real adsorptive Zn content and possible smaller contents out of water-soluble Zn gels. A proof for the presence of adsorptive Zn in the true sense, is further the fact that even at magnification of x 60,000 often only few Zn phases could be observed in the Zn-rich parts (SEM, surface and X-ray distribution). The adsorptive Zn decreases with depth (see diagenesis).

Cu is possible correlated with Zn and appears mainly in chalcopyrite, especially within the sulphide zones. The copper average value is at 0.37%, the maximum at 2.7%; a minimum, however, at only 30 ppm. The SU_1 zone shows the highest Cu contents, a little less Cu is in the SU_2 zone and the amorphous areas (Fig. 4). This also shows the Zn/Cu ratio, since the lowest Zn/Cu quotients are noticed in the lowermost parts of the core (Table 2).

Table 2. Chemism (average values) of the ore sludges. (Groups of samples within one facies range. Dried and de-salted substance)

Group	1	2	3	4	5	6	7	8
Sample No.	1–36	37–70	71–72 78	75–76 77 79–87	73–74	88–89	90–100	101–116
n	36	34	2	3	12	2	11	16
SiO_2	15.4	7.3	11.8	5.7	16.1	21.7	13.8	21.4
TiO_2	0.13	0.07	0.59	0.37	0.29	0.37	0.15	0.21
Al_2O_3	4.6	11.0	9.85	4.9	5.3	7.95	3.1	4.9
Fe_2O_3 [a]	49.0	65.4	9.6	28.0	51.0	30.1	58.3	36.5
MnO	0.66	1.14	2.35	38.0	2.1	0.95	1.8	2.5
CaO	5.4	3.7	27.65	11.8	7.2	12.0	4.15	3.6
MgO	1.64	1.02	2.5	1.06	2.2	1.95	1.21	1.3
Na_2O	1.74	1.29	1.37	0.53	1.21	1.63	3.7	1.32
K_2O	10.0	7.3	10.7	7.9	11.8	15.2	11.0	8.4
P_2O_5	0.17	0.09	0.09	0.10	0.18	0.15	0.17	0.19
CO_2	0.21	0.24	19.5	0.9	0.61	0.91	0.16	0.5
S	7.0	0.87	1.64	0.24	1.06	4.5	1.6	12.9
Zn	2.6	0.17	0.4	0.19	0.21	1.62	0.32	4.5
Cu	0.5	0.066	0.04	0.01	0.17	1.0	0.14	1.06
Σ	99.05	99.08	98.08	99.70	99.43	100.03	99.60	99.28
Range correlation coefficient	1.0 [b]	−0.047	−5.04	−4.11	0.865	−0.174	0.674	0.42
Zn/Cu	5.2	2.57	10	19	1.23	1.62	2.28	4.24
MgO/CaO	0.3	0.27	0.09	0.09	0.3	0.16	0.29	0.36

Group 1: AM and SU_2 zone; 2–7: CO zone (2 and 5: Fe montmorillonite facies; 3: calcite/anhydrite facies; 4: manganite facies; 5 and 7: sulphide montmorllonite facies); 8: SU_2 zone (sulphide facies).

[a] Σ from Fe as Fe_2O_3.
[b] Group 1 given as standard (R = 1.0).

Periods of relatively high copper precipitation were followed by periods which brought the poorer solutions. Especially the oxidic layers which contain higher Cu content in only some thin bands, must be considered as results of such processes. Contrary to Zn "adsorptive" Cu content could only rearely be measured.

3.2 Trace Components

It is evident that the analyzed trace elements appear in some cases in unusually high contents. This fact also indicates a feeding through hydrothermal solutions, since ordinary sea water does not contain such high concentrations of metals of value, and the detritus component in the Atlantis-II-Deep is very low.

A comparison of the elements-of-value content with data from literature could, without knowledge of all the facts, also indicate volcanic action as cause for an enrichment – not considering S (correlating positively with Ca).

The variation diagram, as well as factor analysis go to prove that. The existence of Fe sulphides allows the conclusion that enough S did exist to precipitate all metals of value in solution as sulphides. The enrichment of these metals of value in the sludges can basically be attributed to two factors: increased supply of the metals of value in the solutions and favourable conditions of precipitation.

In the following the trace elements are discussed according to their listing in Table 1.

Ag, identified in some samples by SEM as acanthite/argentite ($Ag_2 S$), has high values of concentration which reach a maximum of 200 ppm. The average value, however, is considerably lower at 19 ppm. Short periods of intensive Ag feeding were followed by longer periods which were Ag-poor.

Cd has to be reagarded as a part of sphalerite because of the strong positive correlation with Zn and S. It reaches a maximum of 990 ppm. Like Ag only very low contents (minimum of 2 ppm) are in the metal of value poor CO zone.

Co is enriched in the sulphide facies (up to 760 ppm); it has tendencies similar to the above-mentioned trace elements. The high contents in the SU_1 zone are striking.

Cr, however, does not show high concentration values; they differ only little in the various ranges of facies.

Ni has a content between 14 and 280 ppm and is enriched in the sulphide zones.

Pb can sometimes be seen as galena. It shows a positive correlation with the trace elements discussed above. In Figure 4, it is evident that Pb shows trends similar to the other shown elements Zn, Cu and Cd: namely higher content in the sulphide zones (up to 2400 ppm), lower content in the silica/goethite zone (partly < 100 ppm).

Rb has maximum values little under 100 ppm, a minimum is at 1 ppm. Contrary to the younger SU_2 zone, especially low content are to be found in the SU_1 zone — an indication of a changed solution feeding.

Sr shows no unusual content (Table 1). The average value is at 100 ppm, the maximum at 3800 ppm, the minimum at 40 ppm. As expected, Sr correlated positively with Ca.

3.3 Discussion

The analysis of the geochemical data leads to a host of indications and results. These can be summarized as follows:

a) The hydrothermally formed sediments can be regarded as Fe sludges in the complete range of the core, excluding some small thin bands of the manganite and calcite/anhydrite facies. The most important Fe minerals are Fe montmorillonite, various Fe hydroxides, hematite and pyrite/markasite. This recent ore deposit is oxidic; sulphides only appear to a smaller extent.

b) Chemically, the samples vary largely. Correlation of rank shows only little conformity, even in samples belonging to the same facies.

c) Fe decreases with increasing S, the correlation is negative. On the other hand S correlates positively with many elements, especially the metals of value (Zn, Cu, Ag, Cd, Co, Ni, Pb). They increase, whereas Fe decreases.

d) The variation diagram of metals of value with S which show ± corresponding trends in all concentration fields, indicates the origin of sulphur as hydrothermal solutions. The formation of sulphur by bacteria can be excluded to a large extent due to geochemical reasons and observations of other sludges of the Atlantis-II-Deep.

e) The high content of metals of value in some areas of the core is due to hydrothermal pulsations and/or changed conditions of precipitation. However, also at times of low feeding of these elements, the values of their ratios usually stay constant. The elements mentioned do not correlate equally in all samples, sometimes opposite tendencies can be observed. This observation cannot be explained so far. Possible reasons are: change in the feeding solutions, change in salinity and temperature of the brines, change of pH and EH, selective filtering qualities of the sludges.

f) Statistical multi-variat methods prove the predominantly hydrothermal-volcanic origin of the metals of value (1st factor of main-component analysis).

This classification of the sludges through the metals of value can only be observed in volcano-hydrothermally formed sediments.

4. Investigations of Diagenesis of the Ore Sludges of the Red Sea

The analyzed ore sludges come from a thin sediment package. Thus the question arises: can early diagenetic processes be determined even with only thin over-lying material? If so, which factors are responsible? The conditions of sedimentation and precipitation in the Atlantis-II-Deep are extraordinary: solutions that are NaCl-rich; hot, carrying heavy-metal complexes; an almost complete absence of oxygen; relatively low pH values; negative EH and the morphological trap structure lead to precipitation of new minerals. These are mainly Fe oxides/hydroxides. Fe montmorillonite and partly also sulphides (see Sect. 2).

The partly high content of diatom-skeletons influences the texture. The sedimentation rates are high, the contents of the pore solutions can reach $> 90\%$.

4.1 Mechanical Diagenesis

Compaction in clay muds effects an intense consolidation of the sediment; the pore volume is decreased through the higher pressure and the pore solution is pressed out. Porosity is a measure for the consolidation of the sediment. It can be measured with various methods. For highly water-bearing muds, it is necessary to determine the exact water content, to calculate thus the pore volume through density measurements of the solid. The ore sludges examined had an average water content of about 68% at the beginning of the research, whereas literature data call for 95% H_2O^- values of fresh cores of the Atlantis-II-Deep (e.g., *Bischoff*, 1969).

Figure 5 shows that the content of solid increases with depth. The two curves show: (average values of the samples put together per 1 m, and average values of the samples of one facies respectively) down to -500 cm the H_2O content remains constant, then an unstable development follows to about -1100 cm, this is probably due to the different minerals in the different facies (it is known that not all minerals react in the same way to the influence of compaction). At -1100 cm and lower, a distinct increase of solid is to be noted, the pore solutions are pressed out more.

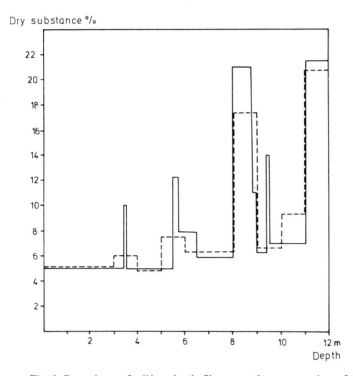

Fig. 5. Dependence of solid on depth. Shown are the average values of the samples per 1 m (————) and the average values of the samples within one facies range (- - - - - -). An increase of solid with depth can be noticed. The unstability in the central part can be due to a change of facies (= changed mineral content)

4.2 Conclusion

Compaction started noticeably at a depth of 5 m of these ore sludges, increased only slightly and extended its influence remarkably at − 11 m. *Von Engelhardt* (1973) put together the porosity of some recent clay sediments depending on depth. Provided that porosity and water content in the ore sludges should bring similar results in respect to compaction, the difference between the two descriptions is that the sediments described by *von Engelhardt* show porosities which decrease linear with depth. This is not true for the ore sludges. The cause is probably their extremely high water content, the salinity of the solutions and the flaky-cloudy precipitates which take a longer time to orientate.

Furthermore, the influence of the OH-rich Fe hydroxides is not known. Compaction causes an orientation of the minerals. For the usually flaky clay minerals this means a texture development from the, to a large extent, unoriented honeycomb texture over the better packaged "house of cards"-texture to the well-oriented mineral plates of the "real" parallel texture.

During SEM research, very distinct parallel textures could sometimes be found. These are mostly due, however, to a participation of nano-fossils. The generally observed lamellation (stressed through colouring) dissolved to forms which are similar to a honeycomb textrue at higher magnifications; the overlying pressure is too low to cause a better orientation of the components.

Down to sample 76 (− 840 cm) layering is generally only little to moderately distinct; then compaction increasingly shows its effects: parallel texture can already be seen at lower magnification in the SEM, the fossils are better oriented, and from − 1100 cm on, parallel texture is reached without participation of fossils.

The content of the spheroids decreases with depth, at the same time the average values of the agglomerate diameter increases (see Sect. 2: spheroids/framboids). This collective reaction is probably a consequence of the increased pressure and maybe of the changed pore solutions (Fig. 6).

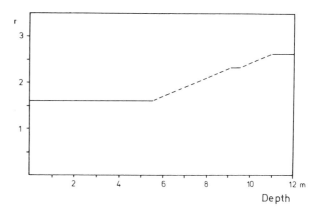

Fig. 6. Dependence of the average agglomerate diameter r (μ) on depth (SEM measurements). Considered are samples of similar chemism in the sulphide zones (1, 2, 6, 8, 17, 19, 34, 109, 111). Explanation see text

From the macroscopic texture descriptions it is evident that the ore-rich layers usually show in surface-near areas indistinct boundary planes to the neighbouring sediment layers. Only in deeper areas are these layers clearly marked off and seem denser.

It is generally observed that samples from deeper layers are better crystallized than those from higher layers: the amount of the interferences (samples of similar chemism) obtained by means of diffractometry is larger, their full width at half maximum is smaller.

Figures 4 and 8 show that Zn contents in comparable sulphide zones have different values. This might partly be due to different conditions of precipitation and different solutions (displacement of the springs), but certainly compaction causes in the lower parts a decrease of pore volume. The sediment now is relatively richer in solids. Average Zn contents (about 3%) are noted in the ore to about 5 m depth, Zn-poorer layers follow till about 11 m, then a strong increase to 7% follows. The picture is similar when considering the facies ranges.

4.3 Chemical Diagenesis

Mechanical and chemical diagenesis overlap. Compaction also plays a part in the following processes: due to the higher pressure and the changed pore solutions new possibilities of reaction between the minerals and solutions are formed. The increase of FeS_2 in depth seems striking (see Sect. 2); the development of the Fe hydroxides is similar; hematite can only be found in the deeper parts. Fe hydroxides (ferrihydrit, goethite, lepidocrocite) only appear closer to the surface. Goethite is the more stable and the same time OH-poorer phase. Two series showing chemical trends can be developed with increasing depth: the development of the series FeS-FeS_2 is linked to a better developed crystallinity of the phases.

Sulphur is drawn out of the pore solutions to form the sulphides. This development confirms the laboratory experiments of *Sweeney* and *Kaplan* (1973).

Besides the effects of chemical and mechanical diagenesis also time has to be considered as an important factor. Also the change of OH-rich Fe hydroxides to goethite, which is more stable under these given conditions (*Bischoff*, 1969) is due to the higher pressure and oxidation, whereas the reaction to hematite only takes place under local temperature anomalies ($> 250°C$; after *Bischoff*, 1969). This assumption cannot be sustained, since hematite also appears next to goethite in areas which certainly have never reached these temperatures.

Zn not only exists as sulphide in the ore sludges, it is partly bound adsorptively (see Sect. 3). The adsorptive Zn contents decrease with depth, the quotient total Zn/adsorptive Zn increases with depth. In Figure 7, the quotient Σ Zn/Zn adsorptive of samples of comparable Zn concentrations is shown towards depth: in several steps an increase of this quotient can be observed an indication for beginning diagenesis. It is to be supposed that the adsorptively bound Zn is gradually changed to crystalline phases with increasing depth.

Often very differing adsorptive Zn contents can be measured within this trend, which cannot be explained by compaction.

Possibe causes can be different; for ZnS-formation unfavourable conditions of precipitation, which result in an adsorption of Zn to hydroxide precipitates or silica. Apart from unfavourable physico-chemical conditions, also insufficient supply of S would show a similar effect.

Fig. 7. Dependence Zn/Zn adsorptive on depth. Considered are samples with Zn contents of > 2.6%. An increase of the quotient and thus an indication of beginning diagenesis can be observed

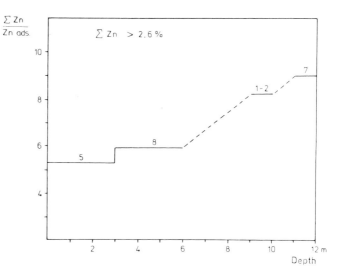

Fig. 8. Trend-behaviour of Zn in depth, dried and de-salted substance. ——— average values of samples per 1 m; - - - - average values of samples considering the facies ranges

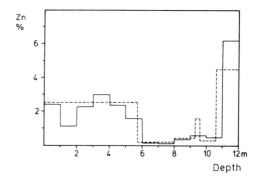

5. The Ore Deposit Atlantis-II-Deep

5.1 Origin of the Hydrothermal Solutions

The hydrothermal-sedimentary ore deposit Atlantis-II-Deep owes its origin to hot NaCl-saturated solutions, which contain, besides other elements, high contents of Fe, Mn, Si, Zn and Cu. The average chemism of the brine is put together in Table 1. The system in the basin is rapidly changing; the temperature for example, is increasing, it has now reached 60°C in present spring area (*Hartmann*, 1973). For comparison: *Degens* (1970) 56.5°C. The brine originates through the entrance of sea water in tectonically caused zones of weakness of evaporites. It is supposed that this takes place laterally of the central trench. Because of the contents of stable isotopes ^{18}O and deuterium which correspond to that of ordinary sea water, *Degens* (1970) is of the opinion that juvenile and continental waters can be excluded as primary sources.

Because of the high salinity (157‰) of the originally normal saliniferous (38 ‰) sea water, its density is increased. Hence, it can sink further down and heat up because of the geothermal anomalies (uprising magmas in the area of the rift zone). An ascent follows and exit of the hydrothermal solutions in submarine springs, especially in the deepest parts of the basin. The different density, compared to ordinary sea water, leads to a separation; the morphological trap structure prevents a mixing of the brine with sea water to a large extent.

Temperature and salinity of the hydrotherms can well be explained. The question as to the origin of the heavy minerals cannot be answered as readily, however. A source for the heavy minerals can be the host rocks through which the solutions flow: schists and especially the young basalts could, attacked by aggressive solutions, have supplied these elements.

The isotope reatios of the dissolved elements ^{87}Sr, ^{86}Sr, ^{208}Pb and ^{13}C/^{12}C (*Degens*, 1970) as well as sulphur-isotope research (*Hartmann* and *Nielsen*, 1966; *Kaplan* et al., 1969) rather indicate a magmatic origin. The good positive correlation of the metals of value with sulphur and the pattern of these elements in the factor analysis (all in 1st factor), as well as the absolute contents of the heavy minerals which correspond rather to volcano-hydrothermal processes, are further indicators for a hydrothermal-volcanigenic mode of origin.

The enrichment factors of the heavy minerals, in respect to the geochemical data of Red Sea basalt, for example (*Schneider* and *Wachendorf*, 1973) differ widely (values between 7 and 1200), which indicates less a leaching, for which more constant values should be expected, but rather a primarily differing supply of solution from depth.

Apart from this, the substantially increased solution capability of the hydrothermal waters must be mentioned, it is considerably increased through the high salinity (*Ellis*, 1968). Thus it does seem very likely that magmatic influences (predominantly) as well as leaching of host rock (to a lesser extent) are a common cause for the enrichment of the heavy minerals in the hydrotherms.

As carrier of the metals, especially chloride complexes can be considered. *Hartmann* (1973) assumes that additionally, dissolved sulphides are in the springs, possibly also free H_2S. These react immediately when existing from the spring with the heavy-metal ions to sulphide precipitates; sulphide ions and H_2) are therefore not traceable in the brines.

Basically important for the often simultaneous precipitation of such different minerals (sulphides, oxides/hydroxides, carbonates, sulphates, silicates) are the two brine bodies which are overlain by a transition zone (to the ordinary sea water). Each of these three zones is, within the undisturbed system, responsible for the formation of certain mineral assemblages.

In the 60°C brine especially the sulphides are formed. The redox potential is so low, that Mn and Fe can appear in bivalent form in the solution.

In the 50°C brine the formation of the Fe oxides/hydroxides takes place (oxidation through oxygen and descending Mn hydroxides which are thus reduced and dissolved) accordingly the brine is poorer in Fe compared to the 60°C brine, but contains high Mn contents.

The transition zone which is enriched in oxygen through mixing with ordinary sea water, is responsible for oxidation of Mn^{II}, which precipitates as oxide/hydroxide.

The precipitates, especially of the upper brine zones, usually do not sink directly, but convective flow cares for a lateral distribution (*Bäcker* and *Richter*, 1973).

In this way, and because of the reduction of the lowest brine, the extensive lack of Mn minerals in the ore sludges can be explained, only a change in the physico-chemical conditions in the basins can lead to a formation of the manganite facies.

5.2 Formation of the Ore Deposit

Requirements for the formation of this recent ore deposit are the hydrothermal heavy-mineral carrying solutions. The brines on the other hand guarantee the special physico-chemical conditions, which lead to precipitation. Almost as important are also the morphological structures ("trap structure": deep, isolated basins), which guarantee the maintainance of the complicated system over longer periods and are also responsible for the high rates of sedimentation (50–100 cm/1000 a).

Even considering the high contents of pore solutions in the sediments, the genesis of the ores takes place very quickly, an indication that formation of ore deposits does not always take geological times. The enrichment of the ores is facilitated by the low detritic content, which furthermore is concentrated in certain layers. A higher detritic content during the precipitation of the sediment would lead to a relative dilution of the ores in the sludges. The Atlantis-II-Deep has become known predominantly as a Zn and Cu ore deposit (with considerable Ag and Cd contents). Correctly, however, it is an Fe deposit, especially of the oxidic type (goethite, hematite); the pyrite markasite content is less important.

Although considerable Zn and Cu concentrations are in the sulphide zone (dried substance), it should be stressed that the average content of these elements in the sediment package tends to be rather low (here: dried substance: Zn 1.31%, Cu 0.31%). The presence of Fe sulphides indicates that the metal-of-value content does not suffice to bind sulphur. *Degens* (1970) supposes that similar to the fossil hydrothermal ore deposits (gang type), a large part of the metals of value is in the fissures and veins of the underlying rock. The genesis of the deposit seems to be certain. Out of salt-rich hydrothermal solutions, stemming in the first place from magma, and enriched with heavy minerals, synsedimentary Fe-, Mn-, Zn- and Cu ores precipitate. Moreover, sulphidically bound Ag, Cd, Pb, Ni and Co content can be measured. Partly high Fe silica content is present. Requirements for these enrichments are the layered brines and the favourable morphology of the basins (trap structure).

References

Allan, T.D.: Magnetic and gravity fields over the Red Sea. Phil. Trans. Roy. Soc. London A 267, 153–180 (1970)

Bäcker, H., Richter, H.: Die rezente hydrothermal-sedimentäre Lagerstätte Atlantis-II-Tief im Roten Meer. Geol. Rdsch. 62, 697–741 (1973)

Bischoff, J.L.: Red Sea geothermal brine deposits: Their mineralogy, chemistry, and genesis. In: Hot Brines and Recent Heavy Metal Deposits in the Red Sea. Degens, E.T., Ross, D.R. (eds.).New York–Heidelberg–Berlin: Springer 1969, pp. 368–401

Carella, R., Scarpa, M.: Geological results of exploration in Sudan by Agip Mineraria. 4th Arab Petrol. Congr. Agip Mineraria, San Donato, Milanese, 1962, p. 23

Degens, E.T.: Sea floor spreading: Lagerstättenkundliche Untersuchungen im Roten und im Schwarzen Meer. Umschau 70, 268–274 (1970)

Ellis, A.: Natural hydrothermal systems and experimental hot water/rock interactions: reactions with NaCl-solutions and trace metal extractions. Geochim. Cosmochim. Acta 32, 1356–1363 (1968)

Engelhardt, W. von: Die Bildung von Sedimenten und Sedimentgesteinen. Teil III. Stuttgart: Schweizerbarth 1973

Girdler, R.W.: A review of Red Sea heat flow. Phil. Trans. Roy. Soc. London A 267, 191–203 (1970)

Haenel, R.: Heat flow measurements in the Red Sea and the Gulf of Aden. Z. Geophys. 38, 1035–1047 (1972)

Hartmann, M.: Untersuchungen von suspendiertem Material in den Hydrothermallaugen des Atlantis-II-Tiefs. Geol. Rdsch. 62, 742–754 (1973)

Hartmann, M., Nielsen, H.: Sulfur isotopes in the hot brine and sediment of Atlantis-II-Deep (Red Sea). Mar. Geol. 4, 305–306 (1966)

Kaplan, J.R., Sweeney, R.E., Nissenbaum, A.: Sulfur isotope studies on the Red Sea geothermal brines and sediments. In: Hot Brines and Recent Heavy Metal Deposits in the Red Sea. Degens, E.T., Ross, D.R. (eds.). New York–Heidelberg–Berlin: Springer 1969, pp. 474–498

Laughton, A.S.: A new bathymetric chart of the Red Sea. Phil. Trans. Roy. Soc. London A 267, 21–22 (1970)

Phillips, J.D., Woodside, J., Bowin, C.O.: Magnetic and gravity anomalies in the central Red Sea. In: Hot Brines and Recent Heavy Metal Deposits in the Red Sea. Degens, E.T., Ross, D.R. (eds.). New York–Heidelberg–Berlin: Springer 1969, pp. 98–113

Schneider, W., Wachendorf, H.: Vulkanismus und Grabenbildung im Roten Meer. Geol. Rdsch. 62, 754–773 (1973)

Sweeney, R.E., Kaplan, I.R.: Pyrite framboid formation: Laboratory synthesis and marine sediments. Econ. Geol. 68, 618–634 (1973)

Subject and Locality Index

Indications in brackets, e.g., (W), refer to economic metal or mineral content.

MINERALIUM DEPOSITA

International Journal for Geology, Mineralogy and Geochemistry of Mineral Deposits

Official Bulletin of the Society for Geology Applied to Mineral Deposits

Edited by G. C. Amstutz, E. Grip, G. L. Krol, G. Kullerud, I. de Magnée, A. Maucher, J. Pereira, P. Ramdohr, P. Routhier

With the cooperation of J. Agard, V. Angelelli, A. Bernard, H. Borchert, J. Bouladon, F. W. Chuchrow, K. C. Dunham, P. Evrard, T. Febrel, N. H. Fisher, P. Haapala, H. D. Holland, E. Ingerson, S. Jankovič, E. Kautzsch, H. King, H. J. Koark, L. Kostelka, J. Kutina, L. J. Lawrence, P. Nicolini, C. H. Smith, R. L. Stanton, W. Uytenbogaardt, F. Vokes, T. Watanabe, A. Wilke

Managing Editor: A. Maucher

This journal introduces basically new observations, principles, and interpretations from the field of applied geology, with emphasis on economic mineral deposits. It contains comprehensive articles, brief original reports, scientific discussions, and news on meetings of importance to mineral research. Emphasis is laid on high quality of content and form of all articles and on an international coverage of subject matter. The journal is intended to be a link between research and application and, consequently, will be of interest to those engaged in mineral research as well as to field geologists.

Mineralium Deposita will, therefore, include investigations from the fields of geochemistry, mineralogy, physical chemistry, tectonics and sedimentology of ore deposits, with occasional digressions into problems of mineral dressing, smelting and other fields of mineral technology and applied geology.

Springer-Verlag
Berlin
Heidelberg
New York

For subscription information and sample copies write to: Springer-Verlag, P.O. Box 105280, D-6900 Heidelberg 1